A BIBLIOGRAPHY OF MEDICAL AND BIOMEDICAL BIOGRAPHY

This new third edition of *A Bibliography of Medical and Biomedical Biography* details readily available sources of information in the English language on significant figures in the history of medicine and the biomedical sciences. Archival collections are also noted and a representative selection of general and specialized histories is included. This new edition features more than 1300 new entries, plus updates and amendments to previous entries. Also included, as far as possible, are the names of individuals recorded in Morton and Moore's *Chronology of Medicine and Related Sciences* (Ashgate, 1997).

This bibliography remains an essential guide to the life and works of individuals who have contributed to the development of medicine and the biomedical sciences.

A BIBLIOGRAPHY OF MEDICAL AND BIOMEDICAL BIOGRAPHY

Third Edition

Leslie T. Morton

FCLIP

FORMERLY
LIBRARIAN, NATIONAL INSTITUTE FOR MEDICAL RESEARCH

and

Robert J. Moore

MBE, BA, MCLIP, MIBiol

FORMERLY
LIBRARIAN, NATIONAL INSTITUTE FOR MEDICAL RESEARCH

ASHGATE

© Leslie T. Morton and Robert J. Moore, 1989, 1994, 2005

All rights reserved. No part of this publication may be reproduced, stored in a retrieval system, or transmitted in any form or by any means, electronic, mechanical, photocopying, recording, or otherwise without the prior permission of the publisher.

The authors have asserted their moral right under the Copyright, Designs and Patents Act, 1988, to be identified as the author of this work.

First edition published 1989
Second edition published 1994
Third edition published 2005

Published by
Ashgate Publishing Limited
Gower House
Croft Road
Aldershot
Hants GU11 3HR
England

Ashgate Publishing Company
Suite 420
101 Cherry Street
Burlington, VT 05402-4405
USA

Ashgate website http://www.ashgate.com

British Library Cataloguing in Publication Data
Morton, Leslie T. (Leslie Thomas)
 A bibliography of medical and biomedical biography
 1. Medical personnel – Great Britain – Biography – Bibliography 2. Medical personnel – United States – Biography – Bibliography
 I. Title II. Moore, Robert J., 1939-
 016.6'1'092

Library of Congress Cataloging-in-Publication Data
Morton, Leslie T. (Leslie Thomas), 1907-2004
 A bibliography of medical and bio-medical biography / Leslie T. Morton and Robert J. Moore.– 3rd ed.
  p. cm.
 Includes bibliographical references and index.
 ISBN 0-7546-5069-3 (hardback : alk. paper)
 1. Medical personnel–Great Britain–Biography–Bibliography. 2. Medical personnel–United States–Biography–Bibliography. 3. Medical scientist–Great Britain–Biography–Bibliography. 4. Medical scientist–United States–Biography–Bibliography.
 [DNLM: 1. Physicians–Bibliography. 2. Biography–Bibliography. 3. History of Medicine, 19th Cent.–Bibliography. 4. History of Medicine, 20th Cent.–Bibliography. 5. Science–Bibliography. ZWZ 112 M889b 2004 I. Title.

 Z6660.5.M67 2004
 [R134]
 016.61'092'2—dc22

2004000357

ISBN 0 7546 5069 3

Typeset in 9pt Times by J.L. & G.A. Wheatley Design, Aldershot.
Printed and bound in Great Britain by T J International, Padstow, Cornwall.

CONTENTS

Introduction to Third Edition	vii
From the Introduction to the Second Edition	ix
List of Abbreviations	xi
Individual Biographies	1
Collective Biographies	412
Short List of Publications on the History of Medicine and Related Subjects	416

INTRODUCTION TO THIRD EDITION

In this third edition we have followed the guidelines laid down for the previous edition. We have again been selective in cases where numerous bibliographies of an individual are available. We have included more references to biographical, autobiographical and bibliographical information published in the periodical literature. We have included as far as possible the names of individuals recorded in our *Chronology of Medicine and Related Sciences* (1997).

We have taken the opportunity to make some amendments and corrections to existing entries. The section on Collective Biographies and the Short List of Books on the History of Medicine have been updated.

The second edition contained biographical information concerning 2368 individuals. In the present edition we have added 1372 biographies, making a total of 3740. Archival Collections have been noted but readers should bear in mind that approaches to librarians at universities, medical and scientific societies, hospitals, etc. could be profitable.

We again gratefully acknowledge the help provided by various libraries, especially those of the Wellcome Library for the History and Understanding of Medicine, the Library of the Royal Society of Medicine, and the Library of the National Institute for Medical Research. John Morton gave considerable technical help in the final preparation of the text. The help and advice of Nigel Farrow, Ann Newell, and Rachel Lynch of Ashgate Publishing Ltd. during production are gratefully acknowledged. Notes of errors and possible additions would be much appreciated.

L.T.M.
R.J.M.

Postscript: My very good friend and collaborator for 42 years died peacefully on the 17th of February 2004, in his 97th year. However, he had the great satisfaction of being able to send the completed text to our publisher on 30th of October 2003. I shall miss him greatly.   R.J.M.

FROM THE INTRODUCTION TO THE SECOND EDITION

The previous edition of this *Bibliography* was restricted to English-language publications in book form, supplemented by references to relevant entries in the *Dictionary of Scientific Biography*, the *Biographical Memoirs of the Royal Society* (and its predecessor, the *Obituary Notices*), the *Biographical Memoirs of the National Academy of Sciences*, Washington, and a few references to the periodical literature.

Following comments in reviews of the first edition, this edition includes references to relevant literature in most modern European languages, enabling us to add biographies of many individuals not previously included and to strengthen with native-language publications the information on some persons previously included.

The *Bibliography* covers a wide spectrum of medico-scientific endeavour and will, we hope, prove of value to a wide variety of users. Readers will appreciate that the biographies of some distinguished persons are missing because they have still to be written or are represented by very brief accounts, and we hope that this *Bibliography* will encourage work to fill such gaps. Sadly, John Thornton, whose *Select bibliography of Medical Biography* (1970) provided a basis for the present work, died on 24 September 1992.

ABBREVIATIONS

BMFRS: Royal Society of London. Biographical Memoirs of Fellows.
BMNAS: National Academy of Sciences, Washington. Biographical Memoirs.
CB: Refers to an item in the Collective Biographies section.

CMAC: Contemporary Medical Archives Centre, Wellcome Library for the History and Understanding of Medicine.

DIEPGEN: Diepgen, P. Unvollendete: vom Leben und Wirken frühverstorbener Forscher und Ärzte. Stuttgart, 1960.

DNB: Dictionary of National Biography. London.

DSB: Dictionary of Scientific Biography. New York, 1970–90.

GREENWOOD: Greenwood, M. The medical dictator. London, 1986.

GROTE: Grote, L.R.R. Die Medizin der Gegenwart in Selbstdarstellungen. Leipzig, 1923–29.

HALE-WHITE: Hale-White, W. Great doctors of the nineteenth century. London, 1935.

MUNK: Royal College of Physicians of London. Munk's Roll.

ONFRS: Royal Society of London. Obituary Notices of Fellows.

PLARR: Royal College of Surgeons of England. Plarr's Lives.

Full references to the above are to be found in the Collective Biographies section.

INDIVIDUAL BIOGRAPHIES

ABANO, Pietro d' *see* **PIETRO d'ABANO**

ABBE, Ernst Karl (1840–1905). German physicist; developed optical microscope; invented Abbe condenser.

GUNTHER, N. Ernst Abbe, Schöpfer der Zeiss-Stiftung. 2te. Aufl. Stuttgart, Wissenschaftliche Verlagsgesellschaft, 1951. 203 pp.
GUNTHER, N. DSB, 1970, **1**, 6–9.
WITTIG, J. Ernst Abbe. Leipzig, Teubner, 1989. 148 pp.

ABBOTT, Maude Elizabeth Seymour (1869–1940). Canadian cardiologist; pioneer in medical education for women in Canada.

MacDERMOT, H.E. Maude Abbott: a memoir. Toronto, Macmillan, 1941. 264 pp.
WAUGH, D. Maudie of McGill: Dr Maude Abbott and the foundations of heart surgery. Toronto, Hannah Institute and Dundurn Press, 1992. 142 pp.

ABDERHALDEN, Emil (1877–1950). Swiss physiologist and biochemist; professor at Halle University.

EMIL ABDERHALDEN zum Gedächtnis. *Nova Acta Leopoldina*, 1952, **141**, No. 103, 143–89 (contributions by E. Schoen, O. Schüter, E. von Skramlik).
GABATHALER, J. Emil Abderhalden: sein Leben und Werk. Wattwil, Abderhalden-Vereinigung; St Gallen, Buchhandlung Ribaux, 1991. 362 pp.
IN MEMORIAM Emil Abderhalden, 1877–1950. Vorträge e. Gedenksymposions. Halle, Universität. 1977.

ABEL, John Jacob (1857–1938). American pharmacologist and biochemist; professor at Johns Hopkins University.

DALE, H.H. ONFRS, 1939, **1**, 577–85.
MacNIDER, W.B. BMNAS, 1946, **24**, 231–57.
PARASCANDOLA, J. The development of American pharmacology. John J. Abel and the shaping of a discipline. Baltimore, Johns Hopkins University Press, 1992. 224 pp.
ROSENBERG, C.E. DSB, 1970, **1**, 9-12.
VOEGTLIN, C. John Jacob Abel 1857–1938. *Journal of Pharmacology and Experimental Therapeutics*, 1939, **67**, 373-406.
Archival material: Welch Memorial Library, Johns Hopkins University, Baltimore, Md.

ABERCROMBIE, John (1780–1844). British physician in Edinburgh; wrote first textbook on neuropathology

–. Account of the late Dr Abercrombie. *Edinburgh Medical and Surgical Journal*, 1845, **63**, 225

COMRIE, J.D. History of Scottish medicine. London, Baillière, 1932. 489–91.

ABERCROMBIE, Michael (1912–1979). British biologist.

MEDAWAR, P. BMFRS, 1980, **26**, 1–15.
Archival material: CMAC.

ABERNETHY, John (1764–1831). British surgeon at St Bartholomew's Hospital, London.

MACILWAIN, G. Memoirs of John Abernethy, with a view to his lectures, his writings, and character; with additional extracts from original documents, now published. 3rd ed., London, Hatchard, 1856. 396 pp. First edition 1853.
THORNTON, J.L. John Abernethy: a biography. London, for the author, 1953. 184 pp.

ABRAHAM, James Johnston (1876–1963). Irish-born surgeon and writer; author of definitive biography of John Coakley Lettsom.

ABRAHAM, J.J. Surgeon's journey: the autobiography of J.Johnston Abraham. London, Heinemann, 1957. 441 pp.
Archival material: Trinity College Dublin.

ABREU, Aleixo de (1568–1630). Portuguese specialist in tropical medicine.

GUERRA, F. Aleixo de Abreu (1568–1630), author of the earliest book on tropical medicine. *Journal of Tropical Medicine*, 1971, **71**, 55–69.
GUERRA, F. DSB, 1970, **1**, 25–6.

ABREU, Manoel de (1892–1962). Brazilian radiologist; introduced mass chest radiography.

BENICIO DOS SANTOS, I. Vida e obra de Manoel de Abreu, o criador da Abreugrafia. Rio de Janeiro, Pengetti, 1963. 256 pp.
BLUNDI, E. Momentos decisivos en la vida de Manoel de Abreu. *Semana Medica* (B. Aires), 1964, **104**, 509–14.

ABT, Isaac Arthur (1867–1955). American paediatrician.

ABT, I.A. Baby doctor [autobiography]. New York, Whittlesea House, [1944]. 310 pp.

ABU BAKR MUHAMMAD IBN ZAKARIYA AL-RAZI, *see* **RHAZES**

ABULCASIS [Abu al-Qasim al Zahrawi] (*c*.936–1013). Arabian physician and pharmacist, born Córdoba.

HAMARNEH, S. DSB, 1976, **14**, 584–5.
HAMARNEH, S. & SONNEDECKER, G. A pharmaceutical view of Abulcasis al Zahrawi in Moorish Spain. Leiden, Brill, 1936. 176pp.

ABUMERON, *see* **AVENZOAR**

ACCUM, Friedrich Christian (1769–1838). German/British chemist; exposed food adulteration.

BING, F.C. Frederick Accum. A biographical sketch. *Journal of Nutrition*, 1966, **89**, 3–8.
COLE, Friedrich Accum: a biographical study. *Annals of Science*, 1951, **7**, 128–43.

ACHARD, Émile Charles (1860–1944). French physician; isolated *Salmonella paratyphi B*; described Achard-Thiers syndrome.

ACHARD, E.C. La confession d'un vieil homme du siècle. Paris, Mercure France, 1943. 430 pp.

LOEPER, M. Émile-Charles Achard. *Bulletin de l'Académie Nationale de Médecine*, 1944, **128**, 504–15; 1952, **136**, 613–19.

ACHILLINI, Alessandro (1463–1512). Italian anatomist; described ducts of submaxillary salivary glands.

FRANCESCHINI, P. DSB, 1970, **1**, 46–7.

MATSEN, H.S. Alessandro Achillini (1463–1512) and his doctrine of "Universals" and "Transcendentals": a study of Renaissance Ockhamism. Lewisburg, Ohio, Bucknell University Press, 1974. 332 pp.

ACLAND, Henry Wentworth Dyke (1815–1900). British physician; Regius Professor of Medicine, Oxford.

ATLAY, J.B. Sir Henry Wentworth Acland, Bart.: a memoir. London, Smith, Elder, 1903. 507 pp.

ACOSTA, Cristóbal (1525–*c*.1594). Portuguese colonial botanist and surgeon; wrote on Indian materia medica.

GUERRA, F. DSB, 1970, **1**, 47–8.

ACOSTA, José de (1539–1600). Spanish traveller; gave early account of altitude sickness.

CARRACIDO, J.R. El P [padre] José Acosta y su importáncia en la literatura cientifico española. Madrid, Sucesores de Rivadenyra, 1899. 163pp.

ADAIR, Gilbert Smithson (1896–1979). British physiologist.

JOHNSON, P. & PERUTZ, M.F. BMFRS, 1981, **27**, 1–27.

ADAMI, John George (1862–1926). British pathologist; professor at McGill University; Vice-Chancellor, University of Liverpool.

ADAMI, M.J. George Adami: a memoir ... and an introduction by Sir Humphry Rolleston. London, Constable, 1930. 179 pp.

Archival material: Library, Wellcome Institute for the History of Medicine, London.

ADAMS, Francis (1796–1861). British physician; made scholarly translations of Greek and Latin medical classics.

SINGER, C. A great country doctor, Francis Adams of Banchory. *Bulletin of the History of Medicine*, 1943, **12**, 1–17.

ADAMS, John (1805–1877). British surgeon; first to distinguish between hypertrophy and carcinoma of the prostate.

PLARR

ADAMS, Robert (1791–1875). Irish physician.

HERRICK, J.B. Robert Adams, surgeon, and his contributions to cardiology. *Annals of Medical History*, 1939, **1**, 45–49.

ADAMS, Zabdiel Boylston (1829–1902). American physician.

PEABODY, C.N. ZAB: Brevet Major Zabdiel Boylston Adams, 1829–1902, physician of Boston and Framingham. Boston, Francis A. Countway Library of Medicine, 1984. 255 pp.

ADDAMS, Jane (1860–1935). American paediatrician.

LINN. J.W. Jane Addams, a biography. New York, Appleton-Century, 1926. 437 pp.

ADDENBROOKE, John (1680–1719). English physician; beqeathed money founding hospital in Cambridge ('Addenbrooke's').

ROOK, A. & MARTIN, L. John Addenbrooke, M.D. (1680–1719). *Medical History*, 1982, **26**, 169–78.

ADDIS, Thomas (1881–1949). American physician; authority on renal function and disease.

LEMLEY, K.V. & PAULING, L. BMNAS, 1964, **63**, 2–46

ADDISON, Christopher (1869–1951). British physician; lecturer on anatomy; politician whose posts included Minister of Health; influenced British Government in the establishment of the Medical Research Committee (afterwards Council).

MINNEY, R.J. Viscount Addison: Leader of the Lords. London, Odhams, 1958. 256 pp.
MORGAN, K. & MORGAN, J. Portrait of a progressive; the political career of Christopher, Viscount Addison. Oxford, Clarendon Press, 1980. 326 pp.

ADDISON, Thomas (1793–1860). British physician at Guy's Hospital, London; described 'Addison's disease' of the adrenal cortex.

BENJAMIN, J.A. DSB, 1970, **1**, 59–61.
GUY'S HOSPITAL REPORTS, 1960, **109** (4), 227–83; special number dedicated to Thomas Addison in commemoration of the centenary of his death.
PALLISTER, G. Thomas Addison, M.D., F.R.C.P., (1793–1860). Newcastle upon Tyne, for the author, 1975. 32 pp.

ADDISON, William (1802–1881). British physician; gave an important description of inflammation and the first description of leucocytosis.

RATHER, L.J. Addison and the white corpuscles: an aspect of nineteenth-century biology. Berkeley, Univ.California Press, 1972. 236 pp.

ADLER, Alfred (1870–1937). Austrian psychiatrist; founder of individual psychology.

BOTTOME, P. Alfred Adler, apostle of freedom. 3rd ed. London, Faber, 1957. 300 pp. First published 1939.
GREY, L. Alfred Adler, the forgotten prophet: a vision for the 21st century. New York, Praeger, 1998. 158pp
RATTNER, J. Alfred Adler. Translated from the German by H. Zohn. New York, Ungar, 1983. 226 pp.
SCHIFERER, H.R. Alfred Adler: eine Bildbiographie. München, E. Reinhart, 1995. 232 pp.

ADLER, Ludwig (1876–1958). Austrian obstetrician; with F. Hitschmann first described cyclical changes in the endometrium.

FROBENIUS, W. Fehldiagnose Endometritis: zur Revision eines wissenschaftlichen Irrtums durch die Wiener Gynäkologen Fritz Hitschmann und Ludwig Adler. Hildesheim, Olms, 1988. 220 pp.

ADLER, Saul (1895–1966). British parasitologist; professor at Hebrew University, Jerusalem.

SHORTT, H.E. BMFRS, 1967, **13**, 1–34.

ADRIAN, Bernard Hume (1927–1995). British electrophysiologist.

HUXLEY, A. BMFRS, 1997, **43**, 13–30.

ADRIAN, Edgar Douglas (1889–1977). British neurophysiologist; professor at Cambridge University; shared Nobel Prize with C.S. Sherrington, 1932, for isolation and functional analysis of the motor cell in the spinal cord.

HODGKIN, A. BMFRS, 1979, **25**, 1–73.
Archival material: Nuffield College, Oxford; Trinity College, Cambridge.

AEGIDIUS CORBOLENSIS, *see* **GILLES DE CORBEIL**

AESCULAPIUS, God of Medicine.

EDELSTEIN, E.J. & EDELSTEIN, L. Asclepius: a collection and interpretation of the testimonies. 2 vols. Baltimore, Johns Hopkins University Press, 1945. 470 + 227 pp.
HART, G.D. Asclepius, the God of Medicine. London, Royal Society of Medicine Press, 2000. 262 pp.
SCHOUTEN, J. The rod and the serpent of Asklepios: symbol of medicine. Amsterdam, Elsevier Publ. Co., 1967. 260 pp.

AETIUS of Amida (502–575). Byzantine physician; compiled *Tetrabiblion*, a collection of writings of physicians that might otherwise have been forgotten.

BRAVOS, S. Das Werk von Aetios v. Amida und seine medizinischen und nichtmedizinischen Quellen. Dissertation... Doktors der Medizin, Universität Hamburg, 1974. 174 pp.
KUDLIEN, P. DSB, 1970, **1**, 68–9.

AGASSIZ, Louis Jean Rodolphe (1807–1873). Swiss/American naturalist and physician.

LURIE, E. Louis Agassiz: a life in science. Chicago, Univ. Chicago Press, 1960. 449 pp.
MARCOU, J. Life, letters and works of Louis Agassiz. 2 vols. Westmead, Farnborough, Gregg International, 1972. 302 + 318 pp. Facsimile of 1896 edition.

AGATHINUS, Claudius (*c*.AD 50). Spartan physician.

KUDLIEN, F. DSB, 1970, **1**, 74–5.

AGNEW, David Hayes (1818–1892). American surgeon; professor at University of Pennsylvania.

ADAMS, J.H. History of the life of D. Hayes Agnew. Philadelphia, London, F.A. Davis Co., 1892. 376 pp.
RADBILL, S.X. David Hayes Agnew, MD, 1818–1892. *Transactions of the College of Physicians of Philadelphia*, 1966, **33**, 252–60.

AGOTE, Luis (1868–1944). Argentinian physician; introduced citrated blood in transfusion 1914.

FIGUEROA ALCORTA, L. Homenaje de la Academia Nacional de Medicina al Doctor Luis Agote en el cincentario des discrubimento de la transfusión sanguínea por el metodo citratado. Buenos Aires, Academia Nacional de Medicina, 1964. 45 pp.

AIKIN, John (1747–1822). British physician and man of letters.

AIKIN, L. Memoir of John Aikin... With a selection of his miscellaneous pieces, biographical, moral and critical. 2 vols., London, Baldwin, Cradock & Joy, 1823. 487 pp.

AIRD, Ian (1905–1962). British surgeon; professor at Royal Postgraduate Medical School, London.

McLEAVE, H. A time to heal. The life of Ian Aird, the surgeon. London, Heinemann, 1964. 278 pp.

AIRY, George Biddell (1801–1892). British astronomer and physicist.

LEVENE, J.R. Sir George Biddell Airy, F.R.S. (1801–1892) and the discovery and correction of astigmatism. *Notes and Records of the Royal Society of London*, 1966, **21**, 180–99.

AKENSIDE, Mark (1721–1770). British physician and poet; physician to Queen Charlotte and to St Thomas' Hospital.

BUCKE, C. On the life, writings and genius of Akenside, with some account of his friends. London, J. Cochrane, 1832. 312 pp.

DIX, R. (ed.) Mark Akenside: a reassessment. Madison, NJ. Fairleigh Dickinson Univ. Press, *c.* 2000. 206 pp.

HOUPT, C. Mark Akenside: a biographical and critical study. PhD thesis. Philadelphia, University of Pennsylvania, 1944. 180 pp.

ALBARRÁN Y DOMINGUEZ, Joaquín Maria (1860–1912). Cuban urological surgeon in Paris.

PAULÍS PAGÉS, J. & MONTEROS-VALDIVIESCO, M.Y. Joaquín Albarrán, genial artifice de la urología. Havan, Museo "Carlos J. Finlay", 1963. 194 pp.

ALBEE, Fred Houdlett (1876–1945). American surgeon; pioneer of living bone-graft surgery.

ALBEE, F. A surgeon's fight to rebuild men. An autobiography. London, Hale, 1950. 270 pp.

ALBERS-SCHÖNBERG, Heinrich Ernst (1865–1921). German radiologist; invented the compression diaphragm.

ALBERS-SCHÖNBERG, E Heinrich Albers-Schönberg, 1865–1921). Eine biographische Skizze. *Röntgenpraxis*, 1965, **18**, 1–7.

ALBERT, César-Alphonse (1810–1862). French surgeon.

VERNEUIL, A. Éloge de Dr C.Alp. Albert. Paris, A. Delahaye, 1864. 96 pp.

ALBERT, Eduard (1841–1900). Bohemian orthopaedic surgeon; introduced joint arthrodesis.

JIRASEK, A. Eduard Albert. Prague, Cesk. Chirurg. Spol., 1946. 468 pp. [in Czech]

ALBERTI, Salomon (1540–1600). German physician; professor of medicine at Wittenberg.

O'MALLEY, C.D. DSB, 1970, **1**, 98.

ALBERTUS MAGNUS [Albert of Bollstädt] (1193–1280). Dominican monk, Bishop of Ratisbon; naturalist.

BALSS, H. Albertus Magnus als Biologe: Werk und Ursprung. Stuttgart, Wissenschaftliche Verlagsgesellschaft, 1947. 307 pp.

WILMS, H. Albert the Great; saint and doctor of the Church. London, Burns, Oates and Washbourne, [1933]. 226 pp. First published in German, München, 1930.

ALBINUS, Bernard Siegfried (1697–1770). German anatomist; professor at Leiden and foremost anatomist of the 18th century.

HALE, R.B. & COYLE, T. (eds.) Albinus on anatomy, with 80 original Albinus plates. New York, Dover, 1988. 208 pp.

PUNT, H. Bernard Siegfried Albinus (1697–1770) and 'human nature': anatomical and physiological ideas in 18th-century Leiden. Amsterdam, B.M. Israel, 1983. 226 pp.

VAN DER PAS, P.W. DSB, 1978, **15**, 4–5.

ALBRIGHT, Fuller (1900–1969). American endocrinologist.

FINK, M.E. DSB, 1990, **17**, 8–11.

LEAF, A. BMNAS, 1976, **48**, 3-22.

ALBUCASIS, *see* **ABULCASIS**

ALCMAEON OF CROTON (*c*.500 BC). Italian philosopher, physician and anatomist; made first known human dissections.

ARCIERI, J.P. Why Alexander of Croton is the father of experimental or scientific medicine. New York, Alcmaeon Editions, 1970. 52 pp.

CODELLAS, P.S. Alcmaeon of Croton: his life, work and fragments. *Proceedings of the Royal Society of Medicine*, 1932, **25**, 1041–6.

ALCOCK, Alfred William (1859–1935). British medical entomologist and zoologist; first professor of medical zoology, London University (at London School of Hygiene and Tropical Medicine).

CALNAN, W.T. & MANSON-BAHR, P. ONFRS, 1932–35, **1**, 119–26.

ALDER, Albert (1888–1980). Swiss haematologist.

SPÄTH, A.V. Der Hämatologe Albert Alder. Zürich, Juris Druck, 1983. 81 pp.

ALDEROTTI, Taddeo (1223–1303). Italian medical teacher; founder of Bolognese school.

SIRAISI, N.G. Taddeo Alderotti and his pupils. Two generations of Italian medical learning. Princeton, N.J., Princeton Univ. Press, (*c*. 1981). 461 pp.

ALDRICH-BLAKE, Louisa (1865-1925). Dean of London (Royal Free Hospital) School of Medicine for Women.

RIDDELL, G.A. Dame Louisa Aldrich-Blake. London, Hodder & Stoughton, 1926. 91 pp.

ALDROVANDI, Ulisse (1522–1600). Italian physician and naturalist; professor at Bologna.

TOSI, A. (ed.) Aldrovandi la Toscani. Carteggio e testimonianze documentarie. Firenze, L.S. Olschi Editore, 1989. 472 pp.

ALEXANDER of Tralles (525–605). Greek physician.

KUDLIEN, F. DSB, 1970, **1**, 121.

PUSCHMANN, T. Alexander von Tralles. 2 vols. Wien, W.Braumüller, 1878–9. Reprinted Amsterdam, 1963.

ALEXANDER, Frederick Matthias (1869–1955). British psychotherapist; invented 'Alexander technique'.

EVANS, J.A. Frederick Matthias Alexander: a family history. Chichester, Phillimore, 2001. 286 pp.
STARING, J. The first 43 years of the life of F. Matthias Alexander. 2 vols. Nijmegen, J. Staring, 1966–97. 448+610 pp.
WESTFELDT, L. F. Matthias Alexander: the man and his work. London, Allen & Unwin, 1964. 163 pp. Munk, 4, 253–4.

ALIBERT, Jean Louis Marc (1768–1837). French dermatologist, the 'Father of French dermatology'.

BRODIER, L. J.L. Alibert, médecin de l'Hôpital Saint-Louis, 1768–1837. Paris, A. Maloine & fils, 1923. 390 pp.
CIVATTE, J. *et al.* Jean Louis Alibert, 1768–1837: fondateur de la dermatologie française, *etc.* Villefranche-de-Rouergue, Société des Amis de Villefranche et du Bas-Rouergue, 1987. 125 pp.
MORTON, L.T. Jean Louis Marc Alibert: a bibliography. *Journal of Medical Biography*, 1993, **1**, 108–12.

ALLBUTT, Thomas Clifford (1836–1925). British physician; Regius Professor of Physic, University of Cambridge.

KEYNES, M. & BUTTERFIELD, J.I. Sir Clifford Allbutt: physician and Regius Professor. *Journal of Medical Biography*, 1993, **1**, 67–75.
ROLLESTON, H.D. The Right Hon. Sir Thomas Clifford Allbutt: a memoir. London, Macmillan, 1929. 314 pp.

ALLCHIN, William Henry (1846–1912). British physician; gave first detailed account of ulcerative colitis.

–. Sir William Henry Allchin, MD, FRCP Lond., FRS Edin. *Lancet*, 1912, **1**, 544–5.

ALLEN, Edgar (1892–1943). American endocrinologist; professor of medicine at Yale University.

COWAN, R.S. DSB, 1970, **1**, 123–4.
EDGAR ALLEN: Curriculum vitae and bibliography. *Yale Journal of Biology and Medicine*, 1944–5, **17**, 2–12.
G., W.U. Edgar Allen (1892–1943). *Yale Journal of Biology and Medicine*, 1942–3, **15**, 641–4.

ALLEN, Willard Myron (1904–1993). American endocrinologist:; co-discoverer of progesterone.

ALLEN, W.M. Recollections of my life with progesterone. *Gynecologic Investigation*, 1974, **5**, 142–82.

ALPINI, Prospero (1553–1617). Italian physician-botanist.
ONGARD, G. Contributi alla biografia di Prospero Alpini. *Acta Medicae Historiae Patavini*, 1961–3, **8–9**, 79–168.

STANNARD, J. DSB, 1970, **1**, 124–5.

ALVAREZ, Walter Clement (1884–1978). American gastroenterologist.

ALVAREZ, W.C. Incurable physician. An autobiography. Englewood Cliffs, Prentice-Hall Inc., 1963. 273 pp.
SCOTT, D.H. Walter C. Alvarez: American man of medicine. New York, Van Nostrand Reinhold, 1976. 380 pp.

ALZHEIMER, Alois (1864–1915). German psychiatrist; described presenile dementia ('Alzheimer's disease').

BERRIOS, G. & FREEMAN, H.L. (eds.) Eponyms in medicine: Alzheimer and the dementias. London, Royal Society of Medicine Services, 1991. 150 pp. [Includes a biographical chapter and thumb-nail biographies of other eponymists in this field.]
MAURER, K. & MAURER, K. Alzheimer: das Leben eine Artzes und die Karriere einer Krankheit. München, Piper, 1998. 319 pp.

AMATUS LUSITANUS [Rodrigues, Joáo] (1511–1568). Portuguese physician and surgeon.

FRIEDENWALD, H. Amatus Lusitanus. *Bulletin of the Institute of the History of Medicine*, 1937, **5**, 603–53.
KELLER, A.G. DSB, 1973, **8**, 554–5.
LEMOS, M. Amato Lusitano: a sua vida e a sua obra. Porto, E. Tavares Martins, 1907. 212 pp.

AMMON, Friedrich August von (1799–1861). German surgeon; his colour plate atlas of the eye (1838–47) is one of the best to appear before the introduction of the ophthalmoscope.

WERNATZ, -. *Annales d'Oculistique*, 1861, **45**, 269–74.
ZEIS, E. Rede zum Gedächtnisse des am 18.Mai 1861 verstorbenen Herrn Dr Friedrich August von Ammon. Dresden, C.G. Ernst am Ende, 1861. 55 pp.

AMOROSO, Emmanuel Ciprian (1901–1982). Trinidad-born endocrinologist; professor of veterinary physiology, University of London.

SHORT, R.V. BMFRS, 1985, **31**, 1–30.
Archival material: CMAC.

AMUSSAT, Jean Zuléma (1796–1856). French surgeon; performed first lumbar colostomy for colonic obstruction, 1835, and established artificial anus, 1839.

LARREY, -. Discours prononcé aux obsequies.... *Bulletin de l'Académie de Médecine*, 1855–6, **21**, 765–78.

ANCEL, Paul Albert (1873–1961). French anatomist and embryologist.

KLEIN, M. DSB, 1970, **1**, 152–3.
WOLFF, E. Le professeur Paul Ancel. *Archives d'Anatomie, d'Histologie et d'Embryologie Normales et Experimentales*, 1961, Suppl. **44**, 5–27.

ANDERSON, Elizabeth Garrett (1836–1917). British physician; pioneer in the movement to include women in the medical profession.

ANDERSON, L.G. Elizabeth Garrett Anderson 1836–1917. London, Faber, 1939. 338 pp.
MANTON, J. Elizabeth Garrett Anderson. London, Methuen, 1965. 382 pp. American
edition, with title *Elizabeth Garrett, M.D.*, New York, Abelard-Schuman, 1960.

ANDERSON, Hugh Kerr (1865–1923). British neurophysiologist.

H., W.B. & SHERRINGTON, C.S. *Proceedings of the Royal Society of London*, B, 1928,
104, xx–xxv.
Archival material: University of Cambridge.

ANDRAL, Gabriel (1797–1876). French physician, professor of pathology, Paris; published
first monograph devoted to haematology.

HUARD, P. & IMBAULT-HUART, M.J. Gabriel Andral (1797–1876). *Revue d'Histoire
des Sciences*, 1982, **35**, 131–53.

ANDREWES, Christopher Howard (1896–1987). British virologist; with P.P. Laidlaw and
W. Smith isolated human influenza virus.

ANDREWES, C.H. Fifty years with viruses. *Annual Review of Microbiology*, 1973, **27**, 1–
11.
TYRRELL, D.A.J. BMFRS, 1991, **37**, 33–54.
Archival material: CMAC.

ANDREWES, Frederick William (1859–1932). British pathologist; professor at St
Bartholomew's Hospital, London.

G., M.H. & P., E.B. ONFRS, 1932–35, **1**, 37–44.
Archival material; Library, Wellcome Institute for the History of Medicine, London.

ANDREWS, Edmund (1824–1904). American anaesthetist; advocated oxygen-nitrous oxide
anaesthesia.

QUINE, W.E. Edmund Andrews. Chicago, Surgical Publishing, 1922. 4 pp.

ANDRY, Nicolas (1658–1742). French physician; wrote first medical parasitology text (1700)
and the first book on orthopaedics, which term Andry himself introduced.

MAUCLAIRE, P. Nicolas Andry, médecin Lyonnais (XVIIe siècle). *Bulletin de la Société
Française d'Histoire de Médecine*, 1933, **27**, 345–9.
NEUHAUS, E. Nicolas Andry, ein zu Unrecht vergessener Arzt des 18. Jahrhunderts.
Opladen, Eiden, 1939. 36 pp.

ANEL, Dominique (1679–1730). French surgeon.

MOULIN, D. de. Dominique Anel and his operation for aneurysm. *Bulletin of the History
of Mdicine*, 1960, **34**, 498–509.

ANFINSEN, Christian Boehmer (1916–1995). American biochemist; shared Nobel Prize
(Chemistry) 1972 for work on ribonuclease.

SCHECHTER, A.N. Christian B. Anfinsen, 1916–1995. *Nature Structural Biology*, 1995,
2, 621–13.

ANGELL, James Rowland (1869–1949). American psychologist; professor at Chicago
University and later President of Yale University.

HUNTER, W.S. BMNAS, 1951, **26**, 191–208.

ANGLE, Edward Hartley (1855–1930). American orthodontist, established orthodontia as a specialty.

HAHN, G.W. Edward Hartley Angle, 1855–1930. *American Journal of Orthodontics*, 1965, **51**, 529–35.

ANICHKOV, Nikolai Nikolaevich (1885–1964). Russian pathologist; produced experimental arteriosclerosis.

SARKISOV, D.S. & POZHARISSKIL, K.M. N.N. Anichkov, 1855–1964. Moskva, Meditsina, 1989. 206 pp.

ANNANDALE, Thomas (1838–1907). British surgeon; professor of clinical surgery, Edinburgh.

PLARR.
DNB, 20th cent, p. 45

ANREP, Gleb Vasilievich (1891–1955). Russian-born physiologist; professor at Cairo.

GADDUM. J.H. BMFRS, 1956, **2**, 19–34.

ANTOMMARCHI, Francesco (1780–1838). Italian anatomist and surgeon; professor of anatomy, Florence; physician to Napoleon I at St Helena.

PAOLI, F. Le Dr Antonmarchi [sic] ou le secret du masque de Napoleon. Paris, Publisud, 1906. 350 pp.

ANTYLLUS (*fl.* 2nd cent. AD). Greek surgeon.

GRANT, R.L. Antyllus and his medical works. *Bulletin of the History of Medicine*, 1960, **34**, 154–74.

APÁTHY, Istan Stephan (1863–1922). Hungarian histologist.

KISS, S. Stephan von Apáthy als Neurolog. *Communicationes ex Bibliotheca Historiae Medicae Hungarica*, 1956, **3**, 1–64.
SAJNER, J. DSB, 1970, **1**, 176–7.

ARBER, Werner (b. 1929). American microbial geneticist; shared Nobel Prize (Physiology or Medicine) 1978, for discovery of restriction enzymes and their application to problems of molecular genetics.

Les Prix Nobel en 1978.

ARBUTHNOT, John (1667–1735). Scottish physician and satirist; inventor of 'John Bull', popular term for the typical Englishman; physician-in-ordinary to Queen Anne and physician to Chelsea Hospital, London.

AITKEN, G.A. The life and works of John Arbuthnot. Oxford, Clarendon Press, 1892. 516 pp. Reprinted New York, Russell, 1968.
BEATTIE, L.M. John Arbuthnot, mathematician and satirist. Cambridge, Mass., Harvard Univ. Press, 1935. 432 pp.
STEENSMA, R.C. Dr John Arbuthnot. Boston, Twayne, 1979.

ARCOLANI, Giovanni (*c.*1390–1460). Italian surgeon; used gold fillings for teeth.

PIERRO, F. Giovanni Arcolano: un chirurgo dimenticato. *Arcispedali S.Anna Ferrara*, 1965, **18**, 183–204.

ARDERNE, John (1307–1380). English surgeon.

BROWN, A. John Arderne, surgeon of early England. *Annals of Medical History*, 1928, **10**, 402–8.
TREATISES of fistula in ano, haemorrhoids and clysters...Edited with introduction, notes, etc. by D'Arcy Power. London, Kegan Paul, 1910. xxxvii, 156 pp. [pp. i–xxxvii biographical and bibliographical]

ARETAEUS of Cappadocia (*c.* 81–138). Greek physician.

KUDLIEN, F. Untersuchungen zu Aretaios von Kappadokien. Mainz, Akademie der Wissenschaften und der Literatur, 1963. 86 pp.
KUDLIEN, F. DSB, 1970, **1**, 234–5.

ARINKIN, Mikhail Innokentievich (1876–1948). Russian pathologist; introduced bone marrow biopsy by needle puncture.

BEIER, V. [In memory of M.I.Arinkin] [in Russian]. *Klinicheskaia Meditsina*, 1976, **54**, 137–40.

ARISTOTLE (384–322 BC). Greek philosopher, biologist and scientist.

GRENE, M. A portrait of Aristotle. Chicago, Univ. Chicago Press, 1963. 271 pp.
LLOYD, G.E.R. Aristotle: the growth and structure of his thought. London, Cambridge Univ. Press, 1968. 324 pp.
OWEN, G.E.L., BALME, D.M., WILSON, L.G. & MINIO-PALEULLO, L. DSB, 1970, **1**, 250–81.

ARKWRIGHT, Joseph Arthur (1864–1944). British bacteriologist, at Lister Institute, London.

MARTIN, C.J. ONFRS, 1945–48, **5**, 127–40.

ARLT, Carl Ferdinand von (1812–1887). Austrian ophthalmologist; professor in Prague and Vienna.

ARLT, F. von. Meine Erlebnisse. Wiesbaden, J.F. Bergmann, 1887. 144 pp.

ARMSTRONG, George (1719–1789). British paediatrician.

MALONEY, W.J. George and John Armstrong of Castleton. Two eighteenth-century medical pioneers. Edinburgh, Livingstone, 1954. 116 pp.

ARMSTRONG, John (1709–1779). British physician, poet and essayist.

MALONEY, W.J. George and John Armstrong of Castleton. Two eighteenth-century medical pioneers. Edinburgh, Livingstone, 1954. 116 pp.

ARMSTRONG, John (1784–1829). British physician to the Fever Institution of London.

BOOTT, F. Memoir of the life and medical opinions of John Armstrong, *etc*. 2 vols., London, Baldwin & Cradock, 1833–34. 616 + 752 pp.

ARNALD OF VILLANOVA (*c*.1240–1311). Spanish physician.

HAVEN, M. La vie et les oeuvres de Maître Arnaud de Villaneuve. Genève, Slatkin Reprints, 1972. 192 pp.
McVAUGH, M. DSB, 1970, **1**, 289–91.
McVAUGH, M.R. Further documents for the biography of Arnaud de Villanova. *Dynamics*, (Granada), 1982, **2**, 363–72.
MENSA I VALLS, J. Arnau de Villanova, espiritual guia bibliographica. Barcelona, Institut d'Estudis Catalans, 1994. 175 pp.

ARNOLD, Julius (1835–1915). German pathologist; described Arnold-Chiari malformation of the cerebellum.

ERNST, P. Julius Arnold. *Folia Haematologica* , 1915, **19**, 220–25.

ARON, Max (1892–1974). French endocrinologist; isolated thyroid-stimulating pituitary hormone.

LEGAIT, E. Le professeur Max Aron (1892–1974) *Annales Médicales de Nancy*, 1975, **14**, 549–50.

ARRHENIUS, Svante August (1859–1927). Swedish chemist, physicist and immunologist; awarded Nobel Prize (Chemistry) 1903.

CRAWFORD, E. Arrhenius from ionic theory to the greenhouse effect. Nantucket MA, Science history Publications, 1996. 320 pp.
KERNBAUER, A. Svante Arrhenius' Beziehungen zu Österreichischen Gelehrten. Graz, Akademische Druck und Verlagsanstalt, 1988. 526 pp.
SNELDERS, H.A.M. DSB, 1970, **1**, 296–302.
Archival material: Royal Swedish Academy of Sciences.

ARSONVAL, Jacques Arsène d' (1851–1940). French biophysicist.

CHAUVOIS, L. D'Arsonval, soixante-cinq ans à travers la science. Paris, Oliven, 1937. 437 pp.
CULOTTA, C.A. DSB, 1970, **1**, 302–5.

ARTELT, Walter (1906–1976). German medical historian.

EULNER, H.H. et al. (eds.) Medizingeschichte in unserer Zeit. Festgabe für Edith Heischkel-Artelt und Walter Artelt zum 65. Geburtstag. Stuttgart, F. Enke, 1971. 491 pp. [Separate bibliographies of E. Heischkel-Artelt's and W. Artelt's publications, pp. 457–77.]

ARTHUS, Nicholas Maurice (1862–1945). French physician; demonstrated 'Arthus phenomenon'.

SIGERIST, H.E. Maurice Arthus. *Bulletin of the History of Medicine*, 1943, **14**, 368–72,

ASCHHEIM, Selmar (1878–1965). German gynaecologist; with B.Zondek isolated the gonadotrophic hormone of the anterior pituitary and introduced the Aschheim-Zondek test for the diagnosis of pregnancy.

HOHLWEG, W. In memoriam Professor Selmar Aschheim. *Zentralblatt für Gynäkologie*, 1965, **87**, 1025–6.

ASCHOFF, Karl Albert Ludwig (1866–1942). German pathologist.

ASCHOFF, L. Ludwig Aschoff: ein Gelehrtenleben in Briefen an die Familie. Freiburg, H.F. Schultz, 1966. 480 pp.

ASCLEPIADES of Bithynia (*fl.* 124–56 BC). Greek physician in Rome.

GREEN, R.M. Asclepiades: his life and writings. A translation of Cocchi's Life of Asclepiades and Gumpert's Fragments of Asclepiades. New Haven, Elizabeth Licht, 1955. 167 pp.
STANNARD, J. DSB, 1970, **1**, 31.

ASCOLI, Alberto (1877–1957). Italian serologist; produced diagnostic test for anthrax.

–. Alberto Ascoli. *Annales de l'Institut Pasteur*, 1957, **93**, 681–2.

ASELLI, Gaspare (1581–1626). Italian anatomist; discovered lacteal vessels.

PREMUDA, L. DSB, 1970, **1**, 115–16.

ASKANAZY, Max (1865–1940). German pathologist; first to associate osteitis fibrosa cystica with parathyroid tumours.

HUEBSCHMANN, P. Max Askanazy, 24.III.1865–23.X.1940. *Verhandlungen der Deutschen Pathologischen Gesellschaft*, 1956, **40**, 359–77.

ASKEW, Anthony (1721–1772). British physician and bibliophile; amassed immense library.

MACMICHAEL: Gold-headed cane (CB).
MUNK, 2, 185–9 (CB).

ASKLEPIOS, *see* **AESCULAPIUS**

ASSALINI, Paolo (1759–1840). Italian surgeon.

LA CAVA, A.F. Paolo Assalini (1759–1840). Trieste, Ziggiotti, 1947. 103 pp.

ASTBURY, William Thomas (1898–1961). British crystallographer; a pioneer of molecular biology.

BERNAL, J.D. BMFRS, 1963, **9**, 1–35.
Archival material: University of Leeds Library.

ASTRUC, Jean (1684–1766). French physician.

DOE, J. Jean Astruc (1684–1766) a biographical and bibliographical study. *Journal of the History of Medicine*, 1960, **20**, 184–97.
HUARD, P. DSB, 1970, **1**, 322–4.
HUARD, P. & IMBAULT-HUART, M.J. Jean Astruc. In Huard, P. (ed.) Biographies médicales et scientifiques, XVIIIe siècle. Paris, R. Dacosta, 1972. pp. 7–30.

ASTWOOD, Edwin Bennett (1909–1976). American endocrinologist; introduced thiourea in treatment of hyperthyroidism.

GREEP, R.O. & GREER, M.A. BMNAS, 1985, **55**, 3–42.

ATKINS, Hedley John Barnard (1905–1983). British surgeon.

ATKINS, H. Memoirs of a surgeon. London, Springwood Books, 1977. 204 pp.

ATKINSON, James (1759–1839). Medical bibliographer.

RUHRÄH, J. James Atkinson and his medical bibliography [pubd. 1834]. *Annals of Medical History*, 1924, **6**, 200–221.

ATLEE, John Light (1799–1885). American ovariotomist.

NEMET, J.H. John Light Atlee and Washington Lemuel Atlee, a biographical sketch. Dissertation. Zürich, 1966. 29 pp.

ATLEE, Washington Lemuel (1808–1878). American ovariotomist.

NEMET, J.H. John Light Atlee and Washington Lemuel Atlee, a biographical sketch. Dissertation, Zürich. 1966. 29 pp.

ATWATER, Wilbur Olin (1844–1907). American agricultural chemist and nutritionist.

DARBY, W.J. Nutrition science: an overview of American genius. *Nutrition Reviews*, 1976, **34**, 1–14.
ROSENBERG, C.E. DSB, 1970, **1**, 325–6.
Archival material: Wesleyan University, Middletown, Conn.

AUENBRUGGER, Joseph Leopold (1722–1809). Austrian physician; introduced chest percussion as a diagnostic method.

BISHOP, P.J. A list of papers, etc. on Leopold Auenbrugger (1722–1809) and the history of percussion. *Medical History*, 1961, **5**, 192–6.
STEUDEL, J. DSB, 1970, **1**, 332–3.
WALSH, J.J. Makers of modern medicine. New York, Fordham Univ. Press, 1907, pp. 55–85.

AUER, John (1875–1948). American physiologist and pharmacologist.

ROTH, G.B. John Auer. *Journal of Pharmacology and Experimental Therapeutics*, 1949, **95**, 285–6.

AUERBACH, Charlotte (b. 1899). German/British geneticist; professor of animal genetics, University of Edinburgh.

AUERBACH, C. A pilgrim's progress through mutation research. *Perspectives in Biology and Medicine*, 1978, **21**, 319–34.
BEALE, G.H. BMFRS, 1995, **41**, 20–42.

AVELING, James Hobson (1828–1892). British obstetrician; employed direct blood transfusion method.

AVELING, J.H. English midwives...Reprint of 1872 edition. With biographical sketch of the author...by J.L. Thornton (pp. xi–xxxi). London, H.K. Elliott, 1967. 186 pp.

AVELLIS, George (1846–1916). German otolaryngologist; recorded recurrent palatal paralysis ('Avellis' syndrome'), 1891.

FINDER, –. *Internationales Zentralblatt für Laryngologie*, 1916, **32**, 199–201.

AVENZOAR [Ibn Zuhr, Abu Marwan 'Abd al-Malik ibn Abi'l'-ata] (*c.*1092–1162). Arabian physician, born Seville; considered the greatest Moslem physician of the Western Caliphate.

COLIN, G. Avenzoar, sa vie et ses oeuvres. Paris, Leroux, 1911. 199 pp.
HAMARNEH, S. DSB, 1976, **14**, 637–9.

AVERROËS [Abu'l-Walid Muhammad ibn Ahmad ibn Muhammad ibn Rushd], (1126–1198). Arabian physician, born Córdoba.

ARNALDEZ, R. & ISKANDAR, A.Z. DSB, 1975, **12**, 1–9.
FAKHRY, M. Averroës (ibn Rushd), his life, works and influence. Oxford, One World, 2000. 187 pp.
LEAMAN, O. Averroes and his philosophy [Rev. ed] London, Curzon Press, [1998]. 204 pp.
URVOY, D. Ibn Rushd (Averroës); translated by O. Stewart. London, Routledge, 1991. 156 pp.
URVOY, D. Averroès. Les ambitions intellectuel musulman. Paris, Flammarion, 1998. 253 pp.

AVERY, Oswald Theodore (1877–1955). Canadian/American bacteriologist at Rockefeller Institute; showed DNA to be the basic material responsible for genetic transmission.

DOCHEZ, A.R. BMNAS, 1958, **32**, 32–49.
DUBOS, R.J. BMFRS, 1956, **2**, 35–48.
DUBOS, R.J. The professor, the Institute, and DNA. New York, Rockefeller Univ. Press, 1976. 238 pp.
KAY, A.S. DSB, 1971, **1**, 342–3.
RUSSELL, N. Oswald Avery and the origin of molecular biology. *British Journal for the History of Science*, 1988, **21**, 393–400.
Archival material: Rockefeller University; Tennessee State Library.

AVICENNA [Ibn Sina] (980–1037). Persian physician; wrote *Liber canonis*, a complete exposition of Galenism.

AFNAN, S.M. Avicenna, his life and work. London, Allen & Unwin, 1958. 298 pp.
GOODMAN, L.E. Avicernna. London & New York, Routledge, 1992. 240 pp.
ISKANDAR, A.Z. DSB, 1978, **15**, 494–501.
SEZGIN, F. Studies on Ibn Sina (d.1037) and his medical works. Collected and reprinted by Fuat Sezgin, 4 vols. Frankfurt am Main, Institut for the History of Arabic-Islamic Science, 1996.
SIRAISI, N.G. Avicenna in Renaissance Italy. The *Canon* and medical teaching in Italian universities after 1500. Princeton, N.J., Princeton Univ. Press, 1987. 410 pp.

AXELROD, Julius (b. 1912). American neuropharmacologist; shared Nobel Prize (Physiology or Medicine) 1970 for studies on the pharmacology of central neurotransmitter substances.

AXELROD, J. An unexplained life in research. *Annual Review of Pharmacology*, 1988, **28**, 1–23.

AXENFELD, Karl Theodor Paul Polykarpos (1867–1930). German ophthalmologist; isolated Morax-Axenfeld bacillus; described metastatic ophthalmia.

BIALASIEWICZ, A.A. et al. Infectious diseases of the eye. Buren, Aeolus Press, [1994?], 15–20.

BABÈS, Victor (1854–1924). Romanian bacteriologist; devised mallein reaction for the diagnosis of glanders.

GHELERTER, I. Victor Babès. *Archivos Iberamericanos de Historia de la Medicina*, 1957, **9**, 227–33.

IGIROSIANU, J. Un grand contemporain de Pasteur: Victor Babès. 1854–1926. *Histoire des Sciences Médicales*, 1974, **8**, 549–58.

BABINGTON, Benjamin Guy (1794–1866). British physician; pioneer in laryngoscopy.

WILSON, T.G. Benjamin Guy Babington. *Archives of Otolaryngology*, 1966, **83**, 72–6.

BABINSKI, Joseph François Felix (1857–1932). French neurologist.

GIJN, J. van. The Babinski sign: a centenary. Utrecht, Universiteit Utrecht, 1996, 176 pp.
TOURNAY, A. La vie de Joseph Babinksi. Amsterdam, Elsevier, 1967. 130 pp.

BABKIN, Boris Petrovich (1877–1950). Russian-born physiologist; professor at McGill University, Montreal.

DALY, I. de B. et al. ONFRS, 1952–3, **8**, 13–23.
Archival material: Osler Library, McGill University, Montreal.

BABUKHIN, Aleksandr Ivanovich (1827–1891). Founder of the Moscow School of Histology and Bacteriology.

METELKIN *et al.* A.I. Babukhin. [In Russian] Moskva, Medgiz, 1955, 305 pp.

BACCELLI, Guido (1832–1916). Italian physician in Rome.

BACCELLI, A. Guido Baccelli, ricordi. Napoli, La Riforma Medica, 1931, 114 pp.

BACON, Roger (*c.*1214–1292). English philosopher and scientist.

EASTON, S.C. Roger Bacon and his research for a universal science. Oxford, Blackwell, 1952. 255 pp.
LITTLE, A.G. Roger Bacon: Essays. Oxford, Clarendon Press, 1914. [Biography: pp. 337–72; bibliography: pp. 373–426]

BACOT, Arthur William (1868–1922). British entomologist; demonstrated rat-flea transmission of the plague bacillus.

GREENWOOD, pp. 109–137 (CB).
SMITH, A.H. Bacot, a martyr to science. *Scientific Monthly*, 1923, **15**, 359–63.

BADHAM, Charles (1780–1845). British physician; intoduced the term 'bronchitis'.

RICHARDS, R. Charles Badham M.D., F.R.S. (1780–1845). *Journal of the History of Medicine*, 1956, **11**, 54–65.

BADO [BALDI] Sebastiano (*fl.* 1640–1676). Italian physician; advocate of cinchona bark in the treatment of malaria.

ROMPEL, J. Der Arzt Baldo und die Chinarinde. *Stimmen der Zeit*, 1929, **47**, 124–36.

BAELZ, Erwin Otto Eduard von (1843–1913). German physician and anthropologist in the Far East; professor of medicine, Tokyo; gave important accounts of tropical diseases.

BÄLZ, T. Erwin Bälz. Das Leben eine deutschen Arzt in erwachenen Japan. 3te. Aufl. Stuttgart, J. Engelhorns Nachf., 1937. 287 pp.
SCHOTTLÄNDER, F. Erwin von Baelz, 1843–1913. Leben und Wirken eines Deutschen Arztes in Japan. Stuttgart, Ausland und Heimat Verlags-Aktiengesellschaft, 1928. 163 pp.

BIBLIOGRAPHY OF MEDICAL AND BIOMEDICAL BIOGRAPHY

VESCOVI, G. Erwin Baelz, Wegbereiter der japanischen Medizin; ein Lebensbild. Stuttgart, Gentner, 1972. 92 pp.

BAER, Karl Ernst von (1792–1876). Estonian biologist and embryologist; discovered mammalian ovum.

BAER, K.E. von. Autobiography... J. M. Oppenheimer, Editor. Canton, Mass, Science History Publications (Watson), 1986. 389 pp. Translated from German edition, Braunschweig, 1886.
OPPENHEIMER, J.M. DSB, 1970, **1**, 385–9.
RAIKOV, B.E. Karl Ernst von Baer, 1792–1876, sein Leben und sein Werk. Leipzig, Barth, 1968. 516 pp.

BÄRENSPRUNG, Friedrich Wilhelm Felix von (1822–1864). German physician, described eczema marginatum,'Bärensprung's disease'.

DIEPGEN, pp. 126–41 (CB).

BAGELLARDO, Paolo (? –1494). Italian paediatrician; author of De infantium aegretudinibus et remediis, 1472, first printed book dealing exclusively with paediatrics.

RUHRÄH, J. *American Journal of Diseases of Children*, 1928, **35**, 289–93.

BAGLIVI, Georgius (1668–1707). Italian biologist and physician; professor of anatomy in Rome.

GRMEK, M.D. DSB, 1970, **1**, 391–2.
SALOMON, D.M. Giorgio Baglivi und seine Zeit. Berlin, A. Hirschwald, 1889. 180 pp.
SCHULLIAN, D.M. (ed.) The Baglivi correspondence from the library of Sir William Osler. Ithaca, N.Y., Cornell Univ. Press, 1974. 531 pp.
Archival material: National Library, Florence; Waller Collection, Uppsala; Osler Collection, McGill University, Montreal.

BAILEY, Hamilton (1894–1961). British surgeon at Royal Northern Hospital, London.

HUMPHRIES, S.V. The life of Hamilton Bailey: surgeon, author and teacher. Beckenham, Ravenswood Press, 1973. 72 pp.
MARSTON, A. Hamilton Bailey: a surgeon's life. London, Greenwich Medical Media, 1999. 157 pp.

BAILEY, Kenneth (1909–1963). British biochemist (protein and muscle biochemistry).

CHIBNALL, A.C. BMFRS, 1964, **10**, 1–13.
Archival material: Cambridge University Library.

BAILEY, Percival (1892–1973). American neurologist and neurosurgeon.

BUCY, P.C. BMNAS, 1989, **58**, 3–46.

BAILEY, Walter, *see* **BAYLEY, Walter**

BAILLIE, Matthew (1761–1823). British pathologist; physician to St George's Hospital London.

BULLOUGH, V.L. DSB, 1970, **1**, 398–9.
CRAINZ, F. The life and works of Matthew Baillie, MD, FRS L&E, FRCP etc (1761–1823) Roma, Peliti Associati, 1995. 195 pp.

MACMICHAEL, W. The gold-headed cane (CB).

RODIN, A.E. The influence of Matthew Baillie's *Morbid anatomy*, Biography, evaluation and reprint. Springfield, C.C. Thomas, 1973. 293 pp.

Archival material: Royal College of Physicians of London; Royal College of Surgeons of England.

BAILLOU, Guillaume de (*c*.1538–1616). French physician; first to describe whooping cough ('tussis quintana').

GOODHALL, E.W. A French epidemiologist of the sixteenth century. *Annals of Medical History*, 1935, N.S.**7**, 405–27.

HUARD, P. DSB, 1970, **1**, 399–400.

BAILY, Walter, *see* **BAYLEY, Walter**

BAKER, George (1722–1809). British physician; demonstrated that lead poisoning in Devonshire was due to contamination of cider in lead casks.

CHILDS, St. J.R. Sir George Baker and the dry belly-ache. *Bulletin of the History of Medicine*, 1970, **44**, 213–40.

McCONAGHEY, R.M. Sir George Baker and the Devonshire colic. *Journal of the History of Medicine*, 1970, **25**, 383–413.

BAKER, John Randal (1900–1984) British cytologist.

BAKER, Peter Frederick (1939–1987). British physiologist; professor at King's College, London.

KNIGHT, D.E. & HODGKIN, A.L. BMFRS, 1990, **35**, 1–35.

BAKER, Sara Josephine (1873–1945). American physician; Chief, Division of Child Hygiene, New York.

BAKER, S.J. Fighting for life. New York, Macmillan, 1939. 260 pp. Reprinted with historical introduction, New York, Krieger, 1980. 272 pp.

WILLMER, E.N. & BRUNET, P.C.J. BMFRS, 1985, **31**, 33–63.

BAKER, William Morrant (1839–1896). British surgeon; described 'Baker's cyst'.

HERSHMAN, M.J. William Morrant Baker (1837–1896). *International Journal of Dermatology*, 1980, **19**, 409–12.

BALARD, Antoine-Jérome (1802–1876). French chemist; discovered bromine, isolated amyl nitrite.

CENTENAIRE de la mort d'Antoine-Jérome Balard. [Contributions by various authors and bibliography of his publications.] *Revue d'Histoire de la Pharmacie*, 1977, **24**, 5–96, 137–46, 203–17.

BALDI, *see* **BADO**

BALEY, Walter, *see* **BAYLEY, Walter**

BALFOUR, Francis Maitland (1851–1882). British embryologist.

CHURCHILL, F.B. DSB, 1970, **1**, 420–22.

F., M. *Proceedings of the Royal Society of London*, 1883, **35**, xx–xxvii.

RUSSELL, E.S. Form and function; a contribution to the history of animal morphology. London, J. Murray, 1916. Reprinted 1982. [Places Balfour's work in historical context (pp. 268–301).]

BALL, Eric Glendinning (1904–1979). British/American biochemist; professor at Harvard Medical School.

BUCHANAN, J.M. & HASTINGS, A.B. BMNAS, 1989, **58**, 49–73.
Archival material: Countway Library, Harvard Medical School.

BALLANCE, Charles Alfred (1856–1936). British surgeon; first President of Society of British Neurological Surgeons.

PLARR, 1930–51, 36–39.

BALLANTYNE, John William (1861–1923). British obstetrician; pioneer advocate of antenatal care.

RUSSELL, H. J.W. Ballantyne, M.D. Edinburgh, Royal College of Physicians of Edinburgh, 1971. 34 pp.

BALLS, Arnold Kent (1891–1966). American enzymologist.

HASSID, W.Z. BMNAS, 1970, **41**, 3-21.

BALTIMORE, David (b.1938). American virologist; shared Nobel Prize (Physiology or Medicine) 1975 for discovery of the interaction between tumour virus and genetic material of the cell.

CROTTY, S. Ahead of the curve: David Baltimore's life in science. Berkeley, Univ. of California Press, 2001. 270 pp.

BALZER, Felix (1848–1929). French dermatologist and syphilologist; first to suggest bismuth in treatment of syphilis.

MENETRIER, –. *Bulletin de l'Académie de Médecine*, Paris, 1929, 3 ser., **101**, 417–25.

BANCROFT, Edward (1744–1821). American physician who spent many years in England; noted transmission of yaws by flies, 1769.

ANDERSON, G.T. & ANDERSON, D.K. Edward Bancroft, 1744–1821, MD, FRS, aberrant 'practitioner of physicke'. *Medical History*, 1973, **17**, 356–67.

BANCROFT, Joseph (1836–1894). British/Australian physician; discovered the parasite of filariasis, *Wuchereria bancrofti*.

FEARN, J. & POWELL, C.W. (eds.) The Bancroft tradition. Brisbane, Amphion Press, 1991. 268 pp.
WINTON, R. From the sidelines of medicine: the Bancroft memorial lecture. *Medical Journal of Australia*, 1964, **2**, 357–69.

BANCROFT, Thomas Lane (1860–1933). Australian naturalist; demonstrated *Aedes aegypti* as a vector of the dengue virus.

FEARN, J. & POWELL, C.W. The Bancroft tradition. Brisbane, Amphion Press, 1991. 268 pp.

BIBLIOGRAPHY OF MEDICAL AND BIOMEDICAL BIOGRAPHY

BANG, Bernhard Laurits Frederik (1848–1932). Danish physician and veterinarian; discovered *Brucella abortus*.

WILLIAMS, T.F. & McKUSICK, V.A. Bernhard Bang: physician, veterinarian, scientist. *Bulletin of the History of Medicine*, 1954, **25**, 60–72.

BANISTER, John Bright (1880–1938) British gynaecologist.

FAIRBAIRN, J.S. *Journal of Obstetrics and Gynaecology of the British Empire*, 1938, **45**, 518–21.

BANISTER, Richard (?1570–1626). English itinerant oculist.

JAMES, R.R. & SORSBY, A. Richard Banister: additional facts in relation to the father of British ophthalmology. *British Journal of Ophthalmology*, 1934, **18**, 156–59.
SORSBY, A. Richard Banister and the beginnings of English ophthalmology. In: Underwood, E.A. (ed.) Science, medicine and history; essays...in honour of Charles Singer. London, Oxford University Press, 1953. vol. **2**, pp. 43–55.
SORSBY, W.J. & BISHOP, W.J. A portrait of Richard Banister. *British Journal of Ophthalmology*, 1948, **32**, 362–66.

BANTI, Guido (1852–1925). Italian pathologist; described 'Banti's disease', 1882, and 'Banti's syndrome', 1894 – splenic and splenomegalic anaemia.

FRANCESCHINI, P. DSB, 1970, **1**, 438-40.
PATRASSI. G. Guido Banti, the hematologist. *Scientia Medica Italica*, 1958, **7**, 13–41.

BANTING, Frederick Grant (1891–1941). Canadian physician; joint isolator of insulin; shared Nobel Prize 1923.

BEST, C.H. ONFRS, 1942–44, **4**, 21–26.
BLISS, M. Banting: a biography. 2nd ed. Toronto, Univ. of Toronto Press, 1993. 336 pp. First published 1984.
STEVENSON, L. G. Sir Frederick Banting. Revised edition, London, Heinemann, 1947. 446 pp. First published 1946.
STEVENSON, L.G. DSB, 1970, **1**, 440-43.
Archival material: University of Toronto.

BÁRÁNY, Robert (1876–1936). Austrian oto-rhino-laryngologist, Nobel Prize 1914.

JOAS, G. Robert Bárány (1876–1936): Leben und Werk. Frankfurt a. Main, Europäischer Verlag der Wissenchaften, 1997. 612 pp.
MAJER, E.H. DSB, 1970, **1**, 446–7.

BARBIER, Charles (1767–1841). Frenchman, devised (1820) a system of raised points, later modified by Braille, to enable the blind to read.

SAKULA, A. That the blind may read. *Journal of Medical Biography*, 1998, **6**, 21–27.

BARBOUR, Henry Gray (1886–1943). American pharmacologist.

KAY, A.S. DSB, 1970, **1**, 449-50.

BARCROFT, Joseph (1872–1947). British physiologist; professor at Cambridge University.

FRANKLIN, K.J. Joseph Barcroft, 1872–1947. Oxford, Blackwell, 1953, 381 pp.

HOLMES, F.L. DSB, 1970, **1**, 452–5.
ROUGHTON, F.J.W. ONFRS, 1948-49, **6**, 315–45.
Archival material: Department of Physiology, University of Cambridge; Medical Research Council, London.

BARD, Archibald (Archibald Philip) American neurophysiologist.

HARRISON, S. BMNAS, 1997, **77**, 15–26.

BARD, Louis Jean Marius (1857–1930). French physician; described 'Bard-Pic syndrome', a complication of pancreatic carcinoma.

COTTIN, E. *et al.* L'oeuvre scientifique du Professeur Louis Bard. Genève, [after 1930], 210 pp.
RAVAUT, P. *Bulletin de l'Académie de Médecine*, Paris, 1930, 3 ser., **103**, 240–44.

BARD, Samuel (1742–1821). American physician.

LANGSTAFF, J.B. Doctor Bard of Hyde Park, the famous physician of revolutionary times, and the man who saved Washington's life. New York, Dutton, 1942. 365 pp.

BARDSLEY, James Lomax (1801–1876). British physician; introduced emetine in amoebiasis.

DNB, 3, 176–7.
Munk, **4**, 98 (CB).

BARFURTH, Dietrich (1849–1927). German anatomist; professor at Dorpat and Rostock.

GROTE, 1923, **2**, 1–22 (CB).

BARGER, George (1878–1939). British chemist; professor at Edinburgh; isolated ergotoxine and other ergot derivatives.

DALE, H.H. ONFRS, 1940, **3**, 63–85.
ROCKE, A.J. DSB, 1978, **15**, 10–11.

BARKER, Horace Albert (b.1907). American biochemist; professor at University of California.

BARKER, H.A. Explorations of bacterial metabolism. *Annual Review of Biochemistry*, 1978, **47**, 1–33.

BARKER, Lewellys Franklin (1867–1943). American physician; professor of medicine at Johns Hopkins Hospital, Baltimore.

BARKER, L.F. Time and the physician: the autobiography of Lewellys F. Barker. New York, Putnam, 1942. 350 pp.
Archival material: Library, Johns Hopkins University Medical School.

BARLOW, George Hilaro (1806–1866). British physician; physician to Guy's Hospital, published early description of subacute bacterial endocarditis.

MUNK, **4**, 28–29.

BARLOW, James (1767–1839). British surgeon; performed first caesarean section in England, from which the mother recovered (1793).

McQUAY, T.A. A Blackburn surgeon: James Barlow, 1767–1839. *Practitioner*, 1965, **195**, 103–8.

BARLOW, Thomas (1845–1945). British physician; distinguished infantile scurvy ('Barlow's disease') from rickets, 1883.

Sir Thomas Barlow, 1845–1945. Three selected lectures and a biographical sketch. London, Dawsons, 1965. 111 pp.
ELLIOTT, T.R. ONFRS. 1945, **5**, 159–67.
Archival material: CMAC.

BARNARD, Christiaan Neethling (1922–2001). South African surgeon; performed first human cardiac transplant, 1967.

BARNARD, C. One life. London, Harrap, 1970. 536 pp.

BARNARD, Joseph Edwin (1870–1949). British physicist; developed ultra-violet microscope.

MURRAY, J.A. ONFRS, 1950, **7**, 3–8.

BARNARDO, Thomas John (1845–1905). British physician and philanthropist; friend of destitute children, for whom he established 'Dr Barnardo's Homes'.

BREADY, J.W. Doctor Barnardo, physician, pioneer, prophet. Child life yesterday and today. London, Allen & Unwin, 1935. 271 pp. First published 1930.
WAGNER, G. Barnardo. London, Weidenfeld & Nicolson, 1979. 344 pp.
WILLIAMS, G. Barnardo the extraordinary doctor. London, Macmillan, 1966. 177 pp. First published 1943.

BARONIO, Giuseppe (1759–1811). Italian surgeon; transplanted tissue and carried out experimental surgery in animals.

ON GRAFTING IN ANIMALS. Degli innesti animali, Giuseppe Baronio. Translated by J.B. Sax, with an introduction by R.M. Goldwyn. Boston, Francis A. Countway Library, 1985. 87 pp.

BARR, Murray Llewellyn (1908–1995). Canadian geneticist; discovered sex chromatin.

POTTER, P. & SOLTAN, H. BMFRS, 1997, **43**, 31–46.

BARRÉ, Jean-Alexandre (1880–1967). French neurologist ('Guillain-Barré syndrome').

ROHMER, F. Jean-Alexandre Barré (1880–1967); l'homme, sa vie et son oeuvre. Conférence de l'Histoire de la Médecine, Lyon. Lyon, Fondation Marcel Mécrieux, 1984. pp. 185–208.

BARRY, James Miranda (*c*.1795–1865). British woman who masqueraded as a man and served as Inspector-General of Army Hospitals, British Army.

HOLMES, R. Scanty particulars: the life of Dr James Barry. London, Viking, 2002. 338pp.
RAE, I. The strange story of Dr. James Barry, Army surgeon, Inspector-General of Hospitals, discovered on death to be a woman. London, Longmans Green, 1958. 124 pp.
ROSE, J. The perfect gentleman. The remarkable life of Dr. James Miranda Barry, the woman who served as an officer in the British Army from 1813 to 1859. London, Hutchinson, 1977. 160 pp.

BARRY, Martin (1802–1855). British embryologist and histologist.

WILLIAMS, W.C. DSB, 1970, **1**, 476–8.

BARTELMEZ, George William (1885–1967). American anatomist and embryologist; professor of anatomy, University of Chicago.

BODIAN, D. BMNAS, 1973, **43**, 1–26.

BARTHEZ, Paul-Joseph (1734–1806). French physiologist.

COWAN, R.S. DSB, 1970, **1**, 478–9.
DULIEU, L. Paul-Joseph Barthez. *Revue d'Histoire des Sciences et leurs Applications*, 1971, **24**, 149–76. Includes bibliography.
REICH, N.S. Paul Joseph Barthez and the impact of vitalism on medicine and psychology. Dissertation submitted... for Doctor of Philosophy in History, University of California, Los Angeles, 1993. 591 pp. UMI Dissertation Services, Ann Arbor.

BARTHOLIN, Caspar Barthelsen (1585–1629). Danish anatomist.

SNORRASON, E. DSB, 1970, **1**, 479-80.

BARTHOLIN, Thomas (1616–1680). Danish anatomist and physiologist; discovered lymphatic system as a separate physiological system.

GARBOE, A. Thomas Bartholin, En bidrag til Dansk natur- og laegevidenskabs historie I det 17.aarhundrede. 2 vols. Kobenhavn, Munksgaard, 1949. 212+203 pp.
O'MALLEY, C.D. DSB, 1970, **1**, 482–3.
SKLAVEM, J.H. The scientific life of Thomas Bartholin. *Annals of Medical History*, 1921, **3**, 67–81.
TERCENTENARY CELEBRATION, *Janus*, 1916, **21**, 271–378.

BARTHOLOW, Roberts (1831–1904). American physician; professor of medicine, Ohio Medical College, Cincinnati.

WALKER, A.E. The development of the concept of cerebral location in the 19th century. *Bulletin of the History of Medicine*, 1957, **31**, 99–121.

BARTISCH, Georg (1536–1606). German ophthalmologic surgeon; wrote the first extensive treatise on ophthalmic surgery, 1583.

SHASTID, T.H. In: American Encyopedia and Dictionary of Ophthalmology, by C.A. Wood, Chicago, 1913, vol. 2, 888–95.
SORSBY, A. Bartisch's contributions to ophthalmology. *Medical History*, 1968, **12**, 205–7.
VERREY, F. Rencontre avec Georg Bartisch, oculiste du XVIe siècle. *Ophthalmologica* (Basel), 1961, **141**, 81–99.

BARTLETT, Frederic Charles (1886–1969). British experimental psychologist.

BROADBENT, D.E. BMFRS, 1970, **16**, 1–13.
Archival material: Library, University of Cambridge.

BARTON, Clara Harlowe (1821–1912). American philanthropist; founded school for nursing in USA, 1873, and the American Red Cross.

OATES, S. B. A woman of valor: Clara Barton and the Civil War. New York, Macmillan, 1994. 527 pp.
PRYOR, E.B. Clara Barton: professional angel. New ed. Philadelphia, University of Philadelphia Press, 1988. 444 pp.

BARTON, William Paul Crillon (1786–1856) American naval surgeon; later professor of botany, Philadelphia.

PLEADWELL, F.L. William Paul Crillon Barton, surgeon United States Navy, a pioneer in American naval medicine (1786–1856). *Annals of Medical History*, 1919, **2**, 367–80.

BARTTER, Frederic Crosby (1914–1983). American physiologist; professor of medicine, San Antonio, Texas; distinguished endocrinologist.

WILSON, J.D. & DELEA, C.S. BMNAS, 1990, **59**, 3–24.

BARUCH, Simon (1840–1921). Polish/American physician, pioneer of hydrotherapy in the USA.

WARD, P.S. Simon Baruch: rebel in the ranks of medicine, 1840–1921. Tuscaloosa, University of Alabama Press, 1994. 399 pp.

BASCH, Samuel Siegfried Karl von (1837–1905). Bohemian physician; devised the earliest sphygmomanometer.

JANTSCH, M. Samuel von Basch: zu seinem 50. Todestag am 25 April 1955. Wien, Brüder Hollinek, 1955. 10pp. From *Wiener Medizinische Wochenschrift*, 1955, **105**, 323–6.

BASEDOW, Carl Adolph von (1799–1854). German physician; gave classical description of exophthalmic goitre.

BROGHAMMER, H. Sanitätsrat Dr Karl Adolf von Basedow (28.3.1799–14.4.1854): Kreisphysikus von Merseburg. Herbolzheim, Centaurus, 2000. 188 pp.
BUCHHEIM, I. Carl Adolph von Basedow. Zu seinem 100. Geburtstag. *Endokrinologie*, 1954, **31**, 129–33.

BASEILHAC, Jean, Frère Côme (1703–1781). French surgeon; advanced the operative technique of lithotomy.

CHEVREAU, A. Un grand chirurgien au XVIIIe siècle, Frère Côme. Mesnil-sur-l'Estrée, Typ. Firmin-Didot; 1912, 92 pp.

BASOV, Vasilii Aleksandrovich (1812–1879). Russian surgeon; established first experimental gastric fistula.

KIPRENSKY, Y.V. [V.A. Basov and his contribution to the development of Russian surgery.] *Sovetskaya Meditsina*, 1963, **26**(4), 145–9 [in Russian].
ZAKHAROV, V.I. V.A. Basov, 1812–1879, Moskva, Medgiz, 1953.

BASS, George (1771–1803). British surgeon and explorer in Australia; Bass Strait and other localities named after him.

BOWDEN, K.M. George Bass (1771–1803): his discoveries, romantic life and tragic disappearance. London, Geoffrey Cumberlege, 1952. 171 pp.

BASSI, Agostino Maria (1773–1856). Italian scientist; founder of the doctrine of parasitic micro-organisms.

ARCIERI, G.P. Agostino Bassi in the history of medical thought. Florence, L.S. Olschki, 1956. 40 pp. First published New York, 1938.
ROBINSON, G. DSB, 1970, **1**, 492–4.

BASSINI, Edoardo (1844–1924). Italian surgeon; devised method for radical cure of inguinal hernia; modified Hahn's nephropexy.

BROWN, R.K. & GALLETTI, G. Bassini's contribution to our understanding of inguinal hernia. *Surgery*, 1960, **47**, 631–32.
SIMPOSIO in honore di Edoardo Bassini. Milano, Università degli Studi, 1987. 201 pp.

BASTIAN, Henry Charlton (1837–1915). British neurologist and bacteriologist; last scientific believer in spontaneous generation.

CLARKE, E. DSB, 1970, **1**, 495–8.
RANG, M. Life and work of Henry Charlton Bastian, 1837–1915. [Thesis]. University College Hospital, London, 1954.
Archival material: Library, School of Medicine, University College London.

BATEMAN, Thomas (1778–1821). British dermatologist; physician to the Public Dispensary and the Fever Institution, London.

[RUMSEY, J.] Some account of the life and character of the late Thomas Bateman. 2nd ed., London, Longman, Rees, Orme, Brown and Green, 1826. 228 pp.

BATES, Marston (1906–1974). American naturalist and epidemiologist.

KIMBER, W.C. DSB, 1990, **17**, 51–3.

BATESON, William (1861–1926). British morphologist and geneticist.

BATESON, B. William Bateson, F.R.S., naturalist: his essays and addresses, together with a short account of his life. London, Cambridge Univ. Press, 1928. 473 pp.
COLEMAN, W. Bateson and chromosomes: conservative thought in science. *Centaurus*, 1970, **15**, 228–314.
COLEMAN, W. DSB, 1970, **1**, 505–6.
Archival material: John Innes Institute, Norwich, England; Royal Society of London; American Philosophical Society, Philadelphia.

BATTELLI, Frédéric (1867–1941). Swiss physiologist.

MORSIER, G. & MOUNIER, M. La vie et l'oeuvre de Frédéric Battelli. Basle, Otto Schwabe, 1977, 130 pp.

BATTEY, Robert (1828–1895). American gynaecologist.

LONGO, D. The rise and fall of Battey's operation: a fashion in surgery. *Bulletin of the History of Medicine*, 1979, **53**, 244–67.

BATTIE, William (1704–1776). British physician; wrote first English textbook on psychiatry.

HUNTER, R.A. & MACALPINE, I. William Battie, M.D., F.R.S. *Practitioner*, 1955, **174**, 208–15.

BATTLE, William Henry (1855–1936) British surgeon.

PLARR, 1930–51.
–. *St. Thomas's Hospital Gazette*, 1937, **35**, 269–73.

BAUDELOCQUE, Jean Louis (1746–1810). French obstetrician; invented a pelvimeter and advanced knowledge of pelvimetry and the mechanics of labour.

LEBEAUD, –. Le Baudelocque des campagnes. Paris, 1835.

LEROUX, –. Discours prononce sur la tombe de M. Baudelocque, le 3 Mai 1810. Paris, Migneret, [1810]. 8 pp.

BAUER, Louis (1814–1898). German/American orthopaedic surgeon; the first exponent of American orthopaedics.

SHANDS, A.R. Louis Bauer, a pioneer in American orthopaedic surgery. *Current Practice in Orthopaedic Surgery*, 1966, **3**, 18–36.

BAUHIN, Caspar (1560–1624). Swiss botanist and anatomist; professor at Basle.

WHITTERIDGE, G. DSB, 1970, **1**, 522–5.

BAUMANN, Eugen Albert (1846–1896). German chemist; initiated research leading to the discovery of thyroxine.

SPAUDE, M. Eugen Albert Baumann (1846–1896): Leben und Werk. Zürich, Juris Druck, 1973, 46 pp.

BAYLE, Antoine Laurent Jessé (1799–1858). French psychiatrist; gave early clear description of dementia paralytica.

MÜLLER, A. Antoine-Laurent Bayle; sein grundlegender Beitrag zur Erforschung der progressiven Paralyse. Zürich, Juris Verlag, 1965, 56 pp.

BAYLE, Gaspard Laurent (1774–1816). French pathologist; published important works on tuberculosis and cancer.

HUARD, P.A. & IMBAULT-HUART, M.J. Bayle, cancérologue. *Histoire des Sciences Médicales*, 1974, **8**, 73–84.

BAYLEY [Bailey, Baley], Walter (1529–1592). English physician to Queen Elizabeth I; published first separate work on ophthalmology printed in England.

HORTON-SMITH, L.G.H. Dr Walter Baily (or Bayley) *c.*1529–92, Physician to Queen Elizabeth: his parentage, his life, and his relatives and descendants. St Albans, Campfield Press, 1952. 115 pp.

POWER, D'A. Dr Walter Bayley and his works, 1529–1592. London, Royal Medico-Chirurgical Society, 1907. Reprinted from *Medico-Chirurgical Transactions*, 1907, **90**, 415–54.

BAYLISS, Leonard Ernest (1900–1964). British physiologist.

WINTON, F.R. DSB, 1970, **1**, 533–5.
Archival material: University College London.

BAYLISS, William Maddock (1860–1924). British physiologist; professor of general physiology, University College London.

BAYLISS, L.E. William Maddock Bayliss, 1860–1924: life and scientific work. *Perspectives in Biology and Medicine*, 1961, **4**, 460–79.

EVANS, C.L. Reminiscences of Bayliss and Starling. Cambridge, Univ. Press, 1964. 17 pp.

EVANS, C.L. DSB. 1970, **1**, 535–8.
Archival material: CMAC; University College London.

BAYNE-JONES, Stanhope (1888–1970). American bacteriologist and medical administrator.

COWDREY, A.E. War and healing; Stanhope Bayne-Jones and the maturing of American medicine. Baton Rouge, Louisiana State Univ. Press, 1991. 230 pp.
HARVEY, A.M. Stanhope Bayne-Jones: the story of a life devoted to country and to medicine. *Johns Hopkins Medical Journal*, 1981, **149**, 150–66.

BAYON, Henry Peter George (1876–1952). South African bacteriologist; induced experimental cancer by injection of tar.

–. Henry Peter Bayon. *British Medical Journal*, 1952, **2**, 1260.

BAZALGETTE, Joseph William (1819–1891). British engineer; planned London sewers and Thames Embankment.

HALLIDAY, S. The great stink of London: Sir Joseph Bazalgette and the cleansing of the Victorian metropolis. Stroud, Sutton, 1999. 210 pp.

BEADLE, George Wells (1903–1989). American geneticist; professor of biology (genetics), Stanford; professor of biology, California Institute of Technology; shared Nobel Prize 1958.

BEADLE, G.W. Recollections. *Annual Review of Biochemistry*, 1974, **43**, 1–13.
HOROWITZ, N.H. BMFRS, 1995, 41, 44–54.
HOROWITZ, N.H. BMNAS, 1990, 26–52, **59**.
Archival material: California Institute of Technology.

BEALE, Lionel Smith (1828–1906). British microscopist.

GEISON, G.L. DSB, 1970, **1**, 539–41.

BEARD, George Miller (1839–1883). American physician; wrote on medical and surgical uses of electricity; first described nervous exhaustion (neurasthenia).

ROSENBERG, C.E. The place of George Miller Beard in nineteenth-century psychiatry. *Bulletin of the History of Medicine*, 1962, **36**, 245–59.

BEATON FAMILY. Scottish medical family of 16th and 17th centuries. Fergus Beaton possessed a family library of ancient medical works, including Hippocrates, Avicenna, Bernard Gordon, etc., translated into Gaelic.

BANNERMAN, J. The Beatons: a medical kindred in the classical Gaelic tradition. Edinburgh, John Donald Publishers, 1986. 161 pp.

BEATSON, George Thomas (1848–1933). British surgeon; treated breast cancer by öophorectomy.

ARAFAN, B.M. Sir George Thomas Beatson, M.D. *Journal of Laboratory and Clinical Medicine*, 1987, **109**, 373.

BEAUMONT, William (1785–1853). American physiologist and physician; investigated digestion and movements of the stomach *in vivo*.

COHEN, I.B. (ed.) The career of William Beaumont and the reception of his discovery. New York, Arno Press, 1980. 642 pp.
HORSMAN, R. Frontier doctor: William Beaumont, America's first great medical scientist. Columbia, MO, Missouri University Press, 1996. 320 pp.

MYER, J.S. William Beaumont, a pioneer American physiologist. A newly edited and completely reillustrated edition of Life and Letters of William Beaumont. With an introduction by Estelle Brodman, to which is added a reprint of the first edition of Experiments and Observations on the Gastric Juice and on the Physiology of Digestion, by William Beaumont. With an introduction by R.E. Shanks. St Louis, C.V. Mosby, 1981. 394 pp. First published 1912; republished 1939.
ROSEN, G. DSB, 1970, **1**, 542–5.
Archival material: University of Chicago; Washington University School of Medicine, St Louis (Index, 1968).

BEAUPERTHUY, Luis Daniel (1803–1871). Born Guadeloupe, W. Indies; first protagonist of the mosquito transmission of yellow fever.

SANABRIA, A. & BEAUPERTHUY DE BENEDETTI, R. Beauperthuy; ensayo biográfico. Español-English, Caracas, Vargas, 1969, 171 pp.
WOOD, C.A. Louis Daniel Beauperthuy, *Annals of Medical History*, 1922, **4**, 166–74.

BECCARI, Nello (1883–1957). Italian anatomist.

FRANCESCHINI, P. DSB, 1970, **1**, 545–6.

BÉCHAMP, Pierre Jacques Antoine (1816–1908). French biochemist and chemist; professor of chemistry and pharmacy, Montpellier and Nancy.

GUÉDON, J.-C. DSB, 1978, **15**, 11–12.
HUME, E.C. Béchamp or Pasteur? A lost chapter in the history of biology. London, Daniel, 1932. 287 pp.
NONCLERCQ, M. Antoine Béchamp 1816–1908: l'homme et le savant originalité de son ouevre. Paris, Maloine, 1982. 249 pp.

BECHTEREV, V.M., *see* **BEKHTEREV, V.M.**

BECK, Theodoric Romeyn (1791–1855). American physician; published first notable American text on forensic medicine.

HAMILTON, F.H. Theodoric Romeyn Beck, 1791–1855. In: Gross, S.D. Lives of eminent American physicians and surgeons. Philadelphia, Lindsay & Blakiston, 1861, 776–95.
HAMILTON, F.H. *Transactions of the Medical Society of New York*, 1856, 3–63.

BÉCLARD, Pierre Augustin (1785–1825). French anatomist; first to excise the parotid.

LABORDE, J.V. Pierre Augustin Béclard et Jules Augustin Béclard: éloge lu à L'Académie de Medecine dans sa séance de 13 décembre 1898. Paris, Masson, 1898. 77 pp.
LE FAILLER, J.M. La vie angevin de P.A. Béclard. Paris, Arnette, 1944. 60 pp.

BÉCLÈRE, Antoine (1856–1939). French endocrinologist and radiotherapist.

BÉCLÈRE, Antoinette. Antoine Béclère, pionnier en endocrinologie; l'un des fondateurs de la virologie et de l'immunologie; fondateur de la radiologie française, 1856–1939. Paris, Baillière, 1973. 475 pp.

BECQUEREL, (Antoine) Henri (1852–1908). French physicist; discovered radioactivity; shared Nobel Prize for Physics in 1903.

ROMER, A. DSB, 1970, **1**, 558–61.

BEDDOE, John (1826–1911). British physician and anthropologist; graduated in Edinburgh, served in the Crimea and practised in Bristol.

> BEDDOE, J. Memories of eighty years. Bristol, Arrowsmith; London, Simpkin Marshall, 1910. 321 pp.
> DANIEL, G. DSB, 1970, **1**, 562–3.

BEDDOES, Thomas (1760–1808). British physician and chemist; invented 'pneumatic' system of therapeutics by inhalation of medicated gases.

> CARTWRIGHT, F.F. The English pioneers of anaesthesia. Bristol, John Wright, 1952, pp. 3–45.
> KNIGHT, D.M. DSB, 1970, **1**, 563–4.
> PORTER, R. Doctor of society: Thomas Beddoes and the sick trade in late-Enlightenment England. London, Routledge, 1992. 238 pp.
> STANSFIELD, D.A. Thomas Beddoes, M.D. 1760–1808; chemist, physician, democrat. Dordrecht, Reidel, 1984. 306 pp.

BEDDOES, Thomas Lovell (1803–1849). British physician, poet and playwright; son of Thomas Beddoes (1760–1808).

> DONNER, H.W. Thomas Lovell Beddoes: the making of a poet. Oxford, Blackwell, 1935. 403 pp.
> SNOW, R.H. Thomas Lovell Beddoes: eccentric and poet. New York, Convici-Friede, 1938. 227 pp.

BEDSON, Samuel Phillips (1886–1969). British virologist.

> DOWNIE, A.W. BMFRS, 1970, **16**, 15–35.

BEER, Georg Joseph (1763–1821). Austrian ophthalmologist; opened first known eye hospital in Europe, 1786.

> ALBERT, D.M. & BLODI, F.C. Georg Joseph Beer: a review of his life and contributions. *Documenta Ophthalmologica*, 1988, **68**, 79–103.

BEERS, Clifford Whittingham (1876–1943). American psychiatrist.

> DAIN, N. Clifford W. Beers: advocate for the insane. Pittsburgh, University of Pittsburgh Press, 1980. 392 pp.

BEESON, Paul Bruce (b.1908) American physician.

> RAPPORT, R. Physician: the life of Paul Beeson. Fort Lee, NJ, Barricade Books, 2001. 277 pp.

BEEVOR, Charles Edward (1854–1908). British neurophysiologist; researched in localization of cerebral function.

> CLARKE, E. DSB, 1970, **1**, 569–70.

BEHRING, Emil von (1854–1917). German bacteriologist; founder of immunology as a science; recipient of first Nobel Prize for Medicine (1901) for his discovery of antitoxins and the introduction of serum therapy.

> BIELING, R. Der Tod hat das Nachsehen; Emil von Behring, Gestalt und Werk. Bielefeld, Bielefelder Verlag, 1954. 275 pp. First published by H. Zeiss and R. Bieling, 1940.

SCHADEWALDT, H. DSB, 1970, **1**, 574–8.

UNGER, F.H.H. Unvergängliches Erbe; das Lebenswerk Emil von Behrings. Oldenberg i.O., Stalling, 1941, 239 pp.

ZEISS, H. & BIELING, R. Behring: Gestalt und Werk. 2te Aufl. Berlin-Grunewald, Bruno Schultz Verlag, 1941. 627 pp.

BEIERWALTES, William Henry (b.1916). American physician.

BEIERWALTES, W.H. Love of life: autobiographical sketches. New York, Vantage Press, *c.* 1996. 215 pp.

BEIJERINCK, Martinus Willem (1851–1931). Dutch virologist and botanist.

HUGHES, S.S. DSB, 1978, **15**, 13–15.

van ITERSON, G. *et al.* Martinus Willem Beijerinck: his life and his work. The Hague, Nijhoff, 1940. 193 pp. Reprinted Madison, Wisconsin, Science Technical Publications, 1983.

BÉKÉSY, Georg von (1899–1972). Hungarian/American otologist; Nobel prizewinner 1961 for work on the inner ear.

DANIEL, J. Békésy, György 1899–1972. Budapest, Akademiai Kaido, 1990. 289 pp.

ELLIS, S.R. DSB, 1990, **17**, 62–4.

RATLIFF, F. BMNAS, 1976, **48**, 25–49.

BEKHTEREV, Vladimir Mikhailovich (1857–1927). Russian neurologist and psychiatrist.

GRIGORYAN, N. DSB, 1970, **1**, 597–9.

GROTE, 1927, **6**, 597–9 (CB).

BELL, Benjamin (1749–1806). British surgeon; surgeon to Royal Infirmary, Edinburgh.

BELL, B. The life, character, and writings of Benjamin Bell, by his grandson. Edinburgh, Edmonston & Douglass, 1868. 170 pp.

BELL, Charles (1774–1842). British surgeon, anatomist, author and artist; made notable contributions to knowledge of the nervous system.

AMACHER, P. DSB, 1970, **1**, 583–4.

GORDON-TAYLOR, G. & WALLS, E.W. Sir Charles Bell: his life and times. Edinburgh, London, Livingstone, 1958. 288 pp.

HALE-WHITE, pp. 42–62. (CB)

Archival material: Royal Army Medical College; University College London; Library, Wellcome Institute for the History of Medicine, London.

BELL, Francis Gordon (1887–1970). New Zealand surgeon; professor at Otago University.

BELL, F.G. Surgeon's saga. Wellington, Reed, 1968. 216 pp.

BELL, John (1763–1820). Scottish surgeon.

WALLS, E.W. John Bell. *Medical History*, 1964, **8**, 69–77.

BELL, Joseph (1837–1911). British surgical teacher in Edinburgh; Arthur Conan Doyle, one of his students, founded the character of Sherlock Holmes on Bell, remarkable for his skill in diagnosis and unexpected deductions.

LIEBOW, E. Dr Joe Bell; model for Sherlock Holmes. Bowling Green, Ohio, Bowling Green Univ. Popular Press, 1982. 269 pp.

SAXBY, J.M.E. Joseph Bell: an appreciation by an old friend. Edinburgh & London, Oliphant, Anderson & Ferrier, 1913. 92 pp.

BELLINI, Lorenzo (1643–1704). Italian anatomist; described 'Bellini's ducts' and 'tubules'

BROWN, T.M. DSB, 1970, **1**, 592–4.

BENACERRAF, Baruj (b.1920). American immunologist; shared Nobel Prize (Physiology or Medicine) 1980.

–. The Nobel Prize for Physiology or Medicine, 1980, awarded to Baruj Benacerraf, Jean Dausset and George D. Snell. *Scandinavian Journal of Immunology*, 1922, **35**, 373–98.
BENACERRAF, B. Reminiscences. *Immunological Reviews*, 1985, **84**, 7–27.

BENCE-JONES, Henry, *see* **JONES, Henry Bence**

BENEDEN, Edouard van (1846–1910). Belgian embryologist and zoologist; described segmentation of the ovum and discovered the centrosome.

FLORKIN, M. DSB, 1970, **1**, 600–602.

BENEDETTI, Alessandro (1460–1512). Italian anatomist and physician.

FERRARI, G. L'esperienza del passato. Alessandro Benedetti, filogo e medica umanista. Firenze, Leo S. Olschki, 1996. 357 pp.
LIND, L.R. Studies in pre-Vesalian anatomy. Philadelphia, American Philosophical Society, 1975, pp. 69–137.
SCHULLIAN, D.M. DSB, 1970, **1**, 603–4.

BENDITT, Earl Philip (1926–1996). American pathologist; carried out important work on amyloid and atherosclerosis.

LAGUNOFF, D. & MARTIN, G.M. BMNAS, 2002, **81**, 24–46.

BENEDICT, Francis Gano (1870–1957). American chemist and physiologist.

CULOTTA, C.A. DSB, 1970, **1**, 609–11.
DUBOIS, E.F. & RIDDLE, O. BMNAS, 1958, **32**, 66–98.
Archival material: Countway Library, Harvard University.

BENEDICT, Stanley Rossiter (1884–1936). American biochemist; professor of physiological chemistry, Cornell University.

McCOLLUM, E.V. BMNAS, 1952, **27**, 155–77.

BENEDIKT, Moritz (1835–1920). Hungarian physician; described unilateral paralysis of oculomotor nerve with trembling on opposite side.

ELLENBERGER, H.F. Moritz Benedikt (1835–1920). *Revue d'Histoire de la Médecine Hebraique*, 1974, **27**, 133–42.

BENIVIENI, Antonio (1443–1502). Italian pathologist; 'Father of pathological anatomy'.

FRANCESCHINI, P. DSB, 1970, **1**, 611–12.

MAJOR, R.II. Antonio di Pagolo Benivieni. *Bulletin of the Institute of the History of Medicine*, 1935, **3**, 739–55.

BENN, Gottfried (1886–1936). Leading German 20th-century poet; medical studies and hospital work provided subject matter for his early poems.

BLUHM, E. & WOLFF, U. Gottfried Benn: eine Bilddokumentation. Munich, Medical Concept, 1981. 121 pp.

BENNET, James Henry (1816–1891). British gynaecologist; pioneer in the study of uterine pathology.

REIS, R.A. James Henry Bennet; the forgotten pioneer. *Annals of Medical History*, 1940, **2**, 234–44.

BENNETT, John Hughes (1812–1875). British physician.

WARNER, J.H. Therapeutic exploration and the Edinburgh bloodletting controversy: two perspectives on the medical meaning of science in the mid-nineteenth century. *Medical History*, 1980, **24**, 241–8.

BENSLEY, Robert Russell (1867–1956). Canadian anatomist.

ROOFE, P.G. DSB, 1970, **1**, 613–14.

BENZER, Seymour (b.1921) American molecular biologist and geneticist.

WEINER, J. Time, love, memory: a great biologist and his quest for the origins of behavior. New York, Knopf, 1999. 300 pp.

BENZI, Ugo, *see* **UGO BENZI**

BÉRARD, Auguste (1802–1846). French surgeon; published first important monograph on parotid tumours.

DENONVILLIERS, –. Élogue du professeur August Bérard. *Bulletin de la Sociéte de Chirurgie de Paris*, 1852, 159–82.
–. *Mémoires de la Société de Chirurgie de Paris*, 1857, **4**, 1–29.

BERDMORE, Thomas (1740–1785). British dentist; published first English textbook on dentistry.

ANDREANA, S. et al. Thomas Berdmore, dentist of his Majesty George III, and dental calculus. *Journal of the History of Dentistry*, 1996, **44**, 115–17.

BERENGARIO DA CARPI, Giacomo (*c*.1460–1530). Italian anatomist; professor of surgery, Padua and Bologna.

O'MALLEY, C.D. DSB, 1970, **1**, 617–21.
PUTTI, V. Berengario da Carpi: saggio biografico e bibliografico sequito della traduzione dell "De fractura calva sive cranei". Bologna, L.Cappelli, 1937. 352 pp.

BERGEL, Franz (1900–1987). Austrian-born medical chemist; professor of chemistry, University of London; Head, Chemistry Dept., Chester Beatty Research Institute, London.

TODD, Lord. BMFRS, 1988, **34**, 1–19.

BERGER, Hans [Johann] (1873–1941). German psychiatrist; founded electroencephalography.

HANS BERGER on the electroencephalogram of man. The fourteen original reports...edited by P. GLOOR. Amsterdam, Elsevier, 1969. 350 pp. (*Electroencephalography and Clinical Neurophysiology*, 1958, Suppl. **28**).

GLOOR, P. Hans Berger and the discovery of the electroencephalogram. *Electroencephalography and Clinical Neurophysiology*, 1966, Suppl. **28**, 1–36.

SCHULTE, B.P.M. DSB, 1973, **2**, 1–2.

BERGER, Paul (1845–1908). French surgeon; introduced interscapulothoracic amputation ("Berger's operation").

FORMESTRAUX, J. de. Paul Berger (1845–1908) *Biographies Médicales*, 1934, **8**, No. 5, 341–56.

BERGMANN, Ernst von (1836–1907). Russian-born surgeon; introduced steam sterilization in surgery.

BUCHHOLTZ, A. Ernst von Bergmann. 4te. Aufl. Leipzig, F.C.W. Vogel, 1925. 642 pp.

FROHNMEYER, K. Ernst von Bergmann: ein Meister der Chirurgie. Stuttgart, 1947. 63 pp.

GENSCHOREK, W. Wegbreiter der Chirurgie: Joseph Lister, Ernst von Bergmann. Leipzig, Hirzel, 1984. 223 pp.

BERGMANN, Max (1886–1944). German/American biochemist and organic chemist.

FRUTON, J.S. DSB, 1978, **15**, 15–16.

Archival material: Library, American Philosophical Society, Philadelphia.

BERGSTRÖM, Sune (b. 1916) Swedish biochemist; shared Nobel Prize (Physiology or Medicine), 1982, for elucidation of the chemical structure of prostaglandins.

Les Prix Nobel en 1982.

BERLIN, Rudolph (1833–1897). German ophthalmologist.

– . *Klinische Monatsblätter für Augenheilkunde*, 1897, **35**, 358–60.

BERNAL, John Desmond (1901–1971). British crystallographer and molecular biologist.

HODGKIN, D.M. BMFRS, 1980, **26**, 17–84.

GOLDSMITH, M. Sage: a life of J.D. Bernal. London, Hutchinson, 1980. 256 pp.

SNOW, C.P. DSB, 1978, **15**, 16–20.

SWANN, B. & APRAHAMIAN, F.J.D. Bernal: a life in science and politics. London, Verso, 1999. 324 pp.

Archival material: Cambridge University Library; Birkbeck College, London.

BERNARD, Claude (1813–1978). French physiologist; a founder of experimental medicine; professor of medicine, Collège de France.

FOSTER, M. Claude Bernard. London, Fisher Unwin, 1899. 245 pp.

GRMEK, M.D. DSB, 1973, **2**, 24–34.

GRMEK, M.D. Le legs de Claude Bernard. Paris, Fayard, 1997. 439 pp.

HOLMES, F.L. Claude Bernard and animal chemistry: the emergence of a scientist. Cambridge, Mass., Harvard Univ. Press, 1974. 541 pp.

OLMSTED, J.M.D., & OLMSTED, E.H. Claude Bernard and the experimental method in medicine. New York, Schuman, 1952. 277 pp.

VIRTANEN, R. Claude Bernard and his place in the history of ideas. Lincoln, Univ. Nebraska Press, 1960. 156 pp.

BERNARD DE GORDON (*fl. c.*1283–1308). French physician; taught at Montpellier; first to mention reading spectacles.

DEMAITRE, L.E. Dr Bernard de Gordon, professor and practitioner. Toronto, Pontifical Institute of Mediaeval Studies, 1980. 236 pp.

BERNAYS, Augustus Charles (1854–1907). American surgeon.

BERNAYS, T. Augustus Charles Bernays. A memoir. St Louis, Mosby, 1912. 309 pp.

BERNHARDT, Martin (1844–1915). German neurologist; described meralgia paraesthetica ('Bernhardt's disease).

JELLIFFE. S.E. *Journal of Nervous and Mental Diseases*, 1917, **44**, 478–80.

BERNHEIM, Hippolyte Marie (1840–1919). French psychologist and psychotherapist.

HUARD, P. DSB, 1973, **2**, 35–6.

BERNOUILLI, Daniel (1700–1782). Dutch physician and mathematician.

HUBER, F. Daniel Bernouilli (1700–1782) als Physiologe und Statistiker. Basel, Schwabe, 1959. 104 pp.
STRAUB, H. DSB, 1973, **2**, 36–46.

BERNSTEIN, Julius (1839–1917). German physiologist; determined characteristics of the action potential.

RUDOLPH, G. Julius Bernstein (1839–1917). In: Founders of experimental physiology, ed. J.W. Boylan. Munich, J.F. Lehmann, 1971, pp. 249–71.
RUDOLPH, G. DSB, 1978, **15**, 20–22.
TSCHERMAK, A. von. Julius Bernstein's Lebensarbeit. Zugleich ein Beitrag zur Geschichte der neueren Biophysik. *Plüger's Archiv für die gesamte Physiologie*, 1919, **174**, 11–89.

BERRILL, Norman Stewart (1903–1996). British/Canadian/American embryologist.

SCRIVER, C.R. BMFRS, 1999, **45**, 21–34.

BERSON, Solomon Aaron (1918–1972). American physician; professor, Mount Sinai School of Medicine; developed radioimmunoassay methods.

RALL, J.E. BMNAS, 1990, **59**, 54–70.

BERT, Paul (1833–1886). French physiologist; pioneer in study of respiratory physiology.

BERT, P. La pression barométrique...facsimile of 1878 edition. Biographical introduction by P. Dejour. Paris, Ed du Centre Nationale de la Recherche Scientifique, 1979. 1183 pp.
DUBREUIL. L. Paul Bert. Paris, Alcan, 1935.
LA BROSSE, P. de. Une des grandes energies françaises: Paul Bert. Hanoi, Extreme-Orient, 1925. 159 pp.
MANI, N. DSB, 1973, **2**, 59–63.

BERTHOLD, Arnold Adolph (1803–1861). German physiologist; founder of hormone research.

KLEIN, M. DSB, 1973, **2**, 72–3.

RUSH, H.P. A biographical sketch of Arnold Adolph Berthold, an early experimenter with ductless glands. *Annals of Medical History*, 1929, N.S. **1**, 208–14.

BERTILLON, Alphonse (1853–1914). French anthropometrist; invented system of identification of persons by measurement of certain parts of the body ('Bertillon system').

BERTILLON, S. Vie d'Alphonse Bertillon, inventeur de l'anthropométrie. 8e. éd. Paris, Gallimard, 1941. 224 pp.

RHODES, H.T.F. Alphonse Bertillon, father of scientific detection. London, Harrap, 1956. 238 pp.

BERTRAND, Gabriel (1867–1962). French biochemist.

KAY, A.S. DSB, 1973, **2**, 86–7.

BESNIER, E. (1831–1909). French dermatologist; described sarcoidosis ('Besnier-Boeck-Schaumann disease'), 1899.

BALSER, F. *Bulletin de la Société Française de Dermatologie et Syphiligraphie*, 1909, **20**, 207–13.

THIBIERGE, G. *Annales* de *Dermatologie et de Syphiligraphie*, 1909, 4 sér., **10**, 353–56.

BESREDKA, Alexandre (1870–1940). Russian bacteriologist and immunologist at Institut Pasteur.

DELAUNAY, A. Vie et oeuvre d'Alexandre Besredka. *Bulletin de l'Institut Pasteur*, 1971, **69**, 73–8.

BEST, Charles Herbert (1899–1978). Canadian biochemist; joint isolator of insulin.

BLISS, M. DSB, 1990, **17**, 80–81.

YOUNG, F.G. & HALES, C.N. BMFRS, 1982, **28**, 1–25.

BESTA, Carlo (1876–1940). Italian neuropsychiatrist.

AROSIO, F. Carlo Besta and the foundation of the National Neurological Institute in Milan. Milan, National Neurological Institute, 1993. 47 pp.

AROSIO, F. Carlo Besta, 1876–1940. Milan, Istituto Nazionale Neurologico "Carlo Besta", 1993. 111 pp.

BETHUNE, Henry Norman (1890–1939). Canadian surgeon, painter, poet, soldier, teacher and medical writer.

GORDON, S. & ALLAN, T. The scalpel the sword: the story of Doctor Norman Bethune. Revised edition. London, Robert Hale, 1954. 271 pp. Revised edition 1973.

HANNANT, L. (ed.) Norman Bethune's writing and art. Toronto, Univ. Taranto Press, 1998. 396 pp.

STEWART, R. Bethune. Toronto, New Press, 1973. 210 pp. Reprinted, Hamden, Conn., Archon Books. 1979.

BETTELHEIM, Bruno (1903–1990). Austrian/American child psychoanalyst.

POLLAK, A. The creation of Dr B. A biography. New York, Simon & Schuster, 1996. 478 pp.

SUTTON, A. Bettelheim: a life and legacy. New York, Basic Books, 1996. 606 pp.

BEURMANN, Charles Lucien de (1851–1933). French dermatologist; with H. Gougerot gave first description of sporotrichosis, 1912.

MASARY, F. *Bulletin de la Société Médical des Hôpitaux de Paris*, 1923, 3 ser. **47**, 1890–92.

BEZOLD, Albert von (1836–1868). German physiologist.

HERRLINGER, R. & KRUPP, I. Albert von Bezold (1836–1868): ein Pionier der Kardiologie. Stuttgart, G. Fischer, 1964. 131 pp.
STEUDEL, J. DSB, 1973, **2**, 110–11.

BIANCHI, Leonardo (1848–1927). Italian neuropsychiatrist.

CAMILLIS, B. de. Leonardo Bianchi. Napoli, 1934. 231 pp.

BICHAT, Marie François Xavier (1771–1802). French histologist, physiologist and surgeon.

CANGUILHEM, G. DSB, 1973, **2**, 122–3.
DOBO, N. & ROLE, A. Bichat: la vié fulgurante d'un génie. Paris, Perrin, 1989. 364 pp.
HAIGH, E. Xavier Bichat and the medical theory of the eighteenth century. *Medical History*, 1984, Suppl. 4. 146 pp; London, Wellcome Institute for the History of Medicine, 1984, 146 pp.
HUNEMAN, P. Bichat, la vie et la mort, Paris, Presses Universitaires de la France, *c*. 1998. 128 pp.

BIDDER, Friedrich Heinrich (1810–1894). Russian neuroanatomist and neurophysiologist.

ACHARD, T. Der Physiologe Friedrich Bidder, 1810–1894. Zürich, Juris, 1969. 57 pp.
CULOTTA, C.A. DSB, 1973, **2**, 123–5.

BIDLOO, Govert (1649–1713). Dutch anatomist and biologist.

BEEKMAN, F. Bidloo and Cowper, anatomists. *Annals of Medical History*, 1935, **7**, 113–29 [W. Cowper plagiarized Bidloo's Anatomia, 1685].
VAN DER PAS, P.W. DSB, 1978, **15**, 28–30.

BIEDL, Artur (1869–1933). Czech endocrinologist; described adiposo-genital dystrophy (Laurence-Moon-Biedl syndrome).

FEHER, L. & KAISER, W. Die Aufänge der modernen Endokrinologie. Pro memoria Artur Biedel. *NTM*, 1970, **7**, Heft 1, 99–108.

BIELSCHOWSKY, Alfred (1871–1940). German ophthalmologist.

KAUFMANN, A.K. Alfred Bielschowsky (1871–1940), ein Leben für die Strabologie. Egelsbach, Washington, Hänsel-Hohenhausen, 1994. 225 pp.

BIER, August Karl Gustav (1861–1949). German surgeon; introduced hyperaemia as an adjuvant in surgical therapy, 1903; introduced cocaine as a spinal anaesthetic, 1899.

SCHMIEDEN, V. *Zentralblatt für Chirurgie*, 1931, **58**, 2931–6.
VOGELER, K. August Bier: Leben und Werk. München, J.F. Lehmann, 1942, 302 pp.

BIERMER, Michael Anton (1827–1892). German physician; gave early description of pernicious ('Biermer's') anaemia.

ALZINGER, S. Der Internist Anton Biermer, 1827–1892. Zürich, Juris, 1992. 120 pp.

BIETT, Laurent Theodor (1780–1840). Swiss dermatologist.

BLUM, G. Laurent Theodor Biett (1780–1840), die erste Schweizer Dermatologe. *Hautartz,* 1985, **36**, 170–72.

BIGELOW, Henry Jacob (1818–1890). American surgeon; professor at Harvard University; surgeon to Massachusetts General Hospital.

[Anon]. A memoir of Henry Jacob Bigelow. Boston, Little Brown, 1900. 297 pp.

BIGGS, Hermann Michael (1859–1923). American physician; Member of Board of Directors, Rockefeller Institute for Medical Research.

WINSLOW, C.E.A. The life of Hermann M. Biggs, physician and statesman of the public health. Philadelphia, Lea & Febiger, 1929. 432 pp.

BIGNAMI, Amico (1862–1929). Italian pathologist; with Grassi demonstrated thaat the malaria parasite reproduces through the *Anopheles* mosquito.

GAMBETTI, F. et al. In memory of Amico Bignami. *Brain Pathology*, 1995, **5**, 105–7.

BILHARZ, Theodor (1825–1862). German parasitologist; discovered causal organism of schistomiasis (bilharziasis).

KNABE, G. Im Kampf gegen die 11. Plage: das Leben der Theodor Bilharz. St Augustin, Stegler Verlag, 1976. 158 pp.

SCHADEWALDT, H. Theodor Biharz (1825–1862): einer der Begründer tropenmedizinischer Forschung. München, J.F. Lehmann, 1962, 12 pp.

SCHADEWALDT, H. DSB, 1973, **2**, 127–8.

SENN, E. Theodor Bilharz: ein deutsches Forschersleben in Ägypten, 1825–1862. Stuttgart, Ausland und Heimat Verlags-Aktiengesellschaft, 1931. 76 pp.

BILLARD, Charles Michel (1800–1832). French paediatrician; pioneer investigator of the pathological anatomy of infants.

BIANCHETTI, Charles Michel Billard und sein Traité des maladies des enfans nouveau-nés et à la mamelle. Zurich, Juris-Verlag, 1963. 89 pp.

OLLIVIER, C.P. Notice historique sur la vie et les travaux de C.-M. Billard. Paris, Baillière, 1832. 30 pp.

BILLINGS, Frank (1854–1932). American physician; professor of medicine, University of Chicago.

HIRSCH, E.F. Frank Billings: the architect of medical education, an apostle of excellence in clinical practice, a leader in Chicago medicine. Chicago, Univ. Chicago Press, 1966. 144 pp.

BILLINGS, John Shaw (1838–1913). American medical bibliographer and librarian, distinguished U.S. Army surgeon; authority on hospital construction; later Director of U.S. Army Surgeon General's Library; founder of *Index Medicus* and *Index Catalog*.

BILLINGS, J. S. Selected papers ... Compiled with a life of Billings, by Frank Bradway Rogers. Chicago, Medical Library Association, 1965. 300 pp.

CHAPMAN, C.B. Order out of chaos: John Shaw Billings and America's coming of age. Boston, Mass, Boston Medical Library at Countway Library of Medicine, 1994. 420 pp.

GARRISON, F.H. John Shaw Billings: a memoir. New York, London, Putnam, 1915. 432 pp.

MITCHELL, S.W. & GARRISON, F.H. BMNAS, 1919, **8**, 375–416.

BILLROTH, Theodor (1829–1894). German surgeon; founder of modern intestinal surgery.

ABSOLON, K.B. The surgeon's surgeon: Theodor Billroth (1829-1894). 4 vols., Lawrence, Kansas, Coronado Press, 1979–89. 952 pp.
ENGEL, H. DSB, 1973, **2**, 129–31.
KOZUSCHEK, W. *et al.* (eds.) Theodor Billroth: ein Leben fur die Chirurgie. Basel, Karger, 1992.

BINET, Alfred (1857–1911). French psychologist; with T. Simon devised tests for measurement of intelligence.

BERTRAND, F.L. Alfred Binet et son oeuvre. Paris, Alcan, 1930. 336 pp.
WOLF, T.H. Alfred Binet. Chicago, University of Chicago Press, 1973. 376 pp.

BINOT, Jean (1867–1909). Chef de Laboratoire, Institut Pasteur, Paris.

VALLERY-RADOT, R. Docteur Jean Binot, chef de laboratoire, Institut Pasteur. Evreux, Imprimerie Ch. Hérissey, [n.d.] 36 pp.

BINZ, Carl (1832–1913). German pharmacologist.

BERTLING, R.M. Der Pharmakologe Carl Binz. [Thesis]. Bonn, Medizinisches Institut, 1969. 90 pp.

BION, Wilfred Ruprecht (1897–1979). British psychoanalyst.

BION, F. The long week-end 1879–1919: part of a life. Abingdon, Fleetwood Press, 1982. 287 pp.
BION, W.R. All my sins remembered; another part of a life; and The other side of genius; family letters. Abingdon, Fleetwood Press, 1985. 243 pp.
BLÉANDONU, G. Wilfred Bion: his life and works 1897–1979. London, Free Association Books, 1994. First publushed in Frebch, 1990.

BIRCHER, Eugen (1882–1957). Swiss physician; introduced laparoscope in arthroscopy.

HELLER, D. Eugen Bircher: Arzt, Militar und Politiker. Ein Beitrag zur Zeitgeschichte. 2te Aufl. Zurich, Verlag Neue Zürcher Zeitung, 1990. 461 pp.

BIRKBECK, George (1776–1842). British physician; founded Mechanics' Institute (later Birkbeck College); began free lectures to workers.

KELLY, G. George Birkbeck, pioneer of adult education. Liverpool, Liverpool Univ. Press, 1957. 380 pp.

BISCHOFF, Theodor Ludwig Wilhelm (1807–1882). German anatomist and physiologist.

ROTHSCHUH, K.E. DSB, 1973, **2**, 160–62.

BISHOP, Ann (1899–1990). British protozoologist and parasitologist.

GOODWIN, L.G. & VICKERMAN, K. BMFRS, 1992, **38**, 29–39.

BISHOP, George Holman (1889–1973). American physiologist; professor at Washington University.

LANDAU, W.M. BMNAS, 1985, **55**, 45–66.
Archival material: Washington University, St Louis.

BISHOP, J. Michael (b. 1936) American microbiologist; shared Nobel Prize (Physiology or Medicine), 1989, for discovery of the cellular origin of retroviral oncogenes.

Les Prix Nobel en 1989.

BIZZOZERO, Giulio Cesare (1846–1901). Italian histologist.

FRANCESCHINI, P. DSB, 1973, **2**, 164–6.
GRAVELA, E. Giulio Bizzozero. Torino, Umberto Allemandi & C, 1989. 196 pp.

BLACK, Davidson (1884–1934). Canadian anthropologist and anatomist; professor of anatomy, Peking Union Medical College; discovered fossil skull known as Peking man.

HOOD, D. Davidson Black: a biography. Toronto, Univ. Toronto Press, 1964. 145 pp.
HOOD, D. DSB, 1973, **2**, 171–2.

BLACK, Douglas Andrew Kilgour (1913–2002). British physician and scientist; professor of medicine, University of Manchester.

BLACK, D. Recollections and reflections. London, Memoir Club, British Medical Journal, 1987. 132 pp.

BLACK, Green Vardiman (1836–1915). American dentist; professor at Chicago and Northwestern University.

PAPPAS, C.N. The life and times of G.V. Black. Chicago, Quintessence Publishing Co., 1980. 128 pp.

BLACK, James White (b. 1924) British pharmacologist; shared Nobel Prize (Physiology or Medicine) 1988, for discovery of important principles for drug treatment.

Les Prix Nobel en 1988.

BLACK, Joseph (1728–1799). British chemist; professor of medicine, Glasgow and later Edinburgh; originated theory of 'specific heat' and 'latent heat'; rediscovered carbon dioxide.

FYFFE, J.G. & ANDERSON, R.G.W. Joseph Black: a bibliography. London, Science Museum, 1992. 125 pp.
GUERLAC, H. DSB, 1973, **2**, 173–83.
SIMPSON, A.D.C. (ed.) Joseph Black 1728–1799; a commemorative symposium. Edinburgh, Royal Scottish Museum, 1982. 69 pp.

BLACKBURN, Luke Pryor (1816–1887). American physician; Governor of Kentucky and prison reformer.

BAIRD, N.D. Luke Pryor Blackburn: physician, governor, reformer. Lexington, Univ. Press of Kentucky, 1979. 128 pp.

BLACKIE, Margery Grace (1899–1981). British homoeopathic physician.

SMITH, C.B. Champion of homoeopathy: the life of Margery Blackie. London, Murray, 1986. 185 pp.

BLACKLEY, Charles Harrison (1820–1900). British physician; showed that pollen could produce hay fever and skin reactions in sensitive persons.

TAYLOR, G. & WALKER, J. Charles Harrison Blackley, 1820–1900. *Clinical Allergy*, 1973, **3**, 103–8.

BLACKMORE, Richard (1653/54–1729). British physician; physician to William III and Queen Anne; voluminous writer in prose and verse.

ROSENBERG, A. Sir Richard Blackmore, a poet and physician of the Augustan age. Lincoln, Univ. Nebraska Press, 1953. 175 pp.

BLACKWELL, Elizabeth (1821–1910). British pioneer medical woman, qualified at Geneva Medical College, New York (1849) and subsequently studied and worked in England.

CHAMBERS, P. A doctor alone: a biography of Elizabeth Blackwell, the first woman doctor. London, Bodley Head, 1956. 191 pp.
GLIMM, A. Elizabeth Blackwell: the first woman doctor of modern times. New York, McGraw-Hill, *c.* 2000. 124 pp.
ROSS, I. Child of destiny: the life story of the first woman doctor. London, Gollancz, 1950. 309 pp.
WILSON, D.C. Lone woman: the story of Elizabeth Blackwell, the first woman doctor. Boston, Little, Brown, 1970. 469 pp.

BLAINVILLE, Henri Marie Ducrotay de (1777–1850). French anatomist and zoologist.

COLEMAN, W. DSB, 1973, **2**, 186–8.

BLAIR, Patrick (1660–1728). Scottish surgeon and botanist; first described hypertrophic pyloric stenosis.

AGNEW, L.R.C. DSB, 1970, **2**, 188–9.

BLAIR, Vilray Papin (1871–1955). American surgeon.

STELNICKI, F.J. et al. Vilray Papin Blair, his surgical descendants and their role in plastic surgical development. *Plastic and Reconstructive Surgery*, 1999, **103**, 1990–2009.

BLAIR-BELL, William (1871–1936). British obstetrician and gynaecologist; founded (Royal) College of Obstetricans and Gynaecologists.

PEEL, J. William Blair-Bell: father and founder. London, Royal College of Obstetricians and Gynaecologists, 1986. 89 pp.

BLALOCK, Alfred (1899–1964). American surgeon; undertook important research into surgical shock; devised operation for treatment of congenital heart block.

HARVEY, A.M. BMNAS, 1982, **53**, 49–81.
LONGMIRE, W.P. Jr. Alfred Blalock: his life and times. Baltimore, Johns Hopkins Hospital. 1992.

BLAND, Edward Franklin (b. 1901). American physician.

FRIEDLICH, A.L. Edward Franklin Bland. Transactions of the Clinical and Climatological Association, 1992, 104, xxxix–xli.
FRIEDLICH, A.L. Profiles in Cardiology. *Clinical Cardiology*, 1990, **13**, 513–15.

BLAND-SUTTON, John (1855–1936). British surgeon to Middlesex Hospital, London.

BETT, W. Sir John Bland-Sutton, 1855–1936. Edinburgh, London, Livingstone, 1956. 100 pp.

BLAND-SUTTON, J. The story of a surgeon. London, Methuen, 4th ed. 1931. 204 pp. First published 1930.

BLANE, Gilbert (1749–1834). Physician to St Thomas' Hospital, to George IV and William IV and to the British Fleet; demonstrated value of citrus fruits as antiscorbutics.

MURRELL, S. Physician extraordinary. London, Hodder & Stoughton, 1949. 318 pp.

BLASCHKO, Alfred (1858–1922). German venereologist.

WEINDLING, P. & SLEVOGT, V. Alfred Blaschko (1858–1922) and the problem of sexually transmitted diseases in Imperial Weimar: a bibliography. Oxford, Wellcome Unit for the History of Medicine, 1992. 150 pp.

BLASCHKO, Hermann Karl Felix [Hugh] (1900–1993). German/British biochemical pharmacologist; professor at Oxford; fundamental work on amine oxidases and dopamine.

BLASCHKO, H.K.F. My path to pharmacology. *Annual Review of Pharmacology*, 1980, **20**, 1–14.

BORN, G.V. & BANKS, P. BMFRS, 1996, **42**, 40–60.

Archival material: CMAC.

BLAUD, Pierre (1774–1858). French physician; treated chlorosis with ferrous sulphate and potassium carbonate ('Blaud's pill').

I, R. Le Docteur Pierre Blaud. *Chronique Médicale*, 1914, **21**, 316–18.

BLEULER, Paul Eugen (1857–1939). German psychiatrist; introduced the concept of schizophrenia.

SCHARFETTER, C. Eugen Bleuler (1857–1839): Studie zu seiner Psychopathologie, Psychologie und Schizophrenielehre. Zürich, Juris Druck, 2001. 119 pp.

BLIZARD, William (1743–1835). British; surgeon at the London Hospital.

COOKE, W. A brief memoir of Sir William Blizard. London, Longman, Rees, Orme, Brown, 1835. 67 pp.

BLOBEL, Günther (b.1936) German/American biochemist; awarded Nobel Prize (Physiology or Medicine) 1999, for the discovery that proteins have intrinsic signals that govern their transport and localization in the cell.

Les Prix Nobel en 1999.

BLOCH, Bruno (1878–1933). Swiss dermatologist; a founder of biologically orientated dermatology.

GUGGENHEIM, F. Bruno Bloch. Biographie und wissenschaftliche Werk. Zürich, Juris, 1969, 75 pp.

BLOCH, Konrad Emil (b.1912). American biochemist; professor at Harvard University; shared Nobel Prize (Physiology) 1964 for work on cholesterol and fatty acid metabolism.

BLOCH, K. Summing up. *Annual Review of Biochemistry*, 1987, **56**, 1–15.
WESTHEIMER, F.H. & LIPSCOMB, W. BMFRS, 2002, **48**, 45–9.

BLOCH, Oscar Thorvald (1847–1926). Danish surgeon.

LUNDSGAARD, K.K.K. *Ugeskrift for Laeger*, 1926, **88**, 606–8.
SCHALDEMOSE, V. *Hospitalstidende*, 1926, **69**, 589–92.

BLOODGOOD, Joseph Colt (1867–1935). American surgeon; professor at Johns Hopkins University.

BLOODGOOD, E.H. & LONG, V.H. Index to the writings of Joseph Colt Bloodwood. Baltimore, Lord Baltimore Press, [1936?]. 51 pp.
MARMON, L.M. The life of Joseph Colt Bloodgood. *Surgery, Gynecology & Obstetrics*, 1993, **177**, 193–200.

BLOOM, William (1899–1972). American histologist.

SINGER, R. BMNAS, 1993, **62**, 16–36.

BLUMENBACH, Baruch Samuel (b. 1925) American medical anthropologist and immunologist; shared Nobel Prize (Physiology or Medicine) 1976, for discoveries concerning new mechanisms for the origin and dssemination of infectious diseases.

Les Prix Nobel en 1976.

BLUMENBACH, Johann Friedrich (1752–1840). German comparative anatomist and biologist; founder of anthropology as a science.

BARON, W. DSB, 1973, **2**, 203–5.

BLUMGART, Herman Ludwig (1895–1977). American surgeon.

WENGER, N.K. The synthesis of humanism and medical science; Herman Ludwig Blumgart, 1895–1977: teacher, administrator, and scientist. *Clinical Cardiology*, 1992, **15**, 308–11.

BLUNDELL, James (1790–1878). British obstetrician; pioneer in blood transfusion.

YOUNG, J.H. James Blundell, 1790–1878; experimental physiologist and obstetrician. *Medical History*, 1964, **8**, 159–69.

BLUNTSCHLI, Hans (1877–1962). Swiss anatomist.

CENER, F. Der Anatom Hans Bluntschli, 1877–1962. Zürich, Juris, 1990. 131 pp.

BOAS, Ismar Isidor (1858–1938). Polish physician; a leading gastroenterologist in Europe.

EICHMANN, A. Ismar Boas (1858–1938) und die Entwicklung der Gastroenterologie als Spezialfach. Zürich, Juris, 1970. 30 pp.
GROTE, 1928, **7**, 51–96. (CB)

BOBBS, John Stough (1809–1870). American surgeon.

SPARKMAN, R.S. Bobbs centennial: the first cholecystotomy. *Surgery*, 1967, **61**, 965–71.

BODINGTON, George (1799–1882). British physician; pioneer advocate of sanatorium treatment in tuberculosis.

KEERS, R.Y. Two forgotten pioneers. James Carson and George Bodington. *Thorax*, 1980, **35**, 483–89.

BOECK, Carl Wilhelm (1808–1875). Norwegian dermatologist and syphilologist; described Norwegian itch ('Boeck's scabies'); with D.C. Danielssen published first modern description of leprosy, 1847.

–. *Dermatologische Wochenschrift*, 1917, **65**, 763–5.
–. *Medicinsk Revue*, Bergen, 1917, **34**, 165–9.
–. *Vierteljahrsschrift für Dermatologie und Syphilis*, 1876, **3**, 135–40.

BÖHLER, Lorenz (1885–1973). Austrian surgeon; professor of traumatology, Vienna.

LEHNE, I. Lorenz Böhler: die Geschichte eines Erfolges. Wien, W. Maudrich, 1991.
LORENS, F. Lorenz Böhler, der Vater der Unfallchirurgie. Wien, W. Maudrich, 1955. 62 pp.

BOENHEIM, Felix (1890–1960). German physician.

RUPRECHT, T.M. Felix Boenheim: Arzt, Politiker, Historiker: ein Biographie. Hildesheim, Georg Olms, 1992. 549 pp.

BOËR, Johan Lukas (1751–1835). Austrian obstetrician; advocate of natural childbirth.

HEILEIN, W.D. Johan Lukas Boër, der Begründer der naturlichen Geburtshilfe (1751–1835). Würzburg, Gebr. Memminger, 1935. 26 pp.

BOERHAAVE, Herman (1668–1738). Dutch physician; eminent medical teacher.

LINDEBOOM, G.A. Herman Boerhaave: the man and his work. London, Methuen, 1968. 452 pp.
LINDEBOOM, G.A. DSB, 1973, **2**, 224–8.
LINDEBOOM, G.A. Bibliographia Boerhaaviana: list of publications written or provided by H. Boerhaave, or based on his works and teaching. Leiden, Brill, 1959. 108 pp.

BOGDANOV [MALINOVSKY], Aleksandr Aleksandrovich (1873–1928). Russian physician and haematologist.

GRAHAM, L. DSB, 1978, **15**, 38–9.

BOHN, Johann (1640–1718). German physiologist; professor of anatomy, Leipzig.

ROTHSCHUH, K. DSB, 1970, **2**, 237–8.

BOIVIN, Marie Anne Victoire (1773–1841). French midwife: published classical account of hydatiform mole, 1827.

–. Notice sur Madame Boivin. *Annales de la Chirurgie Française et Étrangère*, 1841, **2**, 373–7.

BOLK, Lodewijk (1866–1980). Dutch neuroanatomist.

ENGEL, H. DSB, 1970, **2**, 255–6.

BOLLINGER, Otto von (1843–1909). German pathologist; gave first clear description of actinomycosis in cattle.

RÖSSLE, R. Otto v. Bollinger, Nekrolog. *Zentralblatt für allgemeine Pathologie*, 1909, **20**, 961–6.

BOLTON, Charles (1870–1947). British physician; researched mainly on dropsy, on heart failure and on peptic ulcer.

ELLIOTT, T.R. ONFRS, 1948–49, **2**, 23–36.

BOLTON, James (1812–1869). American surgeon.

BECK, J.D.R. & SWAN, K.G. James Bolton MD (1812–1869), reflections on an American surgeon. *Journal of the American College of Surgeons*, 1999, **189**, 324–9.
SCHWARZ, L.L. James Bolton (1812–1869): early proponent of external skeletal fixation. *American Journal of Surgery*, 1944, **66**, 409–13.

BONDT, Jacob, *see* **BONTIUS, Jacobus**

BONET, Théophile (1620–1689). Swiss pathologist.

IRONS, E.E. Théophile Bonet, 1620–1689: his influence on the science and practice of medicine. *Bulletin of the History of Medicine*, 1943, **12**, 623–65.

BONHOEFFER, Karl Ludwig (1868–1948). German neuropsychiatrist.

NEUMÄRKER, K.-J. Karl Bonhoeffer: Leben und Werk eines deutschen Pyschiaters und Neurologen in seiner Zeit. Berlin, Springer, 1990. 232 pp.

BONNER, John Tyler (b.1920). American biologist.

BONNER, J.T. Lives of a biologist: adventure in a century of extraordinary science. Cambridge, MA, Harvard Univ. Press, 2002. 215 pp.

BONNET, Charles (1720–1793). Swiss naturalist and philosopher.

BONNET, C. Mémoires autobiographiques... éd. par R. Savioz. Paris, Vrin, 1948, 414 pp.
MARX, J. Charles Bonnet contre les luminiéres. 2 vols. Oxford, Voltaire Foundation at the Taylor Institution, 1976. 782 pp.

BONNEY, William Francis Victor (1872–1953). British gynaecologist.

CHAMBERLAIN, G. Victor Bonney: the gynaecological surgeon of the twentieth century. London, Parthenon, 200. 140 pp.

BONOMO, Giovan Cosimo (1663–1696). Italian physician; observed the scabies mite and gave first proof of infection of human by microparasite.

FRIEDMAN R. Giovan Cosimo Bonomo (1663–1696): the 250th anniversary of his discovery of the parasitic nature of scabies. *Medical Life*, 1937, **44**, 3–62, 229–51.

BONTIUS [BONDT] Jacobus (1592–1631). Dutch physician and naturalist; gave first modern scientific description of beriberi, 1642.

JEANSELME, E. L'oeuvre de J. Bontius. VI Congrès Internationale de l'Histoire de Médecine, Leyde-Amsterdam, 1927. Antwerp, 1929, 209–28.
RÖMER, L.S.A.M. von. Dr Jacobus Bontius. Batavia, Kolff, 1932. 33 pp.

BOORDE [BORDE] Andrew (?1490–1549). English physician; wrote *The breviary of helthe*, 1547, the first 'modern' work on hygiene.

POOLE, H.E. The wisdom of Andrew Boorde. Edited with an introduction and notes by H. Edmund Poole. Illustrated by A.E. Christopherson. Leicester, Edgar Backus, 1936. 63 pp.

BOOTHBY, Walter Meredith (1880–1953). American physician; professor at Mayo Clinic; devised apparatus for administration of anaesthetics.

VANDAM, L.D. Walter M. Boothby, MD – the wellspring of anesthesiology. *New England Journal of Medicine*, 1967, **276**, 558–63.

BOOTT, Francis (1792–1863). British physician; administered anaesthetic at first major surgical operation in Britain, 21 Dec. 1846.

ELLIS, R.H. The introduction of ether anaesthesia in Great Britain. 2. A biographical sketch of Dr Francis Boott. *Anaesthesia*, 1977, **32**, 197–208.

BORDET, Jules Jean Baptiste Vincent (1870–1961). Belgian bacteriologist and immunologist; Director, Institut Pasteur; Nobel Prize 1919; discovered *Haemophilus pertussis*.

OAKLEY, C.L. BMFRS, 1962, **8**, 19–25.
VIEUCHANGE, J. DSB, 1973, **2**, 300–301.
Archival material: Institut Pasteur du Brabant, Brussels.

BORDEU, Théophile de (1722–1776). French physician; founder of modern hydrotherapy; a pioneer in endocrinology.

DULIEN, L. DSB, 1973, **2**, 301–2.

BORELLI, Giovanni Alfonso (1608–1679). Italian physicist and mathematician; originated neurogenic theory of heart's action.

BARBENSI, G. Borelli. Trieste, Zigiotti Editore, 1947. 141 pp.
SETTLE T.B. DSB, 1973, **2**, 306–14.
Archival material: Biblioteca Nazionale, Florence.

BORGOGNONI OF LUCCA, Theodoric, Bishop of Cervia (*c.* 1205–1298) Italian surgeon; used soporific sponges as anaesthetics.

WALLACE, W.A. DSB, 1973, **2**, 314–15.

BORING, Edward Garrigues (1886–1968). American psychologist; professor at Harvard University.

BORING, E.G. Psychologist at large: an autobiography and selected essays. New York, Basic Books, 1961. 371 pp.
STEVENS, S.S. BMNAS, 1973, **43**, 41–76.

BORODIN, Alexander Porfir'evich (1833–1887). Russian composer, physician and chemist.

FIGUROVSKI, N.A. & SOLOV'EV, T. Alexander Porfir'evich Borodin: a chemist's biography. Translated from Russian. Berlin, Springer-Verlag, *c.*1988. 171 pp.

BORRIES, Bodo von (1905–1956). German physicist; with Ernst Ruska developed first commercially available transmission electron microscope.

SUSSKIND, C. DSB, 1973, **2**, 318.

BOSSI, Luigi Maria (1859–1919). Italian obstetrician and gynaecologist.

CERVELLI, F. & LA TORRE, F. *Clinica Ostetrica*, 1919, **21**, 57–74.

BOSTOCK, John (1773–1846). British physician and medical chemist; gave early classical account of hay fever.

THACKRAY, A. DSB, 1973, **2**, 335–6.

BOSTROEM, Eugen Woldemar (1850–1928). German pathologist; isolated causal organism of actinomycosis.

KRUMHOLZ, W. Eugen Woldemar Bostroem (1850–1928) Lehrer der Pathologie und Förderer der Medizin in Giessen. Giessen, Schmitz, 1983. 311 pp.

BOTALLO, Leonardo (*c*.1519–1587/8). Italian physician; described 'Botallo's duct'.

O'MALLEY, C.D. DSB, 1973, **2**, 336–7.

BOTTAZZI, Filippo (1867–1941). Italian physiologist.

BELLONI, L. DSB, 1973, **2**, 339–40.

BOTTINI, Enrico (1837–1903). Italian surgeon; introduced galvano-cautery for treatment of prostatic hypertrophy and devised operation for radical treatment of prostatic hypertrophy.

ARCIERI, G.P. Enrico Bottini and Joseph Lister in the method of antisepsis. New York, Alcmaeon, 1967. 29 pp.

BOUCHARD, Charles Jacques (1837–1915). French physician; dean, Faculty of Medicine, Paris.

LE GENDRE, P.L. Un médecin philosophique: Charles Bouchard, son oeuvre et son temps. Paris, Masson, 1924. 526 pp.

BOUCHUT, Jean Antoine Eugène (1818–1891). French physician; gave first adequate description of neurasthenia.

–. *France Médicale*, 1891, **38**, 769–71.

BOUILLAUD, Jean Baptiste (1796–1881). French physician; established connection between rheumatic fever and cardiac disease.

ROLLESTON, J.D. Jean Baptiste Bouillaud (1796–1881). A pioneer in cardiology and neurology. *Proceedings of the Royal Society of Medicine*, 1931, **24**, 1253–62.

BOUIN, Pol André (1870–1962). French reproductive endocrinologist.

KLEIN, M. DSB, 1973, **2**, 344–6.

BOURGEOIS, Louise (1563–1636). French accoucheuse; a pioneer of scientific midwifery.

BERNARDAC, M.-L. Louise Bourgeois [translated from the French]. Paris, New York, Flammarion, 1996. 191 pp.
PERKINS, W. Midwifery and medicine in early modern France: Louise Bourgeois. Exeter, University of Exeter Press, 1996. 170 pp.
PFEILSTICKER, W. Louise Burgeois. *Archiv für Geschichte der Medizin*, 1939, **32**, 1–20.

BOURNE, Aleck William (1886–1974). British gynaecologist.

BIBLIOGRAPHY OF MEDICAL AND BIOMEDICAL BIOGRAPHY

BOURNE, A. A doctor's creed. The memoirs of a gynaecologist. London, Gollancz, 1962. 224 pp.
Archival material: CMAC.

BOURNE, Geoffrey (1893–1970). British physician, physician to St Bartholomew's Hospital, London.

BOURNE, G. We met at Bart's; the autobiography of a physician. London, F. Muller, 1963. 288 pp.

BOURNEVILLE, Désiré Magloire (1840–1909). French neurologist, described tuberous sclerosis ('Bourneville's disease').

POIRIER, J. & SIGNORET, L. (eds.) De Bourneville à la sclérose tubérose: un homme, une époque, une maladie. Paris, Médecin-Sciences Flammarion, 1991. 206 pp.

BOVERI, Theodor (1862–1915). German biologist; demonstrated individuality of chromosomes.

BALTZER, F. Theodor Boveri: life and work of a great biologist. Transl. D. Rudnick. Berkeley, Univ. California Press, 1967. 165 pp. First published in German, Stuttgart, 1962. 194 pp.
OPPENHEIMER, J.M. DSB, 1973, **2**, 361–5.
NEUMANN, H.A. Vom Ascaris zum Tumor: Leben und Werk des Biologen Theodor Boveri (1862–1915). Berlin, Blackwell Wissenschafts-Verlag, 1998. 250 pp.

BOVET, Daniel (1907–1992). Swiss pharmacologist; Nobel Prizewinner (Physiology or Medicine) 1957, for his discoveries in chemotherapy.

OLIVERIO, A. BMFRS, 1994, **39**, 60–70.

BOWDITCH, Henry lngersoll (1808–1892). American physician in Boston; specialist in chest disease and opponent of slavery.

BOWDITCH, V.Y. Life and correspondence of Henry Ingersoll Bowditch. 2 vols., Boston, New York, Houghton, Mifflin, 1902.

BOWDITCH, Henry Pickering (1840–1911). American physiologist; professor at Harvard University.

BOWDITCH, H.P. The life and writings of Henry Pickering Bowditch. 2 vols., New York, Arno Press, 1980. Various paginations.
CANNON, W. BMNAS, 1922, **17**, 183–96.
MENDELSOHN, E. DSB, 1973, **2**, 365–8.
Archival material: Countway Library, Harvard Medical School.

BOWDLER, Thomas (1754–1825). British physician and editor; published versions of Shakespeare's works from which all matter which could not "with propriety be read aloud in a family" was deleted, leading to the term 'to Bowdlerize'.

JELLINEK, E.H. Thomas Bowdler: censor, philanthropist, and doctor. *Lancet*, 2001, **358**, 1091–4.

BOWERBANK, Fred Thompson (1880–1960). British/New Zealand cardiologist; Director General of New Zealand Army and Air Medical Services.

48

BOWERBANK, F. A doctor's story. Wellington, Wingfield Press, 1958. 341 pp.

BOWLBY, Edward John Mostyn (1907–1991). British child psychiatrist.

DIJKEN, S. van. John Bowlby: his early life; a biographical journey into the roots of attachment theory. London, Free Association Books, 1998. 214pp.
Archival material: CMAC.

BOWMAN, William (1816–1892). British physiologist and ophthalmologist.

CHANCE, B. Sir William Bowman. *Annals of Medical History*, 1924, **6**, 143–58.
HALE-WHITE, pp. 177–88 (CB).
THOMAS, K.B, DSB, 1973, **2**, 375–7.

BOYCOTT, Arthur Edwin (1877–1938). British pathologist; professor at University College Hospital Medical School, London.

MARTIN, C.J. ONFRS, 1936–39, **2**, 561–71.

BOYD, John Smith Knox (1891–1981). British bacteriologist; President, Royal Society of Tropical Medicine.

GOODWIN, L.G. BMFRS, 1982, **28**, 27–57.
Archival material: CMAC.

BOYD ORR, John (1880–1971). British physiologist and nutritionist; Director, Rowett Research Institute, Aberdeen.

BOYD ORR, J. As I recall. The 1880's to the 1960's. London, MacGibbon & Kee, 1967. 290 pp.
KAY, H.D. BMFRS, 1972, **18**, 43–81.

BOYER, Alexis (1757–1833). French surgeon.

FAIGE, P. Un grand chirurgien limousin: Alexis Boyer. [Strasbourg], Brive, 1941. 119 pp.

BOYLE, Robert (1627–1691). English natural philosopher, physicist and chemist.

FULTON, J.F. A bibliography of the Honourable Robert Boyle, Fellow of the Royal Society. 2nd ed., Oxford, Clarendon Press, 1961. 218 pp.
KAPLAN, B.B. "Divulging of useful truths in physick": the medical agenda of Robert Boyle. Baltimore, Johns Hopkins Univ. Press, 1993. 224 pp.
MADDISON, R.E.W. The life of the Honourable Robert Boyle. London, Taylor & Francis, 1969. 332 pp.
PILKINGTON, R. Robert Boyle: father of chemistry. London, John Murray, 1959. 179 pp.

BOYLSTON, Zabdiel (1680–1766). Pioneer American inoculator against smallpox.

MAGER, G.M. Zabdiel Boylston, medical pioneer of colonial Boston., 1975 Dissertation Abstracts International, 1976, 36, 6270-A. University Microfilms Order No. 76-6853. 255 pp.

BOZEMAN, Nathan (1825–1905). American gynaecologist.

CARMICHAEL, E.B. Nathan Bozeman. *Alabama Journal of Medical Sciences*, 1969, **9**, 233–6.

BOZZINI, Philipp (1773–1809). German physician; first seriously to attempt to make the larynx visible.

MANN, G. Der Frankfürter Lichtleiter. Neues über Philipp Bozzini und sein Endoskop. *Medizinhistorischer Journal*, 1973, **8**, 105–30.

REUTER, H. & REUTER, M.A. Philipp Bozzini and endoscopy in the 19th century. 2 vols. Stuttgart, Max Nitze Museum, 1988.

ROEDIGER, E. Der Frankfurter Arzt Philipp Bozzini, der Erfinder des Lichtleiters, 1773–1809. *Medizinhistorisches Journal*, 1972, **7**, 204–17.

BRACHET, Albert (1869–1930). Belgian embryologist.

CORNER, G.W. DSB, 1973, **2**, 383–5.

HILL, J.P. ONFRS, 1932–35, **1**, 64–70.

BRACHET, Jean (1909–1988). Belgian embryologist and biochemist.

PIRIE, N.W. BMFRS, 1990, **36**, 83–99.

BRADFORD, John Rose (1863–1935). British physician on staff of University College Hospital, London; President, Royal College of Physicians, and Secretary, Royal Society of London.

ELLIOTT, T.R. ONFRS, 1932–35, **1**, 527–35.

BRADWELL, Stephen (1594–1636). English physician; wrote pioneer work on first aid.

GEVITZ, N. "Help for suddain accidents", Stephen Bradwell and the origin of the First Aid guide. *Bulletin of the History of Medicine*, 1993, **67**, 51–73.

BRAID, James (1795–1860). British surgeon in Manchester; a pioneer of scientific hypnotism.

WINK, C.A.S. The life and work of James Braid, surgeon, with special reference to his influence on the development of hypnotism as an orthodox medical procedure. [B.Litt. thesis, Oxford University]. 1970.

BRAILLE, Louis (1809–1852). French teacher of the blind; invented Braille system of reading by touch.

BICKEL, L. Triumph over darkness: the life of Louis Braille. London, Unwin Hyman, 1988. 352 pp. Large-print book.

KUGELMASS, J.A. Louis Braille; windows for the blind. New York, J. Messner, 1951. 160 pp.

BRAIN, Walter Russell (1895–1966). British neurologist and medical statesman.

PICKERING, G.W. BMFRS, 1968, **14**, 61–82.

BRAINARD, Daniel (1812–1866). American surgeon; founder of Rush Medical College.

KINNEY, J. Saga of a surgeon: the life of Daniel Brainard, M.D. Springfield, S.Illinois Univ. School of Medicine, 1987. 233 pp.

BRAMBELL, Francis William Rogers (1901–1970). Irish endocrinologist and embryologist.

OAKLEY, C.L. BMFRS, 1973, **19**, 129–71.

BRAMWELL, Byrom (1847–1931). Scottish physician, Royal Infirmary, Edinburgh.

BRAMWELL, Edwin (1873–1952). Professor of clinical medicine, Edinburgh.

BRAMWELL, John Crichton (1889–1976). Professor of cardiology, Manchester University.

ASHWORTH, B. The Bramwells of Edinburgh: a medical dynasty. Edinburgh, Royal College of Physicians, 1986. 96 pp.

BRANDES, Simon Rudolph (1795–1842). German pharmacist; a founder of *Archiv der Pharmazie*.

ZIMMERMANN, H. Rudolph Brandes (1795–1842): ein bedeutender Apotheker der 19. Jahrhunderts. Stuttgart, Deutscher Apotheker-Verlag, 1985. 278 pp.

BRASHEAR, Walter (1776–1860). American surgeon; first successfully amputated at hip-joint.

McCARTY, A.C. Dr Walter Brashear (1776–1860). *Annals of Medical History*, 1934, **6**, 301–6.

BRAUN, Heinrich Friedrich Wilhelm (1862–1934). German anaesthetist; pioneer in use of local anaesthesia.

GROTE, 1925, **5**, 1–34. (CB)

BRAVO, Francisco (*fl.* 1553–1570). Spanish physician; described Spanish or Mexican typhus.

BRAVO, F. The *Opera medicinalia*. With a biographical and bibliographical introduction by F. Guerra. 2 vols., Folkestone, Dawsons, 1970. 305 leaves. Facsimile reprint of 1570 edition.

BREINL, Anton (1880–1944). Austrian/Australian protozoologist; with H.W. Thomas noted value of the arsenical atoxyl in the treatment of trypanosomiasis, leading to Ehrlich's Salvarsan; later Director of Australian Institute of Tropical Medicine.

DOUGLAS, R.A. Dr Anton Breinl and the Australian Institute of Tropical Medicine. *Medical Journal of Australia*, 1977, **1**, 713–6, 748–51, 784–90.

BREISKY, August (1832–1889). Bohemian obstetrician and gynaecologist; his description of kraurosis vulvae led to term 'Breisky's disease'.

FISCHEL, W. & SCHAUTER, – *Prager Medizinische Wochenschrift*, 1889, **14**, 255–9.

BRENNER, Sydney (b. 1927). South African/British molecular biologist; shared Nobel Prize (Physiology or Medicine), 2002, for discoveries concerning genetic regulation of organ development and programmed cell death.

WOLPERT, L. A life in science: Sydney Brenner as told to Lewis Wolpert. London, BioMedcentral, 2001. 200 pp.

BRETONNEAU, Pierre Fidèle (1778–1862). French physician; worked on diphtheria and typhoid; performed first tracheotomy for croup.

ARON, E. Bretonneau: le médecin de Tours. [Chambray-les-Tours], C.L.D., [1979]. 295 pp.
ROUSSEAU, A. DSB, 1973, **2**, 444–5.

BIBLIOGRAPHY OF MEDICAL AND BIOMEDICAL BIOGRAPHY

TRIAIRE, P. Bretonneau et son correspondents ... publié avec un biographie et des notes par P. Triaire, *etc*. 2 vols. Paris, Alcan, 1892. 599 + 648 pp.

BREUER, Josef (1842–1925). Austrian physiologist and psychopathologist; worked with Freud on the unconscious mind.

CRANEFIELD, P.F. DSB, 1973, **2**, 446–50.
HIRSCHMÜLLER, A. The life and work of Josef Breuer. New York University Press, 1990. 514 pp. First published in German, Bern, 1978.

BRICHETEAU, Isidore (1789–1861). French physician; first to give adequate description of pneumopericardium.

ARLIN, – Éloge du Docteur Bricheteau. Paris, F. Malteste & Cie., 1862. 8 pp.

BRIDGES, Calvin Blackman (1889–1938). American geneticist; discovered non-disjunction of chromosomes.

MORGAN, T.H. BMNAS, 1941–43, **22**, 31–43.
STURTEVANT, A.H. DSB, 1973, **2**, 455–7.

BRIDGES, John Henry (1832–1906). British physician; medical inspector to Local Government Board, London; litterateur.

LIVEING, S. A nineteenth-century teacher, John Henry Bridges. London, Kegan Paul, 1926. 262 pp.

BRIDGES, Robert Seymour (1844–1930). British physician; physician to Great (Royal) Northern Hospital, London, appointed Poet Laureate 1913.

PHILLIPS, C. Robert Bridges: a biography. Oxford, Oxford Univ. Press, 1992. 364 pp.
THOMPSON, E.J. Robert Bridges, 1844–1930. London, Oxford Univ. Press, 1944. 131 pp.

BRIDIE, James, *see* **MAVOR, Osborne Henry**

BRIGGS, Robert William (1911–1983). American geneticist.

DI BERARDINO, M.A. BMNAS, 1999, **76**, 50–63.

BRIGGS, William (1642–1704). English physician and oculist; gave first known description of nyctalopia, 1684, and wrote the first English treatise on the anatomy of the eye.

JAMES, R.R. William Briggs (1650 [sic]–1704. *British Journal of Ophthalmology*, 1932, **16**, 360–68.

BRIGHT, Richard (1789–1858). British physician; described nephritis ('Bright's disease').

BERRY, D. & MACKENZIE, C. Richard Bright 1789–1858: physician in an age of revolution and reform. London, Royal Society of Medicine Services Ltd., 1992. 296 pp.
BRIGHT, P. Dr. Richard Bright, 1789–1858. London, Bodley Head, 1983. 313 pp.
KING, L.S. DSB, 1973, **2**, 463–5.

BRIGHT, Timothie (?1550–1615). British physician; physician to St Bartholomew's Hospital, London; inventor of an early system of shorthand.

CARLTON, W.J. Timothe Bright, doctor of phisicke: a memoir of 'The father of modern shorthand'. London, Stock, 1911. 205 pp.

KEYNES. G.L. Dr Timothe Bright, 1550–1615. A survey of his life with a bibliography of his writings. London, Wellcome Historical Medical Library, 1962. 47 pp.

BRILL, Nathan Edwin (1860–1925). American physician.

OTTENBERG, R. Dr Nathan E. Brill: a profile. *Journal of the Mount Sinai Hospital*, 1956, **23**, 837–41.

BRINTON, William (1823–1867). British physician; an authority on intestinal obstruction; gave first description of linitis plastica.

–. William Brinton. *Lancet*, 1867, **1**, 129.

BRISSEAU, Michel (1676–1743). Belgian surgeon; first to demonstrate the true nature of cataract.

VAN DUYSE, M. "Le Tournasien et le siège de la cataracte". Liber memorialis Tournai, 25 Septembre 1921. Antwerp, J.F. Buschmann, 1922. 31 pp.

BROADBENT, William Henry (1835–1907). British physician; physician to St. Mary's Hospital, London.

BROADBENT, M.E. (ed.) Life of Sir William Broadbent, Physician Extraordinary to H.M. Queen Victoria, Physician in Ordinary to the King and to the Prince of Wales. London, Murray, 1909. 306 pp.

BROCA, Pierre Paul (1824–1880). French surgeon and anthropologist; localized cortical function in the brain.

CLARKE, E. DSB, 1973, **2**, 477–8.
SCHILLER, F. Paul Broca, founder of French anthropology and explorer of the brain. Berkeley, Univ. California Press, 1979. 350 pp. Reprinted Oxford Univ. Press, 1992. 350 pp.

BROCKLESBY, Richard (1722–1797). British physician.

CURRAN, W.S. Dr Brocklesby of London (1722–1797), an 18th-century physician and reformer. *Journal of the History of Medicine*, 1963, **17**, 509–21.

BROCQ, Louis Anne Jean (1856–1928). French dermatologist.

PAUTRIER, L.M. *Annales de Dermatologie*, 1926, 6 ser. **10**, 135–50.

BRODIE, Benjamin Collins (1783–1862). British surgeon, St George's Hospital, London; Sergeant-Surgeon to King William IV; first surgeon to become President of the Royal Society of London (1858–61).

BRODIE, B.C. Autobiography. In: The works of Sir Benjamin Collins Brodie. 2nd ed., 3 vols., London, Longmans Green, 1865, vol. 1, pp. 1–116.
GOODFIELD, C.J. DSB, 1973, **2**, 482–4.
HOLMES, T. Sir Benjamin Collins Brodie. London, Fisher Unwin, 1898. 256 pp.
LeFANU, W.R. Sir Benjamin Brodie, F.R.S. (1783–1862). *Notes and Records of the Royal Society of London*, 1964, **19**, 42–52.
Archival material: Bodleian Library, Oxford; Royal College of Surgeons of England; Library, St George's Hospital Medical School; Library, University of Leicester.

BIBLIOGRAPHY OF MEDICAL AND BIOMEDICAL BIOGRAPHY

BRODMANN, Korbinian (1868–1918). German anatomist; gave comprehensive account of localization of cerebral function.

DANEK, A. & RETTIG, J. Korbinian Brodmann (1868–1919). *Schweizerisches Archiv für Neurologie und Psychiatrie*, 1989, **140**, 555–66.

BRÖDEL, Max (1870–1941). German-born anatomist and medical illustrator; worked at Johns Hopkins University.

CROSBY, R.W. & CODY, J. Max Brödel, the man who put art into medicine. New York, Springer-Verlag, 1991. 352 pp.
McKUSICK, V.A. DSB, 1978, **15**, 64–5.

BRONK, Detlev Wulf (1897–1975). American biophysicist; became influential leader and spokesman for the national scientific establishment.

ADRIAN, Lord. BMFRS, 1976, **22**, 1–9.
BRINK, F. BMNAS, 1979, **50**, 3–87.
INSTITUTE to university. A seventy-fifth anniversary colloquium, 8 June 1976. New York, Rockefeller University, 1977. 108 pp. Contributions by J.T. Flexner, S. Benison, M. McCarty, R. Dubis, H. Gasser, D. Bronk, F. Brink, D. Rockefeller, F. Seitz.
REINGOLD, N. DSB, 1990, **17**, 111–13.
Archival material: Rockefeller University; Johns Hopkins University.

BROOKS, Barney (1884–1952). American physician; specialist in cardiovascular disorders.

ROSENFELD, L. Barney Brooks, MD (1884–1952). Nashville, Tenn., Vanderbilt University Medical Center, 1986. 60 pp.

BROOM, Robert (1866–1951). British/South African physician and palaeontologist.

FINDLAY, G.H. Dr Robert Broom, palaeontologist and physician. Cape Town, A.A. Balkema, 1972. 157 pp.

BROUSSAIS, François Joseph Victor (1772–1838). French physician.

HUARD, P. DSB, 1973, **2**, 507–9.
VALENTIN, M. François Broussais, empereur de la médecine: jeunesse correspondance, vie et oeuvre. Dinard Association des Amis du Musée du Pays de Dinard, 1988. 319 pp.

BROWN, George Lindor (1903–1971). British physiologist.

MacINTOSH, F.C. & PATON. W.D.M. BMFRS, 1974, **20**, 41–72.

BROWN, John (1810–82). British physician in Scotland; author of '*Rab and his friends* and *Horae subsecivae*'.

BROWN, J.T. Dr John Brown. A biography and a criticism. London, Black, 1903. 244 pp.
PEDDIE, A. Recollections of Dr John Brown, author of *Rab and his friends*, etc. With selection from his correspondence. London, Percival, 1893. 197 pp.

BROWN, Michael Stuart (b. 1941). American physician; shared Nobel Prize (Physiology or Medicine), 1985, for discoveries concerning the regulation of cholesterol metabolism.

Les Prix Nobel en 1985.

BROWN, Rachel Fuller (1898–1980). American scientist; jointly with E.L.Hagen isolated nystatin (fungicidin).

BALDWIN, R.S. The fungus fighters: two women scientists and their discovery. Ithaca, Cornell Univ. Press, 1981. 212 pp.

BROWN, Robert (1773–1858). British botanist; discovered cell nucleus.

HATFIELD, G. Robert Brown, A.M., F.R.S.E. (1773–1858). Edinburgh, History of Medicine and Science Unit, Univ. Edinburgh. 1981. 11 pp.

BROWN, Thomas Graham (1882–1965). British neurophysiologist.

ADRIAN, Lord. BMFRS, 1966, **12**, 23–33.

BROWNE, George Buckston (1850–1945). British urological surgeon; benefactor of Royal College of Surgeons of England; endowed Buckston Browne research farm.

DOBSON, J. & WAKELEY, C. Sir George Buckston Browne. Edinburgh, E. & S. Livingstone, 1957. 143 pp.

BROWNE, Stanley George (1907–1986). British physician; Director, Leprosy Study Centre, London; made pioneer studies of leprosy in Africa.

THOMPSON, P. Mister Leprosy: Dr Stanley Browne's fight against leprosy. London, Hodder & Stoughton for Leprosy Mission England & Wales, 1980. 220 pp.

BROWNE, Thomas (1605–1682). British physician, naturalist and antiquary; author of *Religio medici*.

BENNETT, J. Sir Thomas Browne: a man of achievement in literature. Cambridge, Cambridge Univ. Press, 1962. 255 pp.
DONOVAN, D.G., HERMAN, H.G. & IMBRIE, A.E. Sir Thomas Browne and Robert Burton. Boston, Mass., G.K. Hall, 1981. 530 pp.
FINCH, J.S. Sir Thomas Browne, a doctor's life of science and faith. New York, Schuman, 1950. 319 pp.
HUNTLEY, F.L. Sir Thomas Browne: a biographical and critical study. Ann Arbor, Univ. Michigan Press, 1962. 283 pp.
KEYNES, G.L. DSB, 1973, **2**, 522–3.
KEYNES, G.L. A bibliography of Sir Thomas Browne ... Second edition revised and augmented. Oxford, Clarendon Press, 1968. 293 pp. First edition, Cambridge, Cambridge Univ. Press, 1924.

BROWNING, Carl Hamilton (1881–1972). British bacteriologist.

OAKLEY, C.L. BMFRS, 1973, **19**, 173–215.

BROWN-SÉQUARD, Charles-Édouard (1817–1894). French physician; neurologist and endocrinologist, born in Mauritius.

AMINOFF, M.J. Brown-Séquard: a visionary of science. New York, Raven, 1993. 211 pp.
GRMEK, M.D. DSB, 1973, **2**, 524–6.
OLMSTED, J.M.D. Charles-Édouard Brown-Séquard, a nineteenth-century neurologist and endocrinologist. Baltimore, Johns Hopkins Press, 1946. 253 pp.
ROLE, A. La vie étrange d'un grand savant, le professeur Brown-Séquard, (1817–1894). Paris, Plon, 1977. 219 pp.

BRUCE, David (1855–1931). Austrlian/British physician and microbiologist; discovered bacterial cause of Malta fever; made important researches on trypanosomiasis.

B., J.R., ONFRS, 1932–35, **2**, 79–85.
DOLMAN, C.E. DSB, 1973, **2**, 527–30.
Archival material: CMAC; Central Regional Council Archives, Stirling, Scotland.

BRÜCKE, Ernst Wilhelm von (1819–1892). German physiologist; professor in Vienna.

BRÜCKE, E.T. Ernst Brücke. Wien, J. Springer, 1928. 196 pp.
LESKY, E. DSB, 1973, **2**, 530–32.

BRUGSCH, Theodor (1878–1963). German physician and political professor at Halle.

KAISER, W. & HÜBNER, M. (eds.) Theodor Brugsch (1878–1963). Hallesches Brugsch-Symposium 1978. Halle-Wittenberg, Martin Luther-Universität, 1979. 222 pp.
KONERT, J. Theodor Brugsch: Internist und Politiker. Leipzig, S. Hirzel Verlag, 1988. 200 pp.

BRUMPT, Émile (1877–1951). French parasitologist; pioneered study of medical parasitology in France.

GALLIARD, H. DSB, 1973, **2**, 533–4.

BRUN, Rudolph (1885–1969). Swiss neuropsychiatrist.

AESCHLIMAN, J. Rudolph Brun (1885–1969). Leben und Werk des Zürcher Neurologen, Psychoanalytikers und Entomologen. Zürich, 1980. 70 pp.

BRUNFELS, Otto (*c*.1489–1534). German botanist, physician and priest.

ROTH, F.W.E. Otto Brunfels 1489–1534. Ein deutscher Botaniker. *Botanischer Zeitung*, 1900, **2**, 191–232.
STANNARD, J. DSB, 1973, **2**, 535–8.

BRUNNER, Johann Conrad à (1653–1727). Swiss anatomist; demonstrated that polyuria and extreme thirst followed excision of the pancreas.

–. *Acta Physico-Medica Academiae Caesareae Naturae Curiosum Norimbergae*, 1737, **4**, Appendix, 1–14.

BRUNO DA LONGOBURGO (*c*.1200–1286). Italian surgeon; practised in Padua and Verona.

TABANELLI, M. Un chirurgo italiano del 1200: Bruno da Longoburgo. Firenze, Olschki, 1970. 214 pp.

BRUNS, Paul (1846–1916). German surgeon; gave first description of plexiform neurofibroma.

F., M. *Medizinische Korrespondenz-Blatt*, Würtemburg, 1916, **86**, 317–20.

BRUNS, Victor von (1812–1883). German surgeon; professor in Tübingen.

SALTZWEDEL, G. Victor von Bruns (1812–1883): sein Leben und Werk. Tübingen, Mohr, 1977. 214 pp.

BRUNSCHWIG [BRUNSWYCK, BRAUNSCHWEIG], Hieronymus (*c*.1450–1533). Strasbourg surgeon; author of first printed book on surgery in German.

SCHMITZ, R. DSB, 1973, **2**, 546–7.
SIGERIST, H.E. Hieronymus Brunschwig and his work. New York, Argus, 1947. 48 pp.

BRUNTON, Thomas Lauder (1844–1916). British physician and pharmacologist.

BYNUM, W.F. DSB, 1972, **2**, 547–8.
Archival material: British Library; Library of the Wellcome Institute for the History of Medicine; National Library of Scotland.

BUCHAN, William (1729–1805). British physician; wrote *Domestic medicine*, first published in 1769, of which 80,000 copies (19 editions) were sold during the author's lifetime.

BENSLEY, E.H. Medicine laid open. *Medical History*, 1975, **19**, 20–35.
BLAKE, J.B. From Buchan to Fishbein: the literature of domestic medicine. In RISSE, G.B. et al. Medicine without doctors. New York, Science History Publications, 1977, pp. 11–30.

BUCHHEIM, Rudolf (1820–1879). German pharmacologist; professor at Leipzig, Dorpat and Giessen.

BRUPPACHER-CELLIER, M. Rudolf Buchheim (1820–1879) und die Entwicklung einer experimentellen Pharmakologie. Zürich, Juris, 1971. 75 pp.

BUCHINGER, Otto (1878–1966). German dietician.

KLEPZIF, H. Otto Buchinger: ein Leben für das Heilfasten. Friedrichshafen, Gessler, 2000. 168 pp.

BUCHNER, Eduard (1860–1917). German biochemist; Nobel Prize for Chemistry, 1907.

SCHRIEFERS, H. DSB, 1973, **2**, 560–63.

BUCHNER, Hans Ernst (1850–1902). German bacteriologist; discovered complement.

HUEPPE, F. Hans Buchner. *Münchener Medizinische Wochenschrift*, 1902, **49**, 844–7,

BUCK, Gurdon (1807–1877). American surgeon.

CONWAY, H. & STARK, R.B. Plastic surgery at the New York Hospital on hundred years ago, with biographical notes on Gurdon Buck... New York, P.B. Hoeber, 1953. 110 pp.

BUCKE, Richard Maurice (1837–1902). Canadian psychiatrist.

SHORTT, S.E. Victorian lunacy: Richard M. Bucke and the practice of late nineteenth century psychiatry. Cambridge, New York, Cambridge Univ. Press, 1986. 207 pp.

BUCKLAND, Francis Trevelyan (1825–1880). British surgeon and naturalist.

BURGESS, G.H. The curious world of Frank Buckland. London, Baker, 1967. 242 pp.

BUCKNILL, John Charles (1817–1897). British psychiatrist.

LANGLEY, C.E. Sir John Charles Bucknill, 1817–1897: our founder. *British Journal of Psychiatry*, 1980, **137**, 105–10.

BUCKY, Gustav Peter (1880–1963). German/American radiologist; invented Bucky diaphragm.

BORMACHER, K. Gustav Bucky (1880–1963). Bibliographie eines Röntgenologen und Erfinder. Dissertation. Berlin, 1967. 114 pp.

BUDD, George (1808–1892). British physician.

HUGHES, R.E. George Budd (1808–1882) and nutritional deficiency diseases. *Medical History*, 1973, **17**, 127–35.

BUDD, William (1811–1880). British physician and epidemiologist; his work fortified the theory of the water-borne spread of cholera and typhoid.

DOLMAN, C.E. DSB, 1973, **2**, 574–6.

GOODALL, E.W. William Budd, the Bristol physician and epidemiologist. Bristol, Arrowsmith, 1936. 159 pp.

BUDGE, Ludwig Julius (1811–1888). German physiologist; taught at Bonn and Greifswald.

LINDENMEYER, C. Ludwig Julius Budge und der Prioritätstreit um die Inhibitionstheorie. Zürich, Juris, 1967. 44 pp.

BÜLBRING, Edith (1903–1990). German/British pharmacologist; professor, Oxford University.

BOLTON, T.B. & BRADING, A.F. BMFRS, 1993, **38**, 69–95.
Archival material: CMAC.

BÜNGER, Christian Heinrich (1782–1842). German anatomist and surgeon; introduced rhinoplasty.

HEIN, J. Zur Geschichte der Anatomie und Chirurgie: Christian Heinrich Bünger, 1782–1842. Anatom und Chirurg in Marburg. Mannheim, Mannheimer Morgen, 1976. 467 pp.

BÜRGER, Max Theodor (1885–1966). German physician; professor in Bonn.

REIS, W. (ed.) Max Bürger (1885–1966), Internist, Pathophysiologe, Alternforscher. Ausgewahlte Texte. Leipzig, Barth, 1985. 160 pp.

BÜTSCHLI, Otto (1848–1920). German cytologist.

GOLDSCHMIDT, R.B. Otto Bütschli, pioneer of cytology (1848–1920). In: Underwood, E.A. (ed.) Science, medicine and history, *etc.* Oxford, Clarendon Press, 1953, **2**, 223–32.

BUFFON, Georges Louis Leclerc (1707–1788). French naturalist.

ROGER, F. Buffon: a life in natural history. Ithaca, London, Cornell Univ. Press [1997]. 492 pp. First published in French, 1989.

BUIST, John Brown (1846–1915). British bacteriologist in Edinburgh.

MACKIE, T.J. & VAN ROOYEN, C.E. John Brown Buist; an acknowledgement of his early contributions to the bacteriology of variola and vaccinia. *Edinburgh Medical Journal*, 1937, **44**, 72–7.

BULGAKOV, Mikhail Afanasievich (1891–1940). Russian physician, novelist and playwright.

BULGAKOV, M.A. A country doctor's notebook. Translated from Russian. London, Collins, 1975. 155 pp.

EDWARDS, T.R.N. Three Russian writers and the irrational: Zanyatin, Pil'nyakov and Bulgakov. Cambridge, Cambridge Univ. Press, 1982, 220 pp.

BULKLEY, Henry Daggett (1804–1872). American dermatologist.

SYNTEX LABORATORIES INC. Leaders in dermatology: Henry Daggett Bulkley, 1804–1872: the first American dermatologist; the man and his work. Palo Alto, Syntex Laboratories Inc., 1969. 63 pp.

BULLOCH, William (1868–1941). British bacteriologist; professor at London Hospital Medical College.

DOLMAN, C.E. DSB, 1973, **2**, 583–5.
LEDINGHAM, J.C.G., ONFRS, 1939–41, **3**, 819–52.

BULWER, John (*fl.* 1654) First English writer on the teaching of deaf-mutes.

–. A medical psychologist of the seventeenth century. *Journal of Psychological Medicine*, 1860, **13**, 294–314,

BUMKE, Oswald (1877–1950). German psychiatrist; professor in Munich.

BUMKE, O. Erinnerungen und Betrachtungen; der Weg eines deutschen Psychiaters. 2te. Aufl. München, R.Pflaum, 1953. 231 pp.

BUNGE, Gustav von (1844–1920). Estonian physiologist and biochemist.

SCHMIDT, G. Das geistige Vermächtnis von Gustav v. Bunge. Zürich, Juris, 1973. 88 pp.
SCHRIEFERS. H. DSB, 1973, **2**, 585–6.
Archival material: Library, University of Basle.

BURCKHARDT, Gottlieb (1836–1907). Swiss neurosurgeon; performed frontal lobotomy, 1890.

GROSS, D. Der Beitrag Gottlieb Burckhardts (1863–1907) zur Psychochirurgie in medizinischer und ethischer Sicht. *Gesnerus*, 1998, **55**, 221–48.
STONE, J.L. Dr Gottlieb Burckhardt: the pioneer of psychosurgery. *Journal of the History of the Neurosciences*, 2001, **10**, 79–92.

BURDACH, Karl Friedrich (1776–1847). German anatomist and physiologist.

KAY, A.S. DSB, 1973, **2**, 594–6.

BURDENKO, Nicolai Nilovich (1876–1964). Russian neurologist and neurosurgeon.

BAGDASARIAN, S.M. Nikolai Nilovich Burdenko. Moskva, Medgiz, 1954. 246 pp.
GRIGORIAN, N.A. DSB, 1973, **2**, 597–8.
MIRSKII, M.B. Istseliaiushchii skall'pelen akademik N.N. Burdenko. Moskva, Znanie, 1983. 187 pp.

BURDON-SANDERSON, John Scott, *see* **SANDERSON, John Scott Burdon**

BURGER, Herman Carel (1893–1965). Dutch medical physicist.

VAN CITTERT-EYMERS, J.G. DSB, 1973, **2**, 600–601.

BURKITT, Dennis Parsons (1911–1993). British surgeon and pathologist; described 'Burkitt's lymphoma'; pioneer in medical geography.

EPSTEIN,A. & EASTWOOD, M.A. BMFRS, 1995, **41**, 88–102.
GLEMSTER, B. The long safari. London, Bodley Head, 1970. 236 pp.
KELLOCK, B. The fibre man. The life-story of Dr Dennis Burkitt. Tring, Lion Publishing Co., 1985. 208 pp.
Archival material: CMAC.

BURN, Joshua Harold (1892–1981). British pharmacologist.

BÜLBRING, W. & WALKER. J.M. BMFRS, 1984, **30**, 45–89.
BURN, J.H. Essential pharmacology, *Annual Review of Pharmacology*, 1969, **9**, 1–20.
Archival material: University Laboratory of Physiology, Oxford; CMAC.

BURNET, Frank Macfarlane (1899–1985). Australian virologist and immunologist; shared Nobel Prize 1960.

BURNET, [F.] M. Changing patterns: an atypical autobiography. Melbourne & London, Heinemann, 1968. 282 pp.
FENNER, F.J. BMFRS, 1987, **33**, 99–162.
SEXTON, C. Burnet: a life. 2nd edition. Oxford, Oxford University Press, 1999. 327 pp. First published 1991.

BURNS, Allan (1781–1813). British surgeon in Glasgow; pioneer in recognition of mitral stenosis; first to suggest angina pectoris as a sign of coronary obstruction and first to record a case of chloroma.

HERRICK, J.B. Allan Burns, 1781–1813: anatomist, surgeon and cardiologist. *Bulletin of the Society of Medical History of Chicago*, 1928–35, **4**, 457–83.

BURT, Cyril Lodowic (1883–1971). British psychologist.

FLETCHER, R. Science, ideology, and the media; the Cyril Burt scandal. New Brunswick, N.J. Transaction Publishers, 1991. 419 pp.
HEARNSHAW, L.S. Cyril Burt, psychologist. Ithaca, NY, Cornell Univ. Press, 1979. 370 pp.
JOYNSON, R.B. The Burt affair. London, Routledge, 1989. 347 pp.
Archival material: University of Liverpool.

BURTON, Robert (1577–1640). English humanist; compiled first encyclopaedia of psychiatry, *Anatomy of melancholy*, 1621.

DEWEY, N. Robert Burton and the drama. Princeton University Ph.D., 1968. University Microfilms Inc., Ann Arbor, Mich. 245 pp.
DONOVAN, D.G., HERMAN, H.G. & IMBRIE, A.E. Sir Thomas Browne and Robert Burton. Boston, Mass., G.K. Hall, 1981. 530 pp.
JORDAN-SMITH, P. Bibliographia Burtoniana. Stanford, Stanford Univ. Press, 1931. 120 pp.
O'CONNELL, M. Robert Burton. Boston, Twayne Publishers, *c*.1986. 130 pp.
OSLER, W. Robert Burton: the man, his book, his library. *Proceedings and Papers of the Oxford Bibliographical* Society, 1925, **1**, 163–90. Reprinted in Selected writings of William Osler, ed. A.W. Franklin. Oxford, O.U.P., 1951, pp. 65–99.

BUSH, Alice Mary (1914–1974). New Zealand physician.

HERCOCK, F. Alice: the making of a woman doctor, 1914–1974. Auckland, Auckland Univ. Press, 1999. 283 pp.

BUSH, Ian Alcock (1928–1986). British steroid endocrinologist.

BUSH, I.A. Breaking the steroid barrier in partition chromatography. A steroid memoir. *Steroids*, 1985, **45**, 480–96.

BUSCHKE, Abraham (1868–1943). Polish dermatologist; with Otto Busse first described European cryptococcosis ('Busse-Buschke disease').

Gold, G.C. & NURNBERGER, F.J. In memoriam: a tribute to Abraham Buschke. *Journal of the American Academy of Dermatology*, 1992, **26**, 1019–22.

BUSK, George (1807–1886). British microscopist and surgeon.

COOK, G.C. George Busk FRS (1807–1886), nineteenth-century polymath: surgeon. *Journal of Medical Biography*, 1997, **5**, 88–101.

BUSSE, Otto (1867–1922). German physician; described human cryptococcosis.

MEYENBERG, H. von. Otto Busse 6.XII.1867 – 3.II.1922. *Verhandlungen der Deutschen Pathologischen Gesellschaft*, 1939, **31**, 528–31.

BUTENANDT, Adolf Friedrich Johann (1903–1995). German biochemist and endocrinologist; shared Nobel Prize 1939.

AKHTAR, M. & AKHTAR, M.E. BMFRS, 1998, **44**, 79–82.
KARLSON, P. Adolf Butenandt: Biochemiker, Hormonforscher, Wissenschaftspolitiker. Stuttgart, Wissenschaftliche Verlagsgesellschaft, 1990. 336 pp.

BUTLER, Josephine Elizabeth (1828–1906). British Christian worker for social and moral hygiene.

BELL, E.M. Josephine Butler. Flame of fire, London, Constable, 1962. 268 pp.
BUTLER, A.S.G. Portrait of Josephine Butler. London, Faber & Faber, 1954. 222 pp.

BUXTON, Patrick Alfred (1892–1955). British medical entomologist; professor at London School of Hygiene and Tropical Medicine.

WIGGLESWORTH, V.B. BMFRS, 1956, **2**, 69–84.

BUZZARD, Edward Farquhar (1871–1945). British physician; Regius Professor of Medicine, Oxford.

COOKE, A. M. Sir E. Farquhar Buzzard, Bt.: an appreciation. Oxford, for the author, 1975. 36 pp.

BYLON, David (*fl.* 1779). Dutch physician; first definitely to describe dengue.

PEPPER, O.H.P. A note on David Bylon and dengue. *Annals of Medical History*, 1941, **3**, 361–8.

CABANÈS, Augustin (1862–1928). French medical historian.

GOVAERTS, J.R. Un médecin au service de l'histoire, le docteur Augustin Cabanès. Paris, Jouve, 1941. 83 pp.

CABANIS, Pierre-Jean-Georges (1757–1808). French physician and philosopher.

CANGUILHEM, G. DSB, 1971, **3**, 1–3.
ROLE, A. Georges Cabanis, le médecin de Brumaire. Paris, Fernand Lanore, *c*.1994. 420 pp.
STAUM, M.S. Cabanis: enlightenment and medical philosophy in the French Revolution. Princeton, Princeton, Univ. Press, [*c*.1980]. 430 pp.

CADE, John Frederick Joseph (b.1912). Australian psychiatrist.

CADE, J.F. John Frederick Joseph Cade: family memories on the occasion of the 50th anniversary of his discovery of the use of lithium in mania. *Australian and New Zealand Journal of Psychiatry*, 1999, **33**, 615–22.
MITCHELL, P.B. On the 50th anniversary of John Cade's discovery of the anti-manic effect of lithium. *Australian and New Zealand Journal of Psychiatry*, 1999, **33**, 623–8.
MITCHELL, P.B. & HADZI-PAVLOVIC, D. John Cade and the discovery of lithium treatment of manic-depressive illness. *Medical Journal of Australia*, 1999, **171**, 262–4.

CADOGAN, William (1711–97). British paediatrician, physician to the Foundling Hospital, London, and to the Army.

RENDLE-SHORT, M. & RENDLE-SHORT, J. The father of child care: life of William Cadogan (1711–97). Bristol, Wright, 1966. 34 pp.

CAELIUS AURELIANUS (*fl.* AD 400–450). Roman physician; considered the greatest Latin medical writer after Celsus.

DRABKIN, I.E. (ed.) Caelius Aurelianus: on acute diseases and in chronic diseases. Chicago, Chicago Univ. Press, 1950. 1019 pp.

CAESALPINUS, Andreas, *see* **CESALPINO, Andrea**

CAIRNS, Hugh William Bell (1896–1952). Australian neurosurgeon; Nuffield Professor of Surgery, University of Oxford.

FRAENKEL, G.J. Hugh Cairns, first Nuffield Professor of Surgery, University of Oxford. Oxford, Oxford University Press, 1991. 288 pp.

CAIUS, John (1510–1573). British anatomist and surgeon; physician to Henry VIII. Author of first book in English on sweating sickness.

O'MALLEY, C.D. English medical humanists: Thomas Linacre and John Caius. Lawrence, Univ. Kansas Press, 1965. 54 pp.
O'MALLEY, C.D. DSB, 1971, **3**, 12–13.
VENN, J. John Caius, Master of Gonville and Caius College in the University of Cambridge. A biographical sketch. Cambridge, Cambridge Univ. Press, 1910. 78 pp.

CALDANI, Leopoldo Marcantonio (1725–1813). Italian anatomist and physiologist.

HINTZSCHE, E. DSB, 1971, **3**, 15–16.

CALDWELL, Charles (1772–1853). American physician; founded two medical schools in USA.

CALDWELL, C. Autobiography. With a preface, notes and appendix by H.W. Warner. Introduction by Lloyd G. Stevenson. New York, DaCapo, 1968. 454 pp. Reprint of 1855 edition, Philadelphia.

HORINE, E.F. Biographical sketch and guide to the writings of Charles Caldwell, MD. Brooks, Ky., High Acres Press, 1960. 155 pp.

CALDWELL, George Walter (1834–1918). American otolaryngologist; devised 'Caldwell-Luc' operation for abscess of maxillary sinus.

MACBETH, R. Caldwell, Luc and their operation. *Laryngoscope*, 1971, **81**, 1652–7.

CALDWELL, Peter Christopher (1927–1979). British physiologist.

DENTON, E.J. BMFRS, 1981, **27**, 153–72.

CALLOW, Robert Kenneth (1901–1983). British biochemist.

NEUBERGER, A. BMFRS, 1984, **30**, 93–116.

CALMETTE, Léon Charles Albert (1863–1933). French bacteriologist; prepared antivenin sera and BCG antituberculosis vaccine.

BERNARD, N. La vie et l'oeuvre de Albert Calmette. Paris, Editions Albin Michel, 1961. 313 pp.
DELAUNAY, A. DSB, 1971, **3**, 22–3.
KERVRAN, R. Albert Calmette et le B.C.G. Paris, Librairie Hachette, 1962. 222 pp.
MARTIN, C.J. ONFRS, 1934, **1**, 315–25.

CALNE, Roy (b.1930). British surgeon.

CALNE, R. The ultimate gift: the story of Britain's premier transplant surgeon. London, Headline, 1998. 210 pp.

CALVÉ, Jacques (1875–1954). French orthopaedic surgeon; in 1910 described juvenile osteochondritis, afterwards named 'Calvé-Legg-Perthes disease'.

SCHULITZ, K.P. & NIGGEMEYER, O. Jacques Calvé. *Spine*, 1996, **21**, 886–90.

CAMERARIUS, Joachim (1500–1574). German physician and botanist; practised in Nuremberg.

BARON, F. (ed.) Joachim Camerarius (1500–1574): Beiträge zur Geschichte des Humanismus im Zeitalter des Reformation. Essays in the history of humanism during the Reformation. München, Fink, 1978. 255 pp.

CAMERER, Johann Friedrich Wilhelm (1842–1910). German paediatrician.

ZEHENDER, O. Johann Friedrich Wilhelm Camerer und sein Beitrag zur Entstehung der modernen Pädiatrie. Zürich, Juris Verlag, 1969. 39 pp.

CAMERON, Gordon Roy (1899–1966). Australian physician, professor of morbid anatomy, University College Hospital Medical School, London.

OAKLEY, C.L. BMFRS, 1968, **14**, 83–116.
Archival material: Library, University College London.

CAMPBELL, Alfred Walter (1868–1937). Australian neurologist.

EADIE, M.J. A.W. Campbell: Australia's first neurologist. *Clinical and Experimental Neurology*, 1980, **17**, 27–35.

CAMPBELL, Fergus William (1924–1993). British ophthalmologist.

WESTHEIMER, G. BMFRS, 1995, **41**, 104–16.

CAMPER, Petrus (Pieter) (1722–1789). Dutch anatomist, physiologist and surgeon.

LINDEBOOM, G.A. DSB, 1971, **3**, 37–8.
SCHULLER TOT PEURSUM-MEIJER, J. & KOOPS, W.R.H. (eds.) Petrus Camper (1722–1789): onderzoeker van nature. Groningen, Universiteitsmuseum, 1989. 148 pp.

CAMPION, Thomas (1567–1620). English physician, poet and musician.

LOWBURY, E. *et al.* Thomas Campion: poet, composer, physician. London, Chatto & Windus, 1970. 195 pp.

CAMPS, Francis (1905–1972). British pathologist; professor of forensic medicine, London Hospital Medical College.

JACKSON, R. Francis Camps. Famous case histories of the celebrated pathologist. London, Hart-Davis, MacGibbon, 1975. 208 pp.

CANANO, Giovan Battista (1515–1579). Italian anatomist.

MURATORI, G. DSB, 1971, **3**, 40–41.

CANNAN, Robert Keith (1894–1971). American biochemist; professor at New York University.

EDSALL, J.T. BMNAS, 1985, **55**, 107–33.

CANNON, Walter Bradford (1871–1945). American physiologist; professor at Harvard Medical School.

BENNISON, S. & BARGER, A.C. DSB, 1978, **15**, 71–7.
BENNISON, S., BARGER, A.C. & WOLFE, E.L. The life and times of a young scientist. Cambridge, Mass., Harvard Univ. Press, 1987. 506 pp.
BROOKS, C.M. *et al.* The life and contributions of Walter Bradford Cannon 1871–1945: his influence on the development of physiology in the twentieth century. New York, State Univ. of New York, 1975. 264 pp.
CANNON, W.B. The way of an investigator; a scientist's experiences in medical research. New York, W.W. Norton & Co., 1945. 229 pp.
DALE, H.H. ONFRS, 1947, **5**, 407–23.
WOLFE, E.L. *et al.* Walter B. Cannon; science and society. Boston, Harvard Univ. Press, 2000. 644 pp.
Archival material: Countway Library, Harvard Medical School; American Philosophical Society, Philadelphia.

CANSTATT, Carl Friedrich (1807–1850). German physician; wrote important work on geriatrics.

LUKAS, A. Carl Friedrich Canstatt (1807–1850). Leben und Werk eines Klinikers der naturhistorischen Schule. Inaugural-Dissertation ... Freien Universität Berlin, Institut für Geschichte der Medizin, 1986. 109 pp.

CANTLIE, James (1851–1926). British surgeon, Charing Cross Hospital, London; Dean, Chinese School of Medicine, Hong Kong.

CANTLIE, N. & SEAVER, G. Sir James Cantlie: a romance in medicine. London, Murray, 1939. 279 pp.

STEWART, J.C. The quality of mercy: the lives of Sir James and Lady Cantlie, London, Allen & Unwin, 1983. 277 pp.

Archival material: CMAC.

CARCANO LEONE, Giovanni Battista (1536–1606). Italian anatomist.

SCARPA, A. Elogio historico di Giambattista Carcano Leone, professore di anatomia nella Università di Pavia. Milan, Stamperia Reale, 1813. 84 pp.

CARDANO, Girolamo [CARDANUS, Hieronymus] (1501–1576). Italian physician; professor of medicine at Padua and Bologna.

BALDI, M. & CANZIANI, G. Girolamo Cardano: le opere, le fonti, la vita. Milano, Francis Angeli, 1999. 589 pp.

CARDANUS, H. The book of my life (De vita propria liber). Translated from the Latin by Jean Stoner. London, Toronto, Dent, 1931. 331 pp. Reprinted New York, 1963.

FIERZ, M. Girolamo Cardano, 1501–1576; physician, natural philosopher, mathematician, astrologer and interpreter of dreams. Translated by H. Niman. Boston, Basel, *etc.*, Birkhauser, 1983. 202 pp. First published in German, Basel, 1977.

GLIOZZI, M. DSB, 1971, **3**, 64–7.

MORLEY, H. Jerome Cardan. The life of Girolamo Cardano of Milan, physician. 2 vols. London, Chapman & Hall, 1854. 304 + 328 pp.

SIRAISI, N.G. The clock and the mirror: Girolamo Cardano and Renaissance medicine. Princeton, Princeton Univ. Press, 1997. 361 pp.

CAREY, Mathew (1760–1839). Irish/American economist and publisher; published important account of the yellow fever epidemic in Philadelphia in 1793.

GREEN, N.J. Mathew Carey, publisher and patriot. Philadelphia, Library Company of Philadelphia, 1985. 32 pp.

CARLE, Antonio (1854–1927). Italian surgeon; demonstrated the transmissibility of tetanus by inoculation.

DONATI, M. *Archivio Italiano di Chirurgia*, 1927, **47**, 694–7.

UFFREDUZZI, O. *Giornale della Accademia di Medicina di Torino*, 1928, **91**, 182–96.

CARLISLE, Anthony (1768–1840). British surgeon and scientist.

THACKRAY, A. DSB, 1971, **3**, 67–8.

CARLSON, Anton Julius (1875–1956). Swedish-American physiologist; professor at University of Chicago.

DRAGSTEDT, L.R. BMNAS, 1961, **35**, 1–32.

INGLE, D.J. Anton J. Carlson: a biographical sketch. *Perspectives in Biology and Medicine*, 1979, **22**, Suppl., 114–37.

VISSCHER, M.B. DSB, 1971, **3**, 68–70.

CARLSSON, Arvid Per Emil (b. 1923) Swedish pharmacologist; shared Nobel Prize (Physiology or Medicine) 2000, for discoveries concerning signal transduction in the nervous system.

Les Prix Nobel en 2000.

BIBLIOGRAPHY OF MEDICAL AND BIOMEDICAL BIOGRAPHY

CARMICHAEL, Leonard (1898–1973). American physiological psychologist.

PFAFFMANN, C. BMNAS, 1980, **51**, 25–47.

CARPENTER, William Benjamin (1813–1885). British physician and scientist.

THOMAS, K.B. DSB, 1971, **3**, 87–9.

CARREL, Alexis (1873–1944). French surgeon and experimental biologist; Nobel prizewinner 1912.

ANTIER, J.-J. Alexis Carrel: la tentation de l'absolu. Monaco, Editions de Rocher, 1994. 319 pp.
CORNER, G.W. DSB, 1971, **3**, 90–92.
DROUARD, A. Alexis Carrel (1873–1944): De la mémoire à l'histoire. Paris, L'Harmattan, 1995. 262 pp.
LE VAY, D. Alexis Carrel: the perfectibility of man. Rockvelle, MD, Kabel Publishers, 1996. 402 pp.
MALININ, T.I. Surgery and life: the extraordinary career of Alexis Carrel. New York, Harcourt Brace Jovanovich, 1979. 242 pp.
Archival material: Catholic University, Washington; Rockefeller University; Medical Center Library, Georgetown University, Washington, D.C.

CARRIÓN GARCÍA, Daniel Alcides (1857–1885). Peruvian medical student who died after experimental inoculation with blood from patient with Oroya fever ('Carrión's disease').

LASTRES, J.B. Daniel Carrión. Lima, Ed. San Marcos, 1957. 146 pp.
SCHULTZ, M.G. Daniel Carrión's experiment. *New England Journal of Medicine*, 1968, **278**, 1322–6.

CARROLL, James (1854–1907). British/American bacteriologist; with Walter Reed carried out classical studies on aetiology of yellow fever.

BURNHAM, J.C. DSB, 1971, **3**, 94–5.

CARSON, James (1772–1843). British physician; advocate of open pneumothorax in tuberculosis.

KEERS, R.Y. Two forgotten pioneers. James Carson and George Bodington. *Thorax*, 1980, **35**, 483–9.

CARSWELL, Robert (1793–1857). British pathologist; professor of pathological anatomy, University College London.

BEHAN, P.O. & BEHAN, W.M. Sir Robert Carswell: Scotland's pioneer pathologist. In: Rose, F.C. & Bynum, W.F. Historical aspects of the neurosciences. New York, Raven, 1982, pp. 273–92.
RECKERT, H. Das unbekannte Werk des Pathologen Robert Carswell (1793–1857). Köln, 1982, 283 pp. (*Kölner medizinhistorische Beiträge*, Bd. 22).

CARTER, Henry Rose (1852–1925). American epidemiologist; made important studies of yellow fever.

RICHTER, E.D. Henry R. Carter – an overlooked skeptical epidemiologist. *New England Journal of Medicine*, 1967, **277**, 734–8.

CARUS, Carl Gustav (1789–1869). German physician and psychologist.

GENSCHOREK, W. Carl Gustav Carus: Arzt, Kunstler, Naturforscher. 3. Aufl. Leipzig, Hirzel, 1983. 266 pp.

GROSCHE, S. Lebenskunst und Heilkunde bei Carl Gustav Carus (1789–1869). Dissertation ... der Georg-August-Universität zu Göttingen, 1993. 270 pp.

MEFFERT, E. Carl Gustav Carus: Arzt, Künstler-Goethanist. Ein biographische Skizze. Basel, Perseus Verlag, 1999. 144 pp.

CASÁL JULIÁN, Gaspar Roque Francisco Narciso (1680–1759). Spanish physician; gave first clear description of pellagra.

GUERRA, F. DSB, 1971, **3**, 97.

MARTÍNEZ FERNÁNDEZ, J. Perfil de Gaspar Casál. Oviedo, Gráficas Summa, 1961. 122 pp.

VILLA RIO, M.P. Casál en Oviedo: estudio documental de las medicos, cirujanos y boticarios de Oviedo en el siglo XVIII. Oviedo, Instituto de Estudios Asturianos, 1967. 325 pp.

CASH, John Theodore (1854–1930). British pharmacologist; professor of materia medica and therapeutics, Aberdeen.

MARSHALL, C.R. ONFRS, 1936–39, **2**, 295–300.

CASTLE, William Bosworth (1897– 1990). American physician; showed pernicious anaemia to be due to the absence from the gastric juice of haemopoietin ('Castle's intrinsic factor').

KARNAD, A.B. Intrinsic factors: William Bosworth Castle and the development of hematology and clinical investigation at Boston City Hospital. Boston, Francis A. Countway Library of Medicine, 1997. 259 pp.

JANDL, J.H. BMNAS, 1995, **67**, 15–40.

CASSERI [CASSERIO] Giulio (1552–1616). Italian anatomist and surgeon.

PREMUDA, L. DSB, 1971, **3**, 98–100.

STERZI, G. Guilio Casseri, anatomico e chirurgico (*c*.1552–1616) richerche storiche. Venezia, Istituto Veneto di Arti Grafiche, 1909. 167 pp.

CASTELLANI, Aldo (1877–1971). Italian physician; specialist in tropical medicine.

CASTELLANI, A. Microbes, men and monarchs. A doctor's life in many lands. The autobiography of Aldo Castellani. London, Gollancz, 1968. 287 pp.

CASTIGLIONI, Arturo (1874–1953). Italian/American medical historian.

IN MEMORIAM di Arturo Castiglioni, 1874–1953. *Rivista di Storia di Scienze Mediche e Naturali*, 1954, **45**, 1–103. [Special number with articles by A. Corsini, P. Diepgen, J.F. Fulton, R. Neveu, H.E. Sigerist, and L. Belloni.]

CASTLE, William Ernest (1867–1962). American biologist and geneticist; professor of zoology, Harvard University.

ALLEN, G.E. DSB, 1971, **3**, 120–24.

DUNN, L.C. BMNAS, 1965, **38**, 33–80.

Archival material: Library, American Philosophical Society, Philadelphia.

CATCHESIDE, David Guthrie (1907–1994). Australian geneticist.

FINCHAM, J.R.S. & JOHN, B. BMFRS, 1995, **41**, 118–34.

CATHCART, Edward Provan (1877–1951). British physiologist; Regius Professor at University of Glasgow.

WISHART, G.M. ONFRS, 1951, **9**, 35–53.

CATHCART, George Clark (1860–1951). British laryngologist; founded the London Promenade Concerts 1894.

Lancet, 1951, **1**, 98.

CATON, Richard (1842–1926). British electrophysiologist.

COHEN, Lord. Richard Caton (1842–1926) pioneer electrophysiologist. *Proceedings of the Royal Society of Medicine*, 1959, **52**, 645–51.

CATTELL, James McKeen (1860–1944). American psychologist; professor at Columbia University.

PILLSBURY, W.B. BMNAS, 1949, **25**, 1–16.
REINGOLD, N. DSB, 1971, **3**, 130–31.
Archival material: Library of Congress, Washington; Library, American Philosophical Society, Philadelphia.

CAVENTOU, Joseph-Bienaimé (1795–1877). French chemist and pharmacist.

BERMAN, A. DSB, 1971, **3**, 159–60.

CAZENAVE, Pierre Louis Alphée (1795–1877). French dermatologist.

WALLACE, D.J. & LYON, I. Pierre Cazenave and the first detailed modern description of lupus erythematosus. *Seminars in Arthritis and Rheumatism*, 1999, **28**, 305–13.

CÉLINE, Louis Ferdinand [pseudonym of Henri Louis DESTOUCHES] (1894–1961). French physician and writer.

BRASSE, G. Leben und Werk Louis-Ferdinand Célines 1894–1961, unter besonderer Berücksichtigung seiner medizinischen Schriften. Herzogenrath, Murken-Altroge, 2000. 267 pp.
McCARTHY, P. Céline. London, A. Lane, 1975. 352 pp.
NEGRO, F.E. Céline, medico e malato. Milano, Franco Angeli, 2000. 143 pp.
VITROUX, V. La vie de Céline. Paris, B.Grasset, *c.*1988. 596 pp.

CELLI, Angelo (1857–1914). Italian malariologist.

ORAZI, S. Angelo Celli (1857–1914). Roma, Bulzoni Editore, 1992. 270 pp.
PAZZINI, A. The work of Angelo Celli, hygienist, scientist and sociologist. *Scientia Medica Italica* (English ed.), 1958, **7**, 233–41.

CELSUS, Aulus Aurelius Cornelius (25 BC–AD 50). Roman writer on medicine and surgery; author of *De medicina*, oldest medical document after the Hippocratic writings.

KUDLIEN, F. DSB, 1971, **3**, 174–5.
WELLMANN. M.A. Cornelius Celsus, eine Quellenuntersuchung. Berlin, Weidmannsche Buchhandlung, 1913. 138 pp.

BIBLIOGRAPHY OF MEDICAL AND BIOMEDICAL BIOGRAPHY

CERLETTI, Ugo (1877–1963). Italian psychiatrist; introduced electric convulsion therapy.

BARUK, H. Le professeur Hugo Cerletti (1877–1963). *Bulletin de l'Académie de Médecine, Paris*, 1966, **150**, 574–9.

CERMAK, Johann Nepomuk, *see* **CZERMAK.**

CERVELLO, Vincenzo (1854–1919). Italian pharmacologist; introduced paraldehyde as a narcotic.

LAZZARO, C. *Atti della Accademia delle Scienze Mediche, Palermo*, 1920, xiii–xxvi.

CESALPINO, Andrea (1519–1603). Italian physician and botanist.

ARCIERI, G.P. The circulation of the blood and Andrea Cesalpino. New York, Vanni, 1945. 193 pp.
MÄGDEFRAU, K. DSB, 1978, **15**, 80–81.
VIVIANI, U. Vita ed opere di Andrea Cesalpino. Arezzo, Ugo Viviani, 1922. 241 pp.

CESTONI, Giacinto (1637–1718). Italian naturalist; observed the itch mite, *Sarcoptes scabiei*, 1687.

BELLONI, L. DSB, 1971, **3**, 180–81.

CHADWICK, Edwin (1800–1890). British social reformer and statistician.

BRUNDAGE, A. England's "Prussian minister"; Edwin Chadwick and the politics of government growth. University Park, Pennsylvania Univ. Press, 1988. 208 pp.
FINER, S.E. The life and times of Sir Edwin Chadwick. London, Routledge, 1997. 555 pp. First published 1952.
LEWIS, R.A. Edwin Chadwick and the public health movement, 1832–54. London, Longmans Green, 1952. 411 pp.
MARSTON, M. Sir Edwin Chadwick (1800-1890). London, Parsons; Boston, Small, Maynard, 1925. 186 pp.
PERCIVAL, J. The papers of Sir Edwin Chadwick (1800–1890). London, Library, University College London, 1978. 186 pp.

CHAGAS, Carlos Ribeiro Justiniano (1879–1934). Brazilian physician; discovered the causal organism of American trypanosomiasis ('Chagas' disease').

CHAGAS, C. DSB, 1971, **3**, 185–6.
CHAGAS, C., [*filho*]. Meu pai. Rio de Janeiro, Casa de Oswaldo Cruz/Fiocruz, 1993. 293 pp.
LEON, L.A. Carlos Chagas (1879–1934) y la trypanosomiasis americana. Quito, Casa de la Cultura Ecuatoriana, 1980. 101 pp.

CHAIN, Ernst Boris (1906–1979). German-born British biochemist; shared Nobel Prize 1945 for work on penicillin.

ABRAHAM, E. BMFRS, 1983, **29**, 43–91.
CLARK, R.W. The life of Ernst Chain; penicillin and beyond. London, Weidenfeld & Nicolson, 1985. 217 pp.
SWANN, J.P. DSB, 1990, **17**, 148–50.
Archival material: CMAC.

CHAMBERLAND, Charles Edouard (1851–1908). French bacteriologist associated with Pasteur in studies of virus diseases.

DELAUNAY, A. DSB, 1971, **3**, 188–9.

BIBLIOGRAPHY OF MEDICAL AND BIOMEDICAL BIOGRAPHY

CHAMBERLEN Family. Peter, *the elder* (died 1631); **Peter,** *the younger* (1575–1628); **Peter,** *son of the younger* (1601–1683); **Paul** (1635–1717); **Hugh,** *the elder* (1630–?); **Hugh,** *the younger* (1664–1728). Employed midwifery forceps with two distinct blades which interlocked; confined the knowledge to the Chamberlen family.

AVELING, J.H. The Chamberlens and the midwifery forceps. Memorials of the family and an essay on the invention of the instrument. London, Churchill, 1882. 231 pp.
RADCLIFFE, W. The secret instrument. London, Heinemann, 1947. 83 pp.

CHAMBERS, Robert (1802–1871). British publisher and scientist; anticipated the theories of Darwin.

MILLHAUSER, M. Just before Darwin: Robert Chambers and *Vestiges*. Middletown, Conn., Wesleyan Univ. Press, 1959. 246 pp.
WILLIAMS, W.C. DSB, 1971, **3**, 191–3.

CHAMBON DE MONTAUX, Nicolas (1748–1826). French physician; Chief of La Salpetrière.

GÉNÉVRIER, J. La vie et les oeuvres de Nicolas Chambon de Montaux (1748–1836). Paris, G. Steinheil, 1906. 180 pp.

CHAMPIER, Symphorien (1472–1539). French physician; physician to Charles VIII and Louis XIII; medical historian and lexicographer.

ALLUT, P. Étude biographique et bibliographique sur Symphorien Champier. Lyon, Scheuring, 1859. 430 pp.
COPENHAVER, B.P. Symphorien Champier and the reception of the occultist tradition in Renaissance France. The Hague, New York, Mouton, 1978. 323 pp.

CHANG, Min Chueh (1908–1991). Chinese endocrinologist; with Gregory Pincus demonstrated an oral contraceptive.

GREEP, R.O. BMNAS, 1995, **68**, 45–61.

CHANNING, Walter (1786–1876). American obstetrician; early advocate of anaesthesia in childbirth; gave first description (1842) of pernicious anaemia of pregnancy.

KASS, A.M. Midwifery and medicine in Boston: Walter Channing, M.D., 1786–1876). Boston, Northeastern Univ. Press, *c*.2002. 386 pp.

CHANTEMESSE, André (1851–1919). French bacteriologist; introduced anti-typhoid inoculation.

THIEBIERGE, G. *Bulletin de l'Académie de Médecine*, Paris, 1919, **81**, 231–4.

CHAPIN, Charles Value (1856–1941). American pioneer in public health.

CASSEDY, J.H. Charles V. Chapin and the public health movement. Cambridge, Mass., Harvard Univ. Press, 1962. 310 pp.

CHAPMAN, Dennis (1927–1999). British biochemist.

QUINN, P.J. BMFRS, 2001, **47**, 57–66.

CHAPMAN, Nathaniel (1780–1853). American physician; first President of the American Medical Association; founder of *American Journal of the Medical Sciences*.

RICHMAN, I. The brightest ornament: a biography of Nathaniel Chapman. Bellefonte PA., Pennsylvania Heritage, Inc., 1967. 213 pp.

CHARCOT, Jean-Martin (1825–1893). French neuropsychiatrist; a chair of clinical diseases was created for him at the Salpêtrière, Paris.

GOETZ, C.G., BONDUELLE, L. & GELFAND, T. Charcot: constructing neurology. New York, Oxford Univ. Press, 1995. 392 pp. French edition, 1996.
GUILLAIN, G. J.-M. Charcot, 1825–1893: his life – his work. Edited and translated by P. Bailey. London, Pitman, 1959. 202 pp. First published in French, 1955.
OWEN, A.R.G. Hysteria, hypnosis and healing. The work of J.M. Charcot. London. Dennis Dobson, 1971. 252 pp.
TÉTRY, A. DSB, 1971, **3**, 205.
THUILLIER, J. Monsieur Charcot de la Salpétrière. Paris, Laffont, *c*.1993. 309 pp.

CHARGAFF, Erwin (b.1905). Austrian/American biochemist.

ABIR-AM, P. From biochemistry to molecular biology: DNA and the acculturated journey of the critic of science Erwin Chargaff. *History and Philosophy of the Life Sciences*, 1980, **2**, 3–60.
CHARGAFF, E. A fever of recollection: the early way. *Annual Review of Biochemistry*, 1975, **44**, 1–18.
CHARGAFF, E. Heraclitean fire. Sketches from a life before nature. New York, Rockefeller Univ. Press, 1978. 252 pp.

CHARLETON, Walter (1619–1707). English physician and naturalist.

FLEITMANN, S. Walter Charleton (1619–1707). "Virtuoso"; Leben und Werk. Frankfurt am Main, New York, Lang, 1986. 478 pp.

CHARNLEY, John (1911–1982). British orthopaedic surgeon.

NISBET, N.W. & WOODRUFF, M. BMFRS, 1984, **30**, 119–37.
WAUGH, W. John Charnley: the man and the hip. Berlin, New York, Springer Verlag, 1990. 268 pp. Corrected reprint 1991.

CHASE, Merrill Wallace (b.1905). American immunologist.

CHASE, M.W. Immunology and experimental dermatology. *Annual Review of Immunology*, 1985, **3**, 1–29.

CHASSAIGNAC, Édouard Pierre Marie (1804–1879). French surgeon; established surgical drainage on a scientific and methodical basis.

COUÉ, A. Chassaignac: sa vie – son oeuvre. Paris, Marcel Vigne, 1926. 98 pp.

CHATIN, Gaspard Adolphe (1813–1901). French pharmacist; showed value of iodine in endemic goitre and cretinism.

GUIGNARD, L. *Journal de Pharmacie et de Chimie*, 1901, **13**, 151–60.

CHAUFFARD, Anatole Marie Émile (1855–1932). French physician.

HUBER, J. *Semaine des Hôpitaux de Paris*, 1932, **8**, 593–7.
LABBÉ, M. *Bulletin de la Société Médicale des Hôpitaux de Paris*, 1932, 3 ser. **48**, 1387–91.

LE GENDRE, P. Le professeur A. Chauffard (1855–1932). *Bulletin de la Société Française d'Histoire de Médecine*, 1933, **26**, 402–7.

RAVAUT, P. *Bulletin de l'Académie Nationale de Médecine*, Paris, 1932, **108**, 1313–21.

Archival material: American Philosophical Society, Philadelphia.

CHAULIAC, GUY DE, *see* **GUY DE CHAULIAC**

CHAUVEAU, Jean Baptiste Auguste (1827–1917). French comparative physiologist.

McKUSICK, V.A. DSB, 1971, **3**, 219–20.

PITOIS, C. Chauveau: sa vie, son oeuvre anatomique et physiologique. Thèse No. 59, Université Claude Bernard-Lyon I, 1998. École Nationale Vétérinaire, 1998. 227 pp.

CHAVASSE, Noel Godfrey (1884–1917). British physician: Officer in Royal Army Medical Corps in First World War; awarded double Victoria Cross, August 1916 and (posthumously) September 1917; one of only three double recipients of the award.

CLAYTON, A. Chavasse – double V.C. London, Leo Cooper, 1992. 261 pp.

GUMMER, S. The Chavasse twins. London, Hodder & Stoughton, [1963]. 255 pp.

CHEADLE, Walter Butler (1835–1910). British physician.

–. *Lancet*, 1910, **1**, 962–5.

CHEKHOV, Anton Pavlovich (1860–1904). Russian writer and physician.

CALLOW, P. Chekov, the hidden ground: a biography. London, Constable, 1998. 428 pp.

COOPE, J. Doctor Chekhov: a study in literature and medicine. Chale [Isle of Wight], Cross Publishing, 1998. 159 pp.

HINGLEY, R. A new life of Anton Chekhov. London, Oxford University Press, 1976. 352 pp.

CHESELDEN, William (1688–1752). British surgeon; lithotomist to St Thomas' Hospital, Westminster Infirmary and St George's Hospital.

COPE, Z. William Cheselden, 1688–1752. Edinburgh, Livingstone, 1953. 112 pp.

Archival material: Royal Academy of Arts, London (drawings for his *Osteographia*, 1733).

CHEVREUL, Michel Eugène (1786–1889). French organic chemist and gerontologist ; made important studies of animal fats; proved that the sugar in diabetic urine is glucose.

CARMICHAEL, E.B. Michel Eugène Chevreul. Experimental chemist and physicist: lipids and eyes. *Alabama Journal of Medical Sciences*, 1973, **10**, 223–32.

COSTA, A.A. Michel Eugène Chevreul, pioneer of organic chemistry. Madison, State University of Wisconsin, 1962. 116 pp.

ROQUE, M. Michel-Eugène. Chevreul: un savant, des couleurs. Paris, Editions du Muséum Nationale d'Histoire Naturelle, 1997. 277 pp.

CHEYNE, George (1671–1743). British physician; wrote on hypochondria and gout.

BROWN, T.S.M. DSB, 1971, **3**, 244–5.

GUERRINI, A. Obesity and depression in the Enlightenment: the life and times of George Cheyne. Norman, Univ. of Oklahoma Press, 2000. 283 pp.

KING, L.S. George Cheyne, mirror of eighteenth century medicine. *Bulletin of the History of Medicine*, 1974, **48**, 517–39.

CHEYNE, William Watson (1853–1932). British surgeon and bacteriologist.

B.,W. ONFRS, 1932–35, **1**, 26–30.

CHIARI, Hans (1851–1916). German pathologist.

FESTSCHRIFT Hans Chiari aus Anlass seines 25 jährigen Professorenjubiläums gewidnet. Wien, Leipzig, W.Braumüller, 1908. 421 pp.

CHIARUGI, Giulio (1859–1944). Italian anatomist and embryologist.

FRANCESCHINI, P. DSB, 1971, **3**, 245–6.

CHIARUGI, Vincenzo (1739–1820). Italian psychiatrist; one of the first to abandon restraints on patients.

CABRAS, P.L., CAMPANINI, E. & LIFFI, D. Uno psichiatra prima della psichiatria: Vincenzo Chiarugi ed il trattato "Della pazzia in genere, e in specie" (1793–1794). Florence, Scientific Press, 1993. 115 pp.
NEL PRIVATO di una grande vita: l'epistolario di Vincenzo Chiarugi. Manziana, Vecchiarelli, 1992.

CHIBNALL, Albert Charles (1894–1988). British biochemist; professor at universities of London and Cambridge.

CHIBNALL, A.C. The road to Cambridge. *Annual Review of Biochemistry*, 1966, **35**, 1–22.
SYNGE, R.L.M. & WILLIAMS, E.F. BMFRS, 1990, **35**, 55–96.
Archival material: Cambridge University Library.

CHICK, Harriette (1875–1977). British biochemist; conducted important research on aetiology and treatment of rickets.

MORGAN, N. DSB, 1990, **17**, 165–6.

CHISHOLM, Colin (1755–1825). British physician; first to observe the method of transmission of *Dracunculus medinensis*, the Guinea worm.

HOSACK, D. Memoir of the life and writings of the late Colin Chisholm. *American Journal of the Medical Sciences*, 1829, **4**, 394–402.

CHITTENDEN, Russell Henry (1856–1943). American physiological chemist and nutritionist; professor at Yale University.

VICKERY, H.B. BMNAS, 1945, **24**, 59–104.
VICKERY, H.B. DSB, 1971, **3**, 256.
Archival material: Yale University Library, New Haven, Conn.

CHOPART, François (1743–1795). French surgeon; pioneer of urological surgery; introduced method of amputation through the metatarsal joint.

SUE, P. Notice historique sur François Chopart, professeur de pathologie externe à la Faculté de Médecine de Paris. *Journal de Médecine, Chirurgie, Pharmacie*, 1812, **25**, 349–63.

CHOULANT, Johann Ludwig (1791–1861). German physician, medical historian and bibliographer.

BIBLIOGRAPHY OF MEDICAL AND BIOMEDICAL BIOGRAPHY

LAAGE, R.J.Ch.V.ter. Reflections on Johann Ludwig Choulant and his medico-historical bibliographies. In: Smit, P. & Laage, R.J.Ch.ter. Essays on biohistory presented to Frans Verdoon. Utrecht, International Association for Plant Taxonomy, 1970, pp. 115–33.
STEINER, A. Ludwig Choulant und seiner "Anleitung zu dem Studium der Medicin" (1829). Zürich, Juris, 1987. 72 pp.

CHOVET, Abraham (1704–1790). British-born anatomist, moved to Philadelphia, where he taught anatomy and made fine wax models to illustrate his lectures.

MILLER, W.S. Abraham Chovet: an early teacher of anatomy in Philadelphia. *Anatomical Record*, 1911, **5**, 147–72.
MILLER, W.S. Abraham Chovet. *Annals of Medical History*, 1926, **8**, 375–93.

CHRISTIAN, Henry Asbury (1876–1951). American physician.

WARREN, J.V. Henry A. Christian (1876–1951). *Journal of Laboratory and Clinical Medicine*, 1988, **112**, 401–2.

CHRISTISON, Robert (1797–1882). British physician and toxicologist; professor of materia medica at Edinburgh University.

CHRISTISON, R. The life of Sir Robert Christison. Edited by his sons. [Vol. I, Autobiography; II, Memoirs.] 2 vols., Edinburgh, London, Blackwood, 1885–86. 428 + 492 pp.

CHRISTOFFEL, Hans (1888–1959). Swiss psychiatrist and psychoanalyst.

KAISER, W. Leben und Werk des Basler Psychiaters un Psychoanalytikers Hans Christoffel (1888–1959). Zürich, Juris Druck, 1982. 267 pp.

CHRISTOPHERS, Samuel Rickard (1873–1978). British protozoologist.

SHORTT, H.E. & GARNHAM, P.C.C. BMFRS, 1979, **25**, 175–207.
Archival material: CMAC.

CIVIALE, Jean (1792–1867). French physician and lithotomist.

WULFF, O. Contribution à l'histoire de la lithotritie. *Janus*, 1926, **30**, 301–41.

CLAIRMONT [KLARBERG], Paul Johann (1875–1942). Austrian surgeon; professor of surgery in Zürich.

BREITNER, B. Paul Clairmont; das Genie des Lehrens. Basel, Schwabe, 1948. 70 pp.
MÖHR, R.C. Der Chirurg Paul Clairmont (1875–1942). Zürich, Juris, 1986. 99 pp.

CLARK, Alfred Joseph (1885–1941). British pharmacologist; professor in London and Edinburgh.

BARCROFT, J. ONFRS, 1941, **3**, 969–84.
CLARK, D.H. Alfred Joseph Clark 1885–1941: a memoir. [Street, Somerset], C. & J. Clark Ltd Archives, for British Pharmacological Society, 1985. 61 pp.

CLARK, Wilfrid Edward Le Gros (1895–1971). British anatomist and anthropologist; elucidated hypothalamic connections in man.

CLARK, W. Le Gros. Chant of pleasant exploration. Edinburgh, Livingstone, 1968. 250 pp.

74

ZUCKERMAN, Lord. BMFRS, 1973, **19**, 217–33.
Archival material: Bodleian Library, Oxford; General Library, British Museum (Natural History), London.

CLARK, William Mansfield (1884–1964). American biochemist; professor of physiological chemistry at Johns Hopkins University; pioneer in measurement of physicochemical properties.

CLARK, W.M. Notes on a half-century of research, teaching, and administration. *Annual Review of Biochemistry*, 1962, **31**, 1–24.
LEICESTER, H.M. DSB, 1971, **3**, 290.
VICKERY, H.B. BMNAS, 1967, **39**, 1–36.
Archival material: Library, American Philosophical Society, Philadelphia.

CLARKE, Cyril Astley (1907–2000). British geneticist.

WEATHERALL, D. BMFRS, 2002, **48**, 71–85.
Archival material: Liverpool University Library.

CLARKE, Hans Thacher (1887–1972). American biochemist; professor at the College of Physicians and Surgeons, Columbia University.

CLARKE, H.T. Impressions of an organic chemist in biochemistry. *Annual Review of Biochemistry*, 1955, **27**, 1–14.
VICKERY, H.B. BMNAS, 1975, **46**, 3–20.
Archival material: American Philosophical Society, Philadelphia.

CLARKE, John (1761–1815). British physician; gave first account of infantile tetany, 1815.

RUHRÄH, J. *American Journal of Diseases of Children*, 1934, **47**, 184–6.

CLARKE, Robert Henry (1850–1926). British physician and physiologist; introduced stereotactic apparatus for cerebral examination and surgery.

DAVIS, R.A. Robert Henry Clarke, Victorian physician-scholar and pioneer physiologist. *Surgery, Gynecology and Obstetrics*, 1964, **119**, 1333–40.
TEPPERMAN, J. Horsley and Clarke: a biographical medallion. *Perspectives in Biology and Medicine*, 1970, **13**, 293–308.

CLARKE, William Fairlie (1833–1884). British surgeon and general practitioner; assistant surgeon, Charing Cross Hospital, 1871–77.

WALKER, E.A. William Fairlie Clark, MD, FRCS, his life and letters, hospital sketches and addresses. London, Hunt, 1885, 297 pp.

CLAUDE, Albert (1899–1983). Belgian cytologist; shared Nobel Prize 1944 for discoveries concerning organization of the cell.

FLORKIN, M. Pour saluer Albert Claude. *Archives Internationales de Physiologie et de Biochimie*, 1972, **80**, 632–47.
PALADE, G.E. & de DUVE, C. Albert Claude and the beginnings of biological electron microscopy. *Journal of Cell Biology*, 1971, **50**, 5D–55D.

CLELAND, John (1835–1925). British anatomist.

CARMEL, P.W. & MARKESBERY, W.R. Early descriptions of the Arnold-Chiari malformations. The contribution of John Cleland. *Journal of Neurosurgery*, 1972, **37**, 543–7.

CLENDENING, Logan (1884–1945). American physician; professor of medicine and of the history of medicine, University of Kansas.

MAJOR, R.H. Disease and destiny: Logan Clendening. Lawrence, Univ. Kansas Press, 1958. 49 pp.

CLÉRAMBAULT, Gaëtan Henri Alfred Edouard Léonard Murie Gatian de (1872–1934). French psychiatrist.

RENARD, E. Le docteur Gaëtan de Clérambault; sa vie et son oeuvre, 1872–1934. Paris, Le François, 1942. 139 pp.

CLIFT, William (1775–1849). British naturalist; apprentice to John Hunter; conservator of the Hunterian Museum, Royal College of Surgeons of England.

DOBSON, J. William Clift. London, Heinemann, 1954. 144 pp.
Le FANU, W. DSB, 1971, **3**, 323–5.
Archival material: Royal College of Surgeons of England, British Museum (Natural History) in Owen Papers.

CLOUSTON, Thomas Smith (1841–1915). British psychiatrist; first to recognize relationship between general paralysis and congenital syphilis.

–. Sir Thomas Smith Clouston. *British Medical Journal*, 1915, **1**, 744–6, 767.

CLOVER, Joseph Thomas (1825–1882). British anaesthetist; invented apparatus for administration of anaesthetics.

MARSTON, A.D. Life and achievements of Joseph Thomas Clover. *Annals of the Royal College of Surgeons of England*, 1949, **4**, 267–80.
Archival material: Woodward Biomedical Library, University of British Columbia, Victoria, B.C.

CLOWES, William (1544–1604). English surgeon; surgeon to St Bartholomew's Hospital.

CLOWES, W. Selected writings, 1544–1604. Edited, with an introduction and notes, by F.N.L. Poynter. London, Harvey & Blythe, 1948. 179 pp.

COBB, Stanley (1887–1968). American neuropathologist; professor at Harvard University.

WHITE, B.B. Stanley Cobb: a builder of the neurosciences. Boston, Francis A. Countway Library of Medicine, 1984. 445 pp.

COCHRAN, John (1730–1807). American surgeon; Director General of military hospitals, USA; surgeon to George Washington.

SAFFRON, M.H. Surgeon to George Washington, Dr John Cochran 1730–1807. New York, Columbia Univ. Press, 1977. 302 pp.

COCHRANE, Archibald Leman (1909–1988). British epidemiologist; professor at Welsh National School of Medicine, Cardiff.

COCHRANE, A.L. One man's medicine: autobiography of Professor Archibald Cochrane. London, British Medical Journal, 1989. 283 pp.

CODIVILLA, Alessandro (1861–1912). Italian surgeon; first to attempt surgical lengthening of limbs.

FURFARO, D. Alessandro Codivilla chirurgica ortopediste.*Chirurgia degli Organi di Movimento*, 1964, **53**, 75–8.

CODMAN, Ernest Amory (1869–1940). American surgeon.

MALLON, W.J. Ernest Amory Codman: the end result of life in medicine. Philadelphia, W.B. Saunders, 2000. 196 pp.

CODRONCHI, Giovan Battista (1547–1628). Italian physician; wrote first important work on diseases of larynx and forensic medicine, in his De vitiis vocis, 1597.

MAZZINI, G. *Rivista di Storia delle Scienze Mediche*, 1923, **14**, 310–18.
PUCCHINI, C. Il methodus testificandi di G.B. Codronchi. Sala Bolognese, Arnoldo Forni Editore, 1987. 141 pp.
PUCCHINI, C. Per la storia deontologia medica; il contributo di G.B. Codronchi. *Rivista di Storia de la Medicina*, 1991, **22**, 59–68.

COFFEY, Robert Calvin (1869–1922). American surgeon; the modern uretero-intestinal anastomosis is based on his experimental work.

MAYO, W.J. *Surgery, Gynecology & Obstetrics*, 1934, **46**, 671–3.
O'DAY, J. C. *American Medicine*, 1935, **41**, 419–22.

COGHILL, George Ellett (1872–1941). American anatomist and physiologist; professor of comparative anatomy at the Wistar Institute.

MERRICK, C.J. BMNAS, 1941–43, **22**, 251–73.

COHEN, Julius Berend (1859–1935). British chemist; professor at Leeds University; worked on synthesis of antiseptics, including acriflavine.

RAPER, H.S. ONFRS, 1932–35, **1**, 503–13.

COHEN, Philip Pacy (1908–1993). American biochemist.

BURRIS, R.H. BMNAS, 1999, **77**, 34–49.

COHEN, Samuel (b. 1922). American biochemist; shared Nobel Prize (Physiology or Medicine), 1988, for the discovery of growth factors.

Les Prix Nobel en 1988.

COHN, Edwin Joseph (1892–1953). American biochemist; made special studies of proteins.

EDSALL, J.T. BMNAS, 1961, **35**, 47–84.
EDSALL, J.T. DSB, 1971, **3**, 335–6.
SURGENOR, D.M. Edward J. Cohn and the development of protein chemistry: with a detailed account of his work on fractionation of blood during and after World War II. Boston, Harvard Univ. Press, 2002. 434 pp.
Archival material: Medical School, Harvard University.

BIBLIOGRAPHY OF MEDICAL AND BIOMEDICAL BIOGRAPHY

COHN, Ferdinand Julius (1828–1898). German/Polish bacteriologist and botanist; classified bacteria.

GEISON, G.L. DSB, 1971, **3**, 336–41.
DOLLEY, C.S. *Bulletin of the History of Medicine*, 1939, **7**, 49–92.

COHNHEIM, Julius Ferdinand (1839–1884). German pathologist; made important contributions to the knowledge of inflammation and suppuration.

DIEPGEN, pp. 50–55 (CB).
THOM, A. *et al.* Julius Cohnheim und sein Werk. *Zentralblatt für Allgemeine Pathologie*, 1985, **130**, 281–347.

COINDET, Jean François (1774–1834) Swiss physician; introduced iodine treatment of goitre.

RENTCHNIK, P. Iode et goitre. A propos du 150e anniversaire de la mort du Dr Jean-François Coindet. *Médecine et Hygiène*, 1984, **42**, 465–6, 469–70, 472–4, 477–8.

COITER, Volcher (1534–1576). Dutch/German anatomist, physiologist and embryologist; published important work on comparative osteology.

HERRLINGER, R. Volcher Coiter, 1534–1576/ Nürnberg, Edelmann, 1952. 147 pp.
SCHULLIAN, D.M. DSB, 1971, **3**, 342–3.

COLDEN, Cadwallader (1688–1776). American physician and botanist.

HINDLE, B. DSB, 1971, **3**, 343–5.
JARCHO, S. Biographical and bibliographical notes on Cadwallader Colden. *Bulletin of the History of Medicine*, 1958, **32**, 322–34.

COLE, Francis Joseph (1872–1959). British zoologist and scientific historian; professor of zoology, Reading.

FRANKLIN, K.J. BMFRS, 1959, **5**, 37–47.

COLE, Kenneth Stewart (1900–1984). American biophysicist on staff of National Institutes of Health, Bethesda; professor at University of California.

COLE, K.S. Mostly membranes. *Annual Review of Physiology*, 1979, **41**, 1–24.
HUXLEY, A. BMFRS, 1992, **38**, 97–110.

COLE, Rufus Ivory (1872–1966). American physician; pioneer in development of clinical research; first director of the Hospital of the Rockefeller Institute.

MILLER, C.P. BMNAS, 1979, **50**, 119–39.
Archival material: Library, American Philosophical Society, Philadelphia.

COLE, Warren Henry (1898–1990). American surgeon; introduced cholecystography; professor at University of Illinois College of Medicine.

CONNAUGHTON, D. Warren Cole, MD, and the ascent of scientific surgery. Chicago, Warren and Clara Cole Foundation, distributed by University of Illinois Press, 1991. 246 pp.

COLEBROOK, Leonard (1883–1967). British bacteriologist.

NOBLE, W.C. Coli: great healer of men. The biography of Leonard Colebrook, F.R.S. London, Heinemann, 1974. 149 pp.

78

OAKLEY, C.L. BMFRS, 1971, **17**, 91–138.
Archival material: CMAC.

COLLER, Frederick Amasa (1887–1964). American surgeon.

ROBINSON, J.O. Frederick A. Coller: his philosophy, surgical practice, and teachings. Ann Arbor, Mich., National Institute for Burn Medicine, 1986. 257 pp.

COLLES, Abraham (1773–1843). Irish surgeon.

FALLON, M. Abraham Colles, 1773–1843; surgeon of Ireland. London, Heinemann, 1972. 238 pp.

COLLIP, James Bertram (1892–1965). Canadian endocrinologist; isolated the parathyroid hormone and adrenocorticotrophic hormone; improved insulin.

BARR, M.L. & ROSSITER, R.J. BMFRS, 1973, **19**, 235–67.
NOBLE, R.L. DSB, 1971, **3**, 351–4.
Archival material: University of Alberta, Edmonton.

COLOMBO, Realdo (1516–1559). Italian anatomist and physiologist.

BYLEBYL, J. DSB, 1971, **3**, 354–6.

COMBE, Andrew (1797–1847). Physiologist and phrenologist.

COMBE, G. The life and correspondence of Andrew Combe. Edinburgh, Maclachlan & Stewart; London, Longmans & Simpkin Marshall, 1850. 563 pp.

COMROE, Julius Hiram (1911–1984). American physiologist.

KETY, S.S. & FORSTER, R. BMNAS, 2001, **70**, 67–83.

CONCATO, Luigi Maria (1825–1882). Italian physician.

SIMILI, A. Luigi Concato e la catedra torinese di clinica medica. Torino, S.A.F.M.M., 1962. 60 pp. First published in *Minerva Medica*, 1962, **53**, 3069–89.

CONKLIN, Edwin Grant (1863–1952). American embryologist and developmental biologist; professor of biology, Princeton University.

ALLEN, G.F. DSB, 1971, **3**, 389–91.
HARVEY, E.N. BMNAS, 1958, **31**, 54–91.
Archival material: Library, American Philosophical Society, Philadelphia, and Princeton University (see *Journal of the History of Biology*, 1968, **1**, 325–32).

CONN, Jerome W. (1907–1981). American endocrinologist; described primary aldosteronism ("Conn's syndrome").

DAUGHADAY, W.H. BMNAS, 1997, **71**, 3–14.

CONOLLY, John (1788–1870). British psychiatrist; pioneer in humane methods of treatment of the insane.

CLARK. J. A memoir of John Conolly; comprising a sketch of the treatment of the insane in Europe and America. London, Murray, 1869. 298 pp.
SCULL, A. A Victorian alienist, John Conolly, FRCP, DCL (1794–1866). In: Bynum, W.F. *et al.* (eds.) The antomy of madness. London, Tavistock, 1985, vol. 1, pp. 103–50.

CONRING, Hermann (1606–1681). German physician; professor of medicine, Helmstadt.

RESNER, E. Die Bedeutung Hermann Conrings in der Geschichte der Medizin. *Medizinhistorisches Journal*, 1969, **4**, 287–304.
STOLLEIS, M. (ed.) Hermann Conring (1606–1681). Beiträge zur Leben und Werk. Berlin, Duncker & Humblot, 1983. 590 pp.

CONWAY, Edward Joseph (1894–1968). British biochemist and biophysicist.

MAIZELS, M. BMFRS, 1969, **15**, 69–82.

COOK, Albert Ruskin (1871–1951). British medical missionary in Uganda.

FOSTER, W.D. The Church Missionary Society and modern medicine in Uganda. The life of Sir Albert Cook, KCMG, 1871–1951. Newhaven, Sussex, for the author, 1978. 234 pp.
O'BRIEN, B. That good physician. London, Hodder & Stoughton, 1962. 264 pp.
Archival material: CMAC.

COOK, James Wilfred (1900–1975). British chemist; discovered carcinogenic properties of dibenzanthracene compounds.

ROBERTSON, J.M. BMFRS, 1976, **22**, 71–103.

COOKE, Alexander Macdougall (1899–1999). British physician.

COOKE, A. My first 75 years of medicine. London, Royal College of Physicians, 1994. 151 pp.

COOLEY, Denton Arthur (b.1920). American cardiac surgeon.

MINETREE, H. Cooley: the career of a great heart surgeon. New York, Harper's Magazine Press, 1933. 298 pp.
THOMPSON, T. Hearts: DeBakey and Cooley, surgeons extraordinary. London, M. Joseph, 1972. 316 pp.

COOLEY, Thomas Benton (1871–1945). American paediatrician; described erythroblastic anaemia in children.

ZUELZER, W.W. Pediatric profiles: Thomas B. Cooley (1871–1945). *Journal of Pediatrics*, 1956, **49**, 642–50.

COOLIDGE, William David (1873–1975). American physicist; invented Coolidge high vacuum X-ray tube.

LIEBHAFSKY, H.A. William David Coolidge: a centenarian and his work. New York, Wiley, 1974. 96 pp.
SUITS, C.G. BMNAS, 1982, **53**, 141–57.

COONS, Albert Hewett (1912–1978). American immunologist.

KARNOVSKY, M.J. Dedication to Albert H. Coons 1912–1978. *Journal of Histochemistry and Cytochemistry*, 1979, **27**, 1117–18.
McDERMOTT, H.O. BMNAS, 1996, **69**, 27–36.

COOPER, Astley Paston (1768–1841). British surgeon (Guy's Hospital) and teacher; Sergeant-Surgeon to King George IV.

BROCK, R. The life and work of Astley Cooper. Edinburgh, London, Livingstone, 1952. 176 pp.

COOPER, B.B. The life of Sir Astley Cooper, Bart, interspersed with sketches from his notebooks of distinguished contemporary characters. 2 vols., London, Parker, 1843. 448 + 466 pp. Includes an account of the resurrectionists.

HALE-WHITE, 1935, pp. 22–41. (CB)

Archival material: Royal College of Surgeons of England.

COOPER, Lilian Violet (1861–1947). Australian physician and surgeon.

WILLIAMS, L.M. No easy path: the life and times of Lilian Violet Cooper, Australia's first woman surgeon. Brisbane, Amphion Press, 1991. 138 pp.

COPEMAN, Sydney Arthur Monckton (1862–1947). British physician, at St Thomas' Hospital and Ministry of Health; worked on vaccination and cancer.

MacNALTY, A.S. & CRAIGIE, J. ONFRS, 1948–49, **6**, 37–50.

COPP, Douglas Harold (1915–1998). Canadian physiologist.

MacINTYRE, I. BMFRS, 2000, **46**, 49–64.

CORAM, Thomas (1668–1751). British philanthropist; established Foundling Hospital in London.

McCLURE, R.K. Coram's children; the London Foundling Hospital in the eighteenth century. New Haven, Yale Univ. Press, 1981. 321 pp.

CORBOLENSIS, Aegidius, *see* **GILLES DE CORBEIL**

CORDUS, Euricius (1486–1535). German physician; wrote an early account of sweating sickness.

MANN, G. (ed.) Der Englische Schweiss, 1529. Marburg, N.G. Elwert, 1967. 15 pp.

CORDUS, Valerius (1515–1544). German botanist and pharmacist; published first official pharmacopoeia (1546).

GREENE, E.L. Landmarks of botanical history. Washington, Smithsonian Institute, 1909, **1**, 270–314.
SCHMITZ, R. DSB, 1971, **3**, 413–15.

COREY, Robert Brainard (1897–1971). American molecular biologist.

MARSH, R.E. BMNAS, 1997, **72**, 51–68.

CORI, Carl Ferdinand (1896–1984). Czech/American biochemist, born Prague; professor of biological chemistry, Washington University, St Louis; shared Nobel Prize with G.T.R. Cori, 1947.

COHN, M. BMNAS, 1992, **61**, 78–109.
CORI, C.F. The call of science. *Annual Review of Biochemistry*, 1969, **38**, 1–20.
RANDLE, P. BMFRS, 1986, **32**, 67–95.

CORI, Gerty Theresa Radnitz (1896–1957). Czech/American biochemist, born Prague, worked in USA; shared Nobel Prize with husband for work on glycogens, 1947.

CORI, C.F. The call of science. *Annual Review of Biochemistry*, 1969, **38**, 1–20.
LARNER, J. BMNAS, 1992, **38**, 110–35.
SCHMITZ, R. DSB, 1971, **3**, 415–16.

CORMACK, Allan Macleod (b.1924) American biochemist; South African/American physicist; shared Nobel Prize (Physiology or Medicine), 1979, for his contribution to the development of computer-assisted tomography.

Les Prix Nobel en 1979.

CORNER, George Washington (1889–1981). American anatomist, physiologist, and medical historian.

CORNER, G.W. The seven ages of a medical scientist; an autobiography. Philadelphia, Univ. of Philadelphia, 1981. 411 pp.
ZUCKERMAN, Lord. BMFRS, 1983, **29**, 95–112.
Archival material: Library, American Philosophical Society, Philadelphia.

CORNIL, André Victor (1837–1908). French pathologist and bacteriologist.

BIOGRAPHIE de Victor Cornil. Inauguration de la statue de Victor Cornil. Vichy, Imp. Bougarel, 1911. 132 pp.

CORRENS, Carl Franz Joseph Erich (1864–1933).German geneticist and botanist; rediscovered Mendel's laws.

OLBY, R. DSB, 1971, **3**, 421–3.

CORRIGAN, Dominic John (1802–1880). Irish physician; gave classical account of aortic insufficiency.

O'BRIEN, E. Conscience and conflict. A biography of Sir Dominic Corrigan 1802–1880. Dublin, Glendale Press, 1983. 383 pp.

CORTI, Alfonso Giacomo Gaspare (1822–1876). Italian anatomist; made important microscopic studies of inner ear.

HINTZSCHE, E. Alfonso Corti (1822–1876). Eine Biographie auf Grund neu aufgefundener Quellen. Bern, Haupt, 1944. 44 pp.
HINTZSCHE, E. DSB, 1971, **3**, 424–5.
ULLMAN, E.V. Life of Alfonso Corti. *Archives of Otolaryngology*, 1951, **54**, 1–28.

CORVISART, Jean-Nicolas (1755–1821). French physician.

GANIÈRE, P. Corvisart: médecin de Napoléon. Paris, Perrin, 1985. 440 pp. First published 1951.
GANIÈRE, P. DSB, 1971, **3** 426–8.
TOUCHE, M. J.–N.Corvisart; practicien célèbre, grand maître de la médecin. Paris, J.B. Baillière, 1968. 60 pp.
Archival material: Bibliothèque Nationale, Paris; Archives de l'Académie des Sciences, Paris; Archives de l'Académie de Médecine, Paris.

COTUGNO, Domenico Felice Antonio (1736–1822). Italian anatomist and physician.

SCHULLIAN, D.M. DSB, 1971, **3**, 437–8.

COTZIAS, George Constantin (1918–1977). American physician.

DOLE, V.P. BMNAS, 1995, **68**, 63–82.

COUÉ, Emil (1857–1926). French practitioner of autosuggestion as a method of cure of disease.

ORTON, J.L. Emil Coué: the man and his work. London, F. Mott, 1935. 306 pp.

COUNCILMAN, William Thomas (1854–1933). American pathologist at Johns Hopkins Hospital; professor at Harvard University.

CUSHING, H. BMNAS, 1936, **18**, 157–74; also in *Science*, 1933, **77**, 613–18.

COURNAND, André Frédéric (1895–1988). French/American physiologist; shared Nobel Prize (Physiology or Medicine) 1956 for work on cardiac catheterization.

COURNAND, A.F. From roots to late budding. The intellectual adventures of a medical scientist. New York, Gardner Press, 1986. 232 pp.
LEQUIME, J. In memoriam André Cournand (1895–1988). His role in the development of modern cardiology. *Acta Cardiologica*, 1988, **43**, 437–42.
WEIBEL, E.R. BMNAS, 1995, **67**, 65–99.

COURRIER, Robert (1895–1986). French reproductive endocrinologist; professor at Collège de France, Paris.

TATA, J.R. BMFRS, 1990, **36**, 101–23.

COURTOIS, Bernard (1777–1838). French chemist; discovered iodine.

COSTA, A.B. DSB, 1971, **3**, 455.

COUTARD, Henri (1876–1950). French radiologist.

REGATO, J.A. del. Henri Coutard. *International Journal of Radiation* Oncology, 1987, **13**, 433–43.

COUVELAIRE, Roger (1903–1986). French surgeon; constructed artificial bladder (1951).

CERBONNET, F. Roger Couvelaire, 1903–1986. *Chirurgie*, 1989, **115**, 13–25.

COWPER, William (1666–1709). English anatomist and surgeon.

BEEKMAN, F. Bidloo and Cowper. *Annals of Medical History*, 1935, N.S. **7**, 113–29.

COX, Alfred (1866-1954). British medical practitioner; Secretary, British Medical Association.

COX, A. Among the doctors. London, Johnson, 1950. 224 pp.

COX, Herald Rea (1907–1986). American urologist; introduced typhus vaccine.

HARDEN, V.A. Rocky Mountain spotted fever. Baltimore, Johns Hopkins University Press, 1990. pp. 175–96.

COX, William Sands (1801–1875). British physician; founder of the Birmingham Medical School, and Queen's Hospital.

MORRISON, J.T.J. William Sands Cox and the Birmingham Medical School. Birmingham, Cornish, 1926. 240 pp.

COYON, Amand (1871–1928). French physician.

–. *Bulletin de la Société Médicale des Hôpitaux de Paris,*1928, 3 ser., **52**, 1841–4.

CRABBE, George (1754–1832). British physician and poet.

BLACKBURNE, J.R. The restless ocean: the story of George Crabbe, the Aldeburgh poet 1754–1832. Lavenham, Dalton, 1972. 236 pp.
CRABBE, G. The life of George Crabbe, by his son; with an introduction by E. Blunden. London, Cresset Press, 1947. 286 pp.
ZAROFF, L. George Crabbe: his poetry and eighteenth century medicine. Thesis, Stanford Univ., 2000. Ann Arbor, Mich., University Microfilms International, 2001. 271 pp.

CRAFOORD, Clarence (1899–1984). Swedish cardiovascular surgeon; pioneered surgical treatment of coarctation of aorta.

CRAFOORD, J. & OLIN, C. Clarence Crafoord en av seklets stora kirurgiska pionjärer. *Läkartidningen*, 1999, **96**, 2627–32.

CRAIB, William Hofmeyr ('Don') (1896–1982). South African electrocardiographer; professor of medicine, Witwatersrand.

ADAMS, E.B. In search of truth. A portrait of Don Craib. London, Royal Society of Medicine, 1990. 126 pp.

CRAIG, Charles Franklin (1872–1930). American military surgeon and pathologist.

BASS, C.C. Charles Franklin Craig: fifty years of work and service in tropical medicine. *American Journal of* Tropical Medcine *and Hygiene*, 1952, **1**, 5–19.
SIMMONS, J.J. Colonel Charles Franklin Craig, M.C., U.S. Army: salute to the Chief. *American Journal of Tropical Medicine and Hygiene*, 1952, **1**, 20–26.

CRAIG, Lyman Creighton (1906–1974). American chemist; developed analytical methods for alkaloids, peptides and proteins.

MOORE, S. BMNAS, 1978, **49**, 49–77.

CRAIGIE, James (1899–1978). British microbiologist.

ANDREWES, C. BMFRS, 1979, **25**, 233–40.
Archival material: CMAC.

CRANE, Augustus Warren (1868–1937). American roentgenologist; introduced kymography in clinical cardiology.

CASE, J.T. *American Journal of Roentgenology*, 1937, **37**, 684–9.

CRANE, John (1571–1652). 17th-century Cambridge apothecary.

MARTIN, L.C. John Crane, 1571–1652, the Cambridge apothecary and philanthropist. Cambridge, Newton and Denny, 1977. 55 pp.

CREDÉ, Carl Siegmund Franz (1819–1892). German obstetrician at Charité, Berlin.

DICKENMANN, W. Carl S.F. Credé (1819–1892) und seine Hauptleistungen. Zürich, Juris, 1969. 37 pp.

CREIGHTON, Charles (1847–1927). British physician and medical historian.

DOLMAN, C.E. DSB, 1971, **3**, 463–4.

UNDERWOOD, E.A. Charles Creighton: the man and his work. In: A history of epidemics in Britain, Charles Creighton. With additional material by D.E.C. Eversley and Lynda Ovenall. Vol. 1., London, Frank Cass, 1985, pp. 43–135.

CREW, Francis Albert Eley (1886–1973). British physician and geneticist.

HOGBEN, L. BMFRS, 1974, **20**, 135–53.

CRICHTON, Alexander (1763–1856). British physician.

TANSEY, E.M. The life and works of Sir Alexander Crichton, F.R.S. (1763–1856), a Scottish physician to the Imperial Russian Court. *Notes and Records of the Royal Society of London*, 1984, **38**, 241–59.

CRICHTON-BROWNE, James (1840–1938). British physician.

CRICHTON-BROWNE, J. Stray leaves from a physician's portfolio. London, Hodder & Stoughton, 1927. 351 pp.

CRICHTON-BROWNE, J. The doctor's second thoughts. London, E. Benn, 1931. 294 pp.

HOLMES, G.M, ONFRS, 1939, **2**, 519–21.

NEVE, M.R. & TURNER, T. What the doctor thought and did: Sir James Crichton-Browne (1840–1938). Medical History, 1995, **39**, 399–432.

CRICHTON-MILLER, Hugh (1877–1959). British physiologist and analytical psychologist; founded Tavistock Clinic, London.

IRVINE, E.F. A pioneer of the new psychology, Hugh Crichton-Miller, M.A., M.D., F.R.C.P., 1877–1959. Chatham, W. & J. Mackay, 1963. 79 pp.

HUGH CRICHTON-MILLER 1877–1959. A personal memoir by his friends and family. Dorchester, Longmans, 1961. 79 pp.

CRICK, Francis Henry Compton (b.1916). British molecular biologist; shared Nobel Prize 1962 for discovery of the molecular structure of DNA.

CRICK, F. What mad pursuit; a personal view of scientific discovery. London, Weidenfeld & Nicolson, 1988. 182 pp.

WATSON, J.A. The double helix: a personal account of the discovery of the structure of DNA. London, Weidenfeld & Nicolson, 1968. 226 pp.

Archival material: CMAC.

CRILE, George Washington (1864–1943). American surgeon; founder of the Cleveland Clinic Foundation.

CRILE, G.W. George Crile: an autobiography. Edited with sidelights by Grace Crile. 2 vols., Philadelphia, New York, Lippincott, 1947. 624 pp.

ENGLISH, P.C. Shock, physiological surgery and George Washington Crile. Westport Conn., Greenwood, 1980. 271 pp.

Archival material: Western Reserve Historical Society, Cleveland.

CROHN, Burrill Bernard (1884–1983). American gastroenterologist; described regional ileitis.

CROHN, B.B. An oral history; recorded by J.B. Boyle [from three interviews] with a bibliography of Dr Crohn's writings. Wilmington, Delaware, American Gastroenterological Association, 1968. 68 ff.

CROHN, B.B. Notes on the evolution of a medical specialist, 1907–1965. New York, Burrill B. Crohn Research Foundation Inc., Mount Sinai Medical Center, 1984. 79 pp.

CROSSE, John Green (1790–1850). British surgeon; worked at Norfolk and Norwich Hospital, England.

CROSSE, V.M. A surgeon in the early nineteenth century. The life and times of John Green Crosse, 1790–1850. Edinburgh, London, Livingstone, 1968. 210 pp.

CRUIKSHANK, William Cumberland (1745–1800). British surgeon and chemist.

DOBSON, J. DSB, 1971, **3**, 486–8.
Archival material: Royal College of Surgeons of England, London.

CRUMMER, Le Roy (1872–1934). American physician and medical historian.

BEAMAN, A.G. A doctor's Odyssey, a sentimental record of Le Roy Crummer, physician, author, bibliophile, artist in living, 1872–1934. Baltimore, Johns Hopkins Press, 1935. 340 pp.

CRUVEILHIER, Jean (1791–1874). French pathologist.

DELHOMME, L. L'école de Dupuytren. Jean Cruveilhier. Paris, J.B. Baillière, 1937. 309 pp.
HUARD, P. DSB, 1971, **3**, 489–91.

CRUZ, Oswaldo Gonçalves (1872–1917). Brazilian physician and sanitarian; director of the Instituto Oswaldo Cruz.

CERQUIEIRA FALCÃO, E. de. Oswaldo Cruz. Monumenta historica. 3 vols. São Paulo, Empr. Graf, Revista dos Tribunais, 1971–73.
FRAGA, C. Vida e obra de Osvaldo Cruz. Rio de Janeiro, Livraria José, Olympia Editora, 1972. 186 pp.
STEPAN, N. DSB, 1978, **15**, 96–8.
STEPAN, N. Beginnings of Brazilian science: Oswaldo Cruz, medical research and policy 1890–1920. New York, Science History, 1976. 255 pp.

CULLEN, Thomas Stephen (1868–1953). American gynaecologist; eponymized in 'Cullen's sign'.

ROBINSON, J. Tom Cullen of Baltimore. London, O.U.P., 1949. 435 pp.

CULLEN, William (1710–1790). British physician; professor of medicine and outstanding teacher at Edinburgh University.

DOIG, A. ed. William Cullen and the eighteenth century medical world: a bicentenary exhibition and symposium arranged by the Royal College of Physicians of Edinburgh. Edinburgh, Edinburgh Univ. Press, 1993. 256 pp.
THOMSON, J. An account of the life, lectures, and writings of William Cullen. 2 vols. Edinburgh, London, Blackwood, 1859. 668 + 764 pp. Vol. I first issued 1832; Vol. 2 commenced by John and William Thomson and concluded by David Craigie.
WIGHTMAN, W.P.D. DSB, 1971, **3**, 494–5.
Archival material: University of Glasgow Library; Bodleian Library, Oxford; Royal College of Physicians, Edinburgh; Royal College of Physicians, London.

CULPEPER, Nicholas (1616–1654). English physician and astronomer.

CHANCE, B. "Nicholas Culpeper, gent: student in physicke and astrologie", 1616–1653/4. *Annals of Medical History*, 1931, **3**, 394–403.

THULESIUS, O. Nicholas Culpeper: English physician and astrologer. London, Macmillan, 1992. 190 pp.

CURIE, Marie (1867–1934). Polish/French physicist; with her husband Paul isolated radium; shared Nobel Prize for Physics 1903 and awarded Nobel Prize for Chemistry 1911.

CURIE, Eve. Madame Curie, a biography. Translated by V. Sheean. Garden City, New York, Doubleday, Doran, 1937. 393 pp.; London, Heinemann, 1938, 411 pp.

PFLAUM, R. Grand obsession: Madame Curie and her world. New York, Doubleday, 1989. 496 pp.

QUINN, S. Marie Curie: a life. New York, Simon & Schuster, *c.*1995. 509 pp.

WEILL, A.R. DSB, 1971, **3**, 497–503.

CURIE, Pierre (1859–1906). French physicist; shared Nobel Prize for Physics 1903 for work on radioactivity.

CURIE, M. Pierre Curie. Translated by C. & V. Kellogg. New York, Macmillan, 1923 & 1932. 242 pp. Reprinted Dover Publications, 1963.

WYART, J. DSB, 1971, **3**, 503–8.

CURRIE, James (1756–1805). British physician and man of letters in Liverpool; biographer and editor of Robert Burns.

CURRIE, W. W. (ed.) Memoir of the life, writings and correspondence of James Currie, M.D., F.R.S., of Liverpool .... Edited by his son. 2 vols. London, Longman, Rees, Orme, Browne, and Green, 1831. 524 + 503 pp.

THORNTON, R.D. James Currie, the entire stranger, and Robert Burns. Edinburgh, Oliver & Boyd, 1963. 459 pp.

CUSHING, Harvey Williams (1869–1939). American neurosurgeon and bibliophile; professor at Ann Arbor, Michigan.

BLACK, P.M. et al. The surgical art of Harvey Cushing. Park Ridge, Ill., American Association of Neurological Surgeons, *c.*1992. 182 pp.

CANNON, W.B. ONFRS, 1940–41, **3**, 277–90.

FULTON, J.F. & EISENHARDT, L. A bibliography of the writings of Harvey Cushing, prepared on the occasion of his seventieth birthday, 8 April 1939. 3rd ed. Park Ridge, Ill., 1993. 144 pp. First published 1940.

FULTON, J.F. Harvey Cushing; a biography. Oxford, Blackwell, 1946. 755 pp.

MacCALLUM, W.G. BMNAS, 1941–43, **22**, 49–70.

THOMSON, E.H. DSB, 1971, **3**, 516–19.

THOMSON, E.H. Harvey Cushing, surgeon, author, artist. New York, Schuman, 1950. 347 pp. Reprinted New York, Academic Press, 1981.

Archival material: Yale University Medical Library; McGill University, Montreal.

CUSHNY, Arthur Robertson (1866–1926). British pharmacologist and physiologist; professor at Michigan, University College London, and Edinburgh.

ABEL, J.J. Arthur Robertson Cushny 1866–1926 and pharmacology, *Journal of Pharmacology & Experimental Therapeutics*, 1926, **27**, 262–86.

GEISOW, G.L. DSB, 1978, **15**, 99–104.

MacGILLIVRAY, H. A personal biography of Arthur Robertson Cushny. *Annual Review of Pharmacology*, 1968, **8**, 1–24.

Archival material: CMAC; Library, University of Edinburgh.

CUTLER, Elliott Carr (1888–1947). American surgeon.

ZOLLINGER, R.M. Elliott Carr Cutler and the cloning of surgeons. Mont Kisco, NY, Futura Publishing Co., 1988. 235 pp.

CUVIER, Georges (1769–1832). French comparative anatomist.

ARDOUIN, P. Georges Cuvier, promoteur de l'idée évolutionniste et créateur de la biologie moderne. Paris, Expansion Scientifique Française, 1970. 207 pp.
OUTRAM, D. Georges Cuvier: vocation, science, and authority in post-revolutionary France. Manchester, Manchester Univ. Press, c.1984. 299 pp.
SMITH, J.C. (compiler). Georges Cuvier: an annotated bibliography of his published works. Washington, DC, Smithsonian Institution Press, 1993. 251 pp.

CZERMAK, Johan Nepomuk (1828–1873). Czechoslovak laryngologist.

KRUTA, V. DSB, 1971, **3**, 530–31.

CZERNY, Adalbert (1863–1941). German paediatrician.

HARTMANN, H.R. Gesunde Kinder; das Lebenswerk Adalbert Czerny. Berlin, K. Siegismund, 1938. 327 pp.
SCHIFF, E. Pediatric profiles: Adalbert Czerny (1863–1941). *Journal of Pediatrics*, 1956, **48**, 391–9.

CZERNY, Vincenz (1842–1916). Bohemian surgeon; performed first vaginal hysterectomy, 1879.

M., W. *Transactions of the American Surgical Association*, 1917, **35**, xlii–xlv.

DA COSTA, Jacob Mendez (1833–1900). American physician, born in W. Indies; professor at Philadelphia.

JULIAN, D.J. Jacob Mendez Da Costa. *Journal of Medical Biography*, 1993, **1**, 248–52.
LEY, C.F. Jacob Da Costa, medical teacher, clinician, and clinical investigator. *American Journal of Cardiology*, 1982, **50**, 1145–8.

DAKIN, Henry Drysdale (1880–1952). British/American biochemist.

BECHTEL, W. DSB, 1990, **17**, 193–4.
HARTLEY, P. ONFRS, 1952–53, **8**, 129–48.
HAWTHORNE, R.M. Henry Drysdale Dakin, biochemist (1880–1952): the option of obscurity. *Perspectives in Biology and Medicine*, 1983, **26**, 553–66.

DALE, Henry Hallett (1875–1968). British physiologist and pharmacologist; first Director, National Institute for Medical Research; shared Nobel Prize 1936 for work on chemical mediation of nervous impulses.

BYNUM, W.F. DSB, 1978, **15**, 104–7.
DALE, H.H. Adventures in physiology with excursions into pharmacology. London, Pergamon Press, 1953. 652 pp. (Reprints 30 of Dale's major papers with comments by Dale.)
DALE, H.H. An autumn gleaning. London, Pergamon Press, 1954. 225 pp. (Reprints an additional 14 of Dale's general essays, etc.)
FELDBERG, W.S. BMFRS, 1970, **16**, 77–174.

TANSEY, E.M. The early scientific career of Sir Henry Dale FRS (1875–1968). For the Degree of Doctor of Philosophy, University of London. London, University College, 1990. 515 pp.
Archival material: CMAC; National Institute for Medical Research, London; Royal Society of London.

DALLDORF, Gilbert Julius (1900–1979). American virologist; isolated Coxsackie virus from children with poliomyelitis.

RAPP, F. BMNAS, 1994, **65**, 94–105.

DALLOS, Josef (1905–1979). Hungarian ophthalmologist; introduced contact lenses.

MANN, I. Dr Josef Dallos, contact lens pioneer. *Ophthalmic Antiques*, 1995, **50**, 6–7.

DALTON, John (1766–1844). English chemist; gave first description of colour blindness ('Daltonism') 1798.

THACKRAY, A.W. John Dalton: critical assessments of his life and science. Cambridge, Mass., Harvard University Press, 1972. 190 pp.

DALTON, John Call (1825–1889). American physiologist; professor at universities of Buffalo, Vermont, and the College of Physicians and Surgeons, New York.

HOLMES, F.L. DSB, 1978, **15**, 107–10.

DALY, Ivan de Burgh (1893–1974). British physiologist.

BARCROFT, H. BMFRS, 1975, **21**, 197–226.
Archival material: CMAC.

DALZIEL, Keith (1921–1994). British biochemist.

GUTFREUND, H. BMFRS, 1996, **42**, 112–50.

DAM, Carl Peter Henrik (1895–1976). Danish biochemist, professor at the University of Copenhagen; discovered vitamin K; shared Nobel Peace Prize, 1943.

SNELDERS, H.A.M. DSB, 1990, **17**, 196–200.

DAMESHEK, William (1900–1969). Russian/American physician.

GUNZ, F.W. William Dameshek, 1900–1969. *Blood*, 1970, **35**, 577–82.

DANCE, Jean Baptiste Hippolyte (1797–1832). French physician.

ASTRUC, P. *Progrès Médicale*, 1936, Suppl., 17–21.

DANDY, Walter Edward (1886–1946). American neurosurgeon.

FOX, W.L. Dandy of Johns Hopkins. Baltimore, Williams & Wilkins, 1984. 293 pp.

DANFORTH, Charles Haskell (1883–1969). American geneticist; professor of anatomy at Stanford University.

WILLIER, B.H., BMNAS, 1974, **44**, 1–56.

BIBLIOGRAPHY OF MEDICAL AND BIOMEDICAL BIOGRAPHY

DANIELLI, James Frederic (1911–1984). British cytologist.

STEIN, W.D. BMFRS, 1986, **32**, 117–35.

DANILEVSKY, Aleksandr Yakovlevich (1814–1875). Russian bichemist; discovered trypsin.

BULANKIN, I.N. *Biokhimia*, 1950, **15**, 97–104.

DANILEVSKY, Vasili Iakovlevich (1852–1934). Russian biologist, physiologist and protozoologist; discovered malaria parasite in birds.

FINKELSTEIN, E.A. Vasilij Jakovlevich Danilevsky [in Russian]. Moskva, [Academy of Sciences of the USSR], 1955. 290 pp.

DANLOS, Henri Alexandre (1844–1912). French dermatologist; described 'Ehlers-Danlos syndrome' 1908; pioneer in use of radium treatment of lupus erythematosus.

HUDELO, –. *Bulletin de la Société Française de Dermatologie et de Syphiligraphie*, 1912, **23**, 500–508.

DARESTE, Gabriel Madeleine Camille (1822–1899). French biologist and experimental teratologist.

FISCHER, J.L. Leben und Werk von Camille Darest, Schöpfer der experimentellen Teratologie. Halle, Deutsche Akademie der Narturforscher Leopoldina. Halle/Saale, 1994. 283 pp.

DARLING, Samuel Taylor (1872–1925). American pathologist; described histoplasmosis ('Darling's disease').

–. *American Journal of Tropical Medicine*, 1925, **5**, 319–21.

DARLINGTON, Cyril Dean (1903–1981). British geneticist.

LEWIS, D. BMFRS, 1983, **29**, 113–57.
OLBY, R.C. DSB, 1990, **17**, 203–9.
Archival material: Bodleian Library, Oxford.

DARWIN, Charles Robert (1809–1882). British naturalist.

AYDON, C. Charles Darwin. London, Constable, 2000. 360 pp.
BARLOW, N. (ed.) The autobiography of Charles Darwin 1809–1882; with original omissions restored. Edited with appendix and notes by his granddaughter. London, Collins, 1958, 253 pp.
BOWLBY, J. Charles Darwin: a new biography. London, Hutchinson, 1990. 384 pp.
BOWLER, P.J. Charles Darwin: the man and his influence. New edition. Cambridge, Cambridge Univ. Press, 1996. 250 pp.
BRENT, F. Charles Darwin. London, Heinemann, 1981. 536 pp.
BROWNE, E.J. Charles Darwin. 2 vols. London, Cape, 1995–2002. 606+591 pp.
CLARK, R.W. The survival of Charles Darwin: a biography of a man and an idea. London, Weidenfeld & Nicolson, 1985. 480 pp.
DARWIN, F. The life and letters of Charles Darwin: including an autobiographical chapter. 3 vols. London, John Murray, 1887. 395+393+418 pp. Reprinted, 2 vols. New York, Basic Books, 1959.
DE BEER, G. DSB, 1971, **3**, 565–7.
DESMOND, A. & MOORE, J. Darwin. London, Michael Joseph, 1991. 807 pp.
Nicolson, 1984. 450 pp.

FREEMAN, F. The works of Charles Darwin: an annotated bibliographical handlist. 2ⁿᵈ ed. Folkestone, Dawson, 1977. 235 pp.

Archival material: Down House, Downe, Kent; Cambridge University Library (Handlist of Darwin papers, 1960. 72 pp.); American Philosophical Society, Philadelphia.

DARWIN, Erasmus (1731–1802). Physician, scientist, poet; grandfather of Charles Darwin; developed concept of biological evolution to a sophisticated level half a century before the publication of the Origin of species.

COHEN OF BIRKENHEAD, Lord, DSB, 1971, **3**, 577–81.
McNEIL, M. Under the banner of science. Erasmus Darwin and his age. Manchester, University Press, 1987. 307 pp.
DARWIN, C. The life of Erasmus Darwin; edited by D. King-Hele. Cambridge, Cambridge Univ. Press, 2002. 172 pp. Previous (2nd) editon, 1887.
HASSLER, D.M. Erasmus Darwin. New York, Twayne, 1973, 143 pp.
KING-HELE, D. Erasmus Darwin: a life of unequalled achievement. London, Giles de la Mare, 1999. 422 pp.

DAUSSET, Jean Baptiste (b. 1916). French haematologist; shared Nobel Prize (Physiology or Medicine) 1980, for discovery of first histocompatibility antigen.

–. The Nobel Prize for Physiology or Medicine, 1980, awarded to Baruj Benacerraf, Jean Dausset and George D. Snell. *Scandinavian Journal of Immunology*, 1922, **35**, 373–98.

DAVAINE, Casimir (1812–1882). French medical microbiologist.

THÉODORIDES, J. DSB, 1971, **3**, 587–9.
THÉODORIDES, J. Un grand médecin et biologiste: Casimir Joseph Davaine (1812–1882). Oxford, Pergamon Press, 1968. 238 pp.

DAVENPORT, Charles Benedict (1866–1944). American zoologist and geneticist.

RIDDLE, O. BMNAS, 1948, **25**, 75–110.
SHOR, E.N. DSB, 1971, **3**, 589–91.
Archival material: Library, American Philosophical Society, Philadelphia.

DAVIDOFF, Leo Max (1898–1975). Lithuanian/American neurosurgeon; introduced lumbar encephalography, 1932.

FEIRING, E.H. Leo Davidoff. *Surgical Neurology*, 1987, **28**, 173–5.
HORWITZ, N.H. Library: historical perspective Leo M. Davidoff, *Neurosurgery*, 1999, **45**, 194–8.
JACOBSON, H.G. A revisit with Leo M. Davidoff, M.D., a titan of his time; some personal notes (1898–1975). *Neurosurgery*, 1983, **13**, 601–6.

DAVIDSON, James Norman (1911–1972). British biochemist; specialized in biochemistry of nucleic acids.

NEUBERGER, A. BMFRS, 1973, **19**, 281–303.
Archival material: University of Glasgow.

DAVIEL, Jacques (1693–1762). French ophthalmic surgeon.

FIGARELLA, J. Jacques Daviel, maître-chirurgien de Marseille, oculiste du Roi (1693–1762). Marseille, Imprimerie Robert, 1979.

FIGARELLA, J. Jacques Daviel à Marseille. *Histoire des Sciences*, 1979, **13**, 347–62.

VETTER, T. Rencontres avec Jacques Daviel (1693–1762). Paris, Geigy, 1963. 55 pp.

DAVIES, Hugh Morriston (1879–1965). British thoracic surgeon.

WEBB, K. Hugh Morriston Davies: pioneer thoracic surgeon 1879–1965. Ruthin, Caerlion Trust, 1998. 40 pp.

DAVIES, Robert Ernest (1919–1993). British biochemist.

KUSHMERICK, M.J. BMFRS, 2001, **47**, 143–57.

DAVIS, Bernard David (1916–1994). American microbiologist.

MAAS, W.K. BMNAS, 1999, **77**, 50–63.

DAVIS, Hallowell (1896–1992). American neurologist.

GALAMBOS, R. BMNAS, 1998, **75**, 117–37.

DAVIS, John Bunnell (1780–1824). British paediatrician; established dispensary for sick children, London, 1816.

LOUDON, I.S. John Bunnell Davis and the Universal Dispensary for Children. *British Medical Jounal*, 1979, **1**, 1191–4.

DAVIS, Loyal Edward (1896–1982). American surgeon; professor of surgery, Northwestern University, Chicago.

ZOLLINGER, M. Overview biography – Doctor Loyal Davis. *Surgery, Gynecology and Obstetrics*, 1983, **157**, 111–13 [followed by further biographical contributions to p. 163].

DAVIS, Nathan Smith (1817–1904). American physician in Chicago; founder of the American Medical Association.

DANFORTH, I.N. The life of Nathan Smith Davis. Chicago, Cleveland Press, 1907. 193 pp.

DAVISON, Wilburt Cornell (1892–1972). American physician and educator; Dean, Duke University School of Medicine.

ARENA, J.M. & McGOVERN, J.P. (eds.) Davison of Duke: his reminiscences. Durham, Duke Univ. Medical Center, 1980. 283 pp.

DAVY, Humphry (1778–1829). British chemist.

FORGAN, S. (ed.) Science and the sons of genius; studies on Humphry Davy. London, Science Reviews, 1980. 247 pp.

FULLMER, J.Z. Young Humphry Davy; the making of an experimental chemist. Philadelphia, American Philosophical Society, 2000. 385 pp.

HARTLEY, H. Humphry Davy. London, Nelson, 1966, 160 pp. New ed. 1971.

KNIGHT, D. Humphry Davy: science and power. Cambridge, C.U.P., 1996. 218 pp.

DAWSON, Bertrand Edward (1864–1945). British physician-in-ordinary to George V.

WATSON, F. Dawson of Penn. London, Chatto & Windus, 1950. 344 pp.

DeBAKEY, Michael Ellis (b.1908). American cardiovascular surgeon.

THOMPSON, T. Hearts: DeBakey and Cooley, surgeons extraordinary. London, M. Joseph, 1972. 316 pp.

DE BEER, Gavin Rylands (1899–1972). British zoologist and embryologist.

BARRINGTON, E.W.J. BMFRS, 1973, **19**, 65–93.
RIDLEY, M. DSB, 1990, **17**, 213–14.

DEBRÉ, Robert (1882–1978). French paediatrician.

DEBRÉ, R. Hommage au professeur Robert Debré: 1882–1978. Paris, Grasset, 1980.
MARIE, J. et al. Robert Debré (1882–1978). *Bulletin de l'Académie Nationale de Médecine*, 1982, **166**, 1257–93. [Six papers by various authors.]

DEEPING, George Warwick (1877–1950). British physician and novelist.

British Medical Journal, 1950, **1**, 1084.

DEITERS, Otto Friedrich Karl (1834–1863). German neuroanatomist; discovered glial cells.

DIEPGEN, pp. 24–8. (CB)

DÉJERINE, Joseph Jules (1849–1917). French neurologist.

GAUCKLER, E. Le professeur J. Déjerine, 1849–1917. Paris, Masson, 1922. 192 pp.

DÉJERINE-KLUMPKE, Augusta (1859–1927). American/French neuro-anatomist; collaborated with her husband J. J. Déjerine.

THOMAS, A. *et al.* Madame Déjerine, 1859–1927. Paris, Masson, 1929. 106 pp.

DE KRUIF, Paul Henry (1890–1971). American bacteriologist and science writer.

DE KRUIF, P. The sweeping wind: a memoir. New York, Harcourt Brace, 1962. 246 pp.

DELAY, Jean Paul Louis (b.1907). French psychiatrist; introduced chlorpromazine in treatment of psychosis.

–. L'Epee d'Academicien du Professeur Jean Delay. Vendome, Presses Univ. de France, 1960. 38 pp.

DELBRÜCK, Max Ludwig Henning (1906–1981). German/American microbiologist, physicist and geneticist; Nobel Prize for work on genetics, 1969.

FISCHER, E.F. & LIPSON, C. Thinking about science; Max Delbrück and the origins of molecular biology. New York, Norton, 1988. 334 pp.
GLASS, B. DSB, 1990, **17**, 217–19.
HAYES, W. BMFRS, 1982, **28**, 597–90.
HAYES, W. BMNAS, 1993, **62**, 66–117.
Archival material: California Institute of Technology; American Philosophical Society, Philadelphia.

DE LEE, Joseph Bolivar (1869–1942). American obstetrician; founder of the Chicago Lying-in Hospital.

FISHBEIN, M. & DE LEE, S. T. Joseph Bolivar de Lee, crusading obstetrician. New York, Dutton, 1949. 313 pp.

DELORME, Edmond (1847–1929). French surgeon; introduced pulmonary decortication for chronic emphysema.

–. *Bulletin de la Société Française de l', Histoire de la Médecine* 1929, **23**, 35–51.
–. *Bulletin de l'Academie de Médecine*, Paris, 3. sér., **81**, 321–5.
JEANSELME, E. *Annales de Dermatologie et de Syphiligraphie*, 1918–1919, 5 sér., **7**, 233–6.

DELLA PORTA, Giovanni Battista, see **PORTA, Giovanni Battista della**

DELPECH, Jacques Matthieu (1777–1832). French plastic and orthopaedic surgeon.

HUARD, P. & IMBAULT-HUART, M.J. Le Dupuytren montpellierain: Delpech, Jacques Matthieu. *Episteme*, 1973, **7** 199–211.

DEMARQUAY, Jean-Nicolas (1811–1875). French surgeon.

SAINT-GERMAIN, –. *Bulletin et Mémoires de la Société de Chirurgie*, Paris, 1878, n.s. **4**, 53–61.
–. *Gazette Médicale de Paris*, 1878, **4**, 26 Jan. 37–44.

DEMEREC, Milislav (1895–1966). Yugoslavia-born American geneticist; Director, Dept. of Genetics, Carnegie Institution, Washington.

GLASS, B. BMNAS, 1971, **42**, 1–27.
GLASS, B. DSB, 1990, **17**, 217–19.
Archival material: American Philosophical Society, Philadelphia.

DENIS [DENYS], Jean Baptiste (1643–1704). French astronomer and mathematician; first to transfuse blood into a human (1667).

HOFF, H.E. DSB, 1971, **4**, 37–8.
MOORE, P. Blood and justice; the seventeenth-century doctor who made blood transfusion history. London, John Wiley, 2000. 223 pp.
PEUMERY, J.-J. Jean-Baptiste Denis et la recherche scientifique au XVIIe. siècle. Paris, Expansion Scientifique Française, 1970. 271 pp.

DENT, Charles Enrique (1911–1976). British clinical biochemist; professor of human metabolism, University of London.

NEUBERGER, A. BMFRS, 1978, **24**, 15–31.
Archival material: CMAC.

DERCUM, Francis Xavier (1856–1931). American neurologist; first to describe adiposis dolorosa.

BURR, C.W. Memoir of Dr Francis X. Dercum, *Medical Life*, 1932, **39**, 221–8.

DERRICK, Edward Holbrook (1898–1976). Australian pathologist.

DOHERTY, R. Epidemic polyarthritis and the search for its cause. The discovery of Ross River virus. In: Some milestones of Australian medicine. Brisbane, Amphion Press, 1994, pp. 135–52.

DESAULT, Pierre Joseph (1744–1795). French surgeon.

DUBOS, J.C. & GIRARDOT, J. P.J. Desault. Société d'Histoire et d'Archéologie de la Réion de Lure, 1982. 88 pp.

HUARD, P.A. & IMBAULT-HUART, M.J. Pierre Desault. In: Huard, P.A. Biographies médicales et scientifiques, XVIIIe. siècle. Paris, R. Da Costa, 1972, pp. 119–80.

DESCARTES, René du Perron (1596–1650). French philosopher, scientist and physiologist.

ARON, E. Descartes et la médecine. Chambre-Les Tours, C.L.D., 1996. 158 pp.

CARTER, R.B. Descartes' medical philosophy. Baltimore, Johns Hopkins Univ. Press, 1983. 301 pp.

CROMBIE, A.C., MAHONEY, M.S. & BROWN, C.M. DSB, 1971, **4**, 51–65.

LINDEBOOM, G.A. Descartes and medicine. Amsterdam, Rodopi, 1979. 134 pp.

DES GENETTES, René-Nicolas-Dufriche (1762–1837). French physician; Physician-in-Chief to the French army.

GRAZEL, L. Le baron Des Genettes (1762–1837): notes biographiques. Paris, Henry Paulin, 1912. 110 pp.

DESTOUCHES, Henri-Louis, *see* **CÉLINE, Louis Ferdinand**

DETMOLD, Willam Ludwig (1804–1894). German/American surgeon.

SHANDS, A.R. William Ludwig Detmold: America's first orthopaedic surgeon. *Military Medicine*, 1968, **133**, 563–69.

DETWILER, Samuel Randall (1890–1957). American embryologist; professor of anatomy, Columbia University.

NICHOLAS, J.S. BMNAS, 1961, **35**, 85–111.

DEUTSCH, Helene (1884–1982). American psychiatrist and psychoanalyst.

ROAZEN, P. Helene Deutsch: a psychoanalyst's life. New York, Anchor Press/Doubleday, 1985. 384 pp. Reprinted 1992, Transaction Publishers, 371 pp.

DEVENTER, Hendrik van (1651–1724). Dutch obstetrician and orthopaedist.

KOUWER, B.F. Hendrik van Deventer. *Janus*, Amsterdam, 1912, **17**, 425–42.

LAMERS, A.J.M. Hendrik van Deventer, medicinae doctor, 1651–1724, leven en werken. Assen, Van Gorcum, 1946. 267 pp.

DEVERGIE, Marie Guillaume Alphonse (1798–1879). French dermatologist.

BEESON, B.B. *Archives of Dermatology and Syphilis*, 1930, **21**, 1030–32.

LAGNEAU, G. *Bulletin de l'Académie Nationale de Médecine*, Paris, 1879, **8**, 1016–19.

DEWEES, William Potts (1768–1841). American paediatrician and gynaecologist; wrote first American textbooks (1825, 1826) on both these specialties.

HODGE, H.L. An eulogium on William P. Dewees. Philadelphia, Merrihew & Thompson, 1842. 58 pp.

HODGE, H.L. *Philadelphia Journal of Medical and Physical Sciences*, 1843, n.s. **5**, 123–44.

DEWEY, Richard Smith (1845–1933). American psychiatrist.

BIBLIOGRAPHY OF MEDICAL AND BIOMEDICAL BIOGRAPHY

DEWEY, R.S. Recollections of Richard Dewey, pioneer in American psychiatry; an unfinished autobiography, edited by E.L. Dewey. Chicago, University of Chicago Press, 1936, 173 pp. Reprinted New York, Arno Press, 1973.

DÍAZ, Francisco (*c*.1525–1590). Spanish urologist; sometimes called 'the Father of urology'.

BUSH, R.B. & BUSH, I.M. Francisco Díaz and the world of sixteenth century urology. Northridge, Cal., Riker Laboratories, 1970. 12 pp.
RIERA, J. La obra urológica de Francisco Diaz. *Cuadernos de Historia de la Medicina Española*, 1967, 613–59.

DIAZ DE ISLA, Rodrigo Ruiz (1462–1542). Spanish physician; probably first to report West Indian origin of syphilis.

HOLCOMB, R.C. Ruiz Diaz de Isla and the American (Hawaiian) origin of syphilis. *Medical Life*, 1936, **43**, 270–364, 414–70, 487–514.

DICKENS, Frank (1899–1986). British biochemist.

THOMPSON, R.H.S. & CAMPBELL, P.N. BMFRS, 1987, **33**, 187–210.

DICK-READ, Grantly (1890–1959). British obstetrician; pioneer of 'natural childbirth'.

SANDELOWSKI, M. Pain, pleasure, and American childbirth. From the twilight sleep to the Read method, 1914–1960. Westport, Conn., & London, Greenwood Press, 1985. 152 pp.
THOMAS, A.N. Doctor courageous: The story of Grantly Dick-Read. London, Heinemann, 1957. 218 pp.
Archival material: CMAC.

DIEFFENBACH, Johann Friedrich (1792–1847). German plastic surgeon; professor of surgery, Berlin.

GENSCHOREK, W. Wegbereiter der Chirurgie: Johann Friedrich Dieffenbach, Theodor Billroth. Leipzig, Hirzel, 1983. 252 pp.
LAMPE, F.R. Dieffenbach. Leipzig, J.A. Barth, 1934. 219 pp.

DIETL, Jósef (1804–1878). Polish physician; described acute ureteral colic ('Dietl's crisis').

KUCHARZ, E. The life and achievements of Joseph Dietl. *Clio Medica*, 1981, **16**, 25–35.

DIETZ, Johann (1665–1738). German military surgeon.

DIETZ, J. Master Johann Dietz, surgeon in the army of the Great Elector and Barber to the Royal Court. From the old MS in the Royal Library, Berlin. Translated by Bernard Miall. First published by Ernst Consentius [His Introduction dated Berlin, 1914. London, Allen & Unwin, 1923. 315 pp. First published in German.

DIEULAFOY, Georges (1839–1911). French physician; professor at Hôtel-Dieu, Paris.

WIDAL, G.F.L. *et al.* Georges Dieulafoy, 1839–1911. Paris, [1913?]. 46 pp.

DIMSDALE, Thomas (1712–1800). British physician; inoculated Catherine the Great against smallpox.

BISHOP, W.J. Thomas Dimsdale, MD, FRS (1712–1800) and the inoculation of Catherine the Great of Russia. *Annals of Medical History*, 1932, N.S. **4**, 321–38.

96

CLENDENING, P.H. Thomas Dimsdale and smallpox inoculation in Russia. *Journal of the History of Medicine*, 1973, **28**, 109–25.

DINGLE, John Holmes (1908–1973). American physician and epidemiologist.

JORDAN, W.S. BMNAS, 1992, **61**, 136–63.

DIOCLES of Carystos (*fl.* 350 BC). Greek physician.

DANNENFELDT, K.H. DSB, 1971, **4**, 105–7.
JAEGER, W. Diokles von Karystos. Die griechische Medizin und die Schuler der Aristoteles. 2nd ed. Berlin, W. de Gruyter, 1963. 244 pp. First published 1938.

DIONIS, Pierre (1643–1718). French anatomist and surgeon.

BARRITAULT, G. L'anatomie en France au XVIIIe siècle: les anatomistes du Jardin du Roi: Dionis, etc. Paris, Angers, 1940. 88 pp.

DIOSCORIDES [Pedanius Dioscorides of Anazarbus] (*fl.* AD 50–70). Greek physician, pharmacist and botanist; originator of the materia medica.

RIDDLE, J.M. DSB, 1971, **4**, 119–23.
RIDDLE, J.M. Dioscorides on pharmacy and medicine. Austin, University of Texas Press, 1985. 198 pp.

DITTRICH, Franz von (1815–1859). German physician; professor in Prague and Erlangen.

DITTRICH, F. von , 1815–1859, Anatom in Prag und Kliniker in Erlangen. München, Berlin, J.F. Lehmann, 1937. 259 pp.

DIX, Dorothea Lynde (1802–1887). American reformer of treatment of the mentally ill; organized nursing services in American Civil War.

BROWN, T.J. Dorothea Dix: New England reformer. Cambridge MA, Harvard Univ. Press, 1998. 422 pp.
GOLLAHER, D. Voice for the mad: the life of Dorothea Dix. New York, Free Press, 1995. 538 pp.
SCHLAIFER, C. & FREEMAN, L. Heart's work: Civil War heroine and champion of the mentally ill. New York, Paragon House, 1991, 175 pp.

DIXON, Malcolm (1899–1985). British biochemist; professor of enzyme chemistry, Cambridge University.

PERHAM, R.N. BMFRS, 1988, **34**, 97–131.
Archival material: Cambridge University Library.

DOBELL, Cecil Clifford (1886–1949). British protozoologist.

DOLMAN, C.E. DSB, 1971, **4**, 132–3.
HOARE, C.A. & MACKINNON, D.L. ONFRS, 1950, **7**, 35–61.
Archival material: Library, Wellcome Institute for the History of Medicine, London: Cambridge University Library.

DOBSON, Mathew (1735–1784). British physician; discovered hyperglycaemia.

DOBSON, J. Mathew Dobson, Collected papers concerning Liverpool *Medical History*; originally read at 8th British Congress on the History of Medicine, Liverpool, 1971. Liverpool, 1977. 113 leaves. [Typescript in Wellcome Library]

MACFARLANE, I. Mathew Dobson (1735–1784) and diabetes. *Medical Historian; Bulletin of Liverpool Medical History Society*, 1994, **7**, 14–22.

DOBZHANSKY, Theodosius Grigorievich (1900–1975). Russian/American geneticist and evolutionist: professor at Rockefeller Institute (University), New York.

ADAMS, M.B. ed. The evolution of Theodosius Dobzhansky: essays on his life and thought in Russia and America. Princeton, Princeton Univ. Press, *c*.1994. 249 pp.
AYALA, F.J. BMNAS, 1985, **55**, 163–213.
AYALA, F.J. DSB, 1990, **17**, 233–42.
FORD, E.B. BMFRS, 1977, **23**, 59–89.

DOCHEZ, Alphonse Raymond (1882–1964). American microbiologist; professor of medicine, College of Physicians and Surgeons, Columbia University.

HEIDELBERGER, M. *et al.* BMNAS, 1971, **42**, 29–46.
Archival material: Butler Library, Columbia University.

DOCK, George (1860–1951). American physician; professor of medicine, Tulane and Washington universities.

DAVENPORT, H.M. Doctor Dock; teaching and learning medicine at the turn of the century. New Brunswick, NJ, Rutgers Univ. Press, 1987. 342 pp.

DODART, Denis (1634–1707). French physician and botanist.

GRMEK, M.D. DSB, 1971, **4**, 135–6.

DODDRIDGE, Philip (1702–1751). British physician.

CLIFFORD, A.C. The good doctor: Philip Doddridge of Northampton: a tercentenary tribute. Norwich, Charenton Reformed Publishing, 2002. 319 pp.

DODDS, Edward Charles (1899–1973). British physician and medical biochemist.

DICKENS, F. BMFRS, 1975, **21**, 227–67.
Archival material: Royal College of Physicians of London.

DODOENS, Rembert [Dodonaeus] (1516–1585). Netherlands physician and botanist; published first important Belgian herbal.

FLORKIN, M. DSB, 1971, **4**, 138–40.
MEERBEECK, P.J. van. Recherches historiques et critiques sur la vie et les ouvrages de Rembert Dodoens (Dodonaeus). Malines, Hanicq, 1841. 340 pp. Utrecht, Hes Publishers, 1990 [reimpression of original edition].
Archival material: Library, American Philosophical Society, Philadelphia.

DÖEBLIN, Alfred (1878–1957). German psychiatrist, novelist and essayist; wrote *Berlin-Alexanderplatz*.

MÜLLER-SALGET, K. Alfred Döblin: Werk und Entwicklung. Bonn , Bouvier-Verlag H. Grundmann, 1972. 515 pp.
PRANGEL, M. Alfred Döblin. Stuttgart, J.B. Metzler, 1979. 130 pp.

DÖEDERLEIN, Albert Siegmund Gustav (1860–1941). Geman obstetrician and gynaecologist; described 'Döderlein's bacillus'.

FESTSCHRIFT Albert Döderlein, aus Anlass seines 60 Geburtstages am 5.Juli 1920, dargebracht. *Monatschrift für Geburtshilfe und Gynäkologie*, 1920, **53**, 1–430.

DÖELLINGER, Ignaz (1770–1841). German physiologist and embryologist.

RISSE, G.B. DSB, 1971, **4**, 146–7.

DOERR, Robert (1871–1952). Hungarian bacteriologist; related sandfly fever to sandfly *Phlebotomus.*

REUTER, F. In memoriam Robert Doerr. *Wiener Klinische Wochenschrift*, 1952, **64**, 129–30.

DOGLIOTTI, Achille Mario (1897–1966). Italian surgeon.

AGOSTONI, G. Achille Mario Dogliotti (Torino 1897–1966). *Cardiologia Pratica*, 1961, **17**, 111–17.

DOHERTY, Peter C. (b. 1940). Australian immunologist; shared Nobel Prize (Physiology or Medicine) 1996, for discoveries concerning the specificity of the cell-mediated immune defence.

Les Prix Nobel en 1996.

DOHRN, Anton (1840–1909). German morphologist.

HEUSS, T. Anton Dohrn: a life for science. Berlin, Springer. 1991. 440 pp. First published in German, 1962.

DOISY, Edward Adelbert (1893–1986). American biochemist; professor at St Louis University; shared Nobel Prize (Physiology or Medicine) 1943 for work on vitamin K.

DOISY, E.A. An autobiography. *Annual Review of Biochemistry*, 1976, **45**, 1–9.

DOLL, William Richard Shaboe (b.1912). British epidemiologist; Regius Professor of Medicine, University of Oxford.

BECKETT, C. An epidemiologist at work: the personal papers of Sir Richard Doll. *Medical History*, 2002, **46**, 403–21.
DOLL, R. Conversation with Sir Richard Doll. *British Journal of Addiction*, 1991, **86** 365–77.
Archival material: CMAC.

DOMAGK, Gerhard (1895–1964). German physician and pharmacist; introduced prontosil, first drug containing sulphanilamide, in treatment of bacterial infections, and thiosemicarbazone in treatment of tuberculosis; Nobel Prize 1939.

COLEBROOK, L. BMFRS, 1964, **10**, 39–50.
GERHARD DOMAGK, 1895–1964. Lebenserinnerungen in Bildern und Texten...red. R. Alstaedter. Leverkusen, Bayer, 1995. 114 pp.
POSNER, E. DSB, 1971, **4**, 153–6.

DONALD, Archibald (1860–1937). British gynaecologist; devised operation to repair complete prolapse of the uterus.

SHAW, W.F. *Journal of Obstetrics and Gynaecology of the British Empire*, 1937, **44**, 527–38.

DONALDSON, Henry Herbert (1857–1938). American neurologist.

CONKLIN, E.G. BMNAS, 1939, **20**, 229–43.
SAFFRON, M.H. DSB, 1971, **4**, 160–61.
Archival material: Library, American Philosophical Society, Philadelphia.

DONATH, Julius (1870–1950). Austrian immunologist.

SILVERSTEIN, A.M. The Donath-Landsteiner autoantibody: the uncommensurable language of the early immunologic dispute. *Cellular Immunology*, 1986, **97**, 173–88.

DONATH, Willem Frederik (1889–1957). Dutch nutritionist; isolated vitamin B_1.

BOETTCHER, H.M. Das Vitaminbuch. Die Geschichte der Vitaminforschung. Köln. Kiepenheuer & Witsch, 1965, pp. 130–33.

DONATI, Marcello (1538–1602). Italian physician; first to demonstrate gastric ulcer and angioneurotic oedema.

ZANCA, A. Notizie sulla vita e sulle opere di Marcello Donati de Mantova (1538–1602); medico, umanista, uomo di stato. Pisa, Tip. Editrice Giardini, 1964. 63 pp.

DONDERS, Franciscus Cornelis (1818–1889). Dutch physiologist and ophthalmologist; professor of physiology, Utrecht.

CLARKE, E. Problems in the accommodation and refraction of the eye. A brief review of the work of Donders and the progress made during the last fifty years. London, Baillière, Tindall & Cox, 1914. 108 pp.
LAAGE, R.J.Ch.v.ter. DSB, 1971, **4**, 162–4.
LEERSUM, E.C. van. Het levenswerk van Franciscus Cornelis Donders. Haarlem, E.F. Bohn, 1932. 408 pp.

DONNAN, Frederick George (1870–1956). British physical chemist. His work on membrane equilibria provided basis for theories of transport of molecules and ions across the living cell.

FREETH, F.A. BMFRS, 1957, **3**, 23–39.
Archival material: Library, University College London.

DONNÉ , Alfred François (1801–1878). French bacteriologist.

THORBURN, A.L. Alfred François Donné, 1801–1878, discoverer of *Trichomonas vaginalis* and of leukaemia. *British Journal of Venereal Diseases*, 1974, **50**, 377–80.

DONNOLO, Shabbetai [Sabbatai ben Abraham] (913–982). Jewish doctor and cosmologist born in southern Italy when it was part of the Byzantine empire; one of the founders of the medical school at Salerno. His *Antidotarium*, a formulary of some 120 remedies, is the oldest known Hebrew medical work written in Europe (ed. M. Steinschneider, 1868).

SHARF, A. The universe of Shabbetai Donnolo. Forest Grove, Oregon, Aris & Phillips, 1976. 232 pp.

DONOVAN, Charles (1863–1951). British physician in Indian Medical Service; described 'Leishman-Donovan' bodies in kala-azar.

GIBSON, M.E. The identification of kala-azar and the discovery of Leishman-Donovan bodies. *Medical History*, 1983, **27**, 203–13.

KRISHNASWAMI, P. Col. C. Donovan, IMS physician, physiologist and discoverer. *Indian Journal of the History of Medicine*, 1956, **1** , 133–8.
Archival material: CMAC.

DORFMAN, Alfred (1916–1982). American biochemist.

SCHWARZ, N.B. & RODÉN, L. BMNAS, 1997, **72**, 71–87.

DORSEY, John Morris (1900–1978). American psychiatrist; professor, Wayne State University.

DORSEY, J.M. University professor John M. Dorsey. Detroit, Wayne State University Press, 1980. 282 pp.

DORSEY, John Syng (1783–1818). American surgeon.

MIDDLETON, W.S. John Syng Dorsey. *Annals* of *Medical History*, 1930, N.S. **2**, 587–601.

DOTT, Norman McOmish (1897–1973). British neurosurgeon; professor, University of Edinburgh.

RUSH, C. & SHAW, J.F. With sharp compassion: Norman Dott, freeman surgeon of Edinburgh. Aberdeen, University Press, 1990. 314 pp.
Archival material: Library, University of Edinburgh.

DOUBLE, François-Joseph (1776–1848). French physician; introduced and applied auscultation.

FINOT, . – . François-Joseph Double, inventeur de l'auscultation en 1817. *Histoire des Sciences Médicales*, Colombes, 1972, **6**, 14–21.

DOUDOROFF, Michael (1911–1975). Russian/American microbiologist.

BARKER, H.A. BMNAS, 1993, **62**, 118–41.

DOUGLAS, Claude Gordon (1882–1963). British physiologist; professor of general metabolism, Oxford.

CUNNINGHAM, D.J.C. BMFRS, 1964, **10**, 51–74.
Archival material: Bath University.

DOUGLAS, James (1675–1742). British anatomist and physician.

BROCK, C.H. Dr James Douglas's papers and drawings in the Hunterian Collection, Glasgow University Library. A handlist. Glasgow, Wellcome Unit for the History of Medicine, 1994. 170 pp.
THOMAS, K.B. James Douglas of the pouch, and his pupil William Hunter. London, Pitman, 1964. 229 pp.; bibliography and annotated catalogue of his MSS and drawings in Hunterian Library, Glasgow.
THOMAS, K.B. DSB, 1971, **4**, 172–3.
Archival material: Hunterian Library, University of Glasgow.

DOUGLAS, Stewart Ranken (1871–1936). British bacteriologist on staff of National Institute for Medical Research, London.

LAIDLAW, P.P. ONFRS, 1936–39, **2**, 175–82.

BIBLIOGRAPHY OF MEDICAL AND BIOMEDICAL BIOGRAPHY

DOUGLAS, William Wilton (1922–1998). British pharmacologist.

RITCHIE, J.M. BMFRS, 2000, **46**, 145–64.

DOUGLASS, William (1691–1752). American physician; gave first adequate description of scarlet fever.

WEAVER, C.H. The life and writings of William Douglass, M.D., 1691–1752. *Bulletin of the Society* of *Medical History of Chicago*, 1912, **2**, 229–59.

DOVER, Thomas (1660–1742). British physician and privateer, who rescued Alexander Selkirk, the original Robinson Crusoe.

DEWHURST, K. The quicksilver doctor: the life and times of Thomas Dover, physician and adventurer. Bristol, Wright, 1957. 192 pp.
STRONG, L.A.G. Dr Quicksilver, 1660–1742: the life and times of Thomas Dover, MD. London, Melrose, 1955, 184 pp.

DOWLING, Geoffrey Barrow (1891–1946). British dermatologist.

CALNAN, C. The life and times of Geoffrey Barrow Dowling. Oxford, Blackwell Scientific, 1993. 316 pp.

DOWN, John Langdon Haydon (1816–1896). British physician; identified characteristics of 'Down's syndrome'.

WARD, O.C. John Langdon Down; a caring pioneer. London, Royal Society of Medicine Press, 1998. 213 pp.

DOYEN, Eugène Louis (1859–1916). French surgeon.

DIDIER, R. Le docteur Doyen: chirurgien de la belle époque. Paris, Maloine, 1964. 235 pp.

DOYLE, Arthur Ignatius Conan (1859–1930). British physician and writer.

BOOTH, M. The doctor, the detective and Arthur Conan Doyle: a biography of Arthur Conan Doyle. London, Hodder and Stoughton, 1997. 371 pp.
CARR, J.D. The life of Sir Arthur Conan Doyle. London, J. Murray, 1949. 362 pp. Reprinted 1987, 1990
HOWARD, O.D. The quest for Sherlock Holmes. A biographical study of Arthur Conan Doyle. Edinburgh, Dolman Press, 1983. 380 pp.
RODIN, A.E. & KEY, J.D. Medical casebook of Doctor Arthur Conan Doyle: from practitioner to Sherlock Holmes and beyond. Melbourne, Fla., Krieger, 1984. 506 pp.
STASHOWER, D. Teller of tales: the life of Sir Arthur Conan Doyle. New York, Henry Holt, 1999. 472 pp.

DOWNIE, Allan Watt (1901–1988). British bacteriologist; professor at University of Liverpool.

TYRRELL, D.A.J. & McCARTHY, K. BMFRS, 1990, **35**, 97–112.
Archival material: Liverpool University.

DRAGENDORF, Johann Georg Noël (1836–1898). German forensic chemist and toxicologist; professor at Dorpat.

KOKOSKA, U. Johann Georg Noël Dragendorf (20.4.1836–7.4.1898): sein Beitrag zur Gerichtsmedizin, Pharmakologie und Pharmazie am der Universität Dorpat., 1983. 234 pp.

BIBLIOGRAPHY OF MEDICAL AND BIOMEDICAL BIOGRAPHY

DRAGSTEDT, Lester Reynold (1893–1975). American physiologist and surgeon.

WANGENSTEEN, O.H. & WANGENSTEEN, S.O. BMNAS, 1980, **51**, 63–95.

DRAKE, Daniel (1785–1852). American physician; wrote pioneer work on the disease as related to geography.

DRAKE, D. Pioneer life in Kentucky, 1785–1800 ... Edited, from the original manuscript, with introductory comments and a biographical sketch by Emmet Field Horine. New York, Schuman, 1948. First published 1870; reprinted 1907.

HORINE, E.F. Daniel Drake (1785–1852), pioneer physician of the Midwest. Philadelphia, Univ. Pennsylvania Press, 1961. 425 pp.

MANSFIELD, E.D. Memoirs of the life and services of Daniel Drake, M.D., physician, professor and author. Cincinnati, Applegate, 1855. 408 pp. Reprinted New York, Arno Press, 1975. 408 pp.

DRESER, Heinrich (1860–1924). German pharmacologist; introduced acetylsalicylic acid (aspirin) into medicine.

MEYER, H. Heinrich Dreser. *Archiv für Pharmakologie und Experimentelle Pathologie*, 1925, **106**, i–vii.

DRESSLER, Lucas Anton (1815–1896). German physician; described paroxysmal cold haemoglobinuria.

VOSWINCKEL, P. Anton Dressler aus Würzburg: vergesssener Arzt-Pionier an der Schwelle der modernen Hämatologie. *Medizinhistorisches Journal*, 1988, **23**, 132–51.

DREW, Charles Richard (1904–1995). American surgeon; pioneer in blood plasma research.

LICHELLO, R. Pioneer in blood plasma, Dr Charles Richard Drew. New York, Julian Messner, 1968. 190 pp.

LOVE, S. One blood; the death and resurrection of Charles R. Drew. Chapel Hill, Univ. of North Carolina Press, 1996. 373 pp.

MAHONE-LONESOME, R. Charles Drew. New York, Chelsea House, 1990. 109 pp.

WYNES, C.E. Charles Richard Drew: the man and the myth. Urbana, University of Illinois Press, 1988. 132 pp.

DREYER, George (1873–1934). Danish-born bacteriologist; professor at Oxford University.

DOUGLAS, S.R. ONFRS, 1932–35, **1**, 569–76.

DREYER, Margrette. George Dreyer: a memoir by his wife. Oxford, Blackwell, 1937. 249 pp.

DRIESCH, Hans Adolf Eduard (1867–1941). German biologist and embryologist.

DRIESCH, H. Lebenserinnerungen. München, Basel, Reinhardt, 1951. 311 pp.

OPPENHEIMER, J. DSB, 1971, **4**, 186–9.

WENZL, H. (ed.) Hans Driesch. Personlichkeit und Bedeutung für Biologie und Philosophie von Heute. Basel, Reinhardt, 1951. 221 pp.

DRINKER, Philip (1894–1972). American industrial hygienist; with C.F. McKhann introduced artificial respirator ('iron lung').

DRINKER, P.A. & McKHANN, C.F. The iron lung: first practical means of respiratory support. *Journal of the American Medical Association*, 1986, **92**, 1476–80.

BIBLIOGRAPHY OF MEDICAL AND BIOMEDICAL BIOGRAPHY

DRUMMOND, Jack Cecil (1891–1952). British biochemist and nutritionist; professor of biochemistry, University College London.

YOUNG, F.G. ONFRS, 1954, **9**, 99–129.

DRUMMOND, William Henry (1854–1907). Canadian physician and poet.

LYONS, J.B. William Henry Drummond; poet in patois. Markham, Ont., Fitzherbert and Whiteside, 1994. 217 pp.

DRURY, Alan Nigel (1889–1980). British physiologist; developed techniques of blood transfusion.

KEKWICK, R.A. BMFRS, 1981, **27**, 173–98.

DRURY, Victor William Michael (b.1926). British physician: professor of general practice, Birmingham University.

HULL, R. Just a GP: a biography of Professor Sir Michael Drury. Oxford, Radcliffe Medical, c.1994. 144 pp.

DUBINI, Angelo (1813–1902). Italian physician; discovered causal agent of ankylostomiasis.

BELLONI, L. DSB, 1971, **4**, 197–8.

Du BOIS, Eugene Floyd (1882–1959). American physiologist; concentrated on metabolism.

AUB, J.C. BMNAS, 1962, **36**, 125–36.

DUBOIS, Jacques [Jacobus Sylvius] (1478–1555). French physician and teacher of anatomy.

O'MALLEY, C.D. DSB, 1971, **4**, 198–9.

DUBOIS DE CHEMANT, Nicolas (1753–1824). French dentist; manufactured artificial (porcelain) teeth, 1788.

WELLS, E.B. Dubois de Chemant's 'lettre à M. Andouillé': a translation with introduction, notes and a bibliography. *Bulletin of the History of Dentistry*, 1976, **24**, 69–77.

Du BOIS-REYMOND, Emil Heinrich (1818–1896). German physiologist; founder of modern electrophysiology.

CRANEFIELD, P.F. (ed.) Two great scientists of the nineteenth century. Correspondence of Emil Du Bois-Reymond and Carl Ludwig. Baltimore, Johns Hopkins Univ. Press, 1982. 183 pp.
FINKELSTEIN, G.W. Emil Du Bois Reymond: the making of a liberal German scientist (1818–1851). Princeton, Princeton University. Thesis DAI 1997 57(9): 440095-A.
MANN, G. (ed.) Naturwissenschaften und Erkenntnis im 19. Jahrhundert: Emil du Bois-Reymond. Hildesheim, Gerstenberg, 1981. 243 pp.
ROTHSCHUH, K.E. DSB, 1971, **4**, 200–205.
RUFF, P.W. Emil Du Bois-Reymond. Leipzig, Teubner, 1981. 106 pp
Archival material: Staatsbibliothek Preussischer Kulturbesitz, Berlin.

DUBOS, Réne Jules (1901–1982). French physician; professor at Rockefeller University, New York.

HIRSCH, J.G. & MOBERG, C.L. BMNAS, 1989, **58**, 133–61.

DUBOST, Charles (1914–1991). French cardiovascular surgeon.

BLONDEAU, P. Charles Dubost, 1914–1991. *Archives des Maladies du Coeur et des Vaisseux*, 1992, **85**, 483–6.
MERCADIER, M. Éloge de Charles Dubost (1914–1991). *Bulletin de l'Académie Nationale de Médecine*, Paris, 1991, **175**, 1005–115.

DUCHENNE, Guillaume Benjamin Amand (1806–1875). French physician; electrotherapist and neurologisy.

GUILLY, P.J.L. Duchenne de Boulogne. Paris, Baillière, 1936. 240 pp.
ROBINSON, V. Guillaume Benjamin Duchenne. *Medical Life*, 1929, **36**, 287–306.

DUCHESNE, Joseph [Josephus Quercetanus] (1544–1609). French physician and pharmacist.

DEBUS, A.G. DSB, 1971, **4**, 208–10.

DUCLAUX, Èmile (1840–1904). French biochemist; director of Institut Pasteur.

DELAUNAY, A. DSB, 1971, **4**, 210–12.

DUDLEY, Benjamin Winslow (1785–1870). American surgeon; notable lithotomist; performed first cerebral operation (1823) and first successful cataract operation (1836) in U.S.A.

BULLOCK, W.O. Dr Benjamin Winslow Dudley (1785–1870). Annals of *Medical History*, 1935, **7**, 201–13.

DUDLEY, Harold Ward (1887–1935). British biochemist on staff of National Institute for Medical Research, London; worked on insulin, histamine and isolation of acetylcholine.

DALE, H.H. ONFRS, 1932–35, **1**, 595–606.

DUDLEY, Sheldon Francis (1884–1956). British epidemiologist; Medical Director-General, Royal Navy.

BOYD, J.S.K. BMFRS, 1956, **2**, 85–99.

DUEHRSSEN, Alfred (1867–1937). German gynaecologist; introduced vaginal caesarean section.

STRASSMANN, P. *Zentralblatt für Gynäkologie*, 1934, **58**, 145–52.

DUGGAR, Benjamin Minge (1872–1956). American botanist; isolated aureomycin.

WALKER, J.C. BMNAS, 1958, **32**, 113–31.

DUHAMEL, Georges (1884–1966). French physician and novelist; medically qualified 1909; served as surgeon in the First World War; in 1920 made writing his career.

KNAPP, B.L. Georges Duhamel. New York, Twayne Publishers, [1972]. 193 pp.

DUHRING, Louis Adolphus (1845–1913). American dermatologist; author of first American textbook on skin diseases.

PARISH, L.C. Louis A. Duhring: pathfinder for dermatology. Springfield, Ill., Thomas, 1967. 137 pp.

BIBLIOGRAPHY OF MEDICAL AND BIOMEDICAL BIOGRAPHY

DUKE-ELDER, William Stewart (1898–1978). British ophthalmologist.

BELLOWS, J.G. (ed.) Contemporary ophthalmology: honoring Sir Stewart Duke-Elder. Baltimore, Williams & Wilkins, 1972. 554 pp. [Includes biographical sections]
LYLE, T.K., MILLER, S. & ASHTON, N.H. BMFRS, 1980, **26**, 85–105.

DU LAURENS [LAURENTIUS] André (1558–1609). French anatomist and physician.

BYLEBYL, J.F. DSB, 1973, **8**, 53–4.

DULBECCO, Renato (b.1914). Italian/American tumour virologist; shared Nobel Prize (Physiology or Medicine) 1975.

DULBECCO, R. Scienza e avventura. Milano, Sperling & Kupfer, 1989. 310 pp. French edn, 1990.
KEVLES, D.J. Renato Dulbecco and the new animal virology in medicine. *Journal of the History of Biology*, 1993, **26**, 409–42.

DUMAS, Jean Baptiste André (1800–1884). French organic chemist; used defibrinated blood in animal transfusions.

CHAIGNEAU, M. Jean Baptiste Dumas, sa vie, son oeuvre, 1800–1884. Paris, Guy Le Prat, *c.*1984. 434 pp.
KLOSTERMAN, L.J. Studies in the life and work of Jean Baptiste André Dumas ((1800–1884); the period up to 1850. [Ph.D. thesis, Kent University, 1976.]
MATIGNON, C. Jean Baptiste Dumas. *Revue Scientifique*, 1927, 23 Mai, 417–24.

DUN, Patrick (1642–1713). Irish physician; president of the Dublin College of Physicians.

BELCHER, T.W. Memoir of Patrick Dun, Physician-General to the Army, and sometime President of the College of Physicians; including his will, his deed for constituting a professor of physic, and other important records concerning the profession of physic in Ireland never before published. Dublin, Hodges, Smith, 1866. 80 pp.

DUNANT, [Jean] Henri (1828–1910). Swiss philanthropist; founder of the International Red Cross.

HART, E. Man born to live; life and work of Henry Dunant, founder of the Red Cross, London, Gollancz, 1953. 371 pp.
HOLE, B. Ubi caritas: an appreciation of the life of Henri Dunant, 1828–1910. Lewes, Book Guild. 1998. 97 pp.
MOOREHEAD, C. Dunant's dream: war, Switzerland and the history of the Red Cross. London, HarperCollins, 1998. 780 pp.

DUNBAR, Helen Flanders (1902–1959). American psychiatrist.

POWELL, R.C. Helen Flanders Dunbar (1902–1959). and a holistic approach to psychosomatic problems. *Psychiatric Quarterly*, 1977, **49**, 133–52.
POWELL, R.C. Healing and wholesomeness; Helen Flanders Dunbar. Origin of the American psychosomatic movement, 1906–36. Dissertation, Duke University. Dissertation Abstracts International, 1975, **35**, 5313-A. University Microfilms Order No. 75-2415, 354 pp.

DUNCAN, James Matthews (1826–1890). British obstetrician at St. Bartholomew's Hospital, London; associated with Sir James Young Simpson in discovery of anaesthetic properties of chloroform.

NEWLANDS, I. James Matthews Duncan. A sketch for his family. Aberdeen, Privately printed, 1891. 181 pp.

DUNCAN, William Henry (1805–1863). British physician; first Medical Officer of Health for Liverpool.

FRAZER, W.M. Duncan of Liverpool: being an account of the work of Dr W.H. Duncan, Medical Officer of Health of Liverpool, 1847–63, London, Hamish Hamilton, 1947. 163 pp. Reprinted Preston, Carnegie Publications, 1997.

DUNGLISON, Robley (1798–1869). British-born first full-time professor of medicine in USA.; first American author of physiology textbook, medical dictionary and history of medicine.

DUNGLISON, R. The autobiographical ana ... with notes and introduction by S.X. Radbill. Philadelphia, American Philosophical Society, 1963. 212 pp.
RADBILL, S.X. DSB, 1971, **4**, 251–3.

DUNHILL, Thomas Peel (1876–1957). Australian surgeon; pioneer of thyroid surgery.

VELLAR, I.D.A. Thomas Peel Dunhill, the forgotten man of thyroid surgery. *Medical History*, 1974, **18**, 22–50.

DUNN, Leslie Clarence (1893–1974). American geneticist; professor of zoology, Columbia University.

BENNETT, D. L.C. Dunn and his contribution to T-locus genetics. *Annual Review of Genetics*, 1977, **11**, 1–12.
DOBZHANSKY, T. BMNAS, 1978, **49**, 79–104.
GLASS, B. DSB, 1990, **17**, 248–50.
Archival material: Library, American Philosophical Society, Philadelphia; Butler Library, Columbia University.

DUPLAY, Simon Emmanuel (1836–1924). French surgeon; described 'frozen shoulder' syndrome.

LENORMANT, C. *Bulletin de la Société Nationale de Chirurgie*, 1927, **53**, 64–82.

DUPUYTREN, Guillaume (1777–1835). French surgeon; surgeon, Hôtel Dieu, Paris.

BARSKY, H.K. Guillaume Dupuytren: a surgeon in his place and time. New York, Vantage Press, 1984. 295 pp.
MONDER, H.J.J. Dupuytren. 4me ed. Paris, Gaillard, 1945. 312 pp.

DURHAM, Herbert Edward (1866–1945). British bacteriologist; discovered *Salmonella aertrycke* and (jointly) bacterial agglutination.

–. *Lancet*, 1945, **2**, 654–5.

DUROZIEZ, Paul Louis (1826–1897). French physician; first to describe congenital mitral stenosis.

BEAUVAIS, G. de. Éloge de Paul Louis Duroziez. Paris, A. Davy, 1898, 27 pp.

DURRER, Dirk (1918–1984). Netherlands cardiologist.

BURCHELL, H.B. Dirk Durrer; thirty-five years of cardiology in Amsterdam. Utrecht, Wetenschappelijke Uitgeverij Bunge, 1986. 29 pp.

BIBLIOGRAPHY OF MEDICAL AND BIOMEDICAL BIOGRAPHY

MEIJLER, F.L. & BURCHELL, H.B. (eds.) Professor Dirk Durrer: 35 years of cardiology in Amsterdam. A selection of papers and a full bibliography. Amsterdam, North-Holland, 1986. 677 pp.

DUTROCHET, René-Joachim-Henri (1776–1847). French physician; developed cell theory.

ARON, E. Henri Dutrochet, médecin et biologiste, honneur de la Touraine, 1776–1847. Chambray, Editions C.L.D., 1990. 123 pp.
KRUTA, V. DSB, 1971, **4**, 263–5.
RICH, A.R. The place of R.-J.-H. Dutrochet in the development of the cell theory. *Bulletin of the Johns Hopkins Hospital*, 1926, **39**, 330–65.
SCHILLER, J. & SCHILLER, T. Henri Dutrochet (1776–1847). Le matérialisme méchanistique et la physiologie générale. Paris, Blanchard, 1975. 227 pp.

DUTTON, Joseph Everett (1874–1905). British parasitologist; discovered organism causing relapsing fever; died of the disease while investigating it; first to recognize human trypanosomiasis.

BRAYBROOKE, J. & COOK, G.C. Joseph Everett Dutton (1874–1905): pioneer in elucidating the aetiology of West African trypanosomiasis. *Journal of Medical Biography*, 1997, **5**, 131–6.

DUVAL, Mathias Marie (1844–1907). French anatomist, histologist and physiologist; professor of histology, Paris.

COURCY, C. DSB, 1971, **4**, 266–7.

DUVE, Christian de (b. 1917). Belgian cell biologist; shared Nobel Prize (Physiology or Medicine) 1974, for discoveries concerning the structural and functional organization of the cell.

Les Prix Nobel en 1974.

DUVERNEY, Joseph-Guichard (1648–1730). French anatomist and physician; wrote first book on structure, function and diseases of the ear.

ASHERSON, N. A bibliography of editions of Du Verney: Traité de l'organe de l'ouïe published between 1683–1750. London, H.K. Lewis, 1979. 110 pp.
RENAUD, J. Les commaunités de maitres chirurgiens avant la revolution de 1789...avec une notice sur Joseph Guichard Duverney. St Étienne, Ed.du Chevalier, 1946. 154 pp.
WILLIAMS, W.C. DSB, 1971, **4**, 267–8.

DU VIGNEAUD, Vincent (1901–1978). American biochemist; professor at Cornell University Medical College.

HOFMANN, K. BMNAS, 1987, **56**, 543–95.
Archival material: New York Hospital, Cornell Medical Center.

DYER, Rolla Eugen (1886–1971). American surgeon; Director, National Institutes of Health, USA; showed murine typhus to be due to an organism later named Rickettsia mooseri.

TOPPING, N. *Transactions of the Association of American Physicians*, 1973, **86**, 11–12.

DYKE, Cornelius Gysbert (1900–1943). American radiologist; introduced lumbar encephalography.

DAVIDOFF, L.M. Cornelius Gysbert Dyke: pioneer neuroradiologist. *Bulletin of the New York Academy of Medicine*, 1969, **45**, 665–80.

EARLE, Pliny (1809–1892). American physician; pioneer in the treatment of the insane.

SANBORN, F. (ed.) Memoirs of Pliny Earle, M.D., with extracts from his diary and letters (1830–1892), and selections from his professional writings (1839–1891). Boston, Damrell & Upham, 1898. 409 pp.

EAST, Edward Murray (1879–1938). American geneticist.

JONES, D.E. BMNAS, 1944, **22**, 217–42.
PROVINE, W.B. DSB, 1971, **4**, 270–72.

EATON, Monro Davis (1904–1989). American microbiologist; isolated 'Eaton agent' in primary pneumonia.

MARMION, B.P. Eaton agent – science and scientific acceptance: a historical commentary. *Review of Infectious Diseases*, 1990, **12**, 338–53.

EBERTH, Carl Joseph (1835–1926). German pathologist and bacteriologist; professor, Zürich and Halle; discovered typhoid bacillus.

BRIEGER, C.H. DSB, 1971, **4**, 275–7.
SCHEU, Der Pathologe Carl Joseph Eberth 1835–1926, Entdecker des Typhuserregers. Zürich, Juris, 1990. 116 pp.

EBSTEIN, Erich Hugo (1880–1931). German medical historian.

STEINER, J. Erich Hugo Ebstein (1880–1931). Biographie und wissenschaftliche Werk. Zürich, Juris, 1979. 124 pp.

EBSTEIN, Wilhelm (1836–1912). German pathologist and medical historian.

MANN, R.J. & LIE, J.T. The life story of Wilhelm Ebstein (1836–1912) and his almost overlooked description of congenital heart disease. *Mayo Clinic Proceedings*, 1979, **54**, 197–204.

ECCLES, John Carew (1903–1997). Australian neurophysiologist; shared Nobel Prize (Physiology or Medicine) 1963 for discoveries concerning the nerve cell membrane in the peripheral and central nervous systems.

ECCLES, J.C. My scientific Odyssey. *Annual Review of Physiology*, 1977, **39**, 1–18.
CURTIS, D.R. BMFRS, 2001, **47**, 161–7.

ECONOMO, Constantin von (1876–1931). Austrian neuroanatomist and neurologist.

BOGAERT, L. van & THÉDORIDÈS, J. Constantin von Economo: the man and the scientist. Vienna, Österreichisches Akademie der Wissenschaften, 1979. 138 pp.
ECONOMO, Baroness C. & WAGNER-JAUREGG, J. von. Baron Constantin von Economo (1876–1931). His life and work. Translated from the 2nd German edition by R. Spillman. Burlington, Vt., Free Press Interstate Printing Corp., 1937. 126 pp.

EDDY, Mary Baker (1821–1910). American founder of Christian Science.

GILL, G. Mary Baker Eddy. Reading MA, Perseus Books, 1998. 713 pp.

SILBERGER, J. Mary Baker Eddy: an interpretive biography of the founder of Christian Science. Boston, Little, Brown, *c*.1980. 274 pp.

EDEBOHLS, George Michael (1853–1908). American surgeon.

B., H.J. *American Journal of Obstetrics*, 1909, **59**, 835–7.
BOLT, H.J. *Transactions of the American Gynecological Society*, 1909, **34**, 637–40.

EDELMANN, Gerald Maurice (b. 1929). American biochemist; shared Nobel Prize (Physiology or Medicine), 1972, for studies on the chemical structure of antibodies.

Les Prix Nobel en 1972.

EDINGER, Ludwig (1855–1918). German neuroanatomist.

LUDWIG EDINGER (1855–1918). Gedenkschrift an seinem 100. Geburstag und zum 50 Jährigen Bestehen des Neurologischen Institutes (Edinger-Institut) der Universität Frankfurt am Main. Wiesbaden, F. Steiner, 1959. 97 pp.
EMISCH, H. Ludwig Edinger: Hirnanatomie und Psychologie. Stuttgart, G. Fischer, 1991. 254 pp.

EDKINS, John Sydney (1863–1940). British physiologist.

LOWICKE, E.M. The discoverer of gastrin: John Sydney Edkins. *Surgery*, 1965, **58**, 1044–8.

EDMAN, Per Victor (1916–1977). Swedish biochemist.

PARTRIDGE, S.M. & BLOMBÄCK, B. BMFRS, 1979, **25**, 241–65.

EDRIDGE-GREEN, Frederick William (1863–1953). British ophthalmologist; introduced lantern test for colour-blindness.

–. Frederick William Edridge-Green. C.B.E., M.D. Durh., F.R.C.S. *Lancet*, 1953, **1**, 856–7.
PLARR, 1952–64, 118–20 (CB).

EDSALL, David Linn (1869–1945). American physician; Dean, Harvard Medical School and School of Public Health.

AUB, J.C. & HAPGOOD, R.K. Pioneer in modern medicine: David Linn Edsall of Harvard. Cambridge, Mass., Harvard University Press, 1970. 384 pp.

EDSALL, John Tileston (b.1902). American biochemist; professor at Harvard University.

EDSALL, J.T. Some personal history and reflections from the life of a biochemist. *Annual Review of Biochemistry*, 1971, **40**, 1–28.
Archival material: Harvard University.

EDWARDES, David (1502–*c*.1542). English physician; published first work in England devoted solely to anatomy.

O'MALLEY, C.D., DSB, 1971, **4**, 284–5.
O'MALLEY, C.D. & RUSSELL, K.F. David Edwardes, Introduction to anatomy, 1532. A facsimile reproduction with English translation and an introductory essay on anatomical studies in Tudor England. London, Oxford Univ. Press, 1961. 64 pp.
ROOK, A. & NEWBOLD, M. David Edwardes: his activities at Cambridge. *Medical History*, 1975, **19**, 389–92.

EDWARDS, Robert Geoffrey (b.1925). British reproductive biologist; with P.C. Steptoe achieved first successful birth after re-implantation of embryo, 1978.

–. The retirement of Professor Robert Edwards. *Human Reproduction*, 1991, **6**, i–xii.
AUSTIN, C.R. Bob Edwards: a profile. *Human Reproduction*, 1991, **6**, 104.
EDWARDS, R.G. & STEPTOE, P.C. A matter of life; the story of a medical breakthrough. London, Hutchinson, 1980. 188 pp.

EGAS MONIZ, Antonio Caetano de Abreu Freire (1874–1955). Portuguese neurologist and neurosurgeon; professor of neurology, Lisbon; introduced cerebral angiography and psychosurgery (frontal leucotomy); Nobel Prize 1949.

EGAS MONIZ: primeiro centenáro 1874–1974. Coimbra, Museo Nacional de Ciência e de Técnica, 1974. 114 pp.
EGAS MONIZ, A. Confidéncias de um investigador cientifico. Lisboa, Atica, 1949. 626 pp.
PEREIRA, A.L. & RUA PITA, J. Egas Moniz: em livre exame. Coimbra, MinervaCoimbra, 2000. 414 pp.
SWAZEY, J.P. DSB, 1971, **4**, 286–7.

EHLERS, Edvard (1863–1937). Danish dermatologist; described cutis laxa ('Ehlers-Danlos syndrome'), 1901.

DARIER, J. *Bulletin de l'Académie de Médecine*, Paris, 1937, 3. sér. **117**, 626–8.
LIMON, C. *Annales de Dermatologie et de Syphiligraphie*, 1937, 7. sér., **8**, 458–61.

EHRENBERG, Christian Gottfried (1795–1876). German microbiologist; classified microorganisms; professor of medicine, Berlin.

DOBELL, C. C.G. Ehrenberg (1795–1876). A biographical note. *Parasitology*, 1923, **15**, 320–25.
LAUE, M. von. Christian Gottfried Ehrenberg; ein Vertreter deutscher Naturforschung in neunzehnten Jahrhundert. Berlin, J. Springer, 1895. 287 pp.

EHRLICH, Paul (1854–1915). German haematologist and immunologist; founder of chemotherapy; shared Nobel Prize 1908.

APOLANT, H. *et al.* Paul Ehrlich. Eine Darstellung seines wissenschaftlichen Wirkens. Festschrift zum 60. Geburtstage des Forschers. Jena, Gustav Fischer, 1914. 668 pp.
BÄUMLER, E. Paul Ehrlich: scientist for life. New York, Holmes & Meier, 1984. 288 pp. First publshed in German, 1979; 3rd German edition, Frankfurt am Main, Wötzel, 1997. 367 pp.
DOLMAN, C.E. DSB, 1971, **4**, 295–305.
MARQUARDT, M. Paul Ehrlich, London, Heinemann, 1949. 255 pp.
SILVERSTEIN, A.M. Paul Ehrlich's receptor immunology: the magnificent obsession. San Diego, CA, Academic Press, *c.*2002. 202 pp.
Archival material: Paul Ehrlich Institut, Frankfurt; New York Academy of Medicine; Rockefeller University, New York; Leo Baeck Institute, New York.

EICHHOLTZ, Fritz (1889–1967). German pharmacologist.

FLECKENSTEIN, A. Zum Gedächtnis. Professor Dr. med. Fritz Eichholtz. *Archives Internationales de Pharmacodynamie*, 1968, **173**, 259–61.

EICHHORST, Hermann Ludwig (1849–1921). German physician; professor of medicine, Göttingen, and director, Medical Clinic, Zürich.

FASOL, E. Der Internist Hermann Eichhorst, 1849–1921. Zürich, Juris Druck, 1983. 98 pp.

EIJKMAN, Christiaan (1858–1930). Dutch physiologist; pioneer in study of deficiency diseases; discovered cause of beri-beri: Nobel Prize (Physiology or Medicine) 1929.

JANSEN, B.C.P. Het levenswerk van Christiaan Eijkman 1858–1930. Haarlem, Bohn, 1959. 206 pp.
LINDEBOOM, G.A. DSB, 1971, **4**, 310–12.
Archival material: Library, University of Utrecht.

EINHORN, Alfred (1856–1917). German chemist; synthesized cocaine.

UHLFELDER, E. Alfred Einhorn. *Berichte der Deutschen Chemischen Gesellschaft*, 1917, **50**, 669–70.

EINHORN, Max (1862–1953). Russian/American gastroenterologist; introduced gastrodiaphany.

GROTE, 1929, **8**, 1–24. (CB)

EINTHOVEN, Willem (1860–1927). Dutch physiologist; invented string galvanometer and introduced clinical electrocardiography and phonocardiography; Nobel Prize, 1924.

HOOGERWERF, S. DSB, 1971, **4**, 334–5.
WAART, A. de. Het levenswerk van Willem Einthoven, 1860–1927. Haarlem, E.F. Bohn. 258 pp. English summary; list of publications, pp. 237–45.
SNELLEN, H.A. Two pioneers of electrocardiography: the correspondence between Einthoven and Lewis from 1908 to 1926. Rotterdam, Donker, 1983. 140 pp.
SNELLEN, H.A. Willem Einthoven (1860–1927) father of electrocardiography: life and work, ancestors and contemporaries. Dordrecht, Kluwer, *c*.1995. 140 pp.

EISELSBERG, Anton von (1860–1938). Austrian surgeon; professor in Vienna.

EISELSBERG, A. von. Lebensweg eines Chirurgen. Innsbrück, Deutscher Alpenverlag, 1939. 566 pp. Reprinted 1949.

EISENMANN, Johann Gottfried (1795–1867). German physician and politician.

HOFFMANN, H. Johann Gottfried Eisenmann (1795–1867), ein fränkischer Arzt un Freiheitskämpfer. Wurzburg, Freunde Mainfränkischer Kunst und Geschichte E.V., 1967. 131 pp.

ELDERFIELD, Robert Cooley (1904–1979). American organic chemist; made important contributions to chemotherapy.

LEONARD, N.J. BMNAS, 2000, **78**, 1–15.

ELFORD, William Joseph (1900–1952). British physical chemist; introduced first method of measuring viruses and bacteriophage (gradocol membranes).

ANDREWES, C.H. ONFRS, 1952–53, **8**, 149–58.

ELION, Gertrude Belle (1918–1999). American biochemist; shared Nobel Prize (Physiology or Medicine) 1998 for discoveries of important principles for drug therapy.

AVERY, M.E. BMNAS, 2000, **78**, 16–29.
ELION, G.B. The quest for a cure. *Annual Review of Pharmacology*, 1993, **33**, 1–23.

ELIOT, Jared (1685–1763). American physician, pastor and naturalist.

THOMS, H. The doctors Jared of Connecticut. Hamden, Conn., Shoe String Press, 1958. 76 pp. [Deals with Jared Eliot, Jared Potter Kirtland and Jared Potter.]

THOMS, H. Jared Eliot, minister, doctor, scientist, and his Connecticut. Hamden, Conn. Shoe String Press, 1967. 156 pp.

ELLENBOG, Ulrich (1440–1499). German physician; wrote first book on industrial hygiene and toxicology.

ELLENBOG, U. Von den gifftigen Besen, Tempffen und Reuchen. Wiedergabe des ersten Ausburger Druckes mit Biographie ... von Franz Koelsch und Friedrich Zoepfl. München, Münchner Drucke, 1927. xix pp., 6 ff.

ELLER VON BROCKHAUSEN, Johann Theodor (1689–1760). German chemist and physician; Director, Akademie der Wissenschaften, Berlin.

DYCK, D.R. DSB, 1971, **4**, 352–3.

ELLIOT SMITH, Grafton (1871–1937). Australian neurologist, anthropologist and ethnologist; professor of anatomy at Cairo, Manchester and London.

DAWSON, W.R. (ed.) Sir Grafton Elliot Smith; a biographical record by his colleagues. London, Cape, 1938. 272 pp.

ELKIN, A.P. & MACINTOSH, N.W.G. (eds.) Grafton Elliot Smith. The man and his work. Sydney, Sydney University Press, 1974. 232 pp. 4 addresses and 15 papers celebrating the centenary of his birth.

SIMPKINS, D.M. DSB, 1971, **4**, 353–4.

WILSON, J.T. ONFRS, 1936–39, **2**, 323–33.

Archival material: University of Manchester; Sydney University; Australian Academy of Sciences, Canberra.

ELLIOTSON, John (1791–1868). British physician and hypnotist.

RIDGWAY, E.S. John Elliotson (1791–1868); a bitter enemy of legitimate medicine? *Journal of Medical Biography*, 1993, **1**, 191–8; 1994, **2**, 1–7.

WILLIAMS, H. Doctors differ. Five studies in contrast ... London, Cape, 1946, pp. 25–91.

ELLIOTT, Thomas Renton (1877–1961). British physiologist; first to suggest chemical mediation of the nervous impulse, by his work on adrenaline.

DALE, H.H. BMFRS, 1961, **7**, 53–73.

Archival material: CMAC.

ELLIS, Henry Havelock (1859–1939). British psychologist; pioneer in the scientific study of sex.

BROME, V. Havelock Ellis, philosopher of sex: a biography. London, Routledge & Kegan Paul, 1979. 271 pp.

ELLIS, H. My life. Boston, Houghton Mifflin, 1939. 647 pp. London, Toronto, Heinemann, 1940. 542 pp.

GROSSKURTH, P. Havelock Ellis. A biography. London, Allen Lane, 1980. 492 pp.

Archival material: British Library Manuscript Collection.

HAWKINS, M.A. Havelock Ellis on criminology, sexology and moral philosophy. Thesis. Arlington, Univ. of Texas, 2000. 200 pp.

ELOESSER, Leo (1881–1976). American thoracic surgeon; professor at Stanford University.

SHUMACKER, H.B. Leo Eloesser, M.D. Eulogy for a free spirit. New York, Philosophical Library, 1982. 483 pp.

ELSBERG, Charles Albert (1871–1948). American neurosurgeon.

ALEXANDER, E. Charles Albert Elsberg, M.D. (1871–1948), father of spinal cord surgery. *Neurosurgery*, 1987, **20**, 811–14.
HORWITZ, N.H. Charles A. Elsberg (1871–1948). *Neurosurgery*, 1997, **40**, 1315–19.

ELSCHNIG, Anton Philipp (1863–1949). Austrian ophthalmologist.

LIBICKY, H. A personal sketch of Professor Anton Elschnig. *Survey of Ophthalmology*, 1982, **26**, 266–8.

ELVEHJEM, Conrad Arnold (1901–1962). American biochemist; researcher on nutrition, particularly on pellagra.

BURRIS, R.H., BAUMANN, C.A. & POTTER, V.R. BMNAS, 1990, **59**, 134–67.
IHDE, A.J. DSB, 1971, **4**, 357–9.
Archival material: University of Wisconsin.

EMBDEN, Gustav (1874–1933). German physiological chemist; researched on metabolism.

SCHMAUDERER, E. DSB, 1971, **4**, 359–60.

EMMERICH, Rudolph (1852–1914). German microbiologist; prepared pyocyanase *Ps. pyocyanea*.

EMMERICH, R. *Münchener Medizinische Wochenschrift*, 1914, **61**, 2342–3.

EMMET, Thomas Addis (1828–1919). American gynaecologist.

MARR, J.P. Pioneer surgeons of the Woman's Hospital. Philadelphia, F.A. Davis, 1957. pp. 63–101.

EMPEDOCLES (*c.* 490–430 BC). Greek philosopher and physician.

BOLLACK, J. Empédocle. Vol. 1. Paris, Editions du Minuit, 1965. 411 pp.

ENDERS, John Franklin (1897–1985). American microbiologist; shared Nobel Prize 1954 for work on poliomyelitis.

TYRRELL, D.A.J. BMFRS, 1987, **33**, 211–33.
WELLER, T.A. & ROBBINS, C. BMNAS, 1991, **60**, 46–65.

ENGELHARDT, Vladimir Alexandrovich (1894–1984). Russian biochemist; discovered oxidative phosphorylation.

ENGELHARDT, V.A. Life and science. *Annual Review of Biochemistry*, 1982, **51**, 1–19.

ENGELMANN, Theodor Wilhelm (1843–1909). German electrophysiologist.

ENGELMANN, T.W. Some papers and his bibliography, with an introduction by F.L. Meijler. Amsterdam, Rodopi, 1984. 264 pp.
KINGREEN, H. Theodor Wilhelm Engelmann (1843–1909). Münster, Institut für Geschichte der Medizin, 1972. 121 pp.

ROTHSCHUH, K.E. DSB, 1971, **4**, 371–3.
Archival material: Staatsbibliothek, Preussischer Kulturbesitz, Berlin.

ENT, George (1604–1689). English physician; associate of Harvey and distinguished Fellow of Royal College of Physicians of London.

WEBSTER, C. DSB, 1971, **4**, 376–7.
Archival material: Bodleian Library, Oxford; Royal College of Physicians, London.

EPPINGER, Hans (1846–1916). Bohemian physician.

BEITZKE, H. Hans Eppinger. 17.II.1846–12.VIII.1916. *Verhandlungen der Deutschen Pathologischen Gesellschaft*, 1937, **29**, 351–4.

EPPINGER, Hans (1879–1946). Bohemian physician; performed human experimentation in German concentration camps.

SPIRO, H.M. Eppinger of Vienna: scientist and villain. *Journal of Clinical Gastroenterology*, 1984, **6**, 493–7.

ERASISTRATUS (*c.* 304–250 BC). Greek physician; originated anatomical dissection and made physiological experiments.

FUCHS, R. Erasistratus quae in librorum memoria congesta enarrantur. Leipzig, G. Fock, 1892. 32 pp.
DOBSON, J.F. Erasistratus. *Proceedings of the Royal Society of Medicine*, 1927, **20**, 825–32.
LONGRIGG, J. DSB, 1971, **4**, 382–6.

ERB, Wilhelm Heinrich (1840–1921). German neurologist; described 'Erb's palsy'.

SCHULTZE, F. Wilhelm Erb. *Deutsche Zeitschrift für Nervenheilkunde*, 1922, **73**, Heft 1–2, i–xviii.

ERDHEIM, Jacob (1874–1937). Polish pathologist.

HASLHOFER, L. Jacob Erdheim 22.5.1874 bis 21.4.1937. *Verhandlungen der Deutschen Gesellschaft für Pathologie*, 1965, **49**, 370–75.

ERICHSEN, John Eric (1818–1896). British surgeon; wrote popular work on surgery.

JAMES, M.J. The life and times of Sir John Eric Erichsen, Baronet, surgeon extraordinary to the Queen. [n.p., n.d.] 66 leaves [in Wellcome Library].
PLARR, 378–80.

ERICKSON, Milton Hyland (1901–1980). American psychiatrist.

ZEIG, J.K. Experiencing Erickson: an introduction to the man and his work. New York, Brunner/Mazel, 1985. 181 pp.

ERISMANN, Fedor Fedorovich (1842–1915). Swiss-born Russian hygienist; professor at Moscow University.

SEMASHKO, N.A. Friedrich Erismann. The dawn of Russian hygiene and public health. *Bulletin of the History of Medicine*, 1946, **20**, 1–9.
WICK, H. Friedrich Huldreich Erismann (1842–1915): Russischer Hygieniker – Zürcher Stadtrat. Zürich, Juris, 1970. 61 pp.

BIBLIOGRAPHY OF MEDICAL AND BIOMEDICAL BIOGRAPHY

ERLANGER, Joseph (1874–1965). American physiologist; shared Nobel Prize 1944 for work on neurophysiology.

DAVIS, H. BMNAS, 1970, **41**, 111–39.
ERLANGER, J. A physiologist remembers. *Annual Review of Physiology*, 1964, **26**, 1–14.
MONNIER, A.M. DSB, 1971, **4**, 397–9.
Archival material: Medical School, Washington University, St Louis.

ERMENGEM, Emile Pierre Marie van (1851–1932). Belgian bacteriologist; discovered *Clostridium botulinum*.

ZIMMERN, A. *Bulletin de l'Académie Nationale de Médecine*, 1932, **108**, 1465–6.

ESCHERICH, Theodor (1857–1911). German physician; professor of paediatics, Graz, Austria; first described *Eschericha coli*, 1896.

DOLMAN, C.E. DSB, 1971, **4**, 403–6.

ESMARCH, Johann Friedrich von (1823–1908). German surgeon; introduced 'Esmarch bandage' for surgical haemostasis.

DEUTSCHES *Archiv für Geschichte der Medizin*, 1885, **8**, 6–59, 159–64.
ROHLFS, H. Geschichte der Deutschen Medizin. Leipzig, 1885, vol. 4, 353–411.

ESQUIROL, Jean Etienne Dominque (1772–1840). French psychiatrist at the Salpêtrière.

DUMAS, M. Etienne Esquirol; sa famille, ses origines, ses années de formation. Toulouse, Typo-lino, 1972. 121 pp.
CADORET, M. Esquirol et la statistique médicale. Paris, Editions A.G.E.M.P., 1969. 45 pp.
MORA, G. On the bicentenary of the birth of Esquirol. *American Journal of Psychiatry*, 1972, **129**, 562–7.

ESTIENNE, [STEPHANUS] Charles (1504–1564). French anatomist; published first work to include whole external nervous and venous systems (1545).

GRMEK, M.D. DSB, 1971, **4**, 412–13.
RATH, G. Charles Estienne, contemporary of Vesalius. *Medical History*, 1964, **8**, 354–9.

EULENBERG, Michael Moritz (1811–1877). German orthopaedic surgeon.

STÜRZBECHER, M. Michael Moritz Eulenberg, ein Berliner Arzt des 19. Jahrhunderts. *Medizinische Klinik*, 1964, **59**, 1025–8.

EULER, Ulf Svante von (1905–1983). Swedish physiologist; Nobel Prize winner 1970.

BLASCHKO, H.K.F. BMFRS, 1985, **31**, 145–70.
EULER, U.S. von. Pieces in the puzzle. *Annual Review of Pharmacology*, 1970, **10**, 1–12.

EULER-CHELPIN, Hans Karl August Simon von (1873–1964). German-born Swedish chemist; shared Nobel Prize 1929 for work on fermentation.

IHDE, A.J. DSB, 1971, **4**, 485–6.

EURICH, Frederick William (1867–1945). British professor of forensic medicine, Leeds University; physician and bacteriologist in Bradford, where he conducted research on anthrax.

116

BIBLIOGRAPHY OF MEDICAL AND BIOMEDICAL BIOGRAPHY

BLIGH, M. Dr. Eurich of Bradford, London, Clarke, 1960. 296 pp.

EUSTACHI, Bartolomeo (*c.*1500/1510–1574). Italian anatomist; professor at Rome; his fine collection of anatomical plates remained forgotten in Vatican Library until published in 1714.

O'MALLEY, C.D. DSB, 1971, **4**, 486–8.

EVANS, Alice Catherine (1881–1975). American bacteriologist; demonstrated causal organism of Malta fever.

BURNS, V.L. Alice Evans, bacteriologist. Laingsburg, MI, Enterprise Press, 1993. 214 pp.

EVANS, Charles Arthur Lovatt (1884–1968). British physiologist; professor, University College London.

DALY, I. de B. & GREGORY, R.A. BMFRS, 1970, **16**, 233–52.
Archival material: CMAC.

EVANS, David Gwynne (1909–1984). British bacteriologist; Director, National Institute for Biological Standards and Control, London.

DOWNIE, A.W., SMITH, C.E.G., & TOBIN, J. O'H. BMFRS, 1985, **31**, 173–96.

EVANS, Griffith (1835–1935). British veterinary surgeon; discovered pathogenic trypanosome (in horses affected with 'surra').

WARE, J. & HUNT, H. The several lives of a Victorian vet. London, Bachmann & Turner, 1979. 213 pp.

EVANS, Herbert McLean (1882–1971). American anatomist, embryologist and endocrinologist.

AMOROSO, E.C. & CORNER, G.W. BMFRS, 1972, **18**, 83–186.
CORNER, G.W. BMNAS. 1974, **45**, 153–92.
Archival material: University of California, Berkeley; Rockefeller Archives Center, New York.

EVARTS, Edward Vaughan (1926–1985). American neurophysiologist.

THACH, W.T. BMNAS, 2000, **78**, 30–43.

EWALD, Carl Anton (1845–1915). German gastroenterologist.

GUSTEMEYER, I. Carl Anton Ewald (1845–1915), ein Pionier der Gastroenterologie. Dissertation. Köln, 1969. 45 pp.

EWING, James (1866–1943). American pathologist; professor at Cornell Medical College.

ADAIR, F.E. (ed.) Cancer...comprising international contributions to the study of cancer, in honor of J. Ewing. Philadelphia, Lippincott, 1931. 484 pp.
MURPHY, J.B. BMNAS, 1951, **26**, 45–60.
TRIOLO, V.A. DSB, 1971, 4, 498–500.
Archival material: Memorial Sloan-Kettering Cancer Center, New York.

EWINS, Arthur James (1883–1958). British organic chemist; Research Director of May & Baker.

DALE, H.H. BMFRS, 1958, **4**, 81–9.

EY, Henri (1900–1977). French psychiatrist.

CLERVOY, P. Henri Ey, 1900–1997; cinquante ans de psychiatrie en France. Le Plessis-Robinson, Synthélabo, 1997. 303 pp.

EYSENCK, Hans Jurgen (1916–1997). British psychologist.

GIBSON, H.B. Hans Eysenck. The man and his work. London, Owen, 1981. 275 pp.
EYSENCK, H. Rebel with a cause: the autobiography of H.J. Eysenck. Revised and expanded edition. New Brunswick, Transactions Publishers, 1997. 337 pp. First published 1990.

FABER, Knud Helge (1862–1956). Danish physician.

FABER, K.H. Personlige erindringer. København, Gyldendal, 1949. 128 pp.
GROTE, 1929, 8, 29–60 (CB).

FABRICIUS, Hieronymus ab Aquapendente [Fabrizio, Girolamo] (c.1533–1619). Italian professor of anatomy and surgery, Padua; one of the founders of scientific embryology.

FRANKLIN, K.J. ed. De venarum ostiolis (1603). Facsimile edition, etc. Springfield, C.C. Thomas, 1933. 98 pp. Includes biographical sketch.
ZANOBIO, B. DSB, 1971, **4**, 507–12.

FABRICIUS Hildanus, see **FABRY von HILDEN, Wilhelm**

FABRY, Wilhelm, see **FABRY von HILDEN**

FABRY VON HILDEN, Wilhelm [FABRICIUS Hildanus] (1560–1634). German surgeon; 'Father of German surgery'.

HINTZSCHE, E. Gulielmus Fabricius Hildanus, (1560–1634). Hilden, Lindopharm Rönsberg, 1972. 74 pp.
JONES, E. The life and works of Gulielmus Fabricius Hildanus. *Medical History*, 1960, **4**, 112–34, 196–209.
SCHNEIDER-HILTBRUNNER, V. Wilhelm Fabry 1560–1634: Verzeichnis der Werke und des Briefwechsels. Bern, H. Huber, c.1976. 172 pp.
STRANGMEIER, H. Wilhelm Fabry von Hilden: Leben, Gestalt, Wirken. Wuppertal-Eberfeld, Martini und Grüttefien, 1957. 129 pp.
WENNIG, W. (ed.) Fabrystudien. 4 vols. Hilden, Verlag Fr. Peters, 1961–74.

FAGGE, Charles Hilton (1838–1883). British physician.

MANN, W.N. Charles Hilton Fagge. *Guy's Hospital Reports*, 1973, **122**, 196–203.

FÅHRAEUS, Robert [Robin] (1888–1968). Swedish anatomist and pathologist; introduced method of estimating sedimentation in erythrocyte sedimentation rate.

MEMORIAL ISSUE for Robin Fåhraeus (1888–1968). *Biorheology*, 1988, **25**, 823–90.

FAIRLEY, Neil Hamilton (1891–1966). British malariologist.

BOYD, J. BMFRS, 1966, **12**, 123–45.

FALLOPPIO, Gabriele (1523–1562). Italian; professor of anatomy at Pisa and Padua; described Fallopian tubes.

FAVARO, G. Gabriele Falloppio Modenese (MDXXIII–MDLXII). Studio biografico. Modena, Tip. Editrice Immaculata Concezione, 1928. 254 pp.
O'MALLEY, C.D. DSB, 1971, **4**, 519–21.

FALLOT, Etienne-Louis Arthur (1850–1911). French physician; described the congenital heart condition 'Fallot's tetralogy'.

NEILL, C. & CLARK, E.P. Tetralogy of Fallot: the first 300 years. *Texas Heart Institute Journal*, 1994, **21**, 272–9.

FALRET, Jean Pierre (1794–1870). French psychiatrist; first to describe manic-depressive psychosis.

GUARDIA, J.M. Le docteur G. Falret. *Gazette Médicale de Paris*, 1871, **26**, 25–8.

FANCONI, Guido (1892–1979). Swiss paediatrician; with co-workers first described cystic fibrosis.

FANCONI, G. Der Wandel der Medizin, wie ich ihn Erlebte. Bern, Stuttgart, Wien, Hans Huber, 1970. 358 pp.
ROSSI, E. In memoriam Guido Fanconi. *Helvetica Paediatrica Acta*, 1979, **34**, 393–6.

FARR, William (1807–1883). British statistician, epidemiologist and demographer.

EYLER, I.M. Victorian social medicine. The ideas and methods of William Farr. Baltimore, Johns Hopkins Press, 1979. 262 pp.
HUMPHREYS, N.A. (ed.) Vital statistics: a memorial volume of selections from the reports and writings of William Farr. London, Sanitary Institute, 1885. 563 pp. Includes biographical sketch. Reprinted Metuchen, N.J., Scarecrow Press, 1975. 563 pp.

FARRE, John Richard (1775–1862). Barbadian physician and surgeon; joint founder of Royal London Ophthalmic Hospital.

–. *British Medical Journal*, 1862, **1**, 551–2.
MUNK, **3**, 33–5.

FAUCHARD, Pierre (1678–1761). French dental surgeon; wrote first text book on dentistry.

BESOMBES, A. & DAGEN, G. Pierre Fauchard, pére de l'art dentaire moderne...et ses contemporaines. Paris, Société des Publications Médicales et Dentaires, 1961. 150 pp.
WEINBERGER, B.W. Pierre Fauchard, surgeon-dentist. Minneapolis, Pierre Fauchard Academy, 1941. 102 pp.

FAURE-FREMIET, Emmanuel (1883–1971). French protozoologist.

CORLISS, J.O. A man to remember; E. Fauré-Fremiet (1883–1971). Three-quarters of a century of progress in protozoology. *Journal of Protozoology*, 1972, **19**, 389–400.
WILLMER, E.N. BMFRS, 1972, **18**, 187–221.

FAUVEL, Sulpice Antoine (1813–1884). French physician; first to describe the presystolic murmur in mitral stenosis.

BERGERON, –. *Bulletin de l'Académie de Médecine*, Paris, 1884, 2 sér. **13**, 1607–17.

FAY, Temple Sedgwick (1895–1953). American neurosurgeon.

WOLF, J.M. Temple Fay, M.D.: progenitor of the Doman-Delacato treatment procedures. Springfield, Ill., C.C. Thomas, 1968. 258 pp.

FAYRER, Joseph (1824–1907). British surgeon in India; wrote classic account of venomous snakes of India.

> FAYRER, J. Recollections of my life. Edinburgh, London, W. Blackwood, 1900. 500 pp.

FECHNER, Gustav Theodor (1801–1887). German experimental psychologist; developed Fechnr-Weber law on stimulus and sensation.

> ALTMANN, I. Bibliographie Gustav Theodor Fechner. Leipzig, Verlag im Wissenschaftszentrum, 1995. 112 pp.
> ARENDT, H.-J. Gustav Theodor Fechner: ein deutscher Naturwissenschaftler und Philosoph im 19. Jahrhundert. Frankfurt am Main, Peter Lang, c.1999. 285 pp.
> BROZEK, J. & GRUNDLACH, H. (eds.) G.T. Fechner und die Psychologie. Passau, Passavia Universitätsverlag, 1988. 298 pp.
> JAYNES, J. DSB, 1971, **4**, 556–8.
> LENNIG, P. Von der Metaphysik zur Psychophysik: Gustav Theodor Fechner (1801–1887): eine ergobiographische Studie. Frankfurt am Main, New York, P. Lang, 1994. 214 pp.

FEDOROV, Sergei Petrovich (1869–1936). Russian surgeon.

> IVANOVA, A.T. [New data for the biography of S.P. Fedorov.] *Vestnik Khirurgii*, 1968, **101**, (Aug.), 135–9. [In Russian.]
> PAPKO, G.P. [S.P. Fedorov – an outstanding figure in the national surgery (1869–1936).] *Sovetskaya Meditsina*, 1960, **36**, (Feb.), 130–35. [In Russian.]

FEER, Emil (1864–1955). Swiss paediatrician; professor at Heidelberg and Zürich.

> KUNZ, R. Der Kinderarzt Emil Feer (1864–1955). Zürich, Juris, 1987. 84 pp.

FELDBERG, Wilhelm Siegmund (1900–1993). German/British neuropharmacologist.

> BISSET, G.W. & BLISS, T.V.P. BMFRS, 1997, **43**, 143–70.
> FELDBERG, W.S. Fifty years on: looking back on some developments in neurohumoral physiology. Liverpool, Liverpool Univ. Press, 1982, 106 pp.

FELIX, Arthur (1887–1956). Polish-born bacteriologist; typhus research; Weil-Felix reaction.

> CRAIGIE, J. BMFRS, 1957, **3**, 53–79.

FELL, Honor Bridget (1900–1986). British cytologist; Director, Strangeways Research Laboratory, Cambridge, 1929–1970.

> VAUGHAN, J. BMFRS, 1987, **33**, 235–59.
> *Archival material*: CMAC.

FENGER, Christian (1840–1902). Danish/American surgeon and pathologist.

> BUFORD, S.G. Christian Fenger: a biographical sketch. *Bulletin of the Society for Medical History of Chicago*, 1911–16, **1**, 196–204.
> –. *Proceedings of the Institute of Medicine of Chicago*, 1972, **29**, 40–68.

FENN, Wallace Osgood (1893–1971). American physiologist; professor at University of Rochester School of Medicine.

> RAHN, H. BMNAS, 1979, **50**, 141–57.
> WARNER, J.H. DSB, 1990, **17**, 289–91.

FENNER, Frank John (b. 1914). Australian virologist; with F.M. Burnet elucidated acquired immunological tolerance.

FENNER, F. Historical vignette: a life with poxviruses and publishers. *Advances in Virus Research*, 1999, **51**, 1–33.

FERENCZI, Sándor (1873–1933). Hungarian neurologist and psychoanalyst.

ARON, L. *et al.* (eds.) The legacy of Sándor Ferenczi. Hillsdale, NJ, Analysis Press, 1993. 294 pp.
BARANDE, I. Sandor Ferenczi. Paris, Payot & Rivages, 1996. 194 pp. First published 1972.
LORIN, C. Sándor Frenczi: de la médecine à la psychanalyse. Paris, Presses Universitaire de France, 1993. 256 pp.
RACHMAN, A.W. Sándor Ferenczi: the psychotherapist of tenderness and passion. Nothvale, NJ, Jason Aronson, 1997. 464 pp.

FERGUSON, Alexander Hugh (1853–1912). American surgeon; devised 'Ferguson's operation' for hernia.

MacLEAN, N.J. *Surgery, Gynecology & Obstetrics*, 1926, **42**, 721–3.

FERGUSON, Robert George (1883–1964). Canadian physician; director of tuberculosis control, Province of Saskatchewan.

HOUSTON, C.S. R.G. Ferguson: crusader against tuberculosis. Toronto, Hannah Institute, Dundurn Press, 1991. 156 pp.

FERGUSSON, William (1808–1877). British surgeon.

GORDON-TAYLOR, G. Sir William Fergusson, Bt., F.R.C.S., F.R.S. (1801–1877). *Medical History*, 1961, **5**, 1–14.
Archival material: Royal College of Surgeons of England; CMAC.

FERNEL, Jean François (1497–1558). French physiologist, physician, astronomer and mathematician.

FIGARD, L. Un médecin philosophe au XVI siècle. Étude sur la psychologie de Jean Fernel. Paris, Felix Alcan, 1903. 368 pp.
GRANIT, R. DSB, 1971, **4**, 584–5.
SHERRINGTON, C. The endeavour of Jean Fernel, with a list of the editions of his writings. Cambridge, Cambridge University Press, 1946. 223 pp.

FERRARIUS DE GRADIBUS, Johannes Matthaeus (d.1472). Italian physician; professor at Pavia.

FERRARI, H.M. Une chaire de médecine au XVe siècle; un professeur à l'Université de Pavie de 1432 à 1472. Paris, Alcan, 1899. 324 pp.

FERRIER, David (1843–1928). British neurologist; professor of forensic medicine and later of neuropathology, King's College Hospital Medical School, London.

CLARKE, E. DSB, 1971, **4**, 593–5.

FEUCHTERSLEBEN, Ernst von (1806–1849). German psychiatrist; wrote a history of the subject.

RISSMANN, W. Ernst Freiherr von Feuchtersleben (1806–1849). Sein Beitrag zur medizinischen Anthropologie und Psychopathologie. Freiburg im Breisgau, Schulz, 1980, 243 pp.

FEULGEN, Robert Joachim (1884–1955). German biochemist and histochemist.

OLBY, R. DSB, 1971, **4**, 603–4.

FIBIGER, Johannes (1867–1928). Danish oncologist; Nobel Prize, 1926.

SECHER, K. The Danish cancer researcher, Johannes Fibiger, professor in the University of Copenhagen. Copenhagen, Busck; London, Lewis, 1947. 206 pp.

FICK, Adolf Eugen (1829–1901). German physiologist; professor at Zürich; introduced instruments for study of muscle and nerve physiology.

BEZEL, R. Der Physiologe Adolf Fick, 1829–1901; seine Zürcher Jahre. Zürich, Juris, 1979, 85 pp.
ROTHSCHUH, K.E. DSB, 1971, **4**, 614–16.

FIELDS, Bernard N. (1938–1995). American virologist.

SCHLESINGER, S. BMNAS, 1997, **71**, 63–76.

FIESSINGER, Noel (1881–1946). French physician; first (1916) to describe the condition later named 'Reiter's syndrome'.

MONNIER, A.M. DSB, 1971, **4**, 617–18.

FILATOV, Nils Feodorovich (1847–1902). Russian paediatrician; first to describe infectious mononucleosis, 1885.

MOLCHANOV, V.I. N.F. Filatov: [for the centenary of his birth] 1847–1907. Moskva, Medgiz, 1947. 88 pp. [In Russian.]

FILATOV, Vladimir Petrovich (1875–1956). Russian surgeon; introduced pedicle flap in plastic surgery and worked on corneal transplants and tissue therapy.

FILATOV , V.P. My path in science. Moscow, Foreign Languages Publishing House, 1957. 182 pp.
TURBIN, A.M. Naslednicki Filatova. Moskva, Sovestskaia Rossia, 1974. 331 pp.

FILDES, Paul Gordon (1882–1971). British microbiologist; initiated study of bacterial nutrition.

BECHTEL, W. DSB, 1990, **17**, 295–7.
GLADSTONE, G.P., KNIGHT, B.C.J.G. & WILSON, G., BMFRS, 1973, **19**, 317–47.

FILEHNE, Wilhelm (1844–1927). Polish pharmacologist; introduced antipyrin and Pyramidon.

BRUNE, K. Knorr and Filehne in Erlangen. *Agents and Actions*, 1986, **19**, Suppl., 19–29.

FINCH, John (1626–1682). English physician and ambassador in Italy and Turkey.

MALLOCH, A. Finch and Baines: a seventeenth century friendship. Cambridge, Cambridge Univ. Press, 1917. 89 pp.

FINE, Pierre (1760–1814). French surgeon; performed first recorded colostomy for obstruction, 1797.

–. Notice biographique de Pierre Fine. *Journal Générale de Médecine, de Chirurgie et de Pharmacie*, 1815, **54**, 409–14.

FINLAND, Maxwell (1902–1987). Russian/American physician; introduced the sulphonamide, sulphasuxidine.

ROBBINS, F.C. BMNAS, 1999, **76**, 102–13.

FINLAY, Carlos Juan (1833–1915). Cuban physician; first to suggest mosquito as vector of yellow fever.

BEAN, W.B. DSB, 1971, **4**, 619–20.
HART, A.L. Dr Finlay sees it through. 4th ed., New York, Harper, 1942, 370 pp.
RODRIGUEZ, C. Finlay. La Habana, Editorial Librería Selecta, 1951. 340 pp.
SANCHEZ, J.L. Finlay, el hombre y la verdad científica. La Habana, Editorial Científico Tecnica, 1987. 569 pp.

FINLAY, Robert Bannatyne (1842–1929). British physician, lawyer and Chancellor of the Exchequer.

DNB, 1922–30.

FINNEY, John Miller Turpin (1863–1942). American surgeon; professor at Johns Hopkins University.

FINNEY, J.M.T. A surgeon's life. The autobiography of J.M.T. Finney. New York, Putnam's Sons, 1940. 396 pp.

FINSEN, Niels Ryberg (1860–1904). Danish physician; studied therapeutic effects of light and introduced (Finsen) lamp for this therapy; Nobel Prize 1903.

AGGEBO, A.J. Niels Finsen, die Lebensgeschichte eines grossen Arztes und Forschers. Zürich, Rascher, 1947. 330 pp.
LOMHOLT, S. Niels Finsen. København, Gyldendalske Boghandel, 1943. 214 pp.
TRIOLO, V.A. DSB, 1971, **4**, 620–21.

FINSTERER, Hans (1877–1951). Austrian surgeon.

Verzeichnis und Inhaltsgabe der wissenschaftlichen Arbeiten des Dr Hans Finsterer, Assistenten der chirurgischen Universitätsklinik Hofrat Hochenegg in Wien. Wien, Druck von M. Salzer, 1912. 22 pp.

FIORAVANTI, Leonardo (1517–1588). Italian surgeon; follower of Paracelsus.

FURFARO, D. La vita e l'opera di Leonardo Fioravanti. Bologna, Presso l'Azzoguidi Soc. Tip. Editoriale, 1963. 225 pp.

FISCHER, Edmond Henri (b. 1920). American biochemist; shared Nobel Prize (Physiology or Medicine), 1992, for discoveries concerning reversible protein phosphorylation as a biologically regulatory mechanism.

Les Prix Nobel en 1992.

BIBLIOGRAPHY OF MEDICAL AND BIOMEDICAL BIOGRAPHY

FISCHER, Eugen (1874–1967). German anatomist and anthropologist.

GESSLER, B. Leben und Werk des Freiburger Anatomen, Anthropologen und Rassenhygienikers bis 1927. Frankfurt am Main, P. Lang, *c*.2000. 209 pp.
LÖSCH, N.C. Rasse und Konstrukt: Leben und Werk Eugen Fischers. Frankfurt am Main, Peter Lang, *c*.1997. 615 pp.

FISCHER, Hans (1881–1945). German organic chemist; awarded Nobel Prize (Chemistry) 1930, for his work on haemin.

LEICESTER, H.M. DSB, 1978, **15**, 157–8.
WATSON, C.J. Reminiscences of Hans Fischer and his laboratory. *Perspectives in Biology and Medicine*, 1965, **8**, 419–35.

FISCHER, Hermann Otto Laurenz (1888–1960). German biochemist.

BAKER, A.D. DSB, 1972, **5**, 5–7.
FISCHER, H.O.L. Fifty years 'Synthetiker' in the service of biochemistry. *Annual Review of Biochemistry*, 1960, **29**, 1–14.

FISCHER, Isidor (1868–1943). Austrian physician and medical historian.

CASTIGLIONI, A. Dr Isidor Fischer (1868–1943) *Bulletin of the History of Medicine*, 1943, **14**, 114–15.
VOSWINCKEL, P. Das Vermächtnis Isidor Fischers: Chancen und Dilemma der aktuellen Medizin-Biographik. In: BROER, F. (ed.). Eine Wissenschaft emanzipiert sich: die Medizinhistoriographie von der Aufklärung bis zur Postmoderne. Pfaffweiler, Centaurus, 1999, pp. 121–37.

FISHBEIN, Morris (1889–1976). American physician; editor of *Journal of the American Medical Association*.

FISHBEIN, M. Morris Fishbein, M.D.: an autobiography. Garden City, NY, Doubleday & Co., 1969. 505 pp.

FISHER, Ronald Aylmer (1890–1962). British statistician and geneticist; professor at Cambridge.

BOX, J.F. R.A. Fisher: the life of a scientist. Chichester, John Wiley, [*c*.1978]. 512 pp.
GRIDGEMAN, N.T. DSB, 1972, **5**, 7–10.
IN MEMORIAM. Ronald Aylmer Fisher, 1890–1962. *Journal of the Biometric Society*, 1964, **20**, 237–373.
YATES, F. & MATHER, K. BMFRS, 1963, **9**, 91–129.
Archival material: Library, University of Adelaide; Genetics Dept., University of Cambridge.

FITZ, Reginald Heber (1843–1913). American pathologist; professor of theory and practice of physic, Harvard.

MORISON, H. Reginald Heber Fitz. *Bulletin of the History of Medicine*, 1941, **10**, 250–59.

FLAJANI, Giuseppi (1741–1808). Italian anatomist, physician and surgeon; gave one of the earliest accounts of exophthalmic goitre.

CAPPARONI, P. Profili bio-bibliografici di medici, etc. Rome, 1932, vol. 1, 100–102.

FLECHSIG, Paul Emil (1847–1929). German neuro-anatomist and psychiatrist.

124

CLARKE, E. DSB, 1972, **5**, 26–8.

FLEMING, Alexander (1881–1955). British bacteriologist; discovered lysozyme and penicillin; shared Nobel Prize 1945.

COLEBROOK, L. BMFRS, 1956, **2**, 117–27.
DOLMAN, C.E. DSB, 1972, **5**, 28–31.
HARE, R. The birth of penicillin and the disarming of microbes. London, Allen & Unwin, 1970. 236 pp.
MACFARLANE, G. Alexander Fleming:the man and the myth. London, Chatto & Windus, 1984. 304 pp.
SHIPTON, R.C. A bibliography of Sir Alexander Fleming, 1881–1955. London, St Mary's Hospital Medical School, 1993. 26 leaves.

FLEMMING, Carl Friedrich (1799–1880). German psychiatrist.

LEOPOLDT, M. Carl Friedrich Flemming (1799–1880). – ein Vertreter der deutschen Psychiatrie des neunzehnte Jahrhunderts. [Thesis.] Berlin, Institut für Geschichte der Medizin der Freien Universität, 1983. 443 pp.

FLEMMING, Walther (1843–1905). German cytologist; discovered the centrosome.

OLBY, R. DSB, 1972, **5**, 34–6.
PETERS, G. Walther Flemming (1843–1905). Sein Leben und sein Werk. Neumünster, K. Wachholtz, 1967. 111 pp.

FLETCHER, Robert (1823–1912). British-born physician; Principal Assistant Librarian, Surgeon General's Library, Washington; co-editor of *Index-Catalogue* and *Index Medicus*.

BRODMAN, E. Memoirs of Robert Fletcher. *Bulletin of the Medical Library Association*, 1961, **49**, 251–90.
OSLER, W. Robert Fletcher, 1823–1912. *Bristol Medico-Chirurgical Journal*, 1912, **30**, 289–94.

FLETCHER, Walter Morley (1873–1933). British physiologist; secretary to the Medical Research Council, 1914–1933.

ELLIOTT, T.R. ONFRS, 1933, **1**, 153–63.
FLETCHER, M.F. The bright countenance; a personal biography of Walter Morley Fletcher. London, Hodder & Stoughton, 1957. 351 pp.
GEISON, G.L. DSB, 1972, **5**, 36–8.
Archival material: CMAC; Medical Research Council, London; National Institute for Medical Research, London.

FLETT, John Smith (1869–1947). British physician and geologist; Director, Geological Survey of Great Britain and Director, Museum of Practical Geology 1920–35.

DUNHAM, K.C. DSB, 1972, **5**, 38–9.
READ, M.H. ONFRS, 1948, **5**, 689–96.

FLEXNER, Abraham (1866–1959). American educational reformer; expert on medical education.

FLEXNER, A. Abraham Flexner: an autobiography. New York, Simon & Schuster, 1960. 302 pp. (A revised edition of his As I remember, New York, Simon & Schuster, 1940. 414 pp.)

WHEATLEY, S.C. The politics of philanthropy. Abraham Flexner and medical education. Madison, Univ. of Wisconsin Press, 1988, 249 pp.
Archival material: Butler Library, Columbia University, NY.

FLEXNER, Louis Barkhouse (1902–1996). American neurologist.
SPRAGUE, J.M. BMNAS, 1998, **73**, 151–65.

FLEXNER, Simon (1863–1946). American pathologist and bacteriologist; first director of Rockefeller Institute.

CORNER, G.W. DSB, 1972, **5**, 39–41.
FLEXNER, J.T. An American saga: the story of Helen Thomas and Simon Flexner. Boston, Little, Brown & Co., 1984. 494 pp. [Helen Thomas was Mrs Simon Flexner.]
MILLER, M. A guide to selected files of the professional papers of Simon Flexner at the American Philosophical Society Library. Philadelphia, American Philosophical Society Library, 1979. 76 pp. (American Philosophical Society Library, Publ. No. 8.)
ROUS, P. ONFRS, 1948–49, **6**, 409–45.
Archival material: American Philosophical Society, Philadelphia.

FLINT, Austin Sr (1812–1886). American physician.

CHEN, T.S. & CHEN, P.S. The Austin Flints and their contribution to medicine and hepatology. *Surgery, Gynecology and Obstetrics*, 1987, **165**, 367–72.
EVANS, A.S. Austin Flint and his contribution to medicine. *Bulletin of the History of Medicine*, 1958, **32**, 224–41,

FLINT, Austin Jr (1836–1915). American physician.

CHEN, T.S. & CHEN, P.S. The Austin Flints and their contribution to medicine and hepatology. *Surgery, Gynecology and Obstetrics*, 1987, **165**, 367–72.

FLOREY, Howard Walter (1898–1968). Australian pathologist; professor in Oxford; researched on antibacterials, particularly penicillin; shared Nobel Prize (Physiology or Medicine) 1945.

ABRAHAM, E.P. BMFRS, 1971, **17**, 255–302.
BICKEL, L. Rise up to life: a biography of Howard Walter Florey who gave penicillin to the world. London, Angus & Robertson, 1972. 314 pp.
FENNER, F. DSB, 1972, **5**, 41–4.
MACFARLANE, G. Howard Florey: the making of a great scientist. Oxford, Oxford Univ. Press, 1979. 396 pp.
WILLIAMS. T.I. Howard Florey: penicillin and after. Oxford, Oxford Univ. Press, 1984. 404 pp.
Archival material: Royal Society of London.

FLORKIN, Marcel (1900–1979). Belgian biochemist and historian of biochemistry; professor at Liège University.

FLORKIN, M. The call of comparative biochemistry. *Comparative Biochemistry and Physiology*, 1973, **44**, 1–10.
SCHOFFINIELS, E. Marcel Florkin, founding father of comparative biochemistry. *Comparative Biochemistry and Physiology*, 1980, **67B**, 353–8.

FLOURENS, Marie-Jean-Pierre (1794–1867). French physiologist.

KRUTA, V. DSB, 1972, **5**, 44–5.

LEGÉE, G. Pierre Flourens, 1794–1867: physiologiste et historien des sciences: sa place dans l'evolution de la physiologie expérimentale. 2 vols. Abbeville, F. Paillart, 1992. 621 pp.

FLOYER, John (1649–1734). English physician; pioneer in instrumental diagnosis.

SMERDON, G. Four seventeenth-century Oxford medical eccentrics. In: Dewhurst, K. (ed.) Oxford medicine. Oxford, [1970], pp. 14–22.

TOWNSEND, G.L. Sir John Floyer (1649–1734) and his study of pulse and respiration. *Journal of the History of Medicine*, 1967, **22**, 286–316.

FLUDD, Robert (1574–1637). English physician, Rosicrucian and mystic philosopher.

GODWIN, J. Robert Fludd: hermetic philosopher and surveyor of two worlds. London, Thames & Hudson, 1979. 96 pp.

HUFFMAN, W.H. Robert Fludd and the end of the Renaissance. London, Routledge, 1988. 252 pp.

HUTIN, S. Robert Fludd (1574–1637): alchimiste et philosophe rosicrucien. Paris, Omnium Littéraire, 1971. 174 pp.

FLÜCKIGER, Friedrich August (1828–1894). German pharmacist.

HAUG, T. Friedrich August Flückiger (1828–1894): Leben und Werk. Stuttgart, Deutscher Apotheker Verlag, 1985. 405 pp.

FLÜGGE, Carl (1847–1923). German microbiologist; professor of hygiene, Breslau.

HORN, H. & THOM, W. Carl Flügge (1847–1923): Integrator der Hygiene. Wiesbaden, MHP-Verlag GmbH, 1992. 74 pp.

FLYNN, John (1880–1951). Australian physician, notable for his work in the Flying Doctor Service.

McPHEAT. W.S. John Flynn, apostle to the Inland. London, Hodder & Stoughton, 1963. 286 pp.

NORMAN, M. Flying doctor. New York, Day, 1961. 191 pp.

FOERSTER, Otfrid (1873–1941). German neurosurgeon.

ZÜLCH, C.J. Otfrid Foerster: physician and naturalist. Translated from the German by A. Rosenauer and J.P. Evans. Berlin, Springer, 1969. 96 pp. Paperback English translation of original, published in 1966.

FOIX, Charles (1882–1927). French neurologist.

HILLEMAND, P. Charles Foix et son oeuvre 1882–1927. *Clio Medica*, 1976, **11**, 269–87.

FOLCH–PI, Jordi (1911–1979). Spanish/American biochemist.

DEBUS, A.G. DSB, 1972, **5**, 47–9.

LEES, M.B. & POPE, A. BMNAS, 2001, **79**, 135–56.

FOLIN, Otto Knut Olof (1867–1934). Swedish-born/American biochemist; professor at Harvard University.

LEICESTER, H.M. DSB, 1972, **5**, 53.

MEITES, S. Otto Folin: America's first clinical biochemist. Washington, DC, American Association for Clinical Biochemistry, 1989. 428 pp.
SHAFFER, P.A. BMNAS, 1952, **27**, 47–82.

FOLKERS, Karl August (1906–1997). American biochemist.

SHIVE, W. BMNAS, 2002, **81**, 100–114.

FOLLEY, Sydney John (1906–1970). British endocrinologist.

PARKES, A.S. BMFRS, 1972, **18**, 241–65.

FØLLING, Ivan Asbjørn (1888–1973). Norwegian physician; first to describe phenylketonuria.

RINGDAL, N. Kjepper I Vorherres hjul: en bok om Asbjørn Følling og sygdommen fikk hans navn. Oslo, Scanbok, *c.*1992. 182 pp.

FONTANA, Felice (1730–1805). Italian biologist and neurologist.

BELLONI, L. DSB, 1972, **5**, 55–7.
KNOEFEL, P.K. Felice Fontana. Life and works. Trento, Società di Studi Trentini di Scienze Storiche, 1984. 422 pp.
KNOEFEL, P.K. Felice Fontana 1730–1805. An annotated bibliography. Trento, Società di Studi Trentini di Scienze Storiche, 1980. 133 pp.

FONZI, Guiseppangelo (1768–1840). Italian dentist in Paris; introduced porcelain teeth.

GUERINI, V. The life and works of Guiseppangelo Fonzi. Philadelphia, Lea & Febiger, 1925. 136 pp.

FORBES, Alexander (1882–1965). American neurophysiologist; professor at Harvard University.

DAVIS, H. DSB, 1972, **5**, 64–6.
FENN, W.O. BMNAS, 1969, **40**, 113–41.

FORD, Charles Edmund (1912–1999). British geneticist.

LYON, M.F. BMFRS, 2001, **47**, 191–201.

FORD, Edmund Brisco (1901–1988). British geneticist.

CLARKE, B.C. BMFRS, 1995, **41**, 144–68.
Archival material: Bodleian Library, Oxford.

FOREEST [Forestus] Pieter van (1521–1597). Dutch physician; professor at Leyden; called 'the Dutch Hippocrates'.

BOSMAN-JELGERSMA, H.A. (ed.) Petrus Forestus medicus. Amsterdam, Stichting A.D. & L., 1996. 375 pp.

FOREL, Auguste-Henri (1848–1931). Swiss neurologist and psychiatrist; professor of psychiatry, Zürich.

FOREL, A.H. Out of my life and work. New York, Norton, 1937. 352 pp. First published in German, Zürich, 1935.
MEIER, R. August Forel [1848–1931]; Arzt, Naturforscher, Sozialreformer; eine Ausstellung. Zürich, Universität Bern, *c.*1996. 159 pp.

PILET, P.E. DSB, 1972, **5**, 73–4.

WETTLEY, A. August Forel; ein Arztleben in Zwiespalt seiner Zeit. Salzburg, Otto Müller Verlag, 1953. 223 pp.

Archival material: Bibliothèque Cantonale et Universitaire, Lausanne (Inventaire, 1969).

FORESTIER, Jacques (b. 1890). French physician; introduced lipiodol as contrast medium in positive contrast myelography and bronchography.

JAYMAN, M.I.V. & MENCKES, C.J. Jacques Forestier. *Spine*, 1995, **20**, 111–15.

FORESTUS, Petrus, *see* **FOREEST, Pieter van**

FORLANINI, Carlo (1847–1918). Italian surgeon in Padua; originated artificial pneumothorax.

BOTTERO, A. Carlo Forlanini, inventore del pneumotorace artificiale. Milano, U. Hoepli, 1947. 130 pp.

FORMAN, Simon (1552–1611). English astrologer and quack physician.

COOK, J. Dr Simon Forman: a most notorious physician. London, Chatto & Windus, 2001. 228 pp.

TRAISTER, B.H. The notorious astrological physician of London: works and days of Simon Forman. Chicago, Univ. of Chicago Press, 2001. 250 pp.

FORSSELL, Carl Gustav Abrahamson (Gösta) (1976–1950). Swedish radiologist.

AKERLUND, A. Gösta Forssell, 1876–1950. To the memory of his life and work. With a bibliography. *Acta Radiologica*, 1956, Suppl. 131, 50 pp.

NICOU, G. & KOCK, W. (eds.) Från Gösta Forssell och den Svenska röntgenologiens undomestid. Södertälje, Fingraf Tryckeri, 1984. 84 pp.

FORSSMANN, Werner Theodor Otto (1904–1979). German surgeon; first to perform human cardiac catheterization (on himself); shared Nobel Prize with Cournand 1956.

FORSSMANN, W. Experiments on myself. Memoirs of a surgeon in Germany. Translated by H. Davies. Preface by A. Cournand. New York, St Martin's Press, 1975. 352 pp. First published in German, 1972.

FORSTER, John Cooper (1823–1886). British surgeon; performed first gastrostomy in Britain.

RAFFENSPERGER, J.G. John Cooper Forster and the first text-book of paediatric surgery in the English language. *Guy's Hospital Reports*, 1964, **113**, 172–8.

FOSTER, Michael (1836–1907). British physiologist; founder of the Cambridge School of Physiology.

DALE, H.H. Sir Michael Foster, KCB, FRS, a secretary of the Royal Society. *Notes and Records of the Royal Society of London*, 1964, **19**, 10–32.

GEISON, G.L. DSB, 1972, **5**, 79–84.

GEISON, G.L. Michael Foster and the Cambridge School of Physiology: the scientific enterprise in late Victorian society. Princeton, University Press, 1978, 401 pp.

Archival material: Imperial College, London; Royal Society of London.

FOTHERGILL, John (1712–1780). British physician; a Quaker with scientific and philanthropic interests.

FOTHERGILL, J. Chain of friendship: selected letters of Dr John Fothergill of London, 1735–1780. With introduction and notes by B.C. Corner and C.C. Booth. Cambridge, Mass, Harvard Univ. Press, 1971. 538 pp.

FOX, R.H. Dr John Fothergill and his friends: chapters in eighteenth century life. London, Macmillan, 1919. 434 pp.

FOTHERGILL, William Edward (1865–1926). British gynaecologist.

–. *Journal of Obstetrics & Gynaecology of the British Empire*, 1927, n.s. **34**, 102–6.

FOURCROY, Antoine François de (1755–1809). French chemist and physician.

KERSAINT, G. Antoine François de Fourcroy, 1755–1809, sa vie et son oeuvre. Paris, Museum Nationale d'Histoire Naturelle. 1966. 296 pp.

SMEATON, W.A. Fourcroy, chemist and revolutionary. Cambridge, Heffer, 1962. 288 pp.

SMEATON, W.A. DSB, 1972, **5**, 89–93.

FOURNEAU, Ernest (1872–1949). French chemist.

ROUSE, H. DSB, 1972, **5**, 99–100.

FOURNIER, Jean-Alfred (1832–1914). French syphilologist.

GOUGEROT, H. & BRODIER, L. L'Hôpital Saint-Louis et la clinique d'Alfred Fournier. Paris, J. Peyronnet, 1932. 142 pp.

WAUGH, M.A. Alfred Fournier, 1832–1914; his influence on venereology. *British Journal of Venereal Diseases*, 1974, **50**, 232–6.

FOWLER, George Ryerson (1848–1906). American surgeon; introduced 'Fowler position' to facilitate pelvic drainage, 1900.

PILCHER, L.S. George Ryerson Fowler. *Surgery, Gynecology & Obstetrics*, 1924, **39**, 564–7.

FOX, George Henry (1846–1937). American dermatologist; professor, College of Physicians and Surgeons, Columbia University.

FOX, G. Reminiscences. New York, Medical Life Press, 1926. 248 pp.

– LEADERS in dermatology. George Henry Fox: the man, his work. Palo Alta, CA, Syntex Laboratories, 1967. 67 pp.

FOX, Joseph (1775–1816). British dental surgeon.

COHEN, R.A. Joseph Fox of Guy's (1775–1816): dentist and philanthropist. London, Lindsay Society for the History of Dentistry, 195. 32 pp.

HERSCHFELD, J.J. Joseph Fox: pioneer dental investigator and author of "The natural history of the human teeth". *Bulletin of the History of Dentistry*, 1983, **31**, 101–7.

FOX, William Tilbury (1836–1879). British dermatologist; gave original descriptions of several dermatoses.

DNB, 379–80.

Lancet, 1879, **1**, 865.

FRACASTORO, Girolamo (*c.*1478–1553). Italian physician; first to state germ theory of infection.

BIBLIOGRAPHY OF MEDICAL AND BIOMEDICAL BIOGRAPHY

RIDDELL, W.R. Hieronymus Fracastorius and his poetical and prose works on syphilis. With a full glossary of medical and other terms employed by him. Toronto, Canadian Social Hygiene Council, 1928. 136 pp.

SINGER, C. & SINGER, D. The scientific position of Girolamo Fracastoro [?1478–1533] with especial reference to the source, character and influence of his theory of infection. *Annals of Medical History*, 1917–18, **1**, 1–34.

WYNNE-FINCH, H. Fracastor syphilis; or, The French disease, a poem, with... [a biographical] introduction by J.J. Abraham. London, W. Heinemann, 1935. [Abraham's biography is on pp. 1–39.]

ZANOBIO, B. DSB, 1972, **5**, 104–7.

FRAENKEL, Albert (1848–1916). German physician; confirmed role of pneumococcus in pneumonia.

WEISS, G. Albert Fraenkel, Arzt und Forscher. 2nd ed. Mannheim, C.F. Boehringer, 1964. 75 pp.

FRAENKEL, Bernhard (1836–1911). German laryngologist.

BERNHARD Fränkels (zu) 70. Geburtstag, *Zeitschrift für Tuberkulose*, 1906, **10**, 1–16.
KAYSELING, A. *Tuberkulose-Arbeiten*, 1911, **7**, 437–41.

FRAENKEL, Eugen (1853–1925). German pathologist.

SCHOTTMÜLLER, – . *Münchener Medizinischer Wochenschrift*, 1926, **73**, 620–22.
WOHILL, F. *Verhandlungen der Deutschen Pathologischen Gesellschaft*, 1926, **21**, 466–9.

FRANCESCHETTI, Adolphe (1896–1968). Swiss ophthalmologist.

KLEIN, D. In memory of Professor A. Franceschetti (1896–1968). *Genetic Counseling*, 1990, **1**, 91–5.

FRANCIS, Thomas (1900–1969). American epidemiologist; isolated influenza B virus.

PAUL, J.R. BMNAS, 1974, **44**, 57–110.
MacLEOD, C.M. *et al.* Thomas Francis, Jr., M.D., 1900–1969. *Archives of Environmental Health*, 1970, **21**, 226–75.

FRANCO, Pierre (1500–1561). French surgeon and lithotomist; introduced the operation of suprapubic lithotomy and improved the technique of herniotomy.

BAIL, P. Un chirurgien urologiste du seizième siècle, Pierre Franco. Thèse pour le doctorat en médecine. Paris, Jouve, 1932. 113 pp.

FRANÇOIS-FRANCK, Charles-Emile (1849–1921). French physician; performed experimental valvulotomy (1882) and, with Charcot, studied the excitability of the cerebral cortex.

BIANCHON, H. Nos grands médecins. Paris, 1891, pp. 171–9.

FRANK, Alfred Erich (1884–1957). German physician; related diabetes insipidus to posterior pituitary; first reported essential thrombopenia.

STEINITZ, K. Erich Frank. *Deutsche Medizinische Wochenschrift*, 1957, **82**, 1138–9.

FRANK, Fritz (1856–1923). German gynaecologist; introduced suprasymphyseal transperitoneal caesarean section.

-. *Monatsschrift für Geburtshülfe und Gynäkologie*, 1923–1924, **55**, 217–70.

FRANK, Johann Peter (1745–1821). German physician; 'Father of public hygiene'.

BAUMGARTNER, L. & RAMSEY, E.M. Johann Peter Frank and his "System einer vollständigen medizinischen Polizey". *Annals of Medical History*, 1933, N.S. **5b**, 525–32; 1934, N.S. **6**, 697–715.
BREYER, H. Johann Peter Frank. 'Fürst unter den Arzten Europas'. Leipzig, S.Hirzel, 1983. 184 pp.
BIACH, R. Johann Peter Frank, der Wiener Volkshygieniker. Wien, Verlag Notring der Wissenschaftliche Verbände Österreichs. 1962, 160 pp.
FRANK, J.P. Johann Peter Frank: seine Selbstbiographie. Bern, Hans Huber, 1969. 166 pp.

FRANKL, Viktor Emil (1905–1997). Austrian psychiatrist; professor in Vienna.

DOERING, D. Die Logotherapie Viktor Emil Frankls. Köln, Buchhandlung C.E. Kohlhauer, 1981. 272 pp.
HADRUP, G. Viktor E. Frankl. Copenhagen, Forum, 1979. 81 pp.

FRANKLAND, Percy Faraday (1858–1946). British chemist and bacteriologist; professor of chemistry, Dundee and Birmingham.

GEISON, G.L. DSB, 1972, **5**, 127–9.
ONFRS, 1947, **5**, 697–715.

FRANKLIN, Benjamin (1706–1790). American statesman and scientist.

PEPPER, W. The medical side of Benjamin Franklin. Philadelphia, William J. Campbell, 1911. 122 pp.

FRANKLIN, Edward C. (1928–1982). German/American immunologist.

METZGER, H. BMNAS, 2000, **78**, 44–63.

FRANKLIN, Kenneth James (1897–1966). British physiologist; professor at St Bartholomew's Hospital Medical College, London.

DALY, I. de B. & MACBETH, R.G. BMFRS, 1968, **14**, 223–42.

FRANKLIN, Rosalind Elsie (1920–1958). British physical chemist and molecular biologist.

MADDOX, B. Rosalind Franklin: the dark lady of DNA. London, Harper Collins, 2002. 380 pp.
OLBY, R. DSB, 1972, **5**, 139–42.
SAYRE, A. Rosalind Franklin and DNA. New York, Norton, 1975. 221 pp.

FRAPPIER, Armand (b. 1904). Canadian microbiologist; Director of the Institut de Microbiologie de Montréal

STANKÈ, A. & MORGAN, J.L. Ce combat qui n'en finit plus. Montréal, Editions de l'Homme, 1970. 269 pp.

FRASER, Ian James (1901–1999). Irish surgeon; President, Royal College of Surgeons in Ireland.

FRASER, I. Blood, sweat, and cheers. London, Memoir Club, British Medical Journal, 1989. 150 pp.

FRASER, Thomas Richard (1841–1919). British pharmacologist.

C., J.T. Sir Thomas Richard Fraser, 1841–1919. *Proceedings of the Royal Society of London B*, 1921, **92**, xi–xvii.

FRAZIER, Charles Harrison (1870–1936). American surgeon.

GRANT, F.C. *Annals of Surgery*, 1907, **105**, 638–40.
–. *Archives of Neurology and Psychiatry*, 1936, **36**, 1330–2.

FRED, Edwin Broun (1887–1981). American microbiologist; professor of bacteriology, University of Wisconsin.

BALDWIN, I.L. BMNAS, 1985, **55**, 247–90.
JOHNSON, D.O. Edwin Broun Fred, scientist, administrator, gentleman. Madison, Univ. of Madison Press, 1974. 179 pp.

FREDERICQ, Léon (1851–1935). Belgian physiologist; profesor of physiology, University of Liège.

BACQ, Z.M. *et al.* Un pionnier de physiologie, Léon Fredericq. Paris, Masson, 1953. 232 pp.
FLORKIN, M. DSB, 1972, **5**, 148–50.
FLORKIN, M. L'École Liègoise de Physiologie et son Maître Léon Fredericq (1851–1935) pionnier de la zoologie chimique.Liège, Vaillant-Carmanne, 1979. 211 pp.

FREEMAN, Richard Austin (1862–1943). British physician and writer; practised medicine in Accra and Gravesend, Kent; wrote detective stories with Dr Thorndike as his fictitious investigator.

DONALDSON, N. In search of Dr Thorndike; the story of his creator. Bowling Green, University Popular Press, 1971. 288 pp.

FREIND, John (1675–1728). British physician; physician to Queen Caroline (wife of George II).

HALL, M.B. DSB, 1972, **5**, 156–7.
SCHORR, R. Sir [sic] John Freind (1675–1728) M.D.; a pioneer historian of medicine. *Isis*, 1937, **27**, 453–74.

FRENCH, Thomas Rushmore (1849–1929). American laryngologist; obtained first good photographs of the larynx.

–. *Transactions of the American Laryngological Association*, 1929, **51**, 271–81.

FRERICHS, Friedrich Theodor von (1819–1895). German pathologist; gave first important description of multiple sclerosis, 1849 and first description of progressive familial hepatolenticular degeneration, 1861.

NAUNYN, B. *Archiv für experimentelle Pathologie und Pharmakologie*, 1885, **19**, iii–viii.

FREUD, Anna (1895–1982). Austrian child psychoanalyst; worked in London.

DYER, R. Her father's daughter. The work of Anna Freud. New York & London, Jason Aronson, 1983. 323 pp.

BIBLIOGRAPHY OF MEDICAL AND BIOMEDICAL BIOGRAPHY

PETERS, U.A. Anna Freud: a life dedicated to children. London, Weidenfeld & Nicolson, 1985, 281 pp. First published in German, Frankfurt, Fischer, 1984.
YOUNG-BRUEHL, E. Anna Freud: a biography. London, Macmillan, 1998. 528 pp.

FREUD, Sigmund (1856–1939). Austrian psychologist and neuropathologist; founder of psychoanalysis.

AMACHER, P. DSB, 1972, **5**, 171–81.
CLARK, R.W. Freud: the man and the cause. London, Cape and Weidenfeld & Nicolson, 1980. 652 pp.
FREUD, S. An autobiographical study: authorized translation by J. Strachey. New York, W.W. Norton, 1952. 141 pp. First published 1927.
GAY, P. Freud: a life for our time. London, Dent, 1988. 810 pp.
JONES, E. Sigmund Freud: life and work. 3 vols. London, Hogarth Press, 1953–57.
TANSLEY, A.G. ONFRS, 1940–41, **3**, 247–75.
Archival material: Library of Congress, Washington.

FREUND, Jules Thomas (1890–1960). Hungarian/American immunologist ('Freund's adjuvant').

HUMPHTRY, J.H. Jules Freund, M.D. Budapest. *Lancet*, 1960, **1**, 1031–2.

FREUND. Leopold (1868–1943). Bohemian radiologist; first used X rays for deep irradiation therapy.

WEISS, K. In memoriam Leopold Freund. *Wiener Klinische Wochenschrift*, 1947, **59**, 189–90.

FREUND, Wilhelm Alexander (1833–1918). German surgeon, professor in Berlin; introduced abdominal hysterectomy in treatment of cancer.

FREUND, W.A. Leben und Arbeit. Gedanken und Erfahrungen über Schaffen in der Medizin. Berlin, Springer, 1913. 170 pp.

FREY, Hedwig (1877–1938). Swiss anatomist; professor in Zürich.

PLANGGER-VAVRA, M. Die Anatomen Hedwig Frey (1877–1938); erste Professorin der Universität Zürich. Zürich, Juris, 1988. 76 pp.

FREY, Maximilian Ruppert Franz von (1852–1932). German physiologist.

ROTHSCHUH, K.E. DSB, 1972, **5**, 184–5.

FRICK, George (1793–1870). American ophthalmologist; wrote first American textbook on ophthalmology by a specialist ophthalmologist.

FRICKE, Johann Karl George (1790–1841). German plastic surgeon.

RODEGRA, H. Johann Karl Georg Frick (1790–1841): Wegbereiter einer klinische Chirurgie in Deutschland. Herzogenrath, Murken-Altrogge, 1983. 203 pp.

FRIEDENWALD, Harry (1864–1950). American ophthalmologist and medical historian; professor at University of Maryland School of Medicine and College of Physicians and Surgeons of Baltimore.

FRIEDENWALD, H. A medical soldier of fortune. *Bulletin of the History of Medicine*, 1948, **22**, 416–24.

LEVIN, A.L. Vision: a biography of Harry Friedenwald. Philadelphia, Jewish Publication Society of America, 1964. 469 pp.

FRIEDLÄNDER, Carl (1847–1887). German pathologist; isolated *Klebsiella pneumoniae* ('Friedländer's bacillus').

KOEHLER, W. & MOCHMANN, H. Carl Friedländer (1847–1887) und die Entdeckung des "Pneumonecoccus". *Zeitschrift für ärztliche Fortbildung*, 1987, **81**, 615–18.

FRIEDREICH, Nikolaus (1826–1882). German physician;described paramyoclonus multiplex, 'Friedreich's disease', 'Friedreich's ataxia'; first described acute leukaemia.

KUSSMAUL, A. Nikolaus Friedreich. Erinnerungen. *Deutsches Archiv für Klinische Chirurgie*, 1883, **32**, 191–208.

FRIEDRICH, Walter (1888–1968). German radiotherapist.

SCHIERHORN, E. Walter Friedrich. Leipzig, B.G. Teubner, 1983. 100 pp.

FRIEND, Charlotte (1921–1987). American oncologist.

DIAMOND, L. BMNAS, 1994, **63**, 126–48.

FRISCH, Karl von (1886–1982). Austrian zoologist: shared Nobel Prize (Physiology or Medicine) 1973 for discoveries concerning organization and elicitation of individual and social behaviour patterns.

FRISCH, K. von. A biologist remembers. Translated by L.Gombrich. Oxford, Clarendon Press, 1967. 200 pp. First published in German in 1957.

FRITSCH, Gustav Theodor (1838–1927). German physiologist; demonstrated the electrical excitability of the brain.

CLARKE, E. DSB, 1972, **5**, 195–7.

FRITZSCHE, Christian Friedrich (1851–1938). Swiss physician; with E. Krebs described acromegaly before Pierre Marie.

STOFFEL, G. Christian Friedrich Fritzsche, 1851–1938. Erste Glarner Spitalarzt und Beschreiber der Akromegalie. Zürich, Juris, 1983. 66 pp.

FRÖHLICH, Alfred (1871–1953). Austrian physician and pharmacologist; described dystrophia adiposogenitalis ('Fröhlich's syndrome').

BRÜCKE, F. Alfred Fröhlich.. *Wiener Klinische Wochenschrift*, 1953, **65**, 306–7.

FROMM, Erich (1900–1980). German/American psychoanalyst.

CORTINO, M. & MACCOBY, M. Erich Fromm's contribution to psychoanalysis. Northvale, NJ, Aronson, 1996. 461 pp.
KNAPP, G.P. The art of living: Erich Fromm's life and work. New York, Lang, 1989. 270 pp.

FROMM-REICHMANN, Frieda (1889–1957). German/American psychiatrist.

HORNSTEIN, G.A. To redeem one person is to redeem the world. The life of Frieda Fromm-Reichmann. New York, Free Press, *c.*2000. 478 pp.

BIBLIOGRAPHY OF MEDICAL AND BIOMEDICAL BIOGRAPHY

FROSCH, Paul (1860–1928). German pathologist; with A.J.F. Loeffler showed foot-and-mouth disease to be due to a filter-passing virus.

UHLENHUTH, E. Paul Frosch zum Gedächtnis. *Zeitschrift für Immunitätsforschung und Experimentelle Therapie*, 1928, **58**, i–iv.

FRUTON, Joseph Stewart (b.1912). Polish/American biochemist; historian of biochemistry; professor at Yale University.

FRUTON, J. A skeptical biochemist. Cambridge, Mass., Harvard Univ. Press, 1992. 330 pp.
FRUTON, J.S. Eighty years. New Haven, Epikouros Press, *c.*1994. 346 pp.

FUCHS, Ernst (1851–1930). Austrian ophthalmologist; professor in Vienna.

FUCHS, E. Wie ein Augenarzt die Welt sah: Selbstbiographie und Tagebuchblätter, hrsg. von Adalbert Fuchs. Wien, Urban & Schwarzenberg, 1946. 344 pp.

FUCHS, Leonhart (1501–1566). German physician and botanist; published most comprehensive herbal of 16th century (1542); professor, Munich and Tübingen.

EGERTON, F.N. DSB, 1978, **15**, 160–62.
STÜBLER, E. Leonhart Fuchs: Leben und Werk. München, Verlag der Münchner Drucke, 1928. 135 pp.

FÜERBRINGER, Paul Walter (1849–1930). German physician; demonstrated diagnostic value of spinal puncture.

HIS, W. Paul Fürbringer und das Hauptwerk seines Lebens. *Deutsche Medizinische Wochenschrift*, 1931, **57**, 591–3.

FULTON, John Farquhar (1899–1960). American physiologist and medical historian; professor of physiology, Yale University.

BIBLIOGRAPHY of John Farquhar Fulton. *Journal of the History of Medicine*, 1962, **17**, 51–71.
HOFF, H.E. John Fulton's contribution to neurophysiology. *Journal of the History of Medicine*, 1962, **17**, 16–37.
LeFANU, W. John Fulton's historical and bibliographical work. *Journal of the History of Medicine*, 1962, **17**, 38–50.
MUIRHEAD, A. John Fulton – book collector, humanist, and friend. *Journal of the History of Medicine*, 1962, **17**, 2–15.
WALKER, A.E. DSB, 1972, **5**, 207–8.
Archival material: Library, Yale University; Minnesota Historical Society, St Paul.

FUNK, Casimir (1884–1967). Polish-American biochemist; pioneer in vitamin research; introduced the term 'vitamine', later changed to 'vitamin'.

HARROW, B. Casimir Funk, pioneer in vitamins and hormones. New York, Dodd, Mead, 1955. 209 pp.
IHDE, A.J. DSB, 1972, **5**, 208–9.
Archival material: Library of Congress, Washington.

FUNKE, Otto (1828–1879). German physiological chemist; isolated haemoglobin, 1851.

KRIES, J. von. Gedächtnissrede auf Otto Funke bei dessen akademischer Todtenfeier. Freiburg I. Br., H.M. Poppen, 1881. 32 pp.

136

BIBLIOGRAPHY OF MEDICAL AND BIOMEDICAL BIOGRAPHY

FURCHGOTT, Robert E. (b. 1916). American physician; shared Nobel Prize (Physiology or Medicine), 1998, for discoveries concerning nitrous oxide as a signalling molecule in the cardiovascular system.

Les Prix Nobel en 1998.

FURTH, Jacob (1896–1979). Austro-Hungarian/American oncologist.

WEINHOUSE, S. & FURTH, J.J. BMNAS, 1992, **62**, 167–97.

GADDESDEN, John of, *see* **JOHN OF GADDESDEN**

GADDUM, John Henry (1900–1965). British pharmacologist.

FELDBERG, W. BMFRS, 1967, **13**, 57–77.
Archival material: Royal Society of London; CMAC.

GÄRTNER, August (1848–1934). German bacteriologist and hygienist; discovered *Salmonella enteritidis.*

ABEL, R. August Gärtner zur Vollendung des 80. Lebensjahres – 18. April 1928. *Zentralblatt für Bakteriologie*, Ite Abt. Originale, 1928, **107**, i–xvi.

GAFFKY, Georg Theodor August (1850–1918). German bacteriologist.

ROBINSON, G. DSB, 1972, **5**, 219–20.

GAIRDNER, William Tennant (1824–1907). British physician; Regius Professor of the Practice of Medicine, Glasgow University.
GIBSON, G.A. Life of Sir William Tennant Gairdner, Glasgow, Maclehose, 1912. 817 pp.

GAJDUSEK, Daniel Carleton (b. 1923). American paediatrician; shared Nobel Prize (Physiology or Medicine), 1976, for discoveries concerning new mechanisms for the origin and dissemination of infectious diseases.

Les Prix Nobel en 1976.

GALEN (AD 129/130–199/200). Greek physician; prolific writer and dogmatic teacher.

KUDLIEN, F. & WILSON, L.G. DSB, 1972, **5**, 227–37.
SARTON, G.A. Galen of Pergamon. Lawrence, Univ. Kansas Press, 1954. 112 pp.
SIEGAL, R.E. Galen's system of physiology and medicine. Basel, New York, Karger, 1968. 419 pp. [Contains sections on Galen's life and character, and survey of his treatises.]
TEMKIN, O. Galenism; rise and decline of a medical philosophy. Ithaca, Cornell Univ. Press, 1973. 240 pp.

GALL, Franz Joseph (1758–1828). German neuroanatomist and psychologist; introduced phrenology.

ACKERKNECHT, E.H. & VALLOIS, H.V. Franz Joseph Gall, inventor of phrenology and his collection. Translated from the French by C. St Léon. Madison, Dept. of the History of Medicine, Univ. of Wisconsin Medical School, 1956. 86 pp.
LESKY, E. (ed.) Franz Joseph Gall, 1758–1828. Naturforscher und Anthropologe. Wien, Hüber, 1979. 217 pp.
WEGNER, P.C. Franz Joseph Gall 1758–1828. Studien zu Leben, Werk und Wirkung. Hildesheim, Olms, 1991. 201 pp.

YOUNG, R.M. DSB, 1972, **5**, 250–56.

GALLIE, William Edward (1882–1959). Canadian surgeon, professor at Toronto.

HARRIS, R.I. As I remember him: William Edward Gallie, surgeon, seeker, teacher, friend. *Canadian Journal of Surgery*, 1967, **10**, 235–50.

GALLO, Robert (b.1935). American oncologist and virologist; Head, U.S. National Cancer Institute, Laboratory of Tumor Biology; co-discoverer, eith Luc Montagnier (Institut Pasteur) of AIDS virus.

GALLO, R. Virus hunting. New York, Harper-Collins, 1993. 352 pp.

GALTIER, Pierre Victor (1846–1908). French veterinarian; demonstrated transmissibility of rabies virus from dog to rabbit.

ROBIN, Y. La vie et l'oeuvre de P.V. Galtier, savant et professeur de bactériologie à L'Ecole Nationale Vétérinaire de Lyon. Lyon, Annequin, 1957. 91 pp.
THÉODORIDES, J. Un précurseur de Pasteur: Pierre-Victor Galtier, (1846–1908). *Archives Internationales Claude Bernard*, 1972, No. **2**, 167–71.

GALTON, Francis (1822–1911). British anthropologist; founder of eugenics.

FORREST, D.W. Francis Galton. The life and work of a Victorian genius. London, Paul Elek, 1974. 340 pp.
GRIDGEMAN, N.T. DSB, 1972, **5**, 265–7.
KEYNES, M. (ed.) Sir Francis Galton, FRS: the legacy of his ideas. London, Macmillan, 1993. 237 pp.
PEARSON, K. The life, letters and labours of Francis Galton. 3 vols. (in 4), Cambridge, 1914–30. 1787 pp.
Archival material: University College London.

GALVANI, Luigi (1737–1798). Italian anatomist and physiologist; instituted the study of electrophysiology.

BROWN, T.M. DSB, 1972, **5**, 267–9.
GALVANI, L. Commentary on the effect of electricity on muscular motion (translated by M.G. Foley). Norwalk, Conn., Burndy Library, 1954. 176 pp. Includes bibliography of Galvani's books by J.F. Fulton and M.E. Stanton, and biographical information.
MESINI, C. Nuove ricerche Galvani. Bologna, Tamari Editore, 1971. 176 pp.

GAMALEIA, Nikolay Fyodorovich (1859–1949). Russian microbiologist.

GUTINA, V. DSB, 1972, **5**, 269–71.
MILENUSHKIN, I.I. Nikolai Feodorovich Gamaleia. Moskva, Izdatelstvo Akademii Nauk SSSR, 1954. 157 pp. [in Russian].

GAMBLE, James Lawder (1883–1959). American paediatrician; professor at Harvard Medical School.

LOEB, R.F. BMNAS, 1962, **36**, 146–60.
Archival material: Medical School, Harvard University.

GAMGEE FAMILY. British medical family. **Joseph** (1801–1895) veterinary surgeon. **Sampson** (1828–1886) son of Joseph; surgeon; devised the surgical dressing. **John** (1831–1894) brother of Sampson. **Arthur** (1841–1909) brother of Sampson; physiologist and physician.

KAPADIA, H.M. Sampson Gamgee: a great Birmingham surgeon. *Journal of the Royal Society of Medicine*, 2002, **95**, 96–100.

THOMPSON, R.D'A. The remarkable Gamgees: a story of achievement. Edinburgh, Ramsay Head Press, 1974. 216 pp. [Author is daughter of Sir D'Arcy Thompson, the biologist, grandson of Joseph Gamgee.]

GANN, Thomas William Francis (1867–1938). British physician and archaeologist; authority on Mayan architecture.

DNB.

GARCIA, Manuel Patricio Rodriguez (1805–1906). Spanish teacher of singing; inventor of the laryngoscope.

MACKINLAY, M.S. Garcia, the centenarian and his times. Being a memoir of Garcia's life and labours for the advancement of music and science. Edinburgh, Blackwood, 1908. 335 pp.

GARCIA DA ORTA (*c.*1500–1558). Portuguese physician in Goa; wrote on pharmacology and tropical diseases.

BOXER, C.R. Two pioneers of tropical medicine: Garcia d'Orta and Nicolas Monardes. London, Wellcome Historical Medical Library, 1963. 36 pp.

FICALDO, C. de Garcia da Orta e o seu tempo. Lisboa, Imprensa Nacional, 1886. 392 pp. Facsimile reproduction, Lisboa, Temas Portugueses, 1983. 392 pp.

KAPADIA, H.M. Sampson Gamgee: a great Birmingham surgeon. *Journal of the Royal Society of Medicine*, 2002, **95**, 96–100.

KELLER, A.G. DSB, 1974, **10**, 236–8.

REVISTA da Junta de Investigaçoes do Ultramar, 1963, 11, No.4., pp. 615–875. Garcia da Orta: Numero especial commemorativo do quarto centenàrio de publicação do Coloquios dos Simples.

GARCIA TAPIA, Antonio (1875–1950). Spanish otolaryngologist.

BARAJAS GARCIA, J.M. El doctor Don Antonio Garcia Tapia: su vida y su obra. XV Congreso Internacional de Historia de la Medicina, Madrid, Alcala, 22–29 Sep. 1956. 8 pp.

GARDEN, Alexander (1730–1791). British physician, practised in Charlestown, S. Carolina; introduced Virginia pink root as a vermifuge.

BERKELEY, E. & BERKELEY, D.S. Dr Alexander Garden of Charles Town. Chapel Hill, Univ. of North Carolina Press, 1969. 379 pp.

GARLAND, Joseph (1893–1973). American paediatrician; editor of New England Journal of Medicine 1947–1967.

GARLAND, J. A time for remembering. Boston, New England Journal of Medicine, 1972. 203 pp.

GARNHAM, Percy Cyril Claude (1901–1994). British protozoologist.

LAINSON, R. & KILLICK-KENDRICK, R. BMFRS, 1997, **43**, 171–92.
Archival material: CMAC.

GARRISON, Fielding Hudson (1870–1935). American medical bibliographer, historian and librarian.

KAGAN, S.R. Life and letters of Fielding H. Garrison. Boston, Mass., Medico-Historical Press, 1938. 287 pp.

KAGAN, S.R. Fielding H. Garrison: a biography. Boston, Mass., Medico-Historical Press, 1948. 104 pp.

GARROD FAMILY. Alfred Baring (1819–1907); **Archibald Edward** (1857–1936); British physicians.

RUTZ, C. The Garrods. Zürich, Juris Druck-Verlag, 1970. 34 pp.

GARROD, Archibald Edward (1857–1936). British paediatrician; described overt diseases due to congenital biochemical abnormalities; Regius Professor of Medicine, University of Oxford.

BEARN, A.C. Archibald Garrod and the individuality of man. Oxford, Clarendon Press, 1993. 227 pp.

HOPKINS, F.G. ONFRS, 1938, **2**, 225–8.

WILSON, L.G. DSB, 1990, **17**, 333–6.

GARRY, Robert Campbell (1900–1993). British physiologist; Regius Professor of Physiology, University of Glasgow.

GARRY, R.C. Life in physiology. Edited by D. Smith. Glasgow, Wellcome Unit for the History of Medicine, Univ. of Glasgow, 1992. 183 pp.

Archival material: University of Glasgow, CMAC.

GARRY, Thomas Peter (1884–1963). Irish anatomist.

GARRY, J.D. A Dublin anatomist. Tom Garry. An account of the life of Thomas Peter Garry, tutor and prosector in anatomy, Royal College of Surgeons in Ireland. Dublin, Black Cat Press, 1984. 94 pp.

GARTH, Samuel (1661–1719). British physician; physician to George I; Physician-General to the British Army; poet.

COOK, R.I. Sir Samuel Garth. Boston, Twayne Publishers, 1980. 172 pp.

SENA, J.F. The best-natured man: Sir Samuel Garth, physician and poet. New York, AMS Press, 1985. 215 pp.

GASKELL, Walter Holbrook (1847–1914). British morphologist and physiologist.

GEISON, G.L. DSB, 1972, **5**, 279–84.

GASSENDI, Pierre (1592–1655). Dutch anatomist.

BRUNDELL, B. Pierre Gassendi – from Aristotelianism to a new natural philosophy. Dordrecht, D. Reidel, 1987. 251 pp.

JONES, H. Pierre Gassendi 1592–1655: an intellectual biography. Niewkoop, B. de Graaf, 1981. 320 pp.

JOY, L.S. Gassendi the anatomist, advocate of history in an age of science. Cambridge, Univ. Press, 1988. 350 pp.

GASSER, Herbert Spencer (1888–1963). American physiologist; Director, Rockefeller Institute for Medical Research; shared Nobel Prize 1944 for work on functional differentiation of nerve fibres.

BIBLIOGRAPHY OF MEDICAL AND BIOMEDICAL BIOGRAPHY

ADRIAN, E.D. BMFRS, 1964, **10**, 75–82.
CHASE, M.W. & HUNT, C.C. BMNAS, 1995, **67**, 147–77.
HERBERT SPENCER GASSER 1888–1963. Scholar – administrator – Nobel laureate. An autobiographical memoir of a distinguished career in medical science. *Experimental Neurology*, 1964, Suppl. 1, i–vii, 1–38.
LLOYD, D.P.C. DSB, 1972, **5**, 290–91.
Archival material: Rockefeller Archives Center, New York.

GATES, Reginald Ruggles (1882–1962). Canadian geneticist.

ROBERTS, J.A.F. BMFRS, 1964, **10**, 83–106.

GAUCHER, Philippe Charles Ernest (1854–1918). French dermatologist and syphilologist; described familial splenic anaemia ('Gaucher's disease') 1882.

FIAUX, L. Ernest Gaucher. Paris, 1919.

GAUTIER, Emile Justin Armand (1837–1920). French physician and chemist.

LEBON, E. Armand Gautier: biographie, bibliographie analytique des écrits. Paris, Gaulthier-Villars, 1912. 96 pp.

GAUTIER D'AGOTY, Jacques-Fabien (1717–1786). French anatomist.

LOWE, A. ed. Jacques-Fabien Gautier d'Agoty: the anatomical prints, de couleur et grandeur naturelles. Tokyo, Lampoon Press, 1996. 48 pp.

GAY, Frederick Parker (1874–1939). American bacteriologist and pathologist.

DOCHEZ, A.R. BMNAS, 1954, **38**, 99–116.
SAFFRON, M.H. DSB, 1972, **5**, 316–17.

GAYET, Charles Jules Alphonse (1835–1904). French ophthalmologist.

AURAND, I. *Archives d'Ophtalmologie*, 1904, **24**, 629–51.
FRENKEL, H. *Annales d'Oculistique*, 1904, **132**, 161–73.

GEGENBAUR, Carl (1826–1903). German comparative anatomist.

COLEMAN, W. DSB, 1978, **15**, 165–71.

GÉLINEAU, Jean Baptiste Edouard (1828–1906). French physician; first described narcolepsy.

PASSOUANT, P. Le docteur Gélineau (1828–1906). *Histoire des Sciences Médicales*, 1981, **15**, 137–43.

GENGOU, Octave (1875–1957). Belgian bacteriologist; with Bordet devised complement-fixation reaction and discovered *Bordetella pertussis*.

MILLET. M. Eloge académique de Professeur Octave Gengou (1875–1957). *Memoires de l'Académie Royale de Médecine de Belgique*, 1969, **7**, 79–88.

GEOFFROY, Etienne François (1672–1731). French physician and chemist; lecturer on materia medica.

SMEATON, W.A. DSB, 1972, **5**, 352–4.

141

BIBLIOGRAPHY OF MEDICAL AND BIOMEDICAL BIOGRAPHY

GERARD, John (1545–1612). English surgeon and herbalist to James I.

JEFFERS, R.H. The friends of John Gerard (1545–1612), surgeon and botanist. Falls Village, Conn., Herb Grower Press, 1967. 99 pp. Biographical appendix, 1969. 47 pp.
PHELPS, W.H. John Gerard, the herbalist. *Library* (London), 1980, **2**, 76–85.
STEARN, W.T. DSB, 1972, **5**, 361–3.

GERARD, Ralph Waldo (1900–1974). American neurophysiologist.

KETY, S.S. BMNAS, 1982, **53**, 179–210.
NILAN, R-L. Ralph Waldo Gerard, citizen of science: a guide to the Gerard microfiche collection, 1927–1975. Chicago, Univ. of Chicago, 1976. 33 pp.
Archival material: University of California, Irvine.

GERBEC, Markus (1658–1718). Slovene physician; gave early (1692) description of Stokes-Adams syndrome.

GRMEK, M.D. DSB, 1972, **5**, 366–7.
TARTALA, H. Der slowenische Arzt Dr Markus Gerbec (1658–1718); Ein Vorgänger der Fermentationslehre. Internationale Pharmaziegeschichtlichen Kongress, Rotterdam, 1963. Stuttgart, Wissenschaftliche Verlagsgesellschaft, 1965, **26**, 173–82.

GERBEZIUS, Marcus, *see* **GERBEC, Markus**

GERHARD, William Wood (1807–1872). American physician; differentiated typhus from typhoid.

MIDDLETON, W.S. William Wood Gerhard (1807–1872). *Annals of Medical History*, 1935, **7**, 1–18.

GERLACH, Joseph von (1820–1896). German histologist.

ADHAMI, H. Das Erlanger anatom Joseph von Gerlach und seine Bedeutung für die Neurohistologie. *Anatomischer Anzeiger*, 1974, **135**, 277–87.
CONN, H.J. Development of histological staining. *Ciba Symposia*, 1945–6, **7**, 270–300.

GESCHWIND, Norman (1926–1984). American neurologist and behavioural scientist.

SCHACHTER, S.C. & DEVINSKY, O. (eds.) Behavioral neurology and the legacy of Norman Geschwind. Philadelphia, Lippincott-Raven, 1997. 304 pp.

GESELL, Arnold Lucius (1880–1961). American child psychologist; professor at Yale University.

GESELL, A.L. Autobiography. In: Murchison, C. History of psychology in autobiography. Worcester, Mass., Clark Univ. Press, Vol. 4, 1952, pp. 123–42.
MILES, W.R. BMNAS, 1964, **37**, 55–96.
RADBILL, S.K. DSB, 1972, **5**, 377–8.

GESNER, Conrad (1516–1565). Swiss physician, scientist, bibliographer; professor of natural history at Zürich.

FISCHER, H. *et al.* Conrad Gesner, 1516–1565, Universalgelehrter, Naturforscher, Arzt. Zürich, Orell Füssli Verlag, 1967. 234 pp. [Nine papers.]
PILET, P.H. DSB, 1972, **5**, 378–9.

142

WELLISCH, H. Conrad Gesner: a bio-bibliography. Zug, IDC, 1984. 145 pp.

GESSARD, Carle (1850–1925). French bacteriologist; isolated *Ps. aeruginosa (pyocyanea)*.

NAUROY, J. Le carrière militaire et l'oeuvre scientifique de Carle Gessard (1850–1925) *Revue de l'Histoire de la Pharmacie*, 1976, **23**, 175–80.

GHEDINI, Giovanni (1877–1959). Italian physician; introduced bone marrow biopsy.

GELMETTI, P. Giovanni Ghedini e la biopsia medollare. *Minerva Medica*, 1969, **60**, 2963–80.

GHON, Anton (1866–1936). Austrian bacteriologist.

OBER, W.P. Ghon but not forgotten: Anton Ghon and his complex. *Pathology Annual*, 1983, **18** (2), 71–85.

GIANTURCO, Cesare (1905–1995). Italian/American radiologist; pioneer in radio-cinematography.

FISH, R.D. In memoriam: Cesare Gianturco (1905–1995). *Circulation*, 1996, **93**, 1938–9.
WALLACE, S. Cesare Gianturco (1905–1995): a legend in his own time. *Cardiovascular and Interventional Radiology*, 1996, **19**, 59–81.

GIBBON, John Heysham (1903–1973). American cardiovascular surgeon; introduced heart-lung oxygenator.

ROMANE-DAVIS, A. John Gibbon and his heart-lung machine. Philadelphia, Univ. Pennsylvania Press, 1991. 252 pp.
SHUMACHER, H.B. A dream of the heart. The life of John H. Gibbon, Jr.
SHUMACKER, H.B. BMNAS, 1982, **53**, 213–47.

GIBERT, Camille Melchior (1797–1860). French dermatologist; established pityriasis rosea as a clinical entity.

BEESON, B.B. *Archives of Dermatology and Syphilology*, 1934, **30**, 101–3.

GIBSON, William (1788–1868). American surgeon; successfully ligated the common iliac artery.

GERSTER, J.C.A. *Surgery, Gynecology & Obstetrics*, 1931, **52**, 122–4.

GIERKE, Edgar Konrad Otto von (1877–1945). German pathologist; described glycogen storage disease.

BÖHMIG, R. Edgar von Gierke 9.2.1877–21.10.1945. *Verhandlungen der Deutschen Pathologischen Gesellschaft*, 1951, **34**, Suppl., 17–19.

GIGLI, Leonardo (1863–1908). Italian surgeon; introduced a saw used for craniotomy and pubiotomy.

BRUNORI, A. *et al.* Celebrating the centenary (1894–1994): Leonardo Gigli and his wire saw. *Journal of Neurosurgery*, 1995, **82**, 1086–90.

GILBERT, William (1544-1603). English physician and scientist; wrote *De magnete*, 1600.

THOMPSON, S.P. Gilbert, physician; a note prepared for the three hundredth anniversary of the death of William Gilbert, President of the Royal College of Physicians and physician to Queen Elizabeth. London, Chiswick Press, 1903. 31 pp.

SINGER, C. Sir William Gilbert. *Journal of the Royal Naval Medical Service*, 1916, **2**, 494–510.

GILBERTUS ANGLICUS (*fl.* 1245). English exponent of Anglo-Norman medicine. Wrote *Compendium medicinae*.

HANDERSON, H.E. Gilbertus Anglicus. Medicine of the thirteenth century; with a biography of the author. Cleveland, Ohio, Cleveland Medical Library Association, 1918. 77 pp.

GILLES DE CORBEIL [Aegidus Corboliensis] (1165–1213). Physician to Philippe Auguste of France; canon of Nôtre Dame, Paris.

D'IRSAY, S. The life and times of Gilles de Corbeil. *Annals of Medical History*, 1925, **7**, 362–78.

VEILLARD, C. Essai sur la société médicale et religieuse au XIIe siècle, Gilles de Corbeil. Paris, H.Champion, 1908. 456 pp.

GILLES DE LA TOURETTE, Georges Albert Édouard Brutus (1857–1904). French neurologist.

LE GENDRE, L.P. Gilles de la Tourette. Paris, 1905. 53 pp. (*Bulletin et Mémoires de la Société Médicale des Hôpitaux de Paris*, 1904, **21**, 1298–311.)

GILLIAM, David Tod (1844–1923). American gynaecologist.

ZOLLINGER, R.M. David Tod Gilliam. *Surgery, Gynecology and Obstetrics*, 1930, **51**, 873–5.

GILLIES, Harold Delf (1882–1960). New Zealand plastic surgeon.

POUND, R. Gillies, surgeon extra-ordinary: a biography. London, Michael Joseph, 1964. 264 pp.

GILMAN, Albert G. (b. 1941). American biochemist; shared Nobel Prize (Physiology or Medicine), 1994, for discovery of G-proteins and their role in signal transduction in cells.

Les Prix Nobel en 1994.

GILMAN, Alfred (1908–1984). American pharmacologist.

RITCHIE, M. BMNAS, 1996, **70**, 59–80.

GIMBERNAT Y ARBOS, Antonio (1734–1815). Spanish surgeon; wrote on hernia.

GABARRÓ I GARCIA, P. (ed.) Tres traballos trenicati en el homenatge Gimbernat. Barcelona, Laboratorios del Nord d'Espanya, 1936. 213 pp.

MATHESON, N.M. Antonio de Gimbernat, (1734–1816). *Proceedings of the Royal Society of Medicine*, 1949, **42**, 407–10.

GIRARD, Charles (1850–1916). Swiss surgeon; performed first successful hindquarter amputation.

–. *Correspondenz-Blatt für Schweizer Ärzte*, 1916, **46**, 561–4.

GIRDLESTONE, Gathorne Robert (1881–1950). British orthopaedic surgeon.

TRUETA, J. Gathorne Robert Girdlestone. London, Oxford. Univ. Press, 1971. 101 pp.

GLASER, Johann Heinrich (1629–1679). Swiss physician.

PILET, P.E. DSB, 1972, **5**, 418–19.

GLAUBER, Johann Rudolf (1604–1670). German scientist; discovered hydrated sodium sulphate ('Glauber's salt').

GUGEL, K.F. Johann Rudolf Glauber, 1604–1670: Leben und Werk. Würzburg, Freunde Mainfränkischer Kunst und Geschichte, 1955. 71 pp.

GLENNY, Alexander Thomas (1882–1965). British bacteriologist and immunologist.

OAKLEY, C.L. BMFRS, 1966, **12**, 163–80.

GLEY, Marcel-Eugène-Emil (1857–1930). French physiologist; professor of general biology, Collège de France.

GRMEK, M.D. DSB, 1990, **17**, 347–8.

GLISSON, Francis (1597–1677). English physician; Regius Professor of Physic, Cambridge; gave first accurate account of rickets.

TEMKIN, O. DSB, 1972, **5**, 425–7.
WALKER, R.M. Francis Glisson. In: Rook, A. (ed.) Cambridge and its contribution to medicine. London, Wellcome Institute for the History of Medicine, 1971. pp. 35–47.
Archival material: British Library; Royal College of Physicians of London.

GLUCK, Themistokles (1853–1942). German surgeon; chief, Kaiser und Kaiserin Friedrich Krankenhaus, Berlin.

GROTE, 1927, **6**, 89–140 (CB).

GODLEE, Rickman John (1849–1925). British surgeon.

PLARR, 446–50

GODMAN, John Davidson (1794–1830). American anatomist; professor at Rutgers Medical College; professor of surgery, Medical College of Ohio.

MILLER, W.S. John Davidson Godman. *Annals of Medical History*, 1937, N.S., **9**, 293–303.

GOEBEL, Walther Frederick (1899–1993). American biochemist.

McCARTY, M. BMNAS, 1999, **77**, 96–107.

GOETHE, Johann Wolfgang von (1749–1832). German poet and scientist.

MAGNUS, R. Goethe as a scientist. New York, Schuman, 1949. 249 pp. First published in German, 1906.

GOGARTY, Oliver St John (1878–1957). Irish surgeon, writer and poet.

LYONS, J.B. Oliver St John Gogarty. Lewisburg, Pa., Bucknell Univ. Press, 1976. 89 pp.
O'CONNOR, U. Oliver St John Gogarty: a poet of his times. Dublin, O'Brien Press, 2000. 320 pp. First published 1964.

BIBLIOGRAPHY OF MEDICAL AND BIOMEDICAL BIOGRAPHY

GOLDBERGER, Joseph (1874–1929). Austrian/American epidemiologist; discovered cause of pellagra.

BUCHMAN, D.D. The Sherlock Holmes of medicine: Dr Joseph Goldberger. New York, Messner, [c. 1969]. 189 pp.
PARSONS, R.P. Trail to light. A biography of Joseph Goldberger. Indianapolis, Bobbs Merrill, 1934. 353 pp.
ROSENBERG, C. DSB, 1972, **5**, 451–3.
Archival material: Library, University of North Carolina; Library, Vanderbilt University Medical Center.

GOLDBLATT, Harry (1891–1977). American physician; carried out valuable studies on hypertension.

LARAGH, J. Harry Goldblatt (1891–1977). *Transactions of the Association of American Physicians*, 1978, **91**, 24–7.

GOLDING, Benjamin (1793–1863). British physician; founder of Charing Cross Hospital, London.

MINNEY, R.J. The two pillars of Charing Cross. The story of a famous hospital. London, Cassell, 1967. 237 pp.

GOLDMANN, Franz (1895–1970). German/American physician and social hygienist.

ANTONI, C. Sozialhygiene und public health: Franz Goldmann (1895–1970) Husum, Matthiesen Verlag, 1997. 363 pp.

GOLDSCHMIDT, Richard Benedict (1878–1958). German/American geneticist.

GOLDSCHMIDT, R.B. In and out of the ivory tower. The autobiography of Richard B. Goldschmidt. Seattle, Univ. of Washington Press, 1960. 352 pp.
PITERNICK, L.K. Richard Goldschmidt,, controversial geneticist and creative biologist. Basel, Birkhäuser, 1980. 153 pp. (*Experientia*, 1980, Suppl. **35**, 1–153.)
STERN, C. BMNAS, 1967, **39**, 141–92.
Archival material: Bancroft Library, University of California, Berkeley.

GOLDSMITH, Oliver (1730–1774). Irish physician, essayist and poet.

GINGER, J. The notable man: the life and times of Oliver Goldsmith. London, Hamish Hamilton, 1977. 408 pp.
SELLS, A. Oliver Goldsmith: his life and works. London, Allen & Unwin, 1974. 423 pp.

GOLDSTEIN, Joseph Leonard (b. 1940). American physician; shared Nobel Prize (Physiology or Medicine), 1985, for discoveries concerning the regulation of cholesterol metabolism.
Les Prix Nobel en 1985.

GOLGI, Camillo (1843–1926). Italian histologist and pathologist; shared Nobel Prize, 1906.

MAZZARELLO, P. The hidden structure: a scientific biography of Camillo Golgi. Oxford, Oxford University Press, 1999. 407 pp. First published in Italian, 1996.
ZANOBIO, B. DSB, 1972, **5**, 459–61.

GOLL, Friedrich (1829–1903). Swiss physician and physiologist.

LEBRAN, C. Friedrich Goll 1829–1903 als Physiologe und Praktiker. Zürich, Juris, 1971. 64 pp.

BIBLIOGRAPHY OF MEDICAL AND BIOMEDICAL BIOGRAPHY

GOLLA, Frederick Lucien (1878–1968). British neurologist; demonstrated electro-encephalographic changes in epilepsy.

BIRD, J.M. The father of psychophysiology: Professor F.L. Golla and the Burden Neurological Institute. In: 150 Years of psychiatry. London, Athlone Press, 1996, vol. 2, pp. 500–516.

GOLTZ, Friedrich Leopold (1834–1902). Polish physiologist.

EWALD, J.R. Nachruf. *Archiv für die gesamte Physiologie*, 1903, **94**, 1–64.
ROTHSCHUH, K.E. DSB, 1972, **5**, 462–4.

GONIN, Jules (1870–1935). Swiss ophthalmic surgeon.

HUMPF, J. Jules Gonin, inventor of the surgical treatment for retinal detachment. *Survey of Ophthalmology*, 1976, **21**, 276–84.

GOOD, John Mason (1764–1827). British physician and linguist – with good knowledge of 13 languages.

GREGORY, O. Memoirs of the life, writings and character, literary, professional and religious, of the late John Mason Good, M.D. London, H. Fisher, 1828. 472 pp.

GOODPASTURE, Ernest William (1886–1960). American pathologist and virologist; pioneer in virus culture methods.

LONG, E.R. BMNAS, 1965, **38**, 111–44.

GOODSIR, John (1814–1867). British anatomist; professor in Edinburgh.

HEPPELL, D. DSB, 1972, **5**, 469–71.

GORDON, Alexander (1752–1799). British obstetrician in Aberdeen; pioneer investigator into contagiousness of puerperal fever.

PORTER, I. A. Alexander Gordon, M.D., of Aberdeen, 1752–1799. Edinburgh, London, Oliver & Boyd, 1958. 92 pp.

GORDON, Mervyn Henry (1872–1953). British bacteriologist.

GARROD, L.P. ONFRS, 1954, **9**, 153–63.

GORDON-TAYLOR, William Gordon (1878–1960). British surgeon.

HOBSLEY, M. Sir Gordon Gordon-Taylor; a biography and an appreciation. *Journal of Medical Biography*, 1993, **1**, 83–9.
WINDEYER, B. Sir Gordon Gordon-Taylor, surgeon, ambassador, scholar and orator. *Annals of the Royal College of Surgeons of England*, 1965, **36**, 98–115.

GORER, Peter Alfred (1907–1961). British immunologist and geneticist; worked on serological and genetic basis of tissue transplantation.

MEDAWAR, P.B. BMFRS, 1961, **7**, 95–109.

GORGAS, William Crawford (1854–1920). American army surgeon and investigator of sanitation and hygiene; Surgeon General, US Army; investigated spread of yellow fever.

GIBSON, J.M. Physician to the world: the life of General William C. Gorgas. Durham, N.C., Duke Univ. Press, 1950. 315 pp. Reprint with historical introduction by S.W. Wiggins. Tuscaloosa, Univ. Alabama Press, 1989.

GORGAS, M.C. & HENDRICK, B.J. William Crawford Gorgas. Garden City, N.Y., Doubleday Page, 1924. 359 pp.

GORINI, Luigi (1903–1976). American microbial geneticist; professor at Harvard Medical School.

BECKWITH, J. & FRAENKEL, D. BMNAS, 1980, **52**, 203–21.

GORRIE, John (1803–1855). American physician and inventor, born Nevis, British W. Indies; introduced methods of cooling fever patients.

SHERLOCK, V.M. The fever man: a biography of Dr John Gorrie. Tallahassee, Medallion Press, 1982. 152 pp.

GOSIO, Bartolomeo (1863–1944). Italian bacteriologist; first to record the antibacterial effect of a penicillin (from *Penicillium glaucum*).

FERMI, C. *Rivista di Malariologia*, 1944, **23**, 79–82.
JERACI, F. *Rivista di Malariologia*, 1944, **24**, 221–2.

GOTCH, Francis (1853–1913). British neurophysiologist; recorded first correct electroretinograms.

–. Francis Gotch, D.Sc. Oxon & Liverp., LL.D. St And, F.R.S. Wayneflete Professor of Physiology, University of Oxford. *Lancet*, 1913, **2**, 347–51.

GOTTSCHALK, Carl William (1922–1997). American renal physiologist.

BURG, M.B. BMNAS, 1999, **77**, 122–41.

GOUGEROT, Henri Eugène (1881–1955). French dermatologist.

TOURAINE, H. *Bulletin de l'Académie Nationale de Médecine*, Paris, 1955, **129**, 132–5.

GOULD, George Milbry (1848–1922). American ophthalmologist and lexicographer.

WANNARKA, M.B. Dr George Milbry Gould: ophthalmologist and first president of the Medical Library Association. *Bulletin of the History of Medicine*, 1968, **42**, 265–71.

GOWERS, Sir William Richard (1845-1915). British neurologist; physician to National Hospital, London.

CRITCHLEY, M. Sir William Gowers, 1845–1915: a biographical appreciation. London, Heinemann, 1949. 118 pp.

GOYANES CAPDEVILA, José (1876–1964). Spanish surgeon.

BARROS, J.L. Investigaciones sobre los trabajos vasculares del Dr José Goyanes Capdevila. *Cirugiá, Ginecologiá y Urologiá*, 1965, **19**, 1–26.

GRAAF, Regnier de (1641–1673). Dutch physician and physiologist; published pioneer works on the ovary and pancreas.

CATCHPOLE, H.R. Regnier de Graaf 1641–1673. *Bulletin of the History of Medicine*, 1940, **8**, 1261–300.

LINDEBOOM, G.A. Reinier de Graaf. Leven en werken. Delft, Elmar, 1973. 143 pp.

KLEIN, M. DSB, 1972, **5**, 484–5.

GRABAR, Pierre (1892–1986). French chemist; introduced immunoelectrophoresis.

COURTOIS, J.E. Éloge de Pierre Grabar. *Bulletin de l'Académie Nationale de Médecine*, 1986, **170**, 635–9.

GRACE, William Gilbert (1843–1915). British general practitioner in Bristol and famous cricketet.

MIDWINTER, E. W.G. Grace: his life and times. London, Allen & Unwin, 1981. 175 pp.

RAE, S. W.G. Grace: a life. London, Faber & Faber, 1998. 548 pp.

GRADENIGO, Giuseppe (1859–1926). Italian otologist; recorded 'Gradenigo's syndrome' – acute otitis media followed by abductor paralysis, 1904.

BRUZZI, –. *Archivio Italiano di Otologia*, 1926, **37**, 111–17.

S., F. *Archivio Italiano di Otologia*, 1930, **41**, 5–40.

GRAEFE, Alfred Carl (1830–1909). German ophthalmologist; with E.T. Saemisch edited the monumental *Handbuch der gesamten Augenheilkunde*.

SHASTID, H. *American Encyclopaedia and Dictionary of Ophthalmology*, Chicago, 1915, **7**, 5524–626.

GRAEFE, Carl Ferdinand von (1787–1840). German plastic surgeon; professor of surgery, Berlin.

CARL FERDINAND GRAEFE, der Vater. *Deutsches Archiv für Geschichte der Medizin*, 1883, **6**, 305–81.

ROHLFS, H. Geschichte der deutschen Medizin. Leipzig, 1883, vol. 3, 247–324.

ULLMAN, E.V. Albrecht von Graefe: the man in his time. *American Journal of Ophthalmology*, 1954, **38**, 525–43, 695–711, 791–809.

GRAEFE, Friedrich Wilhelm Ernst Albrecht von (1828–1870). German ophthalmic surgeon; one of the founders of modern ophthalmology.

HEYNOLD von GRAEFE, B. Albrecht von Graefe: ein Leben für das Licht. München, Karl Thiemig, 1959. 111 pp.

MÜNCHOW, W. Albrecht von Graefe, Leipzig, B.G. Teubner, 1978. 92 pp.

GRAHAM, Clarence Henry (1906–1971). American experimental psychologist; worked on vision and visual perception.

RIGGS, L.A. BMNAS, 1975, **46**, 71–89.

GRAHAM, Duncan Archibald (1882–1974). Canadian physician; professor of medicine, University of Toronto.

KERR, R.B. & WAUGH, D. Duncan Graham: medical reformer and educator. Toronto, Hannah Institute and Dundurn Press, 1989. 121 pp.

GRAHAM, Evarts Ambrose (1883–1957). American surgeon.

DRAGSTEDT, L.R. BMNAS, 1976, **48**, 221–50.
Archival material: Library, Washington University School of Medicine, St Louis.

GRAM, Hans Christian Joachim (1853–1938). Danish physician; introduced microbiological staining method.

SNORRASON, E. DSB, 1972, **5**, 495–6.

GRANCHER, Jacques Joseph (1843–1907). French phthisiologist.

ROUSILLAT, J. Un patron des hôpitaux de Paris à la Belle Epoque: la vie de Joseph Grancher. Guérat, Société des Sciences Naturelles et Archéologiques de la Creuse, 1989. 133 pp.

GRANIT, Ragnar Arthur (1900–1991). Finnish neurophysiologist; shared Nobel Prize, 1967.

GRILLNER, S. BMFRS, 1995, **41**, 184–97.

GRANT, John Charles Boileau (1886–1973). Canadian anatomist; professor at universities of Montana and Toronto.

ROBINSON, C.L.N. J.C. Boileau Grant, anatomist extraordinary. Markham, Ont., Hannah Institute for the History of Medicine, 1993. 160 pp.

GRANT, Ronald Thomson (1892–1989). British physician on staff of Medical Research Council; consultant physician, Guy's Hospital, London.

THOMPSON, R.H.S. BMFRS, 1991, **37**, 245–62.
Archival material: CMAC.

GRASSI, Benvenuto (*fl*.12th cent.). Italian ophthalmic surgeon; wrote earliest printed book on ophthalmology.

GRASSI, B. De oculis eorumque egretudinibus et curis. Translated with notes and illustrations from the first edition (Ferrara, 1474) by Casey A. Wood, Stanford CA, Stanford Univ. Press, 1929. 101 pp.

GRASSI, Giovanni Battista (1854–1925). Italian parasitologist; made important discoveries on malaria.

FRANCESCHINI, P. DSB, 1972, **5**, 502–4.
NEGHME, A. An appraisal of Giovan Battista Grassi: his work in biology and parasitology. *Experimental Parasitology*, 1964, **15**, 260–78.

GRAUNT, John (1620–1674). English statistician and demographer.

EGERTON, F.N. DSB, 1972, **5**, 506–8.
KEYNES, G.L. A bibliography of Sir William Petty, FRS, and of *Observations on the bills of mortality*, by John Graunt, FRS. Oxford, Clarendon Press, 1971. 103 pp.

GRAVES, Robert James (1797–1853). Irish physician; described 'Graves' disease' (exophthalmic goitre).

TAYLOR, S. Robert Graves; the golden yeara of Irish medicine. London, Royal Society of Medicine, 1989. 160 pp.

GRAWITZ, Paul Albert (1850–1932). German surgeon; made important observations on origin of hypernephroma ('Grawitz tumour').

GROTE, 1923, **2**, 23–75 (CB).

GRAY, Alfred Leftwich (1873–1932). American radiologist; introduced radiotherapy in treatment of carcinoma of the bladder.

BROWN, P. *American Journal of Roentgenology*, 1032, **28**, 679–81.

GRAY, Edward George (1924–1999). British anatomist.

GUILLERY, R.W. BMFRS, 2002, **48**, 153–65.

GRAY, Henry (1825/1827–1861). British anatomist and physician.

GOSS, C.M. A brief account of Gray and his *Anatomy, descriptive and applied*, during a century of its publication in America. Philadelphia, Lea & Febiger, 1959. 51 pp.
NICOL, K.E. Henry Gray, FRCS, FRS. *Friends of the Wellcome Library and Centre for the History of Medicine Newsletter*, 2002, **27**, 8–9.

GRAY, Louis Harold (1905–1965). British radiobiologist.

LOUTIT, J.F. & SCOTT, O.C.A. BMFRS, 1966, **12**, 195–217.

GREATRAKES, Valentine (1629–1683). English faith healer.

LEVER, A.B. Miracles no wonder: the mesmeric phenomena and organic cures of Valentine Greatrakes. *Journal of the History of Medicine*, 1978, **33**, 35–46.

GREENGARD, Paul (b. 1925). American physician; shared Nobel Prize (Physiology or Medicine), 2000, for discoveries concerning signal transduction in the nervous system.

Les Prix Nobel en 2000.

GREENWOOD, Major (1880–1949). British biostatistician; professor of epidemiology and medical statistics, London School of Hygiene and Tropical Medicine.

HOGBEN, L. ONFRS, 1950–51, **7**, 139–54.

GREGG, Alan (1890–1957). American physician; Director, Medical Sciences, Rockefeller Foundation, 1930–51; Vice-president 1951–56; introduced the term 'molecular biology'.

PENFIELD, W. The difficult art of giving: the epic of Alan Gregg. Boston, Little, Brown, 1967. 414 pp.
Archival material: Butler Library, Columbia University.

GREGG, Norman McAlister (1892–1966). Australian ophthalmologist; showed relationship between rubella in early pregancy and congenital defects in infants.

–. *Medical Journal of Australia*, 1966, **2**, 1166–9.

GREGORY, Roderic Alfred (1913–1990). British physiologist.

DOCKRAY, G.J. BMFRS, 1998, **44**, 207–16.

GREIG, David Middleton (1864–1936). British surgeon; first to describe hypertelorism as a separate clinical entity.

–. *Edinburgh Medical Journal*, 1936, **43**, 531–9.

GREN, Friedrich Albrecht Carl (1760–1798). German physician and pharmacologist.

SEILS, M. Friedrich Albrecht Carl Gren in seiner Zeit 1760–1798: Spekulant oder Selbstdenker? Stuttgart, Wissenschaftliche Verlagsgesellschaft, 1995. 261 pp.

GRENFELL, Wilfred Thomason (1865–1940). British medical missionary in Labrador.

GRENFELL, W.T. A Labrador doctor: the autobiography of Wilfred Thomason Grenfell. London, Hodder & Stoughton, 1948. 319 pp. First published 1919.
KERR, J.L. Wilfred Grenfell: his life and work. New York, Dodd, Mead, 1959. 270 pp.
ROMKEY, R. Grenfell of Labrador. Toronto, University Press, 1991. 350 pp.

GREW, Nehemiah (1641–1712). English physician and botanist; comparative anatomist.

LeFANU, W.R. Nehemiah Grew, MD, FRS; a study and bibliography of his writings. Winchester, St Paul's Bibliographies, 1990. 182 pp.

GRIESBACH, Walter Edwin (1888–1968). German physician.

SCHWARZ, V.A. Walter Edwin Griesbach (1888–1968), Leben und Werk: Pharmakologe, Stoffwechselpathologe und Endokinologe/Viola. Frankfurt, Lang, 1999. 183 pp.

GRIESINGER, Wilhelm (1817–1868). German psychiatrist.

ALTSCHULE, M.D. Roots of modern psychiatry. 2nd ed. New York, Grune & Stratton, 1965. 208 pp.
METTE, A. Wilhelm Griesinger, der Begründer der wissenschaftlichen Psychiatrie in Deutschland. Leipzig, Teubner, 1976. 84 pp.
WAHRIG-SCHMIDT, B. Der Jung Wilhelm Griesinger in Spannungsfeld zwischen Philosophie und Psychologie. Tubingen. Narr, *c.*1985. 231 pp.

GRIFFITH, Frederick (1879–1941). British bacteriologist; his work on pneumococcal types led to the discovery that DNA is the basic material responsible for genetic transformation.

DOWNIE, A.W. Pneumococcal transformation – a backward view. *Journal of General Microbiology*, 1972, **73**, 1–11.

GRIFFITH, Harold Randall (1894–1985). Canadian/American anaesthetist; introduced curare into general anaesthesia.

BODMAN, R.L. & GILLIES, D. Harold Griffith: the evolution of modern anaesthesia. Toronto, Hannah Institute; Dundurn Press, 1992. 128 pp.

GRIJNS, Gerrit (1865–1944). Dutch physician; published important work on beri-beri.

LINDEBOOM, G.A. DSB, 1972, **5**, 541–2.

GRITTI, Rocco (1826–1920). Italian surgeon; introduced Gritti-Stokes amputation at the thigh, 1857.

CROSTI, F. La produzione scientifica del Dott. Rocco Gritti. *Ospedale Maggiore*, 1914, 2.ser., **2**, 350–56.
DENT, F. Il Dr.Gritti gli studi oftalmologico. *Ospedale Maggiore*, 1914, 2.ser., **2**, 357–9.

GROBSTEIN, Clifford (1916–1998). American biologist.

WESSELLS, N.K. BMNAS, 2000, **78**, 64–93.

BIBLIOGRAPHY OF MEDICAL AND BIOMEDICAL BIOGRAPHY

GRODDECK, Georg Walter (1866–1934). German pioneer of psychosomatic medicine.

CHEMOUNI. J. Georg Groddeck: psychonalyste de l'imaginaire: psychoanalyse freudienne et psychoanalyse groddeckienne. Paris, Payot, 1984. 347 pp.
GROSSMAN, C.M. & GROSSMAN, Sylva. The wild analyst. The life and work of Georg Groddeck. London, Barrie & Rockliff, 1965. 222 pp.
WILL, H. Die Geburt der Psychosomatik. Georg Groddeck, der Mensch und Wissenschaftler. München, Urban & Schwarzenberg, 1984. 231 pp.

GROEN, Joannes Juda (1903–1990). Dutch physician.

van DAAL, M.J.G.W. & de KNECHT-van-EEKELEN, A. Joannes Juda Groen (1903–1990); een arts op zoek naar het ware welzijn. Rotterdam, Erasmus Publishing, 1994. 240 pp.

GROENVELDT, Jan [GREENFIELD, John] (?1647–1710). Dutch/English lithotomist.

COOK, H.J. Trials of an ordinary doctor: Johannes Groenveldt in seventeenth-century London. Baltimore, Johns Hopkins Univ. Press, c.1994. 301 pp.

GROSS, Robert Edward (1905–1988). American cardiovascular surgeon.

MOORE, F.D. & FOLKMAN, J. BMNAS, 1995, **66**, 130–48.

GROSS, Samuel David (1805–1884). American surgeon.

GROSS, S.D. Autobiography ... With sketches of his contemporaries. Edited by his sons. 2 vols. Philadelphia, Barrie, 1887. 407 + 438 pp. Reprinted New York, Arno Press, 1972.

GROTE, Louis Ruyter Radcliffe (1886–1960). German physician; pioneer in use of music therapy.

ROTHSCHUH, K.E. (ed.) Louis R. Grote: der Arzt im Angesicht von Leben, Krankheit und Tod. Stuttgart, Hippokrates Verlag, 1961. 254 pp.

GRUBBÉ, Emil Herman (1875–1960). American radiologist; pioneered use of lead protection in radiology.

HODGES, P.C. The life and times of Ernest H. Grubbé. The biography of a pioneering Chicago radiologist. Chicago, Univ. Chicago Press, 1964. 135 pp.

GRUBER, Max von (1853–1927). Austrian bacteriologist and hygienist.

FLAMM, H. DSB, 1972, **5**, 563–5.

GRUBY, David (1810–1898). French physician; specialist on medical mycoses.

KISCH, B. Forgotten leaders in modern medicine. *Transactions of the American Philosophical Society*, 1954, N.S. **44**, 193–226.
KRUTA, V. DSB, 1972, **5**, 565–6.

GRÜNEBERG, Hans (1907–1982). German-born geneticist; worked at University College London.

LEWIS, D. & HUNT, D.M. BMFRS, 1984, **30**, 227–47.
Archival material: CMAC.

GRÜNFELD, Joseph (1840–1910). Bohemian physician; successfully catheterized the ureter under endoscopic vision, 1876.

KLEIN, S. *Medizinische Blätter*, 1910, **32**, 286–9.

GRUITHUISEN, Franz Paul von (1774–1852). German urologist and astronomer.

ZAMANN, A.M. Das Leben und Werken des Franz von Paula Gruithuisen (1774–1852). Seine Bedeutung für die Urologie. Mainz, Verlag Murken-Altrogge, 1997. 178 pp.

GRUNDFEST, Harry (1904–1983). Russian/American neurophysiologist.

REUBEN, J.P. BMNAS, 1995, **66**, 150–66.

GUBLER, Adolph Marie (1821–1879). French physician; described "Gubler's paralysis" – crossed hemiplegia.

FRESQUET FEBRER, J.L. Adolph Gubler y el "Journal de Thérapeutique" (1874–1883). *Asclepio*, 1993, **45**, 143–86.

GUDDEN, Bernhard Aloys von (1824–1886). German anatomist and psychiatrist.

SCHROTH, H.H. DSB, 1972, **5**, 569–72.

GUÉRIN, Alphonse-François-Marie (1817–1895). French surgeon.

COURBE, A. Alphonse Guérin: sa vie, son oeuvre. Paris, Jouve et Cie., 1913. 225 pp.

GUÉRIN, Camille (1872–1961). French physician.

SAKULA, A. BCG: Who were Calmette and Guérin? *Thorax*, 1983, **38**, 806–12.

GUGLIELMO DA SALICETTO (*c*.1210–*c*.1280). Italian surgeon; wrote first known treatise on surgical anatomy (*c*.1275), first published in his La ciroxia vulgamnte fata, Book IV, 1474.

HENSCHEL, A.W. *Janus* (Breslau), 1847, **2**, 142–4.

GUIDI, Guido [Vidus Vidius] (1508–1569). Italian anatomist and surgeon.

BROCKBANK, E.M. The man who was Vidius. *Annals of the Royal College of Surgeons of England*, 1956, **19**, 269–95.
GRMEK, M.D. Contribution à la biographie de Vidius (Guido Guidi). *Revue d'Histoire des Sciences*, 1978, **31**, 289–99.
GRMEK, M.D. DSB, 1972, **5**, 580–81.

GUILLAIN, Georges Charles (1876–1961). French physician; described acute infective polyneuritis.

ALAJOUANINE, T. Georges Guillain (1876–1961). *Bulletin de l'Académie Nationale de Médecine*, 1962, **146**, 18–26.

GUILLEMIN, Roger (b.1924). French-born American endocrinologist; shared Nobel Prize 1977 for work on isolation of hypothalamic hormones.

WADE, N. The Nobel duel. Two scientists' 21-year race to win the world's most coveted research prize. Garden City, NY, Anchor Press, Doubleday, 1981. 321 pp.

GUILLOTIN, Ignace Joseph (1738–1814). French physician. The guillotine, of ancient origin, was modified by Guillotin, who suggested its use in executions as being swift and painless.

SOUBIRAN, A. The good Doctor Guillotin and his strange device ... Translated by Malcolm MacGraw. London, Souvenir Press, 1964. 224 pp. First published in French, 1962.

GUINTERIUS, Johannes, of Andernach (1505–1574). German physician.

BREEMSER, F. *et al.* Johann Winter aus Andernach (Ioannes Guinterius Andernacus) 1505–1574): ein Humanist und Mediziner das 16 Jahrhunderts. Andernach, Stadtsmuseum, 1989. 112 pp.

GULL, William Withey (1816–1890). British physician; physician to Guy's Hospital, London.

HALE-WHITE, 1935, pp. 208–26 (CB).
Archival material: Bodleian Library, Oxford; Library, Wellcome Institute for the History of Medicine.

GULLAND, John Masson (1898–1947). British biochemist.

FARRAR, K.R. DSB, 1972, **5**, 589–90.

GULLSTRAND, Alvar (1862–1930). Swedish ophthalmologist; professor of ophthalmology and later physiological and physical optics, Uppsala; Nobel Prize 1911.

HERZBERGER, M.J. DSB, 1972, **5**, 590–91.

GUTHRIE, George James (1785–1856). British military surgeon.

PETTIGREW, vol. 4 (CB).
–. *Lancet*, 1850, **1**, 723–36; 1856, **1**, 519.

GUTHRIE, Samuel (1782–1848). American physician; introduced modern method of manufacturing chloroform.

PAWLING, J.R. Dr Samuel Guthrie, discoverer of chloroform, manufacturer of percussion pellets, industrial chemist (1782–1848). Watertown, NY, Brewster Press, 1947. 106 pp.
ROBINSON, V. Samuel Guthrie, 1782–1848. *Medical Life*, 1927, **34**, 103–52.

GUTTMANN, Ludwig (1899–1980). Polish/British neurologist.

GOODMAN, S. Spirit of Stoke Mandeville. The story of Sir Ludwig Guttmann. London, Collins, 1986. 191 pp.
HARRIS, P. *et al.* [Ludwig Guttmann ... special issue of *Paraplegia*, 1979, **17**, No 1; complete bibliography, pp. 131–8.
WHITTERIDGE, D. BMFRS, 1983, **29**, 227–44.

GUY DE CHAULIAC (*c*.1290–*c*.1367/70). French surgeon.

BULLOUGH, V.L. DSB, 1971, **3**, 218–19.
CHAULIAC, G.A.P.M. Guy de Chauliac, fondateur de la chirurgie didactique, *etc*. Bordeaux, Delmas, 1936. 63 pp.
NICAISE, E. [Introduction to his edition of] *La grand chirurgie*, Paris, Alcan, 1890. pp. lxxxvii–cv; bibliography, pp. cvi–cxci.

GWATHMEY, Jacob Tayloe (1865–1944). American anaesthetist.

COPE, D.K. James Tayloe Gwathmey: seeds of a developing specialty. *Anesthesia and Analgesia*, 1993, **76**, 642–7.

BIBLIOGRAPHY OF MEDICAL AND BIOMEDICAL BIOGRAPHY

GYE, William Ewart (1884–1952). British pathologist; Director, Imperial Cancer Research Fund Laboratories; proponent of viral aetiology of cancer.

ANDREWES, C.H. ONFRS, 1952–53, **8**, 419–30.

HAAB, Otto (1850–1931). Swiss ophthalmologist, invented magnet for extraction of metal particles from eye.

HÜRLIMANN, U. Otto Haab (1850–1931): ein Schweizer Ophthalmologe. Zürich, Juris, 1979. 67 pp.

HAASS, Friedrich Joseph (1780–1853). German physician in Russia; devoted his life to the poor, prisoners, beggars, serfs, etc.

HAMM, A. Der heilige Doktor von Moskau: der Mensch, sein Leben, sein Werk. Berlin, Westkreuz Druckerei, 1979. 136 pp.
KOPELEW, L. Der heilige Doktor Fjodor Petrowitsch. Die Geschichte des Friedrich Joseph Haass, Bad Münstereifel 1780 – Moskau, 1853. 2te. Aufl. Hamburg, Hoffman u. Campe, 1984. 231 pp.

HABICOT, Nicolas (1550–1624). French anatomist and surgeon; successfully performed laryngotomy (4 cases), reported 1620.

VAUCAIRE, R. Étude sur Habicot. L'anatomie at la chirurgie de son temps. Paris, Rueff et Cie., 1891. 244 pp.

HACKER, Victor von (1852–1933). Austrian surgeon; introduced a method of gastrostomy, 1886.

EISELBERG, A. *Archiv für Klinische Chirurgie*, 1933, **175**, i–iv.
SCHMERZ, H. *Chirurg*, 1932, **4**, 833–8.

HADDOW, Alexander (1907–1976). British pathologist; Director, Chester Beatty Cancer Research Institute.

BERGEL, F. BMFRS, 1977, **23**, 133–91.
HADDOW, A. A perspective in time. *Perspectives in Biology and Medicine*, 1975, **18**, 433–55.
Archival material: Chester Beatty Research Institute, London; CMAC.

HADDOW, Alexander John (1912–1978). British medical entomologist.

GARNHAM, P.C.C. BMFRS, 1980, **26**, 225–54.

HADEN, Francis Seymour (1818–1910). British surgeon and painter; founder of (Royal) Society of Painter-Etchers.

PLARR, p. 486.
DNB, Supplement 2.

HADRA, Berthold Ernst (1842–1903). German/American surgeon.

–. *Texas Medical Journal*, 1903–1904, **19**, 11–15.

HAECKEL, Ernst Heinrich Philipp August (1834–1919). German morphologist.

BÖLSCHE, W. Haeckel, his life and work. With introduction and supplementary chapter by the translator, Joseph McCabe. London, T.F. Unwin, 1906. 336 pp.

KEITEL-HOLZ, K. Ernst Haeckel: Forscher, Künstler, Mensch. Eine Biographie. Frankfurt am Main, R.G. Fischer, c.1984. 238 pp.
KRAUSSE, E. Ernst Haeckel. Leipzig, Teubner, 1984. 148 pp.
USCHMANN, G. DSB, 1972, **6**, 6–11.

HAFFKINE, Waldemar Mordecai Wolfe (1860–1930). Russian-born bacteriologist; introduced cholera and plague vaccines.

LUTZKER, E. DSB, 1972, **6**, 11–13.
WAKSMAN, S.A. The brilliant and tragic life of W.M.W. Haffkine, bacteriologist. New Brunswick, N.J., Rutgers Univ. Press, 1964. 86 pp.

HAGEDORN, Hans Christian (1888–1971). Danish physician.

FELIG, P. Landmark perspective: protamine insulin. Hagedorn's pioneering contribution to drug delivery in the management of diabetes. *Journal of the American Medical Association*, 1984, **251**, 389–96.

HAHN, Eugen (1841–1902). German surgeon; introduced nephropexy in treatment of movable kidney.

NEUMANN, A. Deutsche *Zeitschrift für Chirurgie*, 1903, **68**, i–iv.

HAHNEMANN, Christian Friedrich Samuel (1755–1843). German founder of homoeopathic medicine.

COOK, T.M. The founder of homoeopathic medicine, Samuel Hahnemann. Wellingborough, Northamptonshire, Thorsons, 1981. 192 pp.
HAEHL, R. Samuel Hahnemann, his life and work ... Translated from the German by M.L. Wheeler & W.H.R. Grundy. Edited by J.H. Clarke & F.J. Wheeler. 2 vols., London, Homoeopathic Publ. Co., 1931. 443 + 515 pp.
HANDLEY, R. In search of the later Hahnemann. Beaconsfield, UK, Beaconsfield Publishers, 1997. 235 pp.
JOSEF, M. Die Publikationen Samuel Hahnemanns. *Sudoff's Archiv*, 1988, **72**, 14–36.
RITTER, H. Samuel Hahnemann, Begründer der Homöopathie: sein Leben und Werk in neuer Sicht. 2. erw. Aufl. Heidelberg, Haug, 1986. 160 pp.

HALDANE, John Burdon Sanderson (1892–1964). British geneticist; professor of genetics, and later biochemistry, University of London.

CLARK, R.W. The life and work of J.B.S. Haldane. Oxford, Oxford University Press, 1984. 288 pp.
CLARK, R.W. DSB, 1972, **6**, 21–3.
DRONAMRAJU, K.R. Haldane: the life and work of J.B.S. Haldane, with special reference to India. Aberdeen. Aberdeen Univ. Press, 1985. 211 pp.
PIRIE, R.W. BMFRS, 1966, **12**, 219–49.
Archival material: Library, University College London.

HALDANE, John Scott (1860–1936). British physiologist.

CHAPMAN, C.B. DSB, 1972, **6**, 23–5.
DOUGLAS, C.G. ONFRS, 1936, **2**, 115–39.
HALDANE, J.S. The philosophy of a biologist. Oxford, Clarendon Press, 1935. 155 pp.
STURDY, S.W. A co-ordinated whole. The life and work of John Scott Haldane. Univ. Edinburgh dissertation, 1988. 524 pp. Abstr.Int. 1988 49:602A. Univ. Microfilms BRO 81034.

Archival material: National Library of Scotland; Woodward Biomedical Library, University of British Columbia.

HALES, Stephen (1677–1761). British physiologist and public health pioneer; perpetual curate of Teddington, Middlesex.

ALLAN, D.G.C. & SCHOFIELD, R.E. Stephen Hales: scientist and philanthropist. London, Scolar Press, 1980. 220 pp.
CLARK-KENNEDY, A.E. Stephen Hales, D.D., F.R.S.; an eighteenth century biography. Cambridge, Cambridge Univ. Press, 1929. 256 pp. Reprinted, Englewood, NJ, 1965.
GUERLAC, H. DSB, 1972, **6**, 35–48.
Archival material: Royal Society of London.

HALFORD, Henry (1766–1844). British physician, President, Royal College of Physicians of London, 1820–1844; physician to George III, George IV, William IV and Queen Victoria.

MUNK, W. The life of Sir Henry Halford. London, Longmans Green, 1895. 284 pp.

HALL, Granville Stanley (1846–1924). American child psychologist; president of Clark University.

HALL, G.S. Life and confessions of a psychologist. New York, Appleton, 1923. 622 pp.
ROSS, D. G. Stanley Hall: the psychologist as prophet. Chicago, Univ. Chicago Press, 1972. 482 pp.
THORNDIKE, E.L. BMNAS, 1930, **13**, 135–80.

HALL, John (1575–1635).

JOSEPH, H. Shakespeare's son-in-law: John Hall, man and physician ... With a facsimile of the second edition of Hall's *Select observations on English bodies*. Hamden, Conn., Archon Books, 1964. 328 pp.
LANE, J. John Hall and his patients: the medical practice of Shakespeare's son-in-law. Stratford-upon-Avon, The Shakespeare Birthday Trust, 1996. 378 pp.

HALL, John (1795–1866). Chief of medical staff in the Crimean campaign.

MITTRA, S.M. The life and letters of Sir John Hall. London, *etc*, Longmans, Green, 1911. 560 pp.
Archival material: CMAC.

HALL, Marshall (1790–1857). British physician and neurophysiologist.

CLARKE, E. DSB, 1972, **6**, 58–61.
HALE-WHITE, 1935, pp. 85–105 (CB).
HALL, C. Memoirs of Marshall ... by his widow. London, Bentley, 1861. 518 pp.
LEYS, R. From sympathy to reflex: Marshall Hall and his opponents. New York, Garland, 1990. 549 pp.
MANUEL, D.E. Marshall Hall, FRS (1790–1857); a conspectus of his life and work. *Notes and Records of the Royal Society of London*, 1980, **35**, 135–66.

HALLER (Victor) Albrecht von (1708–1777). Swiss anatomist, physiologist, botanist and bibliographer.

BALMER, H. Albrecht von Haller. Bern, Paul Haupt, 1977. 88 pp.

BEER, R.B. Der grosse Haller. Säckingen, Stratz, 1947. 137 pp.
HINTZSCHE, E. DSB, 1972, **6**, 61–7.

HALLERVORDEN, Julius (1882–1965). German neurologist; described Hallervorden-Spatz syndrome, affecting the extrapyramidal system.

SPATZ, H. Erinnerungen an Julius Hallervorden (1882–1965). *Nervenarzt*, 1966, **37**, 477–82.

HALLIBURTON, William Dobinson (1860–1931). British physiologist and biochemist.

MORGAN, N. William Dobinson Halliburton, FRS (1860–1931) – pioneer of British biochemistry. *Notes and Records of the Royal Society of London*, 1983, **36**, 129–45.
MORGAN, N. DSB, 1990, **17**, 377–9.
Archival material: CMAC.

HALLOPEAU, François Henri (1842–1919). French dermatologist.

DENIKER, M.J. *Journal de Chirurgie*, 1925, **25**, 4–6.

HALLPIKE, Charles Skinner (1900–1979). British otologist.

WHITTERIDGE, D. & MERTON, P.A. BMFRS, 1984, **30**, 283–95.

HALSTED, William Stewart (1852–1922). American surgeon; professor at Johns Hopkins University.

CROWE, S.J. Halsted of Johns Hopkins: the man and his men. Springfield, Ill., Thomas, 1957. 247 pp.
MacCALLUM, W.G. William Stewart Halsted, surgeon. Baltimore, Johns Hopkins Press; London, Milford, 1930. 241 pp.
MacCALLUM, W.G. BMNAS, 1937, **17**, 151–70.
OLCH, P.D. DSB, 1972, **6**, 77–8.
OSLER, W. The inner history of the Johns Hopkins Hospital (edited by D.G. Bates & E.H. Bensley). *Johns Hopkins Medical Journal*, 1969, **125**, 184–94.
Archival material: Welch Medical Library, Johns Hopkins Hospital.

HALY ABBAS [al-Majusi, Abu'l Hasan ali ibn 'Abbas] (930–994). Persian physician.

HAMARNEH, S. DSB, 1974, **9**, 40–42.

HAMEY, Baldwin (1568–1640). Flemish physician who settled in London.

KEEVIL, J.J. Hamey the Stranger. London, Bles, 1952. 192 pp.

HAMEY, Baldwin, *the Younger* (1600–1676). English physician; benefactor of the (Royal) College of Physicians, London.

KEEVIL, J.J. The Stranger's son. London, Bles, 1953. 230 pp.
Archival material: Royal College of Physicians of London.

HAMILTON, Alice (1869–1970). American industrial toxicologist.

GRANT, M.P. Alice Hamilton: pioneer doctor in industrial medicine. London, New York, Abelard-Schuman, 1967. 223 pp.
HAMILTON, A. Exploring the dangerous trades: autobiography. Boston, Little, Brown, 1943. 433 pp. Reprinted Clifton, N.J., Kelley.

BIBLIOGRAPHY OF MEDICAL AND BIOMEDICAL BIOGRAPHY

SICHERMAN, B. Alice Hamilton: a life in letters. Cambridge, Harvard Univ. Press, 1984. 460 pp.

HAMILTON, Frank Hastings (1813–1886). American surgeon; pioneered treatment of ulcers by skin grafts.

HAMILTON, F.H. A practical treatise on fractures and dislocations. Reprinted from 1860 edition with a biographical introduction by I.M. Rutkow. San Francisco, Norman Publishing, 1991. 757 pp.

HAMMOND, William Alexander (1828–1900). American neurologist; wrote first American treatise on neurology; surgeon general, US army.

BLUSTEIN, B.E. Preserve your love for science: life of William A. Hammond, American neurologist. Cambridge, University Press, 1991. 289 pp.
KEY, J.D. William Alexander Hammond, M.D. Rochester, Minn., Davies, 1979. 84 pp.

HANAU, Arthur Nathan (1858–1900). German pathologist; first successfuly to transplant tumours in mammals.

BUCHER, H.W. Zur ersten homologen Tumorübertragung in Zürich durch Arthur Hanau 1899. Gesnerus, 1964, **21**, 193–200.
DIEPGEN, pp. 55–6.

HANDLER, Philip (1917–1981). American biochemist, President, National Academy of Sciences for 12 years.

SMITH, E.L. & HILL, R.L. BMNAS, 1985, **55**, 305–53.

HANNOVER, Adolph (1814–1894). Danish pathologist; introduced the term 'epithelioma' although not recognizing its malignant nature.

THOMS, J. Adolph Hannover (1814–1894), en medicin-historisk studie. Thesis, University of Copenhagen, 1978. 62 pp.

HANOT, Victor Charles (1844–1896). French physician.

GILBERT, A. Notice sur les traveux de Hanot. Paris, G. Carré & C. Naud, 1897. 14 pp.

HANSEN, Gerhard Henrik Armauer (1841–1912). Norwegian bacteriologist; discovered *Myco. leprae*, causative agent of leprosy.

HANSEN, G.A. & WATT, F. (eds.) The memories and reflections of Dr G. Armauer Hansen. Würzburg, German Leprosy Relief Association, 1976. 135 pp.
VOGELSANG, T.M. Gerhard Henrik Armauer Hansen, 1841–1912. The discoverer of the leprosy bacillus. His life and work. *International Journal of Leprosy*, 1978, **46**, 257–322.
Archival material: Bergen Leprosy Archives; Regional State Archives; City Archives of Bergen; Leprosy Museum. Bergen. Christopher.Harris & Riksarkiven.dep. no.

HANSON, Emmeline Jean (1919–1973). British physiologist.

RANDALL, J. BMFRS, 1975, **21**, 313–44.
Archival material: King's College, London.

HANSON, Robert Paul (1918–1987). American virologist.

160

YUILL, T.M. & EASTERDAY, B.C. BMNAS, 1996, **70**, 139–54.

HARDEN, Arthur (1865–1940). British chemist; shared Nobel Prize 1929.

HOPKINS, F.G. & MARTIN, C.J., ONFRS, 1942–44, **4**, 3–14.
IHDE, A.A. DSB, 1972, **6**, 110–12.

HARDY, James Daniel (1918–1985). American surgeon.

HARDY, J.D. The world of surgery, 1945–1985; memoirs of one participant. Philadelphia, Univ. Philadelphia Press, 1986. 385 pp.

HARINGTON, Charles Robert (1897–1972). British biochemist; Director, National Institute for Medical Research; synthesized thyroxine.

HIMSWORTH, H. & PITT-RIVERS, R. BMFRS, 1972, **18**, 267–308.

HARINGTON, John (1560–1612). English courtier and poet; invented water closet in which disposal of excreta was mechanically controlled.

ADAMI, J.G. Sir John Harington. *Bulletin of the Johns Hopkins Hospital*, 1908, **19**, 285–95.

HARKEN, Dwight Emary (b.1910). American cardiovascular surgeon.

GONZALEZ-LAVIN, L. Charles P. Bailey and Dwight E. Harken: the dawn of the modern era of mitral valve surgery. *Annals of Thoracic Surgery*, 1993, **53**, 916–19.
SYMBAS, P.N. & JUSTICZ, A.G. Quantum leap forward in the management of cardiac trauma: the pioneering work of Dwight E. Harken. *Annals of Thoracic Surgery*, 1993, **55**, 789–91.

HARLEY, George (1829–1896). British physician and scientist.

TWEEDIE, A. (ed.) George Harley, F.R.S.: the life of a London physician. London, Scientific Press, 1899. 360 pp.

HARPESTRAENG, Henrik (1164–1244). Danish physician and pharmacist.

PEDERSEN, O. DSB, 1972, **6**, 123–4.

HARRIS, Chapin Aaron (1806–1860). American dental surgeon.

ABSELL, M.B. Chapin A. Harris and Horace H. Hayden: an historical review. *Bulletin of the History of Dentistry*, 1969, **17**, 27–31.

HARRIS, Geoffrey Wingfield (1913–1971). British endocrinologist.

VOGT, M. BMFRS, 1972, **18**, 309–29.
Archival material: Bodleian Library, Oxford.

HARRIS, Harry (1919–1994). British biochemical geneticist.

HOPKINSON, D.A. BMFRS, 1996, **42**, 152–70.

HARRIS, Henry (1881–1931). Australian surgeon.

MURPHY, L.J. Harry Harris and his contribution to suprapubic prostatectomy. *Australian and New Zealand Journal of Surgery*, 1984, **54**, 579–88.

HARRIS, Henry (b.1925). Australian pathologist and cell biologist; Regius Professor of Medicine, Oxford.

HARRIS, H. The balance of improbabilities. A scientific life. Oxford, Oxford Univ. Press, 1987. 245 pp.
Archival material: CMAC.

HARRIS, Walter (1647–1732). English paediatrician; wrote standard work on paediatrics.

RUHRÄH, J. Walter Harris, a seventeenth-century pediatrist. *Annals of Medical History*, 1919, **2**, 228–40.

HARRISON, Ross Granville (1870–1959). American biologist; experimental embryologist; pioneer in tissue culture.

ABERCROMBIE, M. BMFRS, 1961, **7**, 111–26.
OPPENHEIMER, J.M. DSB, 1972, **6**, 131–5.
NICHOLAS, A.S. BMNAS, 1961, **35**, 132–63.
Archival material: Yale University Library, New Haven; Johns Hopkins University, Baltimore.

HART, Edwin Bret (1874–1953). American biochemist and nutritionist.

ELVEHJEM, C.A. BMNAS, 1954, **28**, 135–61.
IHDE, A.J. DSB, 1972, **6**, 135–6.
Archival material: University of Wisconsin.

HARTLEY, David (1705–1757). British physician and psychologist.

YOUNG, R.M. DSB, 1972, **6**, 138–40.

HARTLEY, Percival (1881–1957). British biochemist; Director of Department of Biological Standards, National Institute for Medical Research, London.

DALE, H. H. BMFRS, 1957, **3**, 81–100.

HARTLINE, Haldan Keller (1903–1983). American physiologist; Nobel prizewinner 1967 for work on visual and chemical processes of the eye.

GRANIT, R. & RATLIFF, F. BMFRS, 1985, **31**, 263–92.
RATLIFF, R. BMNAS, 1990, **59**, 196–213.

HARTMANN, Arthur (1849–1931). German otolaryngologist; introduced audiometer.

THULLEN, A. Arthur Hartmann zum 120. Geburtstag. *Zeitschrift für Laryngologie, Rhinologie und Otologie*, 1969, **48**, 397–403.

HARTMANN, Frank Alexander (b.1883). American physiologist; professor at Ohio State University; isolated adrernal cortical extract 'cortin'.

HARTMANN, F.A. Biology as a career. *Perpsectives in Biology and Medicine*, 1963, **6**, 280–90.

HARTMANN, Johannes (1568–1631). German iatrochemist and physician.

SCHMITZ, R. DSB, 1972, **6**, 145–6.

HARTMANN, Philipp Karl (1773–1830). German physician.

PAUS, P. Philipp Karl Hartmann, Mensch, Arzt und Philosoph: sein Leben, sein Werk. Ein Beitrag zur Medizingeschichte der Romantik. Inaugural-Disseration...der Rheinischen Friedrich-Wilhelms Universität zu Bonn. Bonn, Friedrich-Wilhelms Universität, 1971. 236 pp.

HARTRIDGE, Hamilton (1886–1976). British physiologist; professor at St Bartholomew's Hospital Medical College; Director, Institute of Ophthalmology, London.

RUSHTON, W.A.H. BMFRS, 1977, **23**, 193–211.
Archival material: CMAC.

HARTWELL, Leland H. (b. 1939). American oncologist; shared Nobel Prize (Physiology or Medicine), 2001, for discovery of key regulators of the cell cycle.

Les Prix Nobel en 2001.

HARVEY, Edmund Newton (1887–1959). American physiologist; professor at Princeton University.

JOHNSON, F.H. BMNAS, 1967, **39**, 193–266.
WARNER, J.H. DSB, 1990, **17**, 383–5.

HARVEY, William (1578–1657). English physician, physiologist, anatomist and embryologist; demonstrated the circulation of the blood through the heart.

BYLEBYL, J. DSB, 1972, **6**, 150–62.
FRANK, R.G. Harvey and the Oxford physiologists: a study of scientific ideas. Berkeley, Univ. of California Press. 1980. 368 pp.
KEELE, K.D. William Harvey: the man, the physician and the scientist, London, Nelson, 1965, 244 pp.
KEYNES, G.L. The life of William Harvey. Oxford, Clarendon Press, 1966. 483 pp.
KEYNES, G.L. A bibliography of the writings of Dr William Harvey, 1578–1657. 3rd ed., revised by G. Whitteridge and C.E. English. Winchester, St Paul's Bibliographies, 1989. 136 pp. First edition 1928.
PAGEL, W. William Harvey's biological ideas. Selected aspects and historical background. Basel, New York, Karger, 1967. 394 pp.
PAGEL, W. New light on William Harvey. Basel, Karger, 1976. 189 pp.
WHITTERIDGE, G. William Harvey and the circulation of the blood. London, Macdonald, 1971. 269 pp.
Archival material: British Library; Royal College of Physicians of London.

HASLAM, John (1764–1844). British physician at Bethlem Royal Hospital; among the first to describe general paralysis of the insane.

LEIGH, D. John Haslam, MD – 1764–1844; apothecary to Bethlem. *Journal of the History of Medicine*, 1955, **10**, 17–44.

HASSALL, Arthur Hill (1817–1894). British physician; founded Royal National Hospital for Diseases of the Chest, Ventnor; compiled first English textbook on microscopic anatomy.

CLAYTON, E.G. Arthur Hill Hassall, physician and sanitary reformer. A short history of his work in public hygiene, and the movement against the adulteration of food and drugs, etc. London, Baillière, Tindall & Cox, 1908. 150 pp.

BIBLIOGRAPHY OF MEDICAL AND BIOMEDICAL BIOGRAPHY

GRAY, E.A. By candlelight: the life of Dr Arthur Hill Hassall, 1817–1894. London, Robert Hale, 1983. 192 pp.

HASTINGS, Albert Baird (1895–1987). American biochemist; professor at Harvard University.

CHRISTENSEN, H.N. BMNAS, 1994, **63**, 172–216.

HASTINGS, A.B. Crossing boundaries: biological, disciplinary, human: a biochemist pioneers for medicine. Ed. by H.N. Christenson. Based on oral histories by P.D. Olch and R.D. Livingstone. Grand Rapids, Michigan, Four Corners Press, 1989. 356 pp.

Archival material: National Library of Medicine, Bethesda, Md.; Countway Library, Harvard Medical School.

HASTINGS, Charles (1794–1866). Founder of the British Medical Association.

McMENEMEY, W.H. The life and times of Sir Charles Hastings, founder of the British Medical Association. Edinburgh, London, Livingstone, 1959. 516 pp.

HATA, Sahachiro (1873–1938). Japanese bacteriologist; co-introducer of salvarsan.

KÖHLER, W. Shibasaburo Kitasato and Sahachiro Hata. Kitasato *Archives of Experimental Medicine*, 1989, **62**, 85–106.

HAÜY, Valentin (1745–1822). French physician; educator of the blind; he originated method of raised letters and embossed paper for teaching the blind to read.

HENRI, P. La vie et l'oeuvre de Valentin Haüy. Paris, Presses Universitaires de France, 1984. 208 pp.

HAUPTMANN. Alfred (1881–1948). German neurologist; introduced phenobarbitone treatment of epilepsy.

STOCKERT, F.G. Alfred Hauptmann. *Archiv für Psychiatrie und Nervenkrankheiten*, 1948, **180**, 529–30.

HAUROWITZ, Felix (1896–1987). Czech/American immunologist.

PUTNAM, F.W. BMNAS, 1994, **64**, 132–63.

HAUSER, Gustav (1856–1935). German pathologist and bacteriologist; isolated *Proteus vulgaris*.

GROTE, 1927, **6**, 141–204. (CB).

HAVERS, Clopton (*c*.1655-1702). English osteologist.

LeFANU, W.R. DSB, 1972, **6**, 183–4.

HAWORTH, Walter Norman (1883–1950). British chemist; Nobel prizewinner (Chemistry) 1937 for his investigations on carbohydrates and vitamin C.

KOPPERL, S.J. DSB, 1972, **6**, 184–6.

HAYEM, Georges (1841–1933). French haematologist.

RIVET, L. Le professeur Georges Hayem. *Sang*, 1927, **1**, 59–68.

HAYES, William (1913–1974). British microbiologist and geneticist.

BRODA, P. & HOLLOWAY, B. BMFRS, 1996, **42**, 172–89.

HAYGARTH, John (1740–1827). British physician and epidemiologist; wrote first monograph on acute rheumatism.

ELLIOTT, J. A medical pioneer: John Haygarth of Chester. *British Medical Journal*, 1913, **1**, 235–42.
WEAVER, G.H. John Haygarth: clinician, investigator, apostle of sanitation. *Bulletin of the Society of Medical History of Chicago*, 1928–35, **4**, 156–200.

HAZEN, Elizabeth Lee (1885–1975). American scientist; isolated nystatin (fungicidin), a soil actinomycete.

BALDWIN, R.S. The fungus fighters: two women scientists and their discovery. Ithaca, Cornell Univ. Press, 1981. 212 pp.

HEAD, Henry (1861–1940). British neurologist.

HENRY HEAD centenary number: *Brain*, 1961, **84**, 529–69.
HOLMES, G.M. ONFRS, 1941, **3**, 665–89.
Archival material: CMAC.

HEAVISIDE, John, *the Elder* (1717/8–1787).

HEAVISIDE, John, *the Younger* (1748–1828). British surgeon.

PEACHEY, G.C. John Heaviside, surgeon. London, St Martin's Press, 1931. 50 pp.

HEBERDEN, William, *the Elder* (1710–1801). British physician.

HEBERDEN, E. William Heberden, physician of the age of reason. London, Royal Society of Medicine Services Ltd., 1989. 246 pp.
ROLLESTON, H. The two Heberdens: **William Heberden** *the Elder* (1710–1801); **William Heberden** *the Younger* (1767–1845). *Annals of Medical History*, 1933, N.S. **5**, 409–27; 566–83.
Archival material: Royal College of Physicians of London.

HEBERDEN, William, *the Younger* (1767–1845). British physician.

BLUMER, G. Some reflections upon the life and accomplishments of William Heberden the Younger (1767–1845). *Bulletin of the History of Medicine*, 1949, **15**, 381–99.
ROLLESTON, H. The two Heberdens: William Heberden *the Elder* (1710–1801); William Heberden *the Younger* (1767–1845). *Annals of Medical History*, 1933, N.S. **5**, 573–83.

HEBRA, Ferdinand von (1816–1880). Austrian dermatologist.

HOLUBAR, K. Ferdinand von Hebra: on the occasion of the centenary of his death. *International Journal of Dermatology*, 1981, **20**, 719–24.

HECHT, Selig (1892–1947). Polish/American biophysicist.

WALD, G. BMNAS, 1991, **60**, 80–100.

HECQUET, Philippe (1881–1737). French physician and scholar.

ROGER, J. Hecquet, docteur régent et ancien doyen de la Faculté de Médecine de Paris. Sa vie, ses oeuvres. Paris, Rétaux-Bray, 1889. 79 pp.

HEGAR, Alfred (1830–1914). German gynaecologist.

MAYER, A. Alfred Hegar und der Gestandwandel der Gynäkologie seit Hegar. Freiburg i. Br., Hans Ferdinand Schulz, 1961. 51 pp.

HEIDELBERGER, Charles (1920–1983). American biochemist and cytopathologist; professor of oncology, McArdle Laboratory, Wisconsin; worked on cancer chemotherapeutic agents.

MILLER, E.C. & MILLER, J.A. BMNAS, 1989, **58**, 259–302.

HEIDELBERGER, Michael (1888–1991). American immunochemist; professor of immunochemistry, Columbia University.

EISEN, H.N. BMNAS, 2001, **80**, 123–40.
HEIDELBERGER, M. Reminiscences – A 'pure' organic chemist's downward path. *Immunological Reviews*, 1984, **81b** 7–19; **82**, 7–27; 1985, **83**, 5–22.
STACEY, M. BMFRS, 1994, **39**, 179–97.
Archival material: National Library of Medicine, Bethesda MD.

HEIDENHAIN, Martin (1864–1949). German microscopic anatomist.

ALPERT, M. DSB, 1972, **6**, 223–4.

HEIDENHAIN, Rudolph Peter Heinrich (1834–1897). German histologist and physiologist.

BROGHAMMER, H. Experimentelle Physiologie und mikroskopische Anatomie in der Nachromantik: Rudolf Peter Heinrich Heidenhain (1834–1897). Aachen, Shaker, 2000. 199 pp.
KRUTA, V. DSB, 1972, **6**, 224–6.

HEIM, Ernst Ludwig (1747–1834). German physician; introduced Jennerian vaccination into Berlin, 1798.

KESSLER, G.W. (ed.) Der alte Heim: leben und Wirken Ernst Ludwig Heims...aus hinterlassen Briefe und Tagebüchern herausgegeben. 2te. Aufl. Leipzig, Brockhaus, 1846. 520 pp.
ROHLFS, H. Geschichte der deutschen Medizin. Stuttgart, Enke, 1875, vol. 1, 480–519.
WILLE, P.F.C. Der Berliner Arzt Ernst Ludwig Heim, genannt "der alte Heim" (1747–1834). *Zeitschrift der ärztliche Forschung*, 1961, **50**, 690–91.

HEINE, Johann Georg (1770–1830). German orthopaedic surgeon.

NAHRATH, H. Johann Georg Heine, der Vater der deutschen Orthopädie. Der Mensch und sein Werk. *Archiv für klinische Chirurgie*, 1927–28, **149**, 476–500.

HEISCHKEL-ARTELT, Edith, *see* **ARTELT, Walter**

HEISTER, Lorenz (1683–1758). German physician, anatomist, and founder of scientific surgery in Germany.

GEIGER, A. Lorenz Heister als Geburtshelfer un gynäkologischer Chirurg. Zürich, Juris Druck, 1983. 140 pp.
KÖRFF, R. Das Berufsethos in der Chirurgie Lorenz Heisters (1683–1758). Zürich, Juris-Verlag, 1975. 94 pp.
VAN DER PAS, P.W. DSB, 1972, **6**, 231–2.

HEKTOEN, Ludwig (1863–1951). American pathologist and microbiologist.

BRIEGER, G.A. DSB, 1972, **6**, 232–3.
CANNON, P.R. BMNAS, 1952–54, **28**, 163–97.

HELLER, Arnold (1840–1913). German pathologist; established syphilis as a cause of aortic aneurysm.

WILKE, E. Arnold Heller. *Münchener Medizinische Wochenschrift*, 1913, **60**, 987–9.

HELLER, Johann Florian (1813–1871). Austrian pathological chemist.

SCHMALHOFER, J. Das Werk von Johann Florian Heller mit besonderer Berücksichtigung der Entstehung des ersten pathologisch-chemischen Laboratoriums am Allgemeinen Wiener Krankenhaus, *etc*. Dissertation. Bonn, 1980. 210 pp.

HELMHOLTZ, Hermann Ludwig Ferdinand von (1821–1894). German physiologist and physicist; invented the ophthalmoscope.

CAHAN, D. (ed.) Hermann von Helmholtz and the foundations of nineteenth-century science. Berkeley, Univ. of California Press, *c*.1993. 666 pp.
EBERT, H. Hermann von Helmholtz. Stuttgart, Wissenschaftliche Verlagsgesellchaft, 1949. 199 pp.
KELLER, A.G. DSB, 1974, **10**, 236–8.
KOENIGSBERGER, L. Hermann von Helmholtz. Translated by F.A. Welby. Oxford, Clarendon Press, 1906. 440 pp. Reprinted, New York, Dover, 1965. 440 pp.
MEULDERS, M. Helmholtz des lumières aux neurosciences. Paris, Editions Odile Jacob, *c*.2000. 313 pp.
TURNER, R.S. DSB, 1972, **6**, 241–53.

HELMONT, Jean Baptiste van (1577–1644). Flemish physiological chemist; founded the Iatrochemical School.

NÈVE de MÉVERGNIES, P. Jean-Baptiste van Helmont: philospopher pour le feu. Paris. Librairie E. Droz, 1935. 232 pp.
PAGEL, W. Jean Baptista van Helmont, reformer of science and medicine. Cambridge, University Press, 1982, 219 pp.
PAGEL, W. DSB, 1972, **6**, 253–9.

HELPERN, Milton (1902–1977). American forensic pathologist.

HELPERN, M. & KNIGHT, B. Autopsy. The memoirs of Milton Helpern, the world's greatest medical detective. New York, St Martin's Press, 1977. 273 pp.

HEMMETER, John Conrad (1864–1931). American physiologist; professor of physiology and clinical medicine, Baltimore.

FESTSCHRIFT. *Medical Life*, 1927, **34**, 154–254.
GROTE, 1924, **3**, 154–254 (CB).

HEMS, Benjamin Arthur (1912–1995). British pharmacologist.

JACK, D. & WALKER, T. BMFRS, 1997, **43**, 215–33.

HENCH, Peter Showalter (1896–1965). American physician: shared Nobel Prize (1950) for introduction of cortisone and ACTH in treatment of rheumatoid arthritis.

FRANÇON, F. La vie exemplaire de P.S. Hench (1896–1965). In memoriam. *Biologie Médicale*, 1967, **56**, 109–23.

HENDERSON, David Willis Wilson (1903–1968). British bacteriologist.

KENT, L.H. & MORGAN, W.T.J. BMFRS, 1970, **16**, 331–41.

HENDERSON, Lawrence Joseph (1878–1942). American physiologist and biochemist; professor at Harvard University.

CANNON, W.B. BMNAS, 1943, **23**, 31–58.
PARASCANDOLA, J. Lawrence J. Henderson and the concept of organized systems. Dissertation, University of Wisconsin, 1968. 259 pp.
PARASCANDOLA, J. DSB, 1972, **6**, 260–62.
Archival material: Harvard University.

HENDERSON, Yandell (1873–1944). American physiologist.

PARASCANDOLA, J. DSB, 1972, **6**, 264–5.
WEST, J.B. BMNAS, 1998, **74**, 145–58.

HENLE, Friedrich Gustav Jacob (1809–1885). German anatomist and pathologist; professor at Göttingen and and Heidelberg.

HINTZSCHE, E. DSB, 1972, **6**, 268–70.
ROBINSON, V. The life of Jacob Henle. New York, Medical Life, 1921. 117 pp.

HENOCH, Eduard Heinrich (1820–1910). German paediatrician.

DUNN, P.M. Dr Eduard Henoch (1820–1910) of Berlin: pioneer of German pediatrics. *Archives of Diseases in Childhood*, 1996, **74**, 149–50.

HENRI DE MONDEVILLE, *see* **MONDEVILLE, Henri de**

HENROTIN, Fernand (1847–1906). Belgian/American surgeon.

SENN, N. Dr Fernand Henrotin a commemorative address. *Surgery, Gynecology and Obstetrics*, 1907, **4**, 91–6.

HENSCHEN, Salomon Eberhard (1847–1930). Swedish neurologist.

GROTE, 1925, **5**, 35–76. (CB).

HENSEN, (Christian Andreas) Victor (1835–1924). German physiologist and pathologist; professor at Kiel.

POREP, R. Der Physiologe und Planktonforscher Victor Hensen (1835–1924): sein Kleben und sein Werk. Neumünster, Karl Wachholtz Verlag, 1970. 147 pp.
ROTHSCHUH, K.E. DSB, 1972, **6**, 287–8.

d'HÉRELLE, Felix (1873–1949). Canadian/French microbiologist; discovered bacteriophage.

SUMMERS, W.C. Felix d'Herelle and the origins of molecular biology. New Haven, Yale Univ. Press, 1999. 236 pp.
THÉODORIDÈS, J. DSB, 1972, **6**, 297–9.

HERFORD, Martin Edward Meakin (b.1908). British medical practitioner.

HALL, M. A doctor at war: the story of Colonel Martin Herford, the most decorated doctor of World War Two, Malvern Wells, Images Publishing, 1995. 220 pp.

HERING, Constantin (1800–1880). German/American homoeopathic physician.

KNERR, C.B. Life of Hering. Philadelphia, Magee, 1940. 317 pp.

HERING, Karl Ewald Konstantin (1834–1918). German physiologist and psychologist.

BAUMANN, C. Der Physiologe Edward [sic] Hering (1834–1918) Curriculum vitae. Frankfurt, Dr Hansel-Hohenhausen, 2002. 171 pp.
KRUTA, V. DSB, 1972, **6**, 299–301.
WILLEBRAND, K. Ewald Hering. Ein Gedenkwort der Psychophysik. Berlin, J. Springer, 1918. 107 pp.

HERMANN, Ludimar (1838–1914). German physiologist.

SCHAWALDER, J.H. Der Physiologe Ludimar Hermann (1838–1914): Berlin-Zürich-Königsberg. Zürich, Juris, 1990. 94 pp.

HERNÁNDEZ, Francisco (1517–1587). Spanish physician and botanist; physician to Philip II of Spain.

VAREY, S. *et al.* (eds.) Searching for the secrets of life: the life and works of Dr Francisco Hernández. Stanford, CA, Stanford Univ. Press, 2000. 299 pp.

HEROPHILUS of CHALCEDON (*fl.* 4th cent. BC). Greek physician and anatomist; the 'Father of human anatomy'.

LONGRIGG, J. DSB, 1972, **6**, 316–19.
VON STADEN, H. Herophilus: the art of medicine in early Alexandria: edition, translation and essays. Cambridge, New York, Cambridge Univ. Press, 1988. 666 pp.

HERRICK, Charles Judson (1868–1960). American comparative neurologist and psychologist.

BARTELMIEZ, G.W. BMNAS, 1973, **43**, 77–108.
O'LEARY, J.L. & BISHOP, C.J. C.J. Herrick, scholar and humanist; a memorial essay written for his centennial. *Perspectives in Biology and Medicine*, 1969, **12**, 492–513.
ROOFE, P.G. DSB, 1972, **6**, 320–22.
Archival material: Spencer Research Library, University of Kansas.

HERRICK, Clarence Luther (1858–1904). American neurobiologist.

WINDLE, W.F. The pioneering role of Clarence Luther Herrick in American neuroscience. Hacksville, NY, Exposition Press, 1979. 140 pp.

HERRICK, James Bryan (1861–1954). American cardiologist.

HERRICK, J.B. Memories of eighty years. Chicago, Univ. Chicago Press, 1949. 270 pp.

HERRLINGER, Robert (1914–1968). German medical historian.

FEINER, E. Bibliographie Robert Herrlinger (1914–1968). Kiel, Institut für Geschichte der Medizin, 1970. 94 pp.

HERSHEY, Alfred Day (1908–1997). American molecular biologist; Nobel Prize (Physiology or Medicine) 1969 for work on replication mechanisms and genetic structure of viruses.

STAHL, F.W. (ed.) We can sleep later: Alfred D.Hershey and the origins of molecular biology. New York, Cold Spring Harbour Laboratory Press, 2000. 359 pp.

HERTER, Christian Archibald (1865–1910). American physician; associated with 'Gee-Herter' disease.

HAWTHORNE, R.M. Christian Archibald Herter, M.D., 1865–1910. *Perspectives in Biology and Medicine*, 1974, **18**, 24–39.

HERTWIG, Karl Wilhelm Theodor Richard von (1850–1937). German embryologist and zoologist.

OPPENHEIMER, J.M. DSB, 1972, **6**, 336–7.

HERTWIG, Wilhelm August Oscar (1849–1922). German embryologist and zoologist.

OLBY, R. DSB, 1972, **6**, 337–40.
WEINDLING, P.J. Darwinism and social Darwinism in Imperial Germany: the contribution of the cell biologist Oscar Hertwig (1849–1922). New York, Stuttgart, Gustav Fischer, 1991. 355 pp.
WEISSENBERG, R. Oscar Hertwig, 1849–1922. Leben und Werk eines deutschen Biologen, Leipzig, Barth. 63 pp.

HERTZ, Carl Hellmuth (b.1920). German physicist; pioneer in the development of medical ultrasonics.

LINDSTROM, K. Carl Hellmuth Hertz. *Ultrasound in Medicine and Biology*, 1991, **17**, 421–4.

HERTZLER, Arthur Emanuel (1870–1946). American surgeon and pathologist.

HASHINGER, E.H. Arthur E. Hertzler: the Kansas horse-and-buggy doctor. Lawrence, Univ. Kansas Press, 1961. 37 pp.
HERTZLER, A.E. The horse and buggy doctor. New York, Harper. 1938. 322 pp.

HESS, Alfred Fabian (1875–1933). American paediatrician and nutritionist.

DARBY, W.J. & WOODRUFF, C.W. Alfred Fabian Hess – a biographical sketch (October 9, 1875–December 5, 1933). *Journal of Nutrition*, 1960, **71**, 3–9.

HESS, Walter Rudolf (1881–1973). Swiss physiologist; shared Nobel Prize 1949 for his discovery of the fundamental organization of the inter-brain as a coordinator of the activities of the internal organs.

HESS, W.R. From medical practice to theoretical medicine: an autobiographical sketch. *Perspectives in Biology and Medicine*, 1963, **6**, 400–424.
JUNG, R. Walter R. Hess (1881–1973). *Review of Physiology, Biochemistry and Pharmacology*, 1981, **88**, 1–21.

HESSING, Johann Friedrich von (1838–1918). German orthopaedic surgeon.

DER ORTHOPÄDE Friedrich von Hessing. München, W. Fritsch, 1970. 72 pp.

BIBLIOGRAPHY OF MEDICAL AND BIOMEDICAL BIOGRAPHY

HEUBNER, Johann Otto Leonhard (1843–1926). German paediatrician; professor at Leipzig, Berlin and Dresden.

GROTE, 1925, **4**, 93–124 (CB).
KLINGENBERG-STRAUB, A. Otto Heubner's Leben und Lehrbuch der Kinderheilkunde. Zürich, Juris-Verlag, 1968. 54 pp.

HEUYER, Georges (1884–1977). French child psychiatrist.

LANG, J.-L. Georges Heuyer: fondateur de la pédo-psychiatrie, un humaniste du Xxe siècle. Paris, Expansion Scientifique, 1997. 185 pp.

HEVESY, George (György) (1885–1966). Hungarian radiochemist; used isotopes as 'tracers' in medical and biological research; Nobel Prize for chemistry 1943.

COCKCROFT, J.D. BMFRS, 1967, **13**, 125–66.
HEVESY, G. Adventures in radioisotope research. 2 vols., Oxford, Pergamon Press, 1962.
LEVI, H, George de Hevesy: life and work; a biography. Bristol, Adam Hilger, 1985, 147 pp.
SZABADVARY, F. DSB, 1972, **6**, 365–7.

HEWITT, Frederic William (1857–1916). British anaesthetist.

BLOOMFIELD, J. *British Journal of Anaesthesia*, 1926–27, **4**, 116–23.

HEWSON, William (1739–1774). British haematologist.

BAILEY, G.H. William Hewson, FRS (1729–1774). An account of his life and works. *Annals of Medical History*, 1923, **5**, 209–24.
LeFANU, W. DSB, 1972, **6**, 367–8.
RICHARDSON, B.W. *Asclepiad*, 1891, **8**, 148–77.
Archival material: Royal Society of London; Royal College of Physician of London; Royal College of Surgeons of England.

HEY, William (1736–1819). British surgeon, founded and worked at Leeds Infirmary.

PEARSON, J. The life of William Hey, Esq., F.R.S. 2 vols. London, Hurst, Robinson & Co., 1822. 139 + 355 pp. 2nd ed., 1827.

HEYFELDER, Johann Ferdinand Martin (1798–1869). German surgeon; introduced ethyl chloride in anaesthesia.

HINTZENSTEIN, U. J.F. Heyfelder (1798–1869): a pioneer of German anaesthesia. In: ATKINSON, R.B. & BOLTON, T.B. (eds.) The history of anaesthesia. London, Royal Society of Medicine, 1988, pp. 502–5.
HINTZENSTEIN, U. & SCHWARZ, W. Frühe Erlanger Beiträge zur Theorie und Praxis der Äther- und Chloroformnarkose. *Anaesthesist*, 1996, **45**, 131–9.
WYSS, H.E. Johann Ferdinand Heyfelder 1798–1869: und seine Beobachtungen über die Krankheiten de Neugeborenen. Zurich, Juris, 1964. 46 pp.

HEYMANS Corneille Jean François (1892–1968). Belgian physiologist; Nobel prizewinner 1968 for work on sinus-aorta mechanism in respiration.

NOTICE biographique sur M. Corneille Heymans, Membre titulaire. *Bulletin de l'Académie Royale de Médecine de Belgique*, 1968, **8**, 591–604.
SCHAEPDRYVER, A.F. de (ed.) Corneille Heymans. A collective biography. Ghent, Heymans Foundation, 1973. 307 pp.

HEYSHAM, John (1753–1834). Practitioner in Carlisle, England; statistician and originator of the 'Carlisle tables'.

LONSDALE, H. The life of John Heysham, MD, and his correspondence with Mr Joshua Milne, relative to the Carlisle Bills of Mortality. London, Longmans Green, 1870. 173 pp.

HIBBS, Russell Aubra (1869–1932). American orthopaedic surgeon.

GOODWIN, G.M. Russell A. Hibbs, pioneer in orthopaedic surgery, 1869–1932. New York, Columbia Univ. Press, 1935. 136 pp.

HICKMAN, Henry Hill (1800–1830). British physician; induced suspended animation by asphyxiation and use of carbon dioxide.

CARTWRIGHT, F.F. The English pioneers of anaesthesia. Bristol, John Wright, 1952, pp. 265–332.
HENRY HILL HICKMAN centenary exhibition at the Wellcome Historical Medical Museum, 54 Wigmore Street, London, W1. London, Wellcome Foundation, 1930. 69 pp.

HICKS, John Braxton (1823–1897). British obstetrician; introduced combined podalic version.

YOUNG, J.H. John Braxton Hicks (1823–1897). *Medical History*, 1960, **4**, 153–62.

HIGHMORE, Nathaniel (1613–1685). English anatomist and physician.

GORDON, J.E. DSB, 1972, **6**, 386–8.
Archival material: British Library.

HILDEGARD of BINGEN, Saint (1098–1179). German abbess; wrote *Physica*, a treatise on medical treatment, etc.

ARIS, M.A. *et al.* (eds.) Hildegard von Bingen: Internationale wissenschaftliche Bibliographie. Mainz, Gesellschaft für Mittelrheinische Kirchengeschichte, 1998. 293 pp.
BERGER, M. Hildegard of Bingen on natural philosophy and medicine. Cambridge, DS. Brewer, 1999. 166 pp.
BETZ, O. Hildegard von Bingen: Gestalt und Werk. München, Kösel, *c.*1996. 246 pp.
FLANAGAN, S. Hildegard of Bingen, 1098–1179; a visionary life. 2nd edition. London, Routledge, 1998. 227 pp.
LAUTER, W. Hildegard-Bibliographie. Wegweiser zur Hildegard-Literatur. 2 vols. Alzey, Rheinhessische Druckwerkstatte, 1970–84. 83+96 pp.
MADDOCKS, F. Hildegard of Bingen: the woman of her age. London, Headline, 2001. 288 pp.

HILL, Archibald Vivian (1886–1977). British physiologist; Jodrell Professor at University College London; shared Nobel Prize 1922.

HILL, A.V. Autobiographical sketch. *Perspectives in Biology and Medicine*, 1970, **14**, 27–42.
HILL, A.V. Memories and reflections. 3 vols., Xerox copy at Churchill Archives Centre, Churchill College, Cambridge. 738 pp.
HILL, A.V. Trails and trials in physiology: a bibliography 1909–1964; with reviews of certain topics and methods and a reconnaissance. London, E. Arnold, 1965. 374 pp.
KATZ, L.N. BMFRS, 1978, **24**, 71–149.
Archival material: Churchill College, Cambridge; CMAC.

HILL, Austin Bradford (1897–1991). British medical statistician; professor at London University.

AUSTIN BRADFORD HILL [a series of papers]. *Stastistics in Medicine*, 1982, **1** (4), 305–59 [bibliography of A.B. Hill, 351–75].
DOLL, R. BMFRS, 1994, **40**, 128–40.

HILL, John (1714–1775). British apothecary and botanist; first to show association between tobacco and cancer.

DOLMAN, C.E. 'That impudent fellow Hill'. *Annals of Science*, 1983, **40**, 281–8.

HILL, John Denis Nelson (1913–1982). British psychiatrist; professor at London University.

REYNOLDS, E.H. & TRIMBLE, M.R. (eds.) The bridge between neurology and psychiatry. London, Churchill Livingstone, 1989. 409 pp. [Chapters 1–2 on Sir Denis Hill.] Reprinted 1990.

HILL, Leonard Erskine (1866–1952). British physiologist.

DOUGLAS, C.G. ONFRS, 1952–53, **8**, 431–43.
HILL, L.E. Philosophy of a biologist. London, Arnold, 1930. 88 pp.

HILLARY, William (1697–1763). English physician; gave early account of sprue.

BOOTH, C.C. William Hillary, a pupil of Boerhaave. *Medical History*, 1963, **7**, 297–316.

HILTON, John (1804–1878). British surgeon at Guy's Hospital, London.

HILTON, J. Rest and pain. Edited by E.W. Wallis and E.E. Philipp. London, G. Bell, 1950. 503 pp. Includes biographical introduction.

HIMSWORTH, Harold Percival (1905–1993). British physician and medical scientist; Secretary of Medical Research Council, 1949–1968.

BLACK, D. & GRAY, J. BMFRS, 1995, **41**, 200–218.
Archival material: CMAC.

HINCKS. Clarence Meredith (1885–1964). Canadian physician; established Canadian National Committee for Mental Hygiene.

ROLAND, C.G. Clarence Hincks: mental health crusader. Toronto, Oxford, Hannah Institute & Dundurn Press, 1990. 128 pp.

HINDLE, Edward (1886–1973). British protozoologist and parasitologist.

GARNHAM, P.C.C. BMFRS, 1974, **20**, 217–34.
Archival material: University of Glasgow.

HINGSON, Robert Andrew (1913–1996). American anaesthesiologist.

ROSENBERG, H. & AXELROD, J.K. Robert Andrew Hingson: his unique contributions to world health as well as to anesthesiology. *American Journal of Anesthesiology*, 1985, **25**, 90–93.

HINSHELWOOD, Cyril Norman (1897–1967). British chemist; shared Nobel Prize 1956 for research on chemical kinetics of bacterial cells.

THOMPSON, H. BMFRS, 1973, **19**, 375–431.
Archival material: Royal Society, London.

HINTON, James (1822–1875). British aural surgeon; first aural surgeon to Guy's Hospital, London.

ELLIS, E.M. James Hinton: a sketch. London, Stanley Paul & Co., 1918. 283 pp.

HINTZSCHE, Gustav Werner (1900–1975). German anatomist and medical historian.

BOSCHUNG, U. Erich Hintzsche, 1900–1975. Nachruf mit Gesamtbibliographie. *Gesnerus*, 1975, **32**, 293–314.

HIPPEL, Arthur von (1841–1917). German ophthalmologist; pioneer in keratoplasty.

SCHIECK, F. *Zeitschrift für Augenheilkunde*, 1917, **37**, 115–19.
SCHLODTMANN, –. *Klinische Monatsblätter für Augenheilkunde*, 1916, **57**, 582–7.

HIPPOCRATES of Cos (460–370 BC). Greek physician, the 'Father of medicine'.

JOLY, R. DSB, 1972, **6**, 419–31.
JOUANNA, J. Hippocrates. Translated by M.B. DeBevoise. Baltimore, Johns Hopkins Univ. Press, 1999. 520 pp. First published in French, 1992.
LEVINE, E.B. Hippocrates. Boston, Twayne, 1971. 172 pp.
MOON, R.O. Hippocrates and his successors in relation to the philosophy of their time. London, Longmans, Green, 1923. 171 pp. [Devoted mainly to the Hippocratic and post-Hippocratic schools, including Pythagoras, Heraclitus, Empedocles, Anaxagoras, etc.]
TEMKIN, O. Hippocrates in a world of pagans and Christians. Baltimore, Johns Hopkins Univ. Press, 1991. 315 pp.

HIRSCH, August (1817–1894). German medical historian; writings include Handbuch der historisch-geographischen Pathologie (1860–64) and Biographisches Lexikon der hervorragenden Aertze (1884–88).

PAGEL, J.L. August Hirsch. Braunschweig, F. Vieweg u. Sohn, 1894. 8 pp.

HIRSCHBERG, Julius (1843–1925). German ophthalmologist and historian of ophthalmology.

HIRSCHBERG, J. Aus jungen Tagen. Berlin, Junk, 1923. 44 pp.
SNYDER, C. Julius Hirschberg, the neglected historian of ophthalmology. *American Journal of Ophthalmology*, 1981, **91**, 664–76.

HIRSCHFELD, Magnus (1868–1935). German physician and sexologist.

HERZER, M. Magnus Hirschfeld: Leven und Werk eines jüdischen, schwulen, und sozialistischen Sexologen. Frankfurt a.M., Campus, 1992.
WOLFF, C. Magnus Hirschfeld: a portrait of a pioneer in sexology. London, Quartet, 1986. 494 pp.

HIRSCHSPRUNG, Harald (1830–1916). Danish paediatrician.

HOLD-PETERSON, K. & ERICHSEN, G. The Danish paediatrician Harald Hirschsprung. *Surgery, Gynecology & Obstetrics*, 1988, **166**, 181–5.

HIRSZFELD, Ludwik (1884–1954). Polish serologist and bacteriologist; discovered heritability of blood groups.

BIBLIOGRAPHY OF MEDICAL AND BIOMEDICAL BIOGRAPHY

JAWORSKI, M. Ludwig Hirzfeld: sein Beitrag zu Serologie und Immunologie. Leipzig, B.G. Teubner, 1980. 91 pp. First published in Polish, Warsaw, 1977.
SCHADEWALDT, H. DSB, 1972, **6**, 432–4.

HIS, Wilhelm *the Elder* (1831–1904). Swiss anatomist, histologist and embryologist,

HIS, W. *the Younger*. Wilhelm His der Aeltere. Lebenserinnerungen und angewälte Schriften. Zusammengestellt und heraugegeben von Eugen Ludwig. Bern, Hüber, 1965. 139 pp.
QUERNER, H. DSB, 1972, **6**, 434–6.

HIS, Wilhelm *the Younger* (1863–1934). Swiss anatomist; described the atrioventricular bundle ('bundle of His').

BAST, T.H. & GARDNER, W.D. Wilhelm His, Jr. and the bundle of His. *Journal of the History of Medicine*, 1949, **4**, 298–318.

HITCHINGS, George Herbert (b.1905). American chemotherapist; shared Nobel Prize (Physiology or Medicine), 1988, for discovery of important principles for drug treatment.

Les Prix Nobel en 1988.

HITSCHMANN, Fritz (1870–1926). Austrian obstetrician.

FROBENIUS, W. Fehlddiagnose Endometritis: zur Revision eines wissenchaftlichen Irrtums durch die Wiener Gynäkologen Fritz Hitschmann und Ludwig Adler. Hildesheim, Olms, 1988. 220 pp.

HITZIG, (Julius) Eduard (1838–1907). German neurophysiologist and psychiatrist; demonstrated muscular responses to electorial stimulation of cerebral cortex.

CLARKE, E. DSB, 1972, **6**, 440–41.

HLAVA, Jaroslav (1855–1924). Bohemian physician; induced experimental amoebiasis.

–. *Bulletin de l'Académie de Médecine*, Paris, 1924, 3 sér., **92**, 1200–1207.
DOBELL, C. Parasitology, 1938, **30**, 239–41.
Archival material: CMAC.

HOARE, Cecil Arthur (1892–1984). British protozoologist.

GOODWIN, L.G. & BRUCE-CHWATT, L.J. BMFRS, 1985, **31**, 295–323.

HOCH, Paul H. (b.1902). American psychiatrist.

LEWIS, N.D.C. & STRAHL, M.O. (eds.) The complete psychiatrist; the achievements of Paul H. Hoch. Albany, State University of New York, 1968. 723 pp.

HOCHE, Alfred Erich (1865–1945). German psychiatrist; professor of psychiatry and neuropathology, Freiburg i. Br.

GROTE, 1923, **1**, 1–24.
HOCHE, A.E. Jahresringe Innenansicht Menschenlebens. München, J.F. Lehmann, 1939. 298 pp.

HODGE, Hugh Lenox (1796–1873). American obsterician and gynaecologist; invented the Hodge pessary.

BIBLIOGRAPHY OF MEDICAL AND BIOMEDICAL BIOGRAPHY

GOODELL, W. Biographical memoir of Hugh L. Hodge. Philadelphia, Collins, 1874. 19 pp.
PENROSE, R.A.F. A discourse commemorative of the life and character of Hugh L. Hodge. Philadelphia, Collins, 1873. 31 pp.

HODGKIN, Alan Lloyd (1914–1998). British biophysicist; shared Nobel Prize (Physiology or Medicine) 1964 for discoveries concerning the ionic mechanism involved in the excitation and inhibition in the peripheral and central nervous portions of the cell membrane.

HODGKIN, A.L. Beginning; some early reminiscences of my early life (1914–1947) *Annual Review of Physiology*, 1983, **45**, 1–16.
HODGKIN, A. Choice and design: reminiscences of science in peace and war. Cambridge, Cambridge Univ. Press, 1992. 412 pp.
HUXLEY, A. BMFRS, 2000, **46**, 219–41.

HODGKIN, Dorothy Mary Crowfoot (1910–1994). British chemist and crystallographer; Nobel Prize, Chemistry, 1961, for determining, by X-ray crystallography, the structure of complex, biologically important, organic molecules.

DODSON, G. BMFRS, 2002, **48**, 181–219.
DODSON, G. *et al.* (eds.) Structural studies on molecules of biological interest. A volume in honour of Dorothy Hodgkin. Historical and biological chapters by M.F. Perutz *et al.* Oxford, Clarendon Press, 1981. 610 pp.
FERRY, G. Dorothy Hodgkin: a life. London, Granta Books, 1998. 423 pp.
Archival material: Bodleian Library, Oxford.

HODGKIN, Thomas (1798–1866). British physician; gave first full description of lymphadenoma ('Hodgkin's disease').

KASS, A.M. & KASS, E.H. Perfecting the world; the life and times of Dr Thomas Hodgkin. Boston, Harcourt Brace Jovanovich, 1988. 642 pp.
ROSE, R. Curator of the dead: Thomas Hodgkin (1798–1866). London, Peter Owen, 1981. 148 pp.
ROSENFELD, L. Thomas Hodgkin: morbid anatomist and social activist. Lanham, MD, Madison, 1993. 0000 pp.

HÖBER, Rudolph Otto Anselm (1873–1953). German/American physiologist.

LENOIR, T. DSB, 1990, **17**, 423–5.

HÖGYES, Endre (1847–1906). Hungarian physician; described rotational nystagmus, 1886.

–. *Orvosi Hetilap*, 1906, **1**, 854–7.

HOFF, Hans (1897–1969). Austrian psychiatrist.

KRAEMER, H.-D. Hans Hoff 1897–1969; Leben und Werk. Inauguraldissertation...der Johannes Gutenberg-Universität Mainz, 1975. Heidesheim Rhein, Ditters Bürodienst, 1976. 91 pp.

HOFFMAN, August Heinrich (1809–1874). German physician, philologist and poet; wrote *Struwwelpeter.*

HOFFMAN, H. Lebenserinnerungen. Frankfurt am Main, Insel Verlag, 1985. 364 pp.

BIBLIOGRAPHY OF MEDICAL AND BIOMEDICAL BIOGRAPHY

HOFFMANN, Erich (1868–1959). German dermatologist; with Schaudinn discovered *Treponema pallidum*, causal organism in syphilis.

HOFFMANN, E. Wollen und Schaffen. Lebenserinnerungen aus eine Wendezeit der Heilkunde. 1868–1932. (Ringen und Vollendung Lebenserinnerungen...1933–46). 2 vols. Hannover, Schmorl und von Seefeld. [1948–49]. 403 + 312 pp.

HOFFMANN, Friedrich (1660–1742). German physician and chemist.

RISSE, G.B. DSB, 1972, **5**, 458–61.
ROTHSCHUH, K.E. Studien zu Friedrich Hoffmann. Sudhoff's Archiv, 1976, **60**, 163–93, 235–70.
WACKERNAGEL, I. Friedrich Hoffmann (1660–1742) und seine Beziehungen zur Zahn, Mund- und Keiferheilkunde. Inaugural-Dissertation...an der Medizinischen Fakultät der Freien Universität, Berlin. Berlin, Medizinischen Fakultät der Freien Universität. 79 pp.

HOFMANN, Albert (b.1906). German chemist; with A. Stoll synthesized lysergic acid diethylamide.

CAMILLA, G. Hofmann: scienziato alchimista. Roma, Stampa Alternativa, 2001. 91 pp.
ULRICH, R.F. & PATTEN, B.M. The rise, decline, and fall of LSD. Perspectives in Biology and Medicine, 1991, **34**, 561–78.

HOFMANN, Klaus (1911–1995). Swiss/American biochemist; worked on ACTH and hormone receptors.

FINN, F.M. & O'MALLEY, B.W. BMNAS, 2002, **81**, 116–34.

HOFMEISTER, Franz (1850–1922). Austro-German biochemist and pharmacologist.

FRUTON, J.S. DSB, 1990, **17**, 430–33.
POHL, J. & SPIRO, K. Franz Hofmeister: sein Leben und Wirken. *Ergebnisse der Physiologie*, 1923, **22**, 1–50.

HOGAN, Albert Garland (1884–1954). American biochemist; isolated vitamin B_c.

RICHARDSON, L.R. Albert Garland Hogan – a biographical sketch. *Journal of Nutrition*, 1969, **97**, 3–7.

HOGBEN, Lancelot Thomas (1895–1975). British biologist.

HARPER, P. Supplementary catalogue of papers and correspondence relating to Lancelot Thomas Hogben FRS (1895–1975): material additional to CSAC 78/2/81 and NCUACS 29/5/91. Bath, University of Bath, 1995. 16 leaves.
HOGBEN, Adrian & HOGBEN, Anne (eds): Lancelot Hogben: scientific humanist – an unauthorized autobiography. Rendlesham, Merlin Press, 1998. 254 pp.
WELLS, G.P. BMFRS, 1978, **24**, 183–221.
Archival material: Library, University of Birmingham.

HOLLAND, Henry (1788–1873). British physician and traveller; physician-in-ordinary to Queen Victoria.

HOLLAND, H. Recollections of past life. London, Longmans, Green, 1872. 346 pp.

HOLLEY, Robert William (b. 1922). American biochemist; shared Nobel Prize (Physiology or Medicine) 1968, for work on RNA.

177

BIBLIOGRAPHY OF MEDICAL AND BIOMEDICAL BIOGRAPHY

Les Prix Nobel en 1968.

HOLMES, Gordon Morgan (1876–1965). British neurologist.

WALSHE, F.M.R. BMFRS, 1966, **12**, 311–19.

HOLMES, Oliver Wendell (1809–1894). American physician; professor of anatomy and physiology, Harvard University; poet and essayist; pointed out, before Semmelweis, the contagious nature of puerperal fever.

CURRIER, T.F. A bibliography of Oliver Wendell Holmes ... Edited by E.M. Tilton for the Bibliographical Society of America. New York, New York Univ. Press, 1953. 708 pp.
HAWTHORNE, H. The happy autocrat: a life of Oliver Wendell Holmes. New York, Longmans Green, 1938. 213 pp.
HOYT, E.P. The improper Bostonian, Dr Oliver Wendell Holmes. New York, William Morrow & Co., 1979. 319 pp.
MORSE, J.T. Life and letters of Oliver Wendell Holmes. 2 vols., Boston, New York, Houghton, Mifflin, 1896. 358 + 335 pp.
TILTON, Eleanor M. Amiable autocrat: a biography of Dr Oliver Wendell Holmes. New York, Schuman, 1947. 470 pp.

HOLMGREN, Alarik Frithiof (1831–1897). Swedish physiologist; investigated colour-blindness.

GRANIT, R. DSB, 1972, **6**, 476–7.

HOLMGREN, Gunnar (1875–1954). Swedish otorhinolaryngologist; devised fenestration operation for otosclerosis.

HAMBERGER, C.A. Gunnar Holmgren (1875–1954). *Archives of Otolaryngology*, 1968, **87**, 214–18.

HOLT, Luther Emmett (1855–1924). American paediatrician.

DUFFUS, R.L. & HOLT, L.E. Jr. L. Emmett Holt, pioneer of a children's century. New York, London, Appleton-Century, 1940. 295 pp.

HOLTFREDER, Johannes (1901–1992). German/American embryologist.

GERHARDT, J. BMNAS, 1998, **73**, 209–24.

HOLTH, Sören (1863–1937). Norwegian ophthalmologist; introduced iridencleisis for treatment of glaucoma.

–. *British Journal of Ophthalmology*, 1937, **21**, 668a–70.

HOLTZ, Friedrich (1898–1967). German pharmacologist; introduced AT10 in treatment of tetany.

MARQUARDT, P. In memoriam: Friedrich Holtz. *Arzneimittelforschung*, 1967, **17**, 919–20.

HOLZKNECHT, Guido (1872–1931). Austrian radiologist; improved dosimetry standards in radiotherapy.

ANGETTER, D. Leben und Werk des Österreichischen Pioniers der Röntgenologie. Wien, Eichbauer, 1998. 187 pp.

WALTHER, K.H. Guido Holzknecht: der grosse Wiener Röntgenarzt. Wien, A. Kaltschmid, 1975. 72 pp.

HOME, Everard (1756–1832). British surgeon and comparative anatomist.

Le FANU, W.R. DSB, 1972, **6**, 478–9.
Archival material: British Library; General Library, British Museum (Natural History); Royal College of Surgeons of England; Royal Society of London.

HOME, Francis (1719–1813). British physician in Edinburgh; inoculated against measles.

ENDERS, J.F. Francis Home and his experimental approach to medicine. *Bulletin of the History of Medicine*, 1964, **38**, 101–12.
HUME, E.E. Francis Home, MD (1719–1813), the Scottish military surgeon who first described diphtheria as a clinical entity. *Bulletin of the History of Medicine*, 1942, **11**, 46–68.

HOOKE, Robert (1635–1703). English physicist; first curator of experiments, Royal Society of London; pioneer microscopist, scientist and architect.

DRAKE, E.T. Restless genius: Robert Hooke and his earthly thoughts. New York, Oxford Univ. Press, 1996. 386 pp.
GUNTHER, R.T. Early science in Oxford Vol. 6, 7, 8, 10, 13. The life and work of Robert Hooke 5 vols., Oxford, for the author, 1930–38.
INWOOD, S. The man who knew too much: the inventive life of Robert Hooke. London, Macmillan, 2002. 485 pp.
JARDINE, L. The curious life of Robert Hooke: the man who measured London. London, HarperCollins, 2003. 352 pp.
KEYNES, G.L. A bibliography of Dr Robert Hooke. Oxford, Clarendon Press, 1960. 115 pp. [Contains lists of his contributions to books, of his papers, letters and manuscripts, with a section on biography and criticism.]
WESTFALL, R.S. DSB, 1972, **6**, 481–8.

HOOKER, Joseph Dalton (1817–1911). British botanist and physician; Director of Kew Gardens, 1865.

DESMOND, R. Sir Joseph Hooker, traveller and plant collector. Woodbridge, Antique Collectors' Club, 1999. 286 pp.
TURRILL, W.B. Joseph Dalton Hooker, botanist, explorer and administrator. London, Thomas Nelson, *c.*1963. 228 pp.

HOPE, James (1801–1841). British cardiologist; physician to St. George's Hospital, London.

FLAXMAN, N. The hope of cardiology, James Hope (1801–1841). *Bulletin of the Institute of the History of Medicine*, 1938, **6**, 1–21.
HOPE, A. Memoir of the late James Hope, M.D. 4th ed. London, Hatchard, 1848. 348 pp. First published 1842.

HOPKINS, Frederick Gowland (1861–1947). British biochemist, 'father of British biochemistry'; shared Nobel Prize 1929.

BALDWIN, E. Gowland Hopkins. A memoir on the discoverer of vitamins. London, Van den Berghs, 1962. 25 pp.
BALDWIN, E. DSB, 1972, **6**, 498–502.
DALE, H.H. ONFRS, 1948, **6**, 115–45.

BIBLIOGRAPHY OF MEDICAL AND BIOMEDICAL BIOGRAPHY

NEEDHAM, J. & BALDWIN, E. (eds.) Hopkins and biochemistry. Cambridge, Heffer, 1949. 361 pp.
Archival material: Churchill College, Cambridge; Library, University of Cambridge.

HOPKINS, Harold Horace (1918–1994). British physician; introduced flexible fibrescope, 1954.

GOW, J.G. Harold Hopkins and optical systems for urology: an appreciation. *Urology*, 1998, **52**, 153–7.

HOPPE-SEYLER, Ernst Felix Emmanuel (1825–1895). German physiological chemist and haematologist.

BAUMANN, E. & KOSSEL, A. Zur Erinnerung am Felix Hoppe-Seyler. *Hoppe-Seylers Zeitschrift für physiologische Chemie*, 1895, **21**, i–lxii.
FRUTON, J.S. DSB, 1972, **6**, 504–6.

HORBACZEWSKI, Jan (1854–1942). Austro-Hungarian (Czech) biochemist.

TEICH, M. DSB, 1972, **6**, 506.

HORDER, Thomas Jeeves (1871–1955). British physician; physician to St Bartholomew's Hospital.

HORDER, M. The little genius; a memoir of the first Lord Horder. London, Duckworth, 1966. 147 pp.
Archival material: CMAC.

HORN, Ernst (1774–1843). German physician.

SCHNEIDER, H. Ernst Horn (1774–1848). Ein ärtzlicher Direktor des Berliner Charité an der Wende zur naturwissenschaftliche Medizin. Inaugural-Dissertation...and der Medizinischen Fachbereichen der Freien Universität Berlin. Berlin, Medizinische Fachbereichen der Freien Universität, 1986. 288 pp.

HORNER, Johann Friedrich (1831–1886). Swiss neurologist; professor in Zürich; described oculopupillary ('Horner's) syndrome – ptosis due to lesion of cervical sympathetic.

KOELBING, H.M. & MÖRGELI, C. (eds.) Johann Friedrich Horner, 1831–1886; der Begründer der Schweizer Augenheilkunde in seiner Autobiographie. Zürich, Hans Rohr, 1986. 113 pp.
OTT, E. Friedrich Horner, 1831–1886. Leben und Werk. Zürich, Juris Druck, 1980. 54 pp.

HORNEY, Karen (1885–1952). German/American psychoanalyst.

PARIS, B.J. Karen Horney: a psychoanalyst's search for self-understanding. New Haven, Yale Univ Press, 1994. 270 pp.
QUINN, S. A mind of her own. The life of Karen Horney London, Macmillan, 1987. 479 pp.
RUBINS, J.L. Karen Horney: gentle rebel of psychoanalysis. New York, Dial Press, 1978. 372 pp.

HORSFALL, Frank Lappin (1906–1971). American virologist; Vice-President of Rockefeller Institute and director of Sloan-Kettering Institute.

HIRST, G.K. BMNAS, 1979, **50**, 233–67.

180

HORSLEY, Victor Alexander Haden (1857–1916). British neurosurgeon, neurophysiologist and pathologist.

CLARKE, E. DSB, 1972, **6**, 518–19.
LYONS, J.B. The citizen surgeon: a biography of Sir Victor Horsley, 1857–1916, London, Peter Downay, 1966. 305 pp.
PAGET, S. Sir Victor Horsley: a study of his life and work. London, Constable, 1919. 358 pp.
Archival material: Library, University College London.

HORSTIUS, Johann Daniel (1616–1685). German balneologist; professor of medicine, Giessen.

PATON, S. Johann Daniel Horstius (1616–1685). Sein Leben und Wirken, sowie seine Bedeutung für die Balneologie des 17, Jahrhunderts. Giessen, Wilhelm Schmitz Verlag, 1989. 252 pp.

HORTON, James Africanus Beale (1835–1883). Born Sierra Leone, qualified in medicine in Britain and served as an officer in the British Army in West Africa.

FYFE, C. Africanus Horton 1835–1883; West African scientist and patriot. New York, Oxford University Press, 1972. 169 pp.

HORVITZ, H. Robert (b. 1947). American molecular biologist; shared Nobel Prize (Physiology or Medicine), 2002, for discoveries concerning genetic regulation of organ development and cell death.

Les Prix Nobel en 2002.

HOSACK, David (1769–1835). American physician; founder of the Elgin Botanic Garden, New York.

ROBBINS, C. C. David Hosack, citizen of New York. Philadelphia, American Philosophical Society, 1964. 246 pp.
Archival material: Library, American Philosophical Society, Philadelphia.

HOUNSFIELD, Godfrey Newbold (b. 1919). British electrical engineer; shared Nobel Prize (Physiology or Medicine), 1979, for introduction of computer-assisted tomography.

Les Prix Nobel en 1979.

HOUSSAY, Bernardo Alberto (1887–1971). Argentinian physiologist; shared Nobel Prize 1947 for his work on the pituitary.

RISSE, G.B. DSB, 1978, **15**, 228–9.
YOUNG, F.G. & FOGLIA, V.G. BMFRS, 1974, **20**, 247–70.

HOUSTON, Alexander Cruikshank (1865–1933). British physician and bacteriologist; Director of Water Examination, Metropolitan (London) Water Board.

G., M.H., ONFRS, 1932–35, **1**, 335–44.

HOUSTOUN [HOUSTON], Robert (1678–1734). Scottish surgeon; first (in 1701) to treat ovarian dropsy by tapping the cyst.

MacKINLAY, C.J. Who is Houston? A biography of Robert Houstoun, M.D., F.R.S. *Journal of Obstetrics and Gyngaecoloy of the British Commonwealth*, 1973, **80**, 193–200.

BIBLIOGRAPHY OF MEDICAL AND BIOMEDICAL BIOGRAPHY

HOWARD, James Griffiths (1927–1998). British immunologist.

MITCHISON, N.A. BMFRS, 2000, **46**, 257–67.

HOWARD, John (1726–1790). British philanthropist and prison reformer.

BAUMGARTNER, L. John Howard (1726–1790); hospital and prison reformer. A bibliography. *Bulletin of the History of Medicine*, 1939, **7**, 486–626.
HOWARD, D.L. John Howard: prison reformer. London, C. Johnson, 1958. 186 pp.

HOWELL, William Henry (1860–1945). American physiologist; professor at Johns Hopkins University.

ERLANGER, J. BMNAS, 1951, **26**, 153–80.
MARCUM, J. William Henry Howell and Jay McLean: the experimental context for the discovery of heparin. *Perspectives in Biology and Medicine*, 1990, **33**, 214–30.
RODMAN, A.C. DSB, 1972, **6**, 525–7.

HOWSE, Neville Reginald (1863–1930). Australian surgeon, awarded Victoria Cross, 1900.

BRAGA, S. Anzac doctor: the life of Sir Neville Howse, Australia's first V.C. Alexandria, NSW, Hale & Iremonger, 2000. 392 pp.
TYQUIN, M.B. Neville Howse, Australia's first Victoria Cross winner. South Melbourne, New York, Oxford Univ. Press, 1999. 212 pp.

HRDLICKA, Aleš (1869–1943). American anthropologist, born Bohemia.

HAJNIŠ, K. DSB, 1972, **6**, 527–8.

HUARTE Y NAVARRO, Juan de Dios (?1529–1588). Spanish physician and psychologist.

READ, M.K. Juan Huarte de San Juan. Boston, Twayne Publishers, 1981. 147 pp.

HUBBARD, John Perry (1903–1990). American paediatrician; with R.E. Goss reported first successful ligation of patent ductus arteriosus, 1934.

LEVIT, E.J. Memoir of John Perry Hubbard, 1903–1990. *Transactions & Studies of the College of Physicians of Philadelphia*, 1991, 5 ser. **18**, 215–19.

HUBEL, David Hunter (b. 1926). Canadian/American neurobiologist; shared Nobel Prix (Physiology or Medicine), 1981, for discoveries concerning information processes in the visual system.

Les Prix Nobel en 1981.

HUBER, Johann Jacob (1707–1778). Swiss anatomist; gave first accurate description of spinal cord.

CLARKE, E. DSB, 1972, **6**, 533–4.

HUEPPE, Ferdinand (1852–1938). German bacteriologist; professor at German University, Prague.

GROTE, 1923, **2**, 11–138.

HUFELAND, Christoph Wilhelm (1762–1836). German professor of pathology at Jena; author of popular work on personal hygiene.

BUSSE, H. Christoph Wilhelm Hufeland: der berühmte Arzt der Goethezeit, Leibarzt der Königin Luise. St Michael, Blaschke, 1982. 334 pp.

GENSCHOREK, W. Christoph Wilhelm Hufeland: der Arzt, der das Leben verlängern half. Leipzig, Wirzel, 1982. 215 pp.

PFEIFFER, K. Medizin der Goethezeit. Christoph Wilhelm Hufeland und die Heilkunst des 18. Jahrhunderts. Cologne, Böhlau, 2000. 293 pp.

HUGGETT, Arthur St George Joseph McCarthy (1897–1968). British physiologist.

BRAMBELL, F.W.B. BMFRS, 1970, **16**, 343–64.

HUGGINS, Charles Brenton (b. 1901). Canadian/American physician; shared Nobel Prize (Physiology or Medicine), 1966, for work on hormonal treatment of cancer.

Les Prix Nobel en 1966.

HUGGINS, Geoffrey, *Viscount Malvern* (1883–1971). Surgeon and Rhodesian statesman; Prime Minister of Rhodesia 1933–56.

GANN, L.H. & GELFAND, M. Huggins of Rhodesia: the man and his country. London, Allen & Unwin, 1964. 285 pp.

HUG-HELLMUTH, Hermine von (1871–1924). German child psychoanalyst.

MACLEAN, G. & RAPPEN, U. (eds.) Hermine Hug-Helmuth; life and selected works. London, Tavistock/Routledge, 1991. 305 pp.

HUGO BENZI, see **UGO BENZI**

HUGO SENENSIS, *see* **UGO BENZI**

HULL, Clark Leonard (1884–1952). American psychologist; professor at Yale University.

BEACH. FA. BMNAS, 1959, **33**, 125–41.

HUME, Edward Hicks (1876–1957). Indian-born American physician, researched on bubonic plague in India; founded and directed the medical school at Yale-in-China, Changsha, Hunan.

HUME, E.H. Doctors east, doctors west. An American physician's life in China. New York, Norton, 1946. 278 pp.

HUMPHREY, John Herbert (1915–1987). British physician; Deputy Director, National Institute for Medical Research; professor of immunology, University of London.

ASKONAS, B.A. BMFRS, 1990, **36**, 273–300.

HUMPHREY, J.H. Serendipity in immunology. *Annual Review of Immunology*, 1984, **2**, 1–21.

HUMPHRY, George Murray (1820–1896). British anatomist and surgeon; professor at Cambridge; founded *Journal of Anatomy and Physiology*.

ROLLESTON, H.D. Sir George Humphry, MD, FRS. New York, P.B. Hoeber, 1927. 11 pp. Reprinted from *Annals of Medical History*, 1927, **9**, 1–11.

HUNAYN IBN ISHĀQ [Johannitius] (808–873). Iraqi physician and translator of Greek scientific works.

ISKANDAR, A.Z. DSB, 1978, **15**, 230–49.
SA'DI, L.M. A bio-bibliographical study of Hunayn ibn Ishāq al-Ibadi (Johannitius) *Bulletin of the Institute of the History of Medicine*, 1934, **2**, 409–46.

HUNT, George Herbert (1884–1926). British physician.

–. *Guy's Hospital Reports*, 1926, **76**, 127–34.

HUNT, James (1833–1869). British anthropologist; wrote on speech disabilities.

SIMPKINS, G.M. DSB, 1972, **6**, 564–5.

HUNT, R. Timothy (b. 1943). British oncologist; shared Nobel Prize (Physiology or Medicine), 2001, for discovery of key regulators of the cell cycle.

Les Prix Nobel en 2001.

HUNT, Reid (1870–1948). American pharmacologist; professor at Harvard Medical School.

MARSHALL, E.K. BMNAS, 1951, **26**, 25–44.
PARASCANDOLA, J. DSB, 1990, **17**, 439–40.
Archival material: Harvard University; Harvard Medical School.

HUNTER, John (1728–1793). British surgeon, anatomy teacher, comparative anatomist and founder of experimental surgery.

DOBSON, J. John Hunter. Edinburgh, London, Livingstone, 1969. 361 pp.
Le FANU, W.R. John Hunter: a list of his books. London, Royal College of Surgeons, 1946. 31 pp.
OPPENHEIMER, J.M. New aspects of John and William Hunter. I. Everard Home and the destruction of the John Hunter manuscripts. II. William Hunter and his contemporaries. New York, Schuman, 1946. 188 pp.
OTTLEY, D. The life of John Hunter, F.R.S. containing a new memoir of the author, his unpublished correspondence with Dr Jenner and Sir Joseph Banks, a short exposition of the Hunterian museum, and many original anecdotes. London, Longman, Orme, Brown, Green and Longman, 1835. 139 pp., bibliography. (Forms vol. I of J.F. Palmer's *Works of John Hunter.*)
PEACHEY, G.C. A memoir of William and John Hunter. Plymouth, W. Brendon, 1924. 313 pp.
QVIST, G. John Hunter, 1728–1793. London, Heinemann Medical, 1981. 216 pp.
Archival material: Royal College of Surgeons of England.

HUNTER, John Irvine (1898–1924). Australian neuroanatomist; professor of anatomy, Sydney University.

BLUNT, M.J. John Irvine Hunter of the Sydney Medical School 1898–1924. Sydney, University Press, 1985. 135 pp.
BRETT, J. The life of Johnny Hunter. Albany, NSW, Wilkinson Printers, 1983. 97 + 8 pp.

HUNTER, Wallace Samuel (1889–1954). American psychologist; professor at Brown University.

GRAHAM, C.H. BMNAS, 1958, **31**, 127–55.

HUNTER, William (1718–1783). British physician, teacher of anatomy, and obstetrician.

BROCK, C.H. Dr William Hunter's papers and drawings in the Hunterian Collection of Glasgow University Library. A handlist. Cambridge, Wellcome Unit for the History of Medicine, 1990. 84 pp.

BYNUM, W.F. & PORTER, R. (eds.) William Hunter and the eighteenth-century medical world. Cambridge, Cambridge Univ. Press, 1985. 424 pp.

ILLINGWORTH, C.F.W. The story of William Hunter. Edinburgh, London, Livingstone, 1967. 134 pp.

OPPENHEIMER, J. New aspects of John and William Hunter. I. Everard Home and the destruction of the John Hunter manuscripts. II. William Hunter and his contemporaries. New York, Schuman, 1946. 188 pp.

PEACHEY, G.C. A memoir of William and John Hunter. Plymouth, W. Brendon, 1924. 313 pp.

Archival material: Library, University of Glasgow; Royal College of Surgeons of England; Royal Society of London.

HUNTINGTON, George (1850–1916). American physician; gave classical description of degenerative (Huntington's) chorea.

STEVENSON, C.S. A biography of George Huntington, MD. *Bulletin of the Institute of the History of Medicine*, 1934, **2**, 53–76.

HUNTINGTON, George Sumner (1861–1927). American comparative anatomist; professor at College of Physicians and Surgeons, Columbia University.

HRDLIČKA, A. BMNAS, 1936, **18**, 245–84.

HURD, Henry Mills (1843–1927). American psychiatrist; professor at Johns Hopkins University.

CULLEN, T.S. Henry Mills Hurd, the first superintendent of the Johns Hopkins Hospital. Baltimore, Johns Hopkins University Press, 1920. 147 pp.

HURLEY, Thomas Ernest Victor (1888–1958). Australian surgeon and medical administrator.

HURLEY, J.V. Sir Victor Hurley: surgeon, soldier and administrator. Hawthorn, Victoria, The Author, 1989. 138 pp.

HURST, Arthur Frederick (1879–1944). British physician; physician to Guy's Hospital.

HURST, A.F. A twentieth century physician: being the reminiscences of Sir Arthur Hurst, London, Arnold, 1949. 200 pp.

RYLE, J.A. Sir Arthur Hurst D.M., F.R.C.P. *Guy's Hospital Reports*, 1945, **94**, 1–11.

HUSCHKE, Emil (1797–1858). German anatomist and physiologist.

SMIT, P. DSB, 1972, **6**, 573–4.

HUTCHINSON, John (1811–1861). British physiologist; researched on respiratory function and invented the spirometer.

BISHOP, P.J. A bibliography of John Hutchinson. *Medical History*, 1977, **21**, 384–96.

CLARKE, E. DSB, 1972, **6**, 575–6.

SPRIGGS, E.A. John Hutchinson, the inventor of the spirometer – his north country background, life in London, and scientific achievements. *Medical History*, 1977, **21**, 357–64.

BIBLIOGRAPHY OF MEDICAL AND BIOMEDICAL BIOGRAPHY

HUTCHINSON, Jonathan (1828–1913). British surgeon; 'the greatest general practitioner in Europe'.

ELLIS, H. Jonathan Hutchinson: life and letters. *Journal of Medical Biography*, 1993, **1**, 11–16.
HUTCHINSON, H. Jonathan Hutchinson: life and letters. London, Heinemann, 1946. 257 pp.
WALES, A.E. The life and work of Sir Jonathan Hutchinson...1828–1913, with an account of his family biography and of some aspects of nineteenth-century life and thought, particularly in science and medicine. 3 vols. Leeds, 1948. Leeds PhD thesis in typescript in the University of Leeds Brotherton Library.

HUTCHISON, Robert (1971–1960). British paediatrician.

MONCRIEFF, A. Pediatric profiles. Sir Robert Hutchison, Bart. *Journal of Pediatrics*, 1961, **58**, 137–9.

HUTTON, James (1726–1797). British physician and geologist; originated the concept of uniformitarianism.

DONOVAN, A. & PRENTISS, J. James Hutton's medical dissertation. Philadelphia, American Philosophical Society, 1980. 57 pp.
EYLES, V.A. DSB, 1972, **6**, 577–89.
PLAYFAIR, J. Biographical account of the late James Hutton, F.R.S. Edin. Transactions of the Royal Society of Edinburgh, 1803, **5**, 39–99.

HUXHAM, John (1692–1768). British physician.

McCONAGHY, R.M.S. John Huxham. *Medical History*, 1969, **13**, 280–87.
WALDRON, H.A. The Devonshire colic. *Journal of the History of Medicine*, 1970, **25**, 383–413.

HUXLEY, Andrew Fielding (b. 1917). British physiologist; shared Nobel Prize (Physiology or Medicine) 1963, for discoveries concerning the ionic mechanisms in excitation and inhibition in peripheral and central portions of the nerve cell membrane.

Les Prix Nobel en 1963.

HUXLEY, Julian Sorell (1887–1975). British biologist.

BAKER, J.R. Julian Huxley, scientist and world citizen 1887–1975. A biographical memoir ... with a bibliography compiled by Jens-Peter Green. Paris, Unesco, 1978. 184 pp.
BAKER, J.R. BMFRS, 1976, **22**, 207–38.
BATES, S.I. & WINKLER, M.G. A guide to the papers of Julian Sorell Huxley. Houston, Woodson Research Center, Fondren Library, Rice University, 1984. 164 pp.
DRONAMRAJU, K.R. If I am to be remembered: the life and work of Julian Huxley, with selected correpondence. Singapore, World Scientific, 1993. 294 pp.
HUXLEY, J. Memories. 2 vols., London, Allen & Unwin, 1970–73. 296 + 269 pp.
WATERS, C.K. & VAN HELDEN, A. (eds.) Julian Huxley: biologist and statesman of science. Houston, Rice Univ. Press, 1992. 344 pp.
Archival material: CMAC; Rice University, Houston, Texas (Guide, 1984).

HUXLEY, Thomas Henry (1825–1895). British biologist and evolutionist.

BIBBY, C. Scientist extraordinary: the life and work of Thomas Henry Huxley, 1825–1895. Oxford, Pergamon, 1972. 208 pp.

186

DESMOND, A. Huxley: evolution's high priest. London, Michael Joseph, 1997. 370 pp. First published in 2 vols: *Devil's disciple*; 1994; *Evolution's high priest*, 1997.

HUXLEY, L. Life and letters of Thomas Henry Huxley. 2 vols. London, Macmillan, 1900. 539 + 541 pp.

JENSEN, J.V. Thomas Henry Huxley: communicating for science. London, Associated University Presses, 1991. 253 pp.

LYONS, S.L. Thomas Henry Huxley: the evolution of a scientist. Amherst, NY, Prometheus Books, 1999. 347 pp.

PINGREE, F. Thomas Henry Huxley: a list of his scientific notebooks, drawings and other papers, preserved in the College Archives. London, Imperial College of Science and Technology, 1968. 94 pp.

WILLIAMS, W.C. DSB, 1972, **6**, 589–97.

Archival material: Library, University of Cambridge; Imperial College, London; Royal Institution, London; Royal College of Physicians of London; Library, American Philosophical Society, Philadelphia.

HYRTL, Joseph (1810–1894). Austrian anatomist.

–. Der Anatom Joseph Hyrtl (1810–1894). Wien, Maudrich, 1991. 202 pp.

MOISSL, R.A. Josef Hyrtl ind sein Vermächtnis. St Pölten, St Pöltner Zeitungs-Verlags-Gesellschaft, 1942. 40 pp.

STEUDEL, J. DSB, 1972, **6**, 618–19.

IBN AL-NAFIS, *see* **NAFIS, ibn al-**

IBN AL-QUFF, see **QUFF, ibn Al-**

IBN-RUSHD, *see* **AVERROËS**

IBN SINA, *see* **AVICENNA**

IBN ZUHR, see **AVENZOAR**

IGNARRO, Louis J. (b.1941). American physician; shared Nobel Prize (Physiology or Medicine), 1998, for discoveries concerning nitrous oxide as a signalling molecule in the cardiovascular system.

Les Prix Nobel en 1963.

IMHOTEP (*c*.3000 BC).

HURRY, J.B. Imhotep: the vizier and physician of King Zoser, and afterwards the Egyptian god of medicine. 2nd ed. Oxford, Humphrey Milford, 1928. 211 pp. First edition 1926.

ING, Harry Raymond (1899–1974). British pharmacological chemist.

SCHILD, H.O. BMFRS, 1976, **22**, 239–55.

INGALS, Ephraim Fletcher (1848–1918). American physician.

–. *Proceedings of the Institute of Medicine, Chicago*, 1918–1919, **2**, 173–8.

–. *Transactions of the American Laryngological Association*, 1918, **40**, 185–94.

INGLE, Dwight Joyce (1907–1978). American endocrinologist.

VISSCHER, M.B. BMNAS, 1992, **61**, 246–55.

BIBLIOGRAPHY OF MEDICAL AND BIOMEDICAL BIOGRAPHY

INGLIS, Elsie (1864–1917). British physician; founded Scottish Women's Hospitals Units to serve in 1914–1918 war.

LAWRENCE, M. Shadow of swords. A biography of Elsie Inglis. London, Michael Joseph, 1971. 370 pp.

LENEMAN, L. In the service of life: the story of Elsie Inglis and the Scottish Women's Hospitals. Mercat Press, 1994. 274 pp.

INGRASSIA, Giovanni Filippo (1510–1580). Italian anatomist and physician.

O'MALLEY, C.D. DSB, 1973, **7**, 16–17.

ISAACS, Alick (1921–1967). British virologist; discovered interferon.

ANDREWES, C.H. BMFRS, 1967, **13**, 205–21.

ISAACS, Charles Edward (1811–1860). American physician; contributed to the knowledge on kidney function.

BIETER, R.N. Charles Edward Isaacs, a forgotten American kidney physiologist. *Annals of Medical History*, 1929, N.S. **1**, 363–77.

RADBILL, S.X. DSB, 1973, **7**, 23–4.

ISAEFF (ISAYEFF), Vasiliy Isayevich (1854–1911). Russian microbiologist; with Pfeiffer recorded bacteriolysis of cholera vibrio under certain conditions ('Pfeiffer phenomenon').

Morskoi Vrach, St Petersburg, 1912, i–viii.

ISRAEL, James (1848–1926). German mycologist; first to describe human actinomycosis.

BLOCH, P. *et al.* James Israel, 1848–1926. Wiesbaden, Steiner, 1983. 256 pp.

ISIDORE OF SEVILLE [ISIDORUS HISPALENSIS] (*c.* 560–636). Archbishop of Seville. Book IV of his encyclopaedic *Etymologiae* includes a survey of contemporary medical knowledge.

BREHAULT, E. An encyclopedist of the Dark Ages. New York, Longman, 1912. 274 pp.

ISSELS, Hermannn Joseph (b.1907). German physician; advocate of controversial cancer treatment.

THOMAS, G. Issels: the biography of a doctor. London, Hodder & Stoughton, 1975. 352 pp.

ITARD, Jean Marie Gaspard (1774–1838). French otologist; published first 'modern' textbook on diseases of the ear (1821); introduced new hearing tests.

CASTEX, A. Jean Itard; notes sur la vie et son oeuvre. *Bulletin d'Oto-Rhino-Laryngologie*, 1920. Sept., 15 pp.

IUDIN, Sergei Sergeivich (1891–1954). Russian surgeon; used cadaver blood in transfusion.

D'IACHENKO, P.K. [In memory of S.S.Udin.] [In Russian.] *Vestnik Khirurgija*, 1967, **98**, Jan., 125–30.

NUVAKOV, B.S. *et al.* Sergei Iudin: etiudy biografii. Moskva, Novvostu, 1991. 102 pp. [In Russian.]

SWAN, H. S.S. Yudin; a study in frustration. *Surgery*, 1965, **58**, 572–85.

188

IVANOVSKII, Dmitri Iosifovich (1864–1920). Russian pioneer virologist; a discoverer of the filterable nature of viruses.

GUTINA, DSB, 1973, **7**, 34–6.
LECHEVALIER, H. Dmitri Iosovich Ivanowski (1864–1920). *Bacteriological Reviews*, 1972, **36**, 135–45.

IVY, Robert Henry (1881–1974). British/American plastic surgeon, born in England.

IVY, R.H. A link with the past. Baltimore, Williams and Wilkins, 1962. 148 pp.

IZQUIERDO, José (1887–1975). Mexican physician and medical historian.

PLAZA IZQUIERDO, F. José Izquierdo: vida y obra. Vol. 1. Biografia. Caracas, Congreso de la República, 1984. 460 pp.

JABOULAY, Mathieu (1860–1913). French surgeon; first to describe the interilio-abdominal operation.

LERICHE, R. Un grand précurseur oublié de la neurochirurgie, Mathieu Jaboulay. *Histoire de la Médecine*, 1951, no. 3, 35–40.

JACKSON, Chevalier (1865–1958). American laryngologist.

JACKSON, C. The life of Chevalier Jackson. An autobiography. New York, Macmillan 1938. 229 pp.

JACKSON, Dennis Emerson (1878–1980). American anaesthetist; introduced trichlethylene as anaesthetic.

MORRIS, L.E. D.E. Jackson 1878–1980: a perspective. In: ATKINSON, R.B. & BOULTON, T.B. (eds.)The history of anaesthesia. London, Royal Society of Medicine, 1988, pp. 571–4.
RENDELL-BAKER, L. Dennis E. Jackson, William B. Neff, Robert A. Higson, unsung heroes of anaesthesia. *Anesthesia History Association Newsletter*, 1944, **12**, 8–21.

JACKSON, Ernest Sandford (1860–1938). Australian surgeon.

PARKER, N. & PEARN, J. (eds.) Ernest Sandford Jackson – the life and times of a pioneer Australian surgeon. Brisbane, Child Health Publishing Fund, Dept. of Child Health, Royal Children's Hospital, 1987. 354 pp.

JACKSON, Hall (1739–1797). American physician and surgeon; introduced digitalis to American medical practice.

ESTES, J.W. Hall Jackson and the purple foxglove: medical practice and research in Revolutionary America 1760–1820. Hanover, Univ. Press of New England, 1890. 291 pp.

JACKSON, James (1777–1867). American physician; professor of clinical medicine, Harvard Medical School.

PUTNAM, J.J. A memoir of Dr James Jackson: with sketches of his father Hon. Jonathan Jackson, and his brothers Robert Henry Charles, and Patrick Tracy Jackson; and some account of their ancestry. Boston, New York, Houghton, Mifflin, 1906. 456 pp.

JACKSON, John Hughlings (1835–1911). British clinical neurologist and neurophysiologist.

CLARKE, E. DSB, 1973, 7, 46–50.

CRITCHLEY, M. & CRITCHLEY, E.A. John Hughlings Jackson: father of English neurology. Oxford, Oxford Univ. Press, 1998. 228 pp.

HEAD, H. Hughlings Jackson on aphasia and kindred affections of speech; together with a complete bibliography of Dr Jackson's publications on speech and a reprint of some of the most important parts. *Brain*, 1915, **28**, 1–190.

LASSEK, A.M. The unique legacy of John Hughlings Jackson. Springfield, Ill, C.C. Thomas, 1970. 146 pp.

JACOB, Arthur (1790–1874). British ophthalmologist; remembered for 'Jacob's membrane' and 'Jacob's ulcer'.

JAMES, R.R. *British Journal of Ophthalmology*, 1927, **11**, 257–63.

JACOB, François (b.1920). French molecular biologist; shared Nobel Prize 1965 with A.M. Lwoff and J. Monod for devising operon theory of the regulation of gene expression.

JACOB, F. The statue within. An autobiography. Translated from the French edition (Paris, 1987) by Franklin Philip. New York, Basic Books, 1988. 321 pp.

JACOBAEUS, Hans Christian (1879–1937). Swedish surgeon; pioneer in the development of the thoracoscope.

EHRSTRÖM, R. *Finska Läkaresällskapets Handlingar*, 1938, **81**, 1–8.

JACOBI, Abraham (1830–1919). German-born American paediatrician; professor of children's diseases, New York Medical College.

ROBINSON, V. The life of Abraham Jacobi. *Medical Life*, 1928, **35**, 214–58.

TRUAX, R. The doctors Jacobi. Boston, Little, Brown, 1952. 270 pp.

JACOBI, Carl Wigand Maximilian (1775–1858). German psychiatrist; director of asylum in Siegburg.

HERTING, J. Carl Wigand Maximilian Jacobi, ein deutscher Arzt (1775–1858). Ein Lebensbild nach Briefen und anderen Quellen. Görlitz, Starke, 1929. 218 pp.

SCHULTE, H. von Maximilian Jacobi – Leben und Lehre. *Sudhoffs Archiv für Geschichte der Medizin*, 1961, **45**, 351–69.

JACOBI, Mary Corinna Putnam (1842–1906). Pioneer American physician; professor of children's diseases, Medical College for Women, New York; born in England.

PUTNAM, Ruth. Life and letters of Mary Putnam Jacobi. New York, London, Putnam, 1925. 381 pp.

MARY PUTNAM JACOBI, M.D., a pathfinder in medicine. With selections from her writings and a complete bibliography. Edited by the Women's Medical Association of New York City. New York, London, Putnam, 1925. 521 pp.

TRUAX, R. The doctors Jacobi. Boston, Little, Brown, 1952. 270 pp.

JACOBS, Walter Abraham (1883–1967). American chemist; worked on chemistry of natural products and chemotherapy.

ELDERFIELD, R.C. BMNAS, 1980, **51**, 247–78.

Archival material: Rockefeller Archives Center.

JACOBSON, Leon Orris (1912–1992). American physician.

GOLDWASSER, E. BMNAS, 1996, **70**, 191–202.

JADASSOHN, Josef (1863–1936). German dermatologist and venereologist.

MILAN, G. *Revue Française de Dérmatologie et de Venérologie*, 1928, **4**, 450–53.
SULZBERGER, M.B. *American Journal of Dermatopathology*, 1985, **7**, 31–6.

JAEGER, Eduard (1818–1884). Austrian ophthalmologist; introduced sight-test types, 1854; an accomplished painter, he illustrated his own works.

–. Eduard Jäger. *Klinische Monatsblätter der Augenheilkunde*, 1884, **22**, 277–85.

JAKOB, Alfons Maria (1884–1931). German neurologist; described spastic pseudosclerosis ('Creutzfeldt-Jacob disease').

JOSEPHY, –. *Zeitschrift für die gesamte Neurologie und Psychiatrie*, 1932, **138**, 165–8.

JAKSCH, Rudolph von (1855–1947). Bohemian paediatrician; gave classic description of infantile pseudoleukaemic anaemia ('von Jaksch's disease').

LOWY, J. *Wiener Medizinische Wochenschrift*, 1925, **75**, 1629–35.

JAMES, Robert (1705–1776). British physician and medical lexicographer.

BURR, C.W. Dr Robert James (1705–1776) and his medical dictionary. *Annals of Medical History*, 1929, **1**, 180–90.

JAMES, Sydney Price (1870–1946). British malariologist.

CHRISTOPHERS, S.R. ONFRS, 1945–48, **5**, 507–23.

JAMES, William (1842–1910). American psychologist.

ALLEN, G.W. William James: a biography. New York, Viking Press, 1967. 556 pp.
FERRELL, S., DSB, 1973, **7**, 67–9.
MYERS, G.E. William James: his life and thought. New Haven, Yale Univ. Press, 1986. 628 pp.
PERRY, R.B. The thought and character of William James. 2 vols. Boston, Little, Brown, 1935.

JAMESON, Leander Starr (1853–1917). South African physician, politician and administrator; led 'Jameson Raid' to overthrow the government of Paul Kruger, 1895.

British Medical Journal, 1917, **2**, 742–3.
COLVIN, I.D. Life of Jameson. 2 vols. London, Arnold, 1922.

JAMESON, William Wilson (1885–1962). British physician; public health officer.

GOODMAN, N.M. Wilson Jameson, architect of national health. London, Allen & Unwin, 1970. 216 pp.

JAMIESON, Edward Bald (1876–1956). British anatomist.

EASTWOOD, J. & EASTWOOD, M. E.B. Jamieson: anatomist and Shetlander. Lerwick, Shetland Times Ltd., 1999. 80 pp.

JANES, Robert Meredith (1894–1966). Canadian surgeon; with N.S. Shenstone introduced hilar tourniquet in lung surgery.

DELANEY, R.J. Dr Robert Meredith Janes (1894–1966) professor of surgery, University of Toronto. *Canadian Journal of Surgery*, 1969, **12**, 2–11.

JANET, Pierre Marie Félix (1859–1947). French psychiatrist.

MAYO, E. The psychology of Pierre Janet. London, Routledge & Kegan Paul, 1952. 132 pp.
PRÉVOST, C.M. La psycho-philosophie de Pierre Janet. Paris, Payot, 1973. 348 pp.

JANSEN, Barend Coenraad Petrus (1884–1962). Dutch nutritionist; jointly isolated vitamin B_1 (thiamine) in crystalline form.

EYS, J. van. Barend Coenraad Petrus Jansen – a biographical sketch (1884–1962). *Journal of Nutrition*, 1970, **100**, 485–90.

JANSKY, Jan (1873–1921). Czech physician; first to classify blood into four groups.

ZOTIKOV, E.A. & DONSKOV, S.I. I. IANSKI [on the centenary of his birth] [In Russian] *Problemy Gematologiia*, 1974, **19** (7), 58–60.

JARISCH, Adolf (1850–1902). Austrian dermatologist; introduced Jarisch-Herxheimer reaction in syphilis therapy.

RILLE, J. Adolf Jarisch. *Deutsche Medizinische Wochenschrift*, 1902, **28**, 267–8.

JARVIS, Edward (1803–1884). American physician, psychiatrist and statistician.

DAVICO, R. The autobiography of Edward Jarvis (1803–1884). London, Wellcome Institute for the History of Medicine, 1992. 162 pp. (*Medical History*, Suppl. 12).
GROB, G.N. Edward Jarvis and the medical world of nineteenth-century America. Knoxville, Univ. Tennessee Press, 1978. 300 pp.

JAVAL, Louis Émile (1839–1907). French ophthalmologist.

LEVENE, J.R. The true inventors of the keratoscope and photokeratoscope. *British Journal of the History of Science*, 1965, **11**, 324–42.

JEFFERSON, Geoffrey (1886–1961). British neurosurgeon.

SCHURR, P.H. So that was life. A biography of Sir Geoffrey Jefferson: master of neurosciences and man of letters. London, Royal Society of Medicine, *c*.1997. 358 pp.
WALSHE, F.M.R. BMFRS, 1961, **7**, 127–35.

JELLIFFE, Smith Ely (1866–1945). American neurologist and psychiatrist.

BURNHAM, J.C. Jelliffe: American psychoanalyst and physician, and his correspondence with Sigmund Freud and C.G. Jung. Edited by W. Maguire. Chicago, Univ. Chicago Press, 1983. 324 pp.
KRASNER, D. Smith Ely Jelliffe and the development of American psychosomatic medicine. [Ph.D. thesis, Bryn Mawr College, 1984]. Ann Arbor, University Microfilms International, 1986. 322 pp.

JENNER, Edward (1749–1823). British physician and naturalist, in general practice in Berkeley, Gloucestershire; introduced vaccination against smallpox.

BARON, J. The life of Edward Jenner, M.D., with illustration of his doctrines and selections from his correspondence. 2 vols., London, Colburn, 1827–38. 624 + 471 pp. Vol. 1 reissued with a new vol. 2, 1838.

BIBLIOGRAPHY OF MEDICAL AND BIOMEDICAL BIOGRAPHY

FISHER, R.B. Edward Jenner. London, Deutsch, 1991. 361 pp.

FISK, D. Dr Jenner of Berkeley. London, Heinemann, 1959. 288 pp.

LeFANU, W. A bibliography of Edward Jenner. 2nd ed. Winchester, St Paul's Bibliographies, 1985. 160 pp. (Includes surveys of most unpublished MS collections and memorabilia at various archives and institutions).

SAUNDERS, P. Edward Jenner: the Cheltenham years 1795–1823. Hanover, Univ. Press of New England, 1982. 469 pp.

WILSON, L.G. DSB, 1973, **7**, 95–7.

Archival material: Wellcome Institute, London; Royal College of Physicians of London; Royal College of Surgeons of England; Medical Library, Duke University, Durham, NC; Welch Memorial Library, Johns Hopkins University, Baltimore.

JENNER, William (1815–1898). British physician; differentiated aetiology of typhus and typhoid.

–. *Transactions of the Epidemiological Society of London*, 1898–99, n.s. **18**, 571–4.

JENNINGS, Herbert Spencer (1868–1947). American biologist and geneticist; director of zoological laboratory, Johns Hopkins University.

SONNEBORN, T.M. BMNAS, 1974, **47**, 143–223.

Archival material: Library, American Philosophical Society, Philadelphia.

JEPHSON, Henry (1798–1878). British physician and philanthropist.

BAXTER, E.G. Dr Jephson of Leamington Spa. Leamington Spa, Local History Society, 1980. 104 pp.

JERNE, Niels Kaj (1911–1994). British-born immunologist; shared Nobel Prize (Physiology or Medicine) 1994 for theories concerning specificity in development and control of immune system.

ASKONAS, B.A. & HOWARD, J.G. BMFRS, 1997, **43**, 235–51.

SÖDERQVIST, T. The troubled life of Niels Jerne. New Haven, Yale Univ. Press, 2003. 259 pp.

JESTY, Benjamin (1737–1816). British farmer of Worth Matravers, Dorset.

WALLACE, E.M. The first vaccinator, Benjamin Jesty of Worth Matravers and his family. Wareham & Swanage, Anglebury-Bartlett, 1981. 20 pp.

JEX-BLAKE, Sophia Louisa (1840–1912). British physician, qualified in Dublin, 1877; founded London School of Medicine for Women, 1874.

ROBERTS, S. Sophia Jex-Blake: a pioneer in nineteenth-century medical reform. London, Routledge, 1993. 207 pp.

TODD, M. Life of Sophia Jex-Blake. London, Macmillan, 1981. 574 pp.

JOANNES ACTUARIUS (*fl.* 13th cent.). Byzantine physician; wrote De urinis, a mediaeval treatise on urinoscopy, first published Venice, 1519.

KUDLIAN, F. Empiric und Theorie in der Harnlehre des Johannes Aktuarios. *Clio Medica*, 1973, **8**, 19–30.

JOANNES DE KETHAM (d. *c.* 1490). German physician; wrote Fasciculus medicinae, a series of medical writings, 1491, the first medical book with woodcuts.

193

BIBLIOGRAPHY OF MEDICAL AND BIOMEDICAL BIOGRAPHY

SINGER, C. & SUDHOFF, K. (eds). Joannes de Ketham Alemanus. Fasciculus medicinae, 1491. Fascimile...with an historical introduction. Milan, Lier, 1924. 55 pp.

JOBERT DE LAMBALLE, Antoine Joseph (1799–1867). French surgeon.

LE SAGE, R. Und grand chirurgien breton, Jobert de Lamballe, 1799–1867. Paris, M. Vigné, 1938. 46 pp.

JOETTEN, Karl Wilhelm (1886–1958). German physician; jointly introduced suramin in treatment of trypanosomiasis.

REPLOH, H. K.W. Joettlin zum Gedächtnis. *Archiv für Hygiene*, 1958, **142**, 339–41.

JOHANNES DE MIRFELD (d.1407). Author of an encyclopaedia of contemporary medical knowledge.

HARTLEY, P. H.-S. & ALDRIDGE, H.R. Johannes de Mirfeld of St Bartholomew's Smithfield: his life and works, Cambridge, C.U.P., 1936. 191 pp.

JOHANNITIUS, *see* **HUNAYN IBN ISHAQ**

JOHANNSEN, Wilhelm Ludvig (1857–1927). Danish biologist.

CHURCHILL, F.B. William Johannsen (1857–1927) and the genotype concept. *Journal of the History of Biology*, 1974, **7**, 5–30.
DUNN, L.C. DSB, 1973, **7**, 113–15.

JOHN OF ARDERNE, *see* **ARDERNE, John**

JOHN OF GADDESDEN [Johannes Anglicus] (?1280–1361). Fourteenth-century English physician who wrote on contemporary medicine.

CHOLMELEY, H.P. John of Gaddesden and the Rosa medicinae. Oxford, Clarendon Press, 1912. 184 pp.

JOHN OF MIRFIELD, *see* **JOHANNES DE MIRFIELD**

JOLIOT-CURIE, Frédéric (1900–1958). French nuclear physicist.

BIQUARD, P. Frédéric Joliot-Curie: the man and his theories. London, Souvenir Press, 1965. 224 pp. First published in French 1961.
GOLDSMITH, M. Frédéric Joliot-Curie; a biography. London, Lawrence & Wishart, 1976. 260 pp.

JOLLY, William Tasker Adam (1878–1939). British physiologist.

BELONJE, P.C. William Jolly – father of medical science in South Africa. *South African Medical Journal*, 1991, **80**, 156–8.

JONES, Alfred Ernest (1879–1958). British psychoanalyst.

BROME, V. Ernest Jones, Freud's alter ego. London, Caliban Books, 1982. 250 pp.
JONES A.E. Free associations: memoirs of a psychoanalyst. New York, Basic Books, 1959. 263 pp.

JONES, Frederic Wood, *see* **WOOD JONES, Frederic**

JONES, Henry Bence (1813–1873). British physician.

COLEY, N.G. Henry Bence-Jones MD., FRS. *Notes & Records of the Royal Society of London*, 1973, **28**, 31–56.
JONES, H.B. An autobiography with elucidation at later dates. Privately printed. Tottenham, London, Crusha & Son Ltd., 1929. 33 pp.

JONES, Joseph (1833–1896). American physician and medical educationist; surgeon in the Confederate army.

BREEDON, J.O. Joseph Jones, D.M.: scientist of the old South. Lexington, Univ. Press of Kentucky, 1975. 293 pp.

JONES, Mary Ellen (1922–1996). American biochemist.

TRAUT, T.W. BMNAS, 2001, **79**, 183–201.

JONES, Robert (1858–1933). British orthopaedic surgeon.

ORR, H.W. On the contributions of Hugh Owen Thomas of Liverpool, Sir Robert Jones of Liverpool and London, John Ridlon, M.D., of New York and Chicago, to modern orthopedic surgery. Springfield, C.C. Thomas, 1949. 253 pp.
WATSON, F. Life of Sir Robert Jones. London, Hodder & Stoughton, 1934. 327 pp.

JONES, Thomas Wharton (1808–1891). British ophthalmologist.

GODLEE, R.J. Thomas Wharton Jones F.R.S. Reprinted from *British Journal of Ophthalmology*, March–April 1921. 36 pp.

JONES, Walter Jennings (1865–1935). American biochemist; professor at Johns Hopkins University; pioneer work on nucleic acids.

CLARK, W.M. BMNAS, 1939, **20**, 79–139.
OLBY, R. DSB, 1990, **17**, 447–8.

JONSTON, John (1603–1675). Polish physician of Scottish extraction.

CRELLIN, J.K. DSB, 1973, **7**, 164–5.

JORDAN, Edwin Oakes (1866–1936). American bacteriologist; professor at University of Chicago.

BURROWS, W. BMNAS, 1939, **20**, 197–228.
CASSEDY, J.H. DSB, 1973, **7**, 170–71.
Archival material: University of Chicago Library; American Society for Microbiology, Washington, DC.

JORDAN, Joseph (1804–1863). British surgeon.

JORDAN, F.W. Life of Joseph Jordan, surgeon, and an account of the rise and progress of medical schools in Manchester, etc. London & Manchester, Sherratt & Hughes, 1904. 185 pp.

JOSEPH, Eugen (1879–1933). German urologist.

MAHRENHOLZ, M.A. Eugen Joseph: Biobibliographie eines Berliner Urologen. Thesis. Berlin, Institut zur Geschichte der Medizin, 1998. 119 pp. Facsimile.

JOSEPH, Jacques (1865–1934). German plastic surgeon; 'Father of modern rhinoplasty'.

NATVIG, P. Jacques Joseph, surgical sculptor. Philadelphia, Saunders, 1982. 264 pp.

JOSLIN, Elliott Proctor (1869–1962). American nutritionist.

HOLT, A.C. Elliott Proctor Joslin: a memoir, 1869–1962. Worcester, Mass., Asa Bartlett Press, 1969. 68 pp.

JOSUE, Otto (1869–1923). Belgian physician.

MASSARY, P. *Bulletin et Mémoires de la Société Médicale des Hôpitaux de Paris*, 1923, 3 sér., **47**, 1878–81.

JUDET, Robert Louis (1909–1980). French orthopaedic surgeon; with his brother devised an acrylic prosthesis for hip arthroplasty.

GOURSOLAS, F. Contribution à la biographie de Professeur Robert Judet. *Histoire des Sciences Médicales*, 1988, **22**, 249–56.

JUNCKER, Johann (1679–1759). German chemist and physician.

FICHMAN, DSB, 1973, **7**, 188.
KAISER, W. & HUEBNER, H. (eds.). Johann Juncker (1679–1759). und seine Zeit. 3 vols. Halle, Universität, 1979, 255 pp.

JUNG, Carl Gustav (1875–1961). Swiss analytical psychologist.

BROME, V. Jung. London, Macmillan, 1978. 327 pp.
CAZENAVE, M. (ed.) Carl Gustav Jung. Paris, L'Herne, 1984. 516 pp.
FORDHAM, M. DSB, 1973, **7**, 189–93.
HANNAH, B. Jung: his life and work. A biographical memoir. London, Michael Joseph, 1977. 376 pp.
HAYMAN, R. A life of Jung. London, Bloomsbury Publishing, 1999. 522 pp.
McLYNN, F. Carl Gustav Jung. London, Bantam, 1996. 623 pp.
RAFF, J. Jung and the alchemical imagination. York Beach, ME, Nicolas-Hays (Samuel Weiser), 277 pp.

JUNG-STILLING, Johann Heinrich (1740–1817). German ophthalmic surgeon; a pioneer in cataract extraction.

JUNG-STILLING: his biography, translated from the German by R.O. Moon. London, G.T. Foulis, 1938. 251 pp.
JUNG-STILLING, H. Lebensgeschichte: Vollständige Ausgabe. Mit Anmerkungen hrsg. von G.A. Benrath. Darmstadt, Wissenschaftliche Buchgesellschaft, 1976. 784 pp.
PROPACH, G. Johann Heinrich Jung-Stilling (1740–1817) als Arzt. Köln, Universität zu Köln, 1983. 410 pp.

JURIN, James (1684–1750). British physician.

RUSNOCK, A.A. The correspondence of James Jurin (1684–1750). Physician and Secretary to the Royal Society. Amsterdam/Atlanta, GA, Rodopi, Clio Medica, 1996. 577 pp.

KABAT, Elvin Abraham (b.1914). American immunochemist; professor of microbiology, Columbia University.

KABAT, E.A. Getting started 50 years ago – experiences, perspectives, and problems of the first 21 years. *Annual Review of Immunology*, 1983, **1**, 1–32.

KABAT, E.A. Before and after. *Annual Review of Immunology*, 1988, **6**, 1–24.

KAEMPFER, Engelbert (1651–1716). German physician, scientist, linguist, artist and traveller.

HABERLAND, D. Engelbert Kaempfer, 1651–1715: a biography. London, British Library, 1996. 158 pp. First published in German.

KAHLBAUM, Karl Ludwig (1828–1899). German psychiatrist; wrote classical work on catatonia.

KATZENSTEIN, R. Karl Ludwig Kahlbaum und sein Beitrag zur Entwicklung der Psychiatrie. Zürich, Juris, 1963. 42 pp.

KAHLER, Otto (1849–1893). Bohemian physician; gave first complete description of syringomyelia.

–. Professor Kahler, *Zeitschrift für Heilkunde*, 1893, **14**, 3–12.

KAHN, Reuben Leon (b. 1887). Lithuanian/American immunologist and serologist; introduced precipitation test for syphilis.

COBB, W.M. Reuben Leon Kahn. *Journal of the National Medical Association*, 1971, **63**, 388–94.

KALCKAR, Herman Moritz (1908–1991). Danish/American biochemist.

KENNEDY, E.P. BMNAS. 1996, **69**, 149–64.

KALDEN, Clemens von (1859–1903). German pathologist; first to describe ovarian granulosa-cell tumour, 1895.

FISCHER, G. & ZIEGLER, F. *Centralblatt für Allgemine Pathologie*, 1903, **14**, 209–11.

KALK, Heinrich Otto (1895–1973). German gastroenterologist.

WILDHURT, T. Heinrich Otto Kalk (1895–1973): Lebensbild eines Gastroenterologen und Hepatologen. 2te. Aufl. Freiburg, Kalk Foundation, 1996. 55 pp.

KAMEN, Martin David (b.1913). Canadian/American biochemist; discovered carbon 14-radioisotopes used as tracers in biochemistry.

KAMEN, M.D. Radiant science, dark politics: a memoir of the nuclear age. Berkeley, Univ. California Press, 1985. 348 pp.
KAMEN, M.D. A cupful of luck, a pinch of sagacity. *Annual Review of Biochemistry*, 1986, **55**, 1–34.

KANDEL, Eric R. (b. 1929). American physician; shared Nobel Prize (Physiology or Medicine), 2000, for discoveries concerning signal transduction in the nervous system.

Les Prix Nobel en 2000.

KANE, Elisha Kent (1820–1857). American surgeon; first notable American Arctic explorer.

CORNER, G.W. Doctor Kane of the Arctic seas. Philadelphia, Temple Univ. Press, 1972. 306 pp.

VILLAREJO, O.M. Dr Kane's voyage to the Polar lands. Philadelphia, Univ. of Pennsylvania Press, 1965. 220 pp.

Archival material: Library, American Philosophical Society, Philadelphia.

KAPLAN, Nathan Oram (1917–1986). American biochemist; isolated coenzyme A.

McELROY, W.D. BMNAS, 1994, **63**, 246–91.

KAPOSI [KOHN], Moritz (1837–1902). Hungarian dermatologist.

FESTSCHRIFT gewidmet Moritz Kaposi zum 20 jahrigen Professorenjubiläum. *Archiv für Dermatologie und Syphilis*, 1900, Ergänzungsband. 924 pp.

KARRER, Paul (1887–1971). Russian/Swiss chemist; synthesized vitamins; shared Nobel Prize for Chemistry 1937.

ISLER, C. BMFRS, 1978, **24**, 245–321.
LEICESTER, H.M. DSB, 1978, **15**, 257–8.

KARTAGENER, Manes (1897–1975). Swiss physician.

SCHRIML, S. Studien zur Geschichte des Kartagener-Syndroms und des Immotile-Cilia Syndroms. Würzburg, Institut für Geschichte der Medizin der Universität Würzburg, 1991. 147 pp.

KATO, Gen-ichi (1890–1979). Japanese physiologist at Keio University; investigated nerve conduction.

KATO, G. The road a scientist followed. Notes of Japanese physiology as I myself experienced it. *Annual Review of Physiology*, 1970, **32**, 1–20.

KATZ, Bernard (1911–2003). German/British physiologist; shared Nobel Prize (Physiology or Medicine) 1970 for researches into chemical neurotransmission.

KATZ, B. Reminiscences of a physiologist, 50 years after. *Journal of Physiology*, 1981, **370**, 1–12.

KAUFFMANN, Fritz Josua (1897–1978). German bacteriologist; classified Salmonella, based on antigen structure.

KAUFFMANN, F. Erinnerungen eine Bakteriologen; Zur Geschichte der Enterobacteriaceen-Forschung. Kopenhagen, Munksgaard, 1969, 366 pp.
ROHDE, R. & WINKLE, S. In memoriam Prof. Dr. med. Fritz Kauffmann. *Zentralblatt für Bakteriologie*, Orig. A., 1979, **243**, 141–6.

KAY, Herbert Davenport (1893–1976). British biochemist.

BLAXTER, K. BMFRS, 1977, **23**, 283–310.

KEATS, John (1795–1821). British poet who qualified as a physician in 1816.

GOELLNICHT, D.C. The poet-physician: Keats and medical science. Pittsburgh, Univ. of Pittsburgh Press, 1984. 291 pp.
HALE-WHITE, W. Keats as doctor and patient. London, Oxford Univ. Press, 1938. 96 pp.
SMITH, H. Keats and medicine. Newport, Isle of Wight, Cross Publishing, 1995. 127 pp.

KEEN, William Williams (1837–1932). American surgeon; professor at Jefferson Medical College.

GEIST, D.C. The writings of William Williams Keen, M.D., Hon. F.R.C.S.; a selected annotated bibliography. *Transactions and Studies of the College of Physicians of Philadelphia*, 1976, **43**, 337–71.
GEIST, D.C. William Williams Keen, M.D. (1837–1932), surgeon and author. *Transactions and Studies of the College of Physicians of Philadelphia*, 1977, **44**, 182–93.
JAMES, W.W. Keen (ed.). The memoirs of William Williams Keen MD. Doylestown, Pa., W.W. Keen James, 1990. 337 pp.
STONE, J.L. W.W. Keen: America's pioneer neurological surgeon. *Neurosurgery*, 1985, **17**, 997–1010.
Archival material: CMAC.

KEHR, Hans (1862–1916). German surgeon; successfully ligated hepatic artery, 1903.

ABE, H.R. [three papers] *Beiträge zur Geschichte der Universität Erfurt*, 1879–83, **19**, 233–64.

KEIBEL, Franz Karl Julius (1861–1929). German embryologist.

NAUCK, E.T. Franz Keibel, zugleich ein Untersuchung über das Problem des wissenschaftlichen Nachwuchses. Jena, G.Fischer, 1937. 112 pp.

KEILIN, David (1887–1963). Russian-born Polish parasitologist and biochemist; worked in Cambridge, England.

FRUTON, J. S. DSB, 1973, **7**, 272–4.
MANN, T. BMFRS, 1964, **10**, 183–205.
Archival material: Cambridge University Library.

KEILL, James (1673–1719). Scottish anatomist and physiologist.

VALADEZ, F.M. DSB, 1973, **7**, 278–9.
VALADEZ, F.M. & O'MALLEY, C.D. James Keill of Northampton; physician, anatomist and physiologist. *Medical History*, 1971, **15**, 317–35.

KEITH, Arthur (1866–1955). British anatomist and anthropologist.

CLARK, W.E. Le G. BMFRS, 1955, **1**, 145–62.
KEITH, A. An autobiography. London, Watts, 1950. 721 pp.
LeFANU, W.R. DSB, 1973, **7**, 278–9.
Archival material: Royal College of Surgeons of England.

KEITH, Norman Macdonnell (1885–1976). Canadian physician.

PRUITT, R.D. *Transactions of the Association of American Physicians*, 1976, **89**, 23–5.

KEKWICK, Ralph Ambrose (1908–2000). British biophysicist.

CREETH, J.H., VALLET, L. & WATKINS, W.M. BMFRS, 2002, **48**, 235–49.

KELLAWAY, Charles Halliley (1889–1952). Australian physiologist and pathologist; Director, Walter & Eliza Hall Institute of Pathology, Melbourne, and Wellcome Research Institute, London.

DALE, H.H. ONFRS, 1952–53, **8**, 503–21.

BIBLIOGRAPHY OF MEDICAL AND BIOMEDICAL BIOGRAPHY

KELLY, Howard Atwood (1858–1943). American gynaecologist; professor at Pennsylvania and Johns Hopkins universities.

CULLEN, T.S. Dr Howard A. Kelly, Professor of Gynecology in the Johns Hopkins University and Gynecologist-in-Chief to the Johns Hopkins Hospital. *Johns Hopkins Hospital Bulletin*, 1919, **30**, 287–93 [bibliography 293–302 (M.W. Blogg)].
DAVIS, Audrey W. Dr Kelly of Hopkins: surgeon, scientist, Christian. Baltimore, Johns Hopkins Press, 1959. 242 pp.

KELSO, John Joseph (1864–1935). Canadian paediatrician.

JONES, A. In the children's aid: J.J. Kelso and child welfare in Ontario. Toronto, Buffalo, University of Toronto Press, 1981. 210 pp.

KENDALL, Edward Calvin (1886–1972). American biochemist and endocrinologist; isolated crystalline thyroxine; shared Nobel Prize 1950.

INGLE, D. BMNAS, 1974, **47**, 249–90.
KENDALL, E.C. Cortisone. New York, Scribner, 1971. 175 pp. [Autobiography of Kendall.]
KOHLER, R.E. DSB, 1978, **15**, 258–9.
Archival material: Library, Princeton University.

KENDREW, John Cowdery (1917–1997). British molecular biologist; shared Nobel Prize (Chemistry) in 1962 for studies in myoglobin and other globular proteins.

HOLMES, K.C. BMFRS, 2001, **47**, 313–32.
NARDONE, A. Supplementary catalogue of the papers and correspondence of Sir John Cowdery Kendrew FRS (1917–1997). Bath, National Cataloguing Unit for the Archives of Contemporary Scientists. NCUACS Catalogue No. 78/7/98.

KENNAWAY, Ernest Laurence (1881–1958). British cancerologist.

COOK, J.W. BMFRS, 1958, **4**, 139–54.
HADDOW, A. Sir Ernest Laurence Kennaway FRS 1881–1958. Chemical cause of cancer then and today. *Perspectives in Biology and Medicine*, 1974, **17**, 543–91.
Archival material: CMAC.

KENNEDY, Robert Foster (1884–1952). British neurologist; emigrated to USA and became professor of neurology, Cornell University.

BUTTERFIELD, I.K. (ed.) The making of a neurologist: the letters ... to his wife. Edited with a memoir by I.K. Butterfield. Cambridge, Heffer, 1981. 115 pp.

KENNY, Elizabeth (1886–1952). Australian nurse; introduced method of physical treatment in infantile paralysis.

COHN, V. Sister Kenny: the woman who challenged the doctors. Minneapolis, Univ. of Minnesota Press, 1975. 302 pp.

KENT, Alfred Frank Stanley (1863–1958). British physiologist; described the 'bundle of Kent'.

ANDERSON, R.H. & BECKER, A.E. Stanley Kent and accessory atrioventricular connections. *Journal of Thoracic and Cardiovascular Surgery*, 1981, **81**, 649–58.

KERMACK, William Ogilvy (1898–1970). British biochemist.

McCREA, W.H. BMFRS, 1971, **17**, 399–429.

KERNER, Justinus Andreas Christian (1786–1862). German physician, toxicologist, poet and clairvoyant.

BERGER-FIX, A. (ed.) Justinus Kerner. Nur wenn Mann von Geistern spricht. Briefe und Klecksorgraphien. Stuttgart, K. Thienemanns Verlag, 1986. 240 pp.
GRÜSSER, O.J. Justinus Kerner (1786–1862) – Arzt, Poet, Geisterseher. Berlin, Springer, 1987. 382 pp.

KETTLE, Edgar Hartley (1882–1936). British pathologist; professor and director of the Institute of Pathology, (Royal) Postgraduate Medical School, London.

MURRAY, J.A. ONFRS, 1936–39, **2**, 301–5.

KEY, Axel (1832–1901). Swedish pathologist; professor at Karolinska Institutet, Stockholm.

LJUNGGREN, B. & BRUYN, G.W. The Nobel Prize in Medicine and the Karolinska Institute: the story of Axel Munthe and Alfred Nobel. Basel, Karger, 2001. 232 pp.

KEYNES, Geoffrey Langdon (1887–1982). British surgeon and bibliographer.

KEYNES, G. The gates of memory. Oxford, Clarendon Press, 1981. 428 pp.

KHORANA, Har Gobind (b. 1922). Indian/American chemist; shared Nobel Prize (Physiology or Medicine) 1968, for establishment of techniques for the synthesis of polynucleotides.

Les Prix Nobel en 1968.

KIDD, John (1775–1851). British anatomist and chemist.

EDMONDS, J.M. DSB, 1973, **7**, 365–6.

KIELLAND, Christian (1871–1941). African-born obstetrician.

PARRY-JONES, E. Kielland's forceps. London, Butterworth, 1952. 211pp.

KIELMEYER, Carl Friedrich (1765–1844). German anatomist and comparative physiologist.

COLEMAN, W. DSB, 1973, **7**, 366–9.
KANZ, K.T. Kielmeyer-Bibliographie. Verzeichnis der Literatur von und über den Naturforscher Carl Friedrich Kielmeyer (1765–1844) Stuttgart, Verlag für Geschichte der Naturwissenchaften und Technik, 1991. 161 pp.

KIKUTH, Walther (1896–1968). German physician; introduced the antimalarial mepacrine; with co-workers introduced Miracil D in treatment of bilharziasis.

GRÜN, L. In memoriam W. Kikuth. *Medizinische Welt*, 1968, (31 Aug.), 1875–76.

KILIAN, Hermann Friedrich (1800–1863). German obstetrician; drew attention to pelvic deformities.

LENTZ, H. Der Bonner Geburtshelfer Hermann Friedrich Kilian. [Thesis]. Bonn, Medizinische Fakultät, 1969. 557 pp.

KILLIAN, Gustav (1860–1921). German laryngologist, professor in Berlin; invented a laryngoscope and a bronchoscope.

KILLIAN, H. Gustav Killian, sein Leben, sein Werk. Remscheid-Lennepi, Dustri-Verlag, 1958. 267 pp.
ZÖLLNER, F. Gustav Killian, father of bronchoscopy. *Archives of Otolaryngology*, 1965, **82**, 656–9.

KIMBALL, Gilman (1804–1892). American gynaecological surgeon; pioneered hysteromyomectomy (for fibromyoma).

DAVENPORT, F.H. In memoriam, G. Kimball. *American Journal of Obstetrics*, 1892, **26**, 560–65.

KIMMELSTIEL, Paul (1900–1970). American pathologist; described 'Kimmelstiel-Wilson syndrome'.

WELLMAN, K.F. In memorian: Paul Kimmelstiel, M.D., 1900–1970. *American Journal of Clinical Pathology*, 1971, **56**, 117–19.

KIMURA, Motoo (1924–1994). Japanese geneticist.

CROW, J.F. BMFRS, 1997, **43**, 253–65.

KING, Albert Freeman Africanus (1841–1914). American physician; supported belief in transmission of malaria by mosquito, 1883.

CHARLES, S.T. Albert F.A. King (1841–1914), an armchair scientist. *Journal of the History of Medicine*, 1969, **24**, 22–6.

KING, Frederic Truby (1858–1938). New Zealand paediatrician.

KING, M. Truby King the man. London, Allen & Unwin, 1948. 355 pp.

KING, Harold (1887–1956). British chemist and pharmacologist.

HARINGTON, C.R. BMFRS, 1956, **2**, 157–71.
LESCH, J.E. DSB, 1990, **17**, 472–4.

KING, Helen Dean (1869–1955). American geneticist.

BOGIN, M. DSB, 1990, **17**, 474–8.
Archival material: Wistar Institute, Philadelphia.

KINMONTH, John Bernard (1916–1982). British surgeon; introduced lymphangiography, 1952.

CORNELIUS, E.H. John Bernard Kinmonth. PLARR, 1974–82, 221–3 (CB).

KINSEY, Alfred Charles (1894–1956). American social scientist.

CHRISTENSON, C.V. Kinsey, a biography. Bloomington, Indiana Univ. Press, 1971. 241 pp.
GATHORNE-HARDY, J. Alfred C. Kinsey. Sex: the measure of all things: a life of Alfred Kinsey. London, Chatto & Windus, 1998. 514 pp.
JONES, J.H. Alfred C. Kinsey: a public/private life. New York, London, Norton, *c*.1997. 937 pp.

BIBLIOGRAPHY OF MEDICAL AND BIOMEDICAL BIOGRAPHY

KIRCHER, Athanasius (1602–1680). German priest, physicist, mathematician and microscopist; probably first to use microscope in investigating cause of disease and to state explicitly the theory of contagion by animalculae as cause of infectious disease.

GODWIN, A. Athanasius Kircher: a Renaissance man and the quest for lost knowledge. London, Thames & Hudson, 1979. 76 pp.

REILLY, P.C. Athanasius Kircher SJ: master of a hundred arts, 1602–1680. Wiesbaden, Rome, Ed. del Mondo, 1974. 207 pp.

KIRK, John (1832–1922). British physician and explorer.

LIEBOWITZ, D. The physician and the slave trade. John Kirk, the Livingstone expeditions, and the crusade against slavery in East Africa. New York, W.H. Freeman, 1998. 314 pp.

KIRKBRIDE, Thomas Story (1809–1883). American psychiatrist.

TOMES, N. A generous confidence: Thomas Story Kirkbride and the art of asylum keeping, 1840–1883. Cambridge, Cambridge Univ. Press, 1984. 387 pp.

KIRKES, William Senhouse (1823–1864). British physician.

POWER, D'A. *St Bartholomew's Hospital Reports*, 1910–1911, **18**, 166–8.

KIRSCHNER, Martin (1879–1942). German surgeon; introduced Kirschner wire for skeletal traction and bone fragment stabilization.

ROMM, S. The person behind the name: Martin Kirschner. *Plastic and Reconstructive Surgery*, 1983, **72**, 104–7.

KIRSTEIN, Alfred (1863–1922). German laryngologist; first to use direct-vision laryngoscopy.

FICKEN, C. von. In memorian Alfred Kirstein. *Zeitschrift für Hals-, Nasen- und Ohrenheilkunde*, 1922–1923, **4**, 275–83.

KIRTLAND, Jared Potter (1793–1877). American physician and naturalist.

THOMS, H. The doctors Jared of Connecticut. Hamden, Conn, Shoe String Press, 1958. 76 pp. [Deals with Jared Potter Kirtland, Jared Eliot and Jared Potter.]

KISCH, Bruno (1890–1966). Czech-born physiologist and biochemist; professor of physiology, Yeshiva University, N.Y.

KISCH, B. Wanderungen und Wandlungen: die Geschichte eines Arztes in 20. Jahrhundert. Köln, Greven, 1966. 360 pp.

STEVENSON, L.G. In memoriam Bruno Kisch (1890–1966). *Journal of the History of Medicine*, 1967, **22**, 47–53.

KITASATO, Shibasaburo (1852–1931). Japanese bacteriologist; cultured tetanus bacillus; joint discoverer of antitoxic immunity.

FUJINO, T. DSB, 1973, **7**, 391–3.

KÖHLER, W. Shibasaburo Kitasato and Sahachiro Hata. *Kitasato Archives of Experimental Medicine*, 1989, **62**, 85–106.

MIYAJIMA, M. The life of Kitasato. Tokyo, Herald Press, 1935. 44 pp.

KIWISCH, Franz (1814–1851). Bohemian gynaecologist and obstetrician; professor in Prague and Würzburg.

203

MUELLER, B. Franz Kiwisch Ritter von Rosenau, 1814–1851. [Thesis.] Würzburg, Universität, 1980.

KLEBS, Arnold Carl (1870–1943). Swiss physician, bibliophile and medical historian; son of T.A.E. Klebs.

BAUMGARTNER, L. Arnold Carl Klebs, 1870–1943. *Bulletin of the History of Medicine*, 1943, **14**, 201–16.
ARNOLD C. KLEBS issue of *Bulletin of the History of Medicine*, 1940, **8**, 317–532; includes Bibliography of the writings of Arnold C.Klebs, by A. Lang, pp. 523–32.

KLEBS, Theodor Albrecht Edwin (1834–1913). German pathologist and bacteriologist; professor of pathology at Berne, Prague, Würzburg, Zürich and Chicago; first to see typhoid and diphtheria bacilli.

RÖTHLIN, O.M. Edwin Klebs (1834–1913): ein früher Vorkämpfer der Bakteriologie und seine Irrfahrten. Zürich, Juris-Verlag, 1962. 38 pp.

KLECZKOWSKI, Alfred Alexander Peter (1908–1970). Polish/Russian immunologist and photobiologist.

BAWDEN, F.C. BMFRS, 1971, **17**, 431–40.

KLEIN, Carl Christian (1772–1825). German surgeon.

TOELLNER, R. Carl Christian von Klein: ein Wegbereiter wissenschaftlicher Chirurgie in Württemberg. Stuttgart, Gustav Fischer, 1965, 112 pp.

KLEIN, Edward Emanuel (1844–1925). Austrian histologist and pathologist.

BULLOCH, W. Emanuel Klein 1844–1925. *Journal of Pathology and Bacteriology*, 1925, **28**, 684–97.

KLEIN, Melanie (1882–1960). Austrian psychoanalyst.

ARNOUX, D.J. Melanie Klein. Paris, Presses Universaires de France, 1997. 127 pp., A.J. Melanie Klein. New York, Columbia Univ. Press, 2002. 304 pp.
LIKIERMAN, M. Melanie Klein: her work in context. London, Continuum, 2001. 202 pp.
GROSSKURTH, P.P. Melanie Klein: her world and her work. New York, Knopf, 1986. 516 pp. Reprinted Northvale, N.J., J. Aronson, 1995.
SEGAL, H. Melanie Klein. Brighton, Harvester Press, 1979. 194 pp.

KLEINE, Friedrich Karl (1869–1950). German parasitologist.

JANITSCHKE, K. Kleine-Koch-Orenstein: the German connection with medical research in South Africa. *Adler Museum Bulletin*, 1985, **11**, No. 2, 16–18.

KLEMPERER, Georg (1865–1946). German physician; introduced antipneumococcus serum.

HAUSEN, K. Georg Klemperer. *Deutsche Medizinische Wochenschrift*, 1947, **72**, 362–3.

KLENCKE, Hermann Philipp Friedrich (1813–1881). German pathologist; showed possibility of transmission of tuberculosis to man by cow's milk.

HIRSCH, H. Hermann Philipp Friedrich Klencke, 1813–1881, zwischen Romantik und Experimentalpathologie. Zürich, Juris-Verlag, 1964. 51 pp.

KLENCKE, H. Aus dem Leben eines Arztes. Hrsg. von Barbara und Günter Albrecht. Berlin, Buchverlag der Morgen, 1968. 379 pp.

KLENK, Ernst (1896–1971). German biochemist.

MORGAN, N. DSB, 1990, **17**, 484–5.

KLIENEBERGER-NOBEL, Emmy (1892–1985). German bacteriologist.

KLIENEBERGER-NOBEL, E. Memoirs. London, Academic Press, 1980. 141 pp. First published in German, Stuttgart, 1977.

KLINEFELTER, Harry Fitch (1912–1990). American physician; described 'Klinefelter syndrome', an endocrine disorder (1942).

–. *Lancet*, 2000, **356**, 333–35.

KLIPPEL, Maurice (1858–1942). French neurologist.

PATEL, P.R. & LAUERMAN, W.C. Maurice Klippel. *Spine*, 1995, **20**, 2157–60.

KLUYVER, Albert Jan (1888–1956). Dutch microbiologist and biochemist.

KAMP, A.F. *et al.* (eds.) Albert Jan Kluyver: his life and work. Amsterdam, New York, Interscience, 1959. 567 pp.
SMIT, P. DSB, 1973, **7**, 405–7.
WOODS, D.D. BMFRS, 1957, **3**, 109–28.

KNAPP, Herman Jacob (1832–1911). German/American ophthalmologist.

SNYDER, C. Our ophthalmic heritage. London, J. & A. Churchill, 1967, pp. 9–12.
Archival material: CMAC.

KNEIPP, Sebastian (1821–1897). German pastor and hydrotherapist.

SCHOMBURG, E. Sebastian Kneipp, 1821–1897. Bad Wörishofen, Sanitas Verlag 1976. 154 pp.

KNIGHT, John (1600–1680). English surgeon; sergeant surgeon to King Charles II.

CALVERT, E.M. & Calvert, R.T.C. Sergeant Surgeon John Knight, Surgeon General 1664–1680. London, Heinemann, 1939. 111 pp.

KNIGHTON, William (1776–1836). British physician; physician to the Prince Regent (George IV).

KNIGHTON, D. The memoirs of Sir William Knighton, Bart ... including his correspondence with many distinguished persons. 2 vols., London, R. Bentley, 1838. 423 + 488 pp.

KNOWLES, James Sheridan (1784–1852). Irish physician and dramatist.

MEEKS, L.H. Sheridan Knowles and the theatre of his time. Bloomington, Ind., Principia Press, 1933. 239 pp.

KNOX, Robert (1791–1862). British anatomist in Edinburgh; suffered notoriety as a result of his connexion with the resurrectionists.

BULLOUGH, V.L. DSB, 1973, **7**, 414–16.

LONSDALE, H. A sketch of the life and writings of Robert Knox, the anatomist. London, Macmillan, 1970. 420 pp.

RAE, I. Knox the anatomist. Edinburgh, London, Oliver & Boyd, 1964. 164 pp.

KOBERT, Rudolf (1854–1918). German pharmacologist; professor at Rostock.

–. Zur Geschichte der Pharmakologie und Toxikologie. Rudolf Kobert und seine Zeit. Rostock, Universität Rostock, Medizinische Fakultät, 1992. 112 pp.

KOCH, Gerhard (b.1913). German geneticist.

KOCH, G. Inhaltsreiche Jahr eines Humangenetikers. Mein Lebensweg in Bildern und Dokumenten. Erlangen, Perimed Fachbuchgesellschaft, 1982. 472 pp.

KOCH, Heinrich Hermann Robert (1843–1910). German bacteriologist, particularly notable for his work on cholera, tuberculosis and anthrax; Nobel Prize 1905.

BARLOW, C. & BARLOW, P. Robert Koch. Geneva, Editio-Servis S.A., distributed by Heron Books, 1971. 392 pp.

BROCK, T.D. Robert Koch, a life in medicine and bacteriology. Madison, Science Tech Publishers; Berlin, New York, Springer-Verlag, 1988. 364 pp. Reprinted Washington DC, ASM Press, 1999.

DOLMAN, C.E. DSB, 1973, **7**, 420–25.

GENSCHOREK, W. Robert Koch. Selbstloser Kampf gegen Seuchen und Infektionskrankheiten. Leipzig, S. Hirzel, 1981. 224 pp.

MÖLLERS, B. Robert Koch: Personlichkeit und Lebenswerk. Hannover, Schmorl & von Seefeld, 1950. 756 pp.

Archival material: Robert Koch Institut, Berlin; Institute of the History of Medicine, Johns Hopkins University.

KOCHER, Emil Theodor (1841–1917). Swiss surgeon, professor at Bern; pioneer thyroidectomist; Nobel Prize 1909.

BOSCHUNG, U. (ed.) Theodor Kocher (1841–1917): Beiträge zur Würdigung von Leben und Werk. Bern, Stuttgart, Huber, 1991. 135 pp.

TRÖHLER, U. Der Nobelpreisträger Theodor Kocher, 1841–1917, auf dem Weg zur physiologischen Chirurgie. Basel, Boston, Birkhäuser, 1984. 240 pp.

KODICEK, Egon Hynek (1908–1982). Czech biochemist; worked especially on vitamins.

FRASER, D.R. & WIDDOWSON, E.M. BMFRS, 1983, **29**, 297–331.

Archival material: CMAC.

KOEBERLÉ, Eugène (1828–1915). French gynaecologist; introduced ovariotomy into France.

HUARD, P.A. Eugène Koberlé, pionnier de l'asepsie, de l'hémostase, de la gynécologie et de la chirurgie modernes. *Episteme*, 1968, **2**, 219–43.

PECHEVIN, R. Le docteur Koberlé et son oeuvre. Strasbourg, [n.p.], 1914. 146 pp.

KÖHLER, Alban (1874–1947). German radiologist; introduced teleradiography of the heart, 1905.

FREYSCHMIDT, Alban Köhler, geb. 1874, gest. 1947. Beurteilung eines Pioniers der klinischen Radiologie aus heutiger Sicht. *Fortschritte auf dem Gebiete der Röntgenstrahlen*, 1995, **163**, 463–8.

KÖHLER, Georges Jean Franz (1946–1995). German immunologist; shared Nobel Prize (Physiology or Medicine), 1984, for discovery of the principle of mononuclear antibodies.

Les Prix Nobel en 1984.

KÖHLER, Wolfgang (1887–1967). Estonian pyschologist.

JAEGER, S. Wolfgang Köhler (1887–1967) zum 100. Geburtstag am 21 Januar: biographische Daten und Publikationen. *Geschichte der Psychologie*, 1987, **4**. Heft. 1, Nr. 10, 5–24.
NEISSER, I. BMNAS,2002, **81**, 186–97.

KÖLLIKER, Rudolph Albert von (1817–1905). Swiss anatomist, embryologist, histologist and physiologist; professor in Würzburg.

HINTZSCHE, E. DSB, 1973, **7**, 437–40.
KÖLLIKER, R.A. Erinnerungen aus meinem Leben. Leipzig, W.Engelmann, 1899. 399 pp.

KÖNIG, Emanuel (1658–1731). Swiss physician and natural historian.

FICHMAN, M. DSB, 1973, **7**, 458–9.

KOENIG, Franz (1832–1910). German orthopaedic surgeon.

KOENIG, F. Lebenserinnerungen. Berlin, Hirschwald, 1912. 155 pp.

KÖRTE, Werner (1853–1937). German surgeon; first successfully to treat bronchiectstasis surgically.

EISELBERG, A. *Archiv für Klinische Chirurgie*, 1933, **176**, 401–6.
NORDMANN, O. *Chirurg*, 1933, **5**, 769–74.

KOLFF, Willem Johann (b. 1911). Dutch/American physician; introduced artificial kidney. 1944.

FRIEDMAN, E.A. Willem J. Kolff, MD: accomplishments beyond standard measure. *ASAIO Journal*, 1997, **43**, 263–7.
FRIEDMAN, E.A. Kolff's heritage renewed. *Artificial Organs*, 1998, **18**, 44–6.
KOLFF, W.J. The early years of artificial organs at the Cleveland Clinic, *ASAIO Journal*, 1998, **44**, 3–11.
WEISSE, A.B. Turning bad luck into good: the alchemy of Willem Kolff, the first successful artificial kidney, and the artificial heart. *Hospital Practice (Office ed.)*, 1992, **27**, 108–10, 115–18, 121 *passim.*

KOLISKI, Alexander (1847–1918). Austrian gynaecologist; with C. Breus published classical description and classification of pelvic deformities.

STOERCK, O. *Wiener Klinische Wochenschrift*, 1918, **31**, 522–4.

KOLLATH, Werner (1892–1970). German nutritionist and artist.

KOLLATH, E. Werner Kollath, Forscher, Arzt und Künstler. München, J.F. Lehmanns Verlag, 1973. 384 pp.

KOLLE, Wilhelm (1868–1935). German bacteriologist.

HETSCH, H. Wilhelm Kolle. *Deutsche Medizinische Wochenschrift*, 1935, **61**, 849–50.

KOLLER, Carl (1857–1944). Bohemian/American ophthalmologist; introduced local anaesthesia.

LILJESTRAND, G. Carl Koller and the development of local anaesthesia. *Acta Physiologica Scandinavica*, 1967, Suppl. 299, 1–30.

KOLTZOFF, Nikolai Konstantinovich (1872–1940). Russian cytologist and geneticist.

ZALKINE, S.Y. DSB, 1973, **7**, 454–7.

KON, George Armand Robert (1892–1951). British chemist; professor at Chester Beatty Research Institute, Royal Cancer Hospital, London.

LINSTEAD, R.P. ONFRS, 1952–53, **8**, 171–92.

KONCHALOVSKII, Maksim Petrovich (1875–1942). Russian physician.

GUKASIAN, A.G. Maksim Petrovich Konchalovskii i ego kliniko-teoreticheskie vagliady. Moskva, Medgiz, 1956. 153 pp.

KONOPKA, Stanislaus (1896–1982). Polish physician, medical librarian and historian.

–. *Clio Medica*, 1983, **18**, 239–41.
DUSINSKA, H. Stanislaw Konopka 1896–1982, Zarys monograficzny. Warsawa, Główna Bibliotika Lekarska, 1995. 363 pp.

KOPLIK, Henry (1858–1927). American paediatrician; drew attention to 'Koplik's spots' as a diagnostic sign in measles.

BASS, M.H. Pediatric profiles: Henry Koplik. *Journal of Pediatrics*, 1955, **46**, 119–25.

KOPROWSKI, Hilary (b. 1916). Polish/American microbiologist; introduced poliomyelitis immunization; pioneer of vaccines and of the therapeutic use of monoclonal antibodies.

KAPLAN, M.M. Three decades of co-operation with Hilary Koprowski. *Journal of Cellular Physiology*, 1982, Suppl. 2, 1–6.
KRITCHEVSKY, D. Hilary Koprowski and the Wistar Institute. *Journal of Cellular Physiology*, 1982, Suppl. 2, vi–viii.
VAUGHAN, R. Listen to the music: the life of Hilary Koprowski. New York, Springer, 2000. 295 pp.

KORÁNYI, Sándor (1866–1944). Hungarian physician; established cryoscopy of urine as kidney function test.

GROTE, 1924, 3, 63–88.
MAGYAR, I. Korányi Sándor. Budapest, Akademiai Kiado, 1970. 239 pp.

KORCZAK, Janusz [Goldszmit, Henryk] (1878 or 1879–1942). Jewish Polish paediatrician; killed voluntarily while accompanying 200 orphanage children exterminated at Treblinka.

LIFTON, B. The king of children: a biography of Janusz Korczak. New York, Farrar, Straus & Giroux, 1988. 404 pp.

KORNBERG, Arthur (b.1918). American biochemist; professor at Washington University and Stanford University; shared Nobel Prize 1959 with S. Ochoa.

KORNBERG, A. For the love of enzymes: the odysssey of a biochemist. Cambridge, Mass., Harvard Univ. Press, 1989. 336 pp.

BIBLIOGRAPHY OF MEDICAL AND BIOMEDICAL BIOGRAPHY

KORNBERG, A. Never a dull enzyme. *Annual Review of Biochemistry*, 1989, **58**, 1–30.

KOROTKOV [KOROTKOFF], Nikolai Sergeievich (1874–1920). Russian physician; introduced the method of applying the stethoscope to the brachial artery during blood pressure examination with the sphygmomanometer.

KOROTKOFF, N.S. Experiments for determining the efficiency of arterial collaterals. Preface, biographical notes and editing of translation from Russian by H.N. Segall. Montreal, H.N. Segal, 1980. 265 pp. [M.D. thesis, Imperial Military Medical Academy, St Petersburg, 1910].

MULTANOVSKY, M.P. The Korotkov's method. History of its discovery and clinical and experimental interpretation, and contemporary appraisal of its merits. *Cor et Vasa* (Prague), 1970, **12**, 1–7.

KORSAKOFF, Sergei Sergeivich (1854–1900). Russian neuropsychiatrist; drew attention to alcoholic polyneuritis ('Korsakoff's syndrome').

VICTOR, M. & YAKOVLEV, P.I. S.S. Korsakoff's psychic disorder in conjunction with peripheral neuritis. A translation...with brief comments on the author and his contribution to clinical medicine. *Neurology*, 1955, **5**, 394–406, 509.

KORTUM, Karl Arnold (1745–1824). German physician and writer.

CARL ARNOLD KORTUM, 1745–1824: Arzt, Forscher, Literat: 250 Geburtstag. Essen, Verlag Peter Pomp, Bottrup, 1995. 277 pp.

KOSSEL, Karl Martin Leonhard Albrecht (1853–1927). Swiss physiological chemist; noted for work on cytochemistry; Nobel Prize (Physiology or Medicine) 1910.

JONES, M.E. Albrecht Kossel: a biographical sketch. *Yale Journal of Biology and Medicine*, 1953, **7**, 466–8.

OLBY, R. DSB, 1973, **7**, 466–8.

KRAEPELIN, Emil (1856–1926). German psychiatrist; pioneer of experimental psychiatry.

KRAEPELIN, E. Memoirs. Edited by H. Hippius *et al.* Translated by C. Wording-Deane. New York, Springer, 1987. 270 pp. First published in German, Berlin, 1983.

STEINBERG, H. Kraepelin in Leipzig: eine Begegnung von Psychiatrie und Psychologie. Bonn, Edition Das Narrenschiff im Psychiatrie-Verlag, 2001. 382 pp.

KRAFFT, Charles (1863–1921). Swiss surgeon.

GASSER, P. Charles Krafft (1863–1921), ein Pionier der Appendektomie und der Krankenpflege in Europa. Basel, Schwabe, 1977. 121 pp.

KRAFFT-EBING, Richard (1840–1902). German psychiatrist.

HAUSER, R.I. Sexuality, neurasthenia and the law. Richard von Krafft-Ebing (1840–1926). London, PhD thesis, University College London. 1992. 471 leaves.

OOSTERHUIS, H. Stepchildren of nature: Krafft-Ebing, psychiatry and the meaning of sexual identity. Chicago, Chicago Univ. Press, 2000. 321 pp.

KRASILNIKOV, Nikolai Aleksandrovich (1896–1973). Russian microbiologist.

GUTINA, V.N. N.A. Krasilnikov [1896–1973]. Moskva, Nauka, 1982.

BIBLIOGRAPHY OF MEDICAL AND BIOMEDICAL BIOGRAPHY

KRAUS, Rudolf (1868–1932). Austrian bacteriologist and immunologist; discovered precipitins.

EISLER, M. Rudolf Kraus. *Wiener Klinische Wochenschrift*, 1932, **45**, 1072–73.

KRAUSE, Fedor (1857–1927). German neurosurgeon.

KAHLENDAHL, H. Anfänge der Neurochirurgie in Deutschland: Fedor Krause. *Zeitschrift für Neurologie*, 1973, **204**, 159–63.

KRAYER, Otto (1899–1982). German-born pharmacologist; professor at Harvard Medical School.

GOLDSTEIN, A. BMNAS, 1987, **57**, 150–225.

KREBS, Edwin Gerhard (b. 1918). American biochemist; shared Nobel Prize (Physiology or Medicine) 1992 for discoveries concerning reversible protein phosphorylation as a biological regulatory mechanism.

KREBS, E.G. An accidental biochemist. *Annual Review of Biochemistry*, 1998, **67**, xiii–xxxii.
Archival material: Library, University of Cambridge; University of Sheffield.

KREBS, Hans Adolf (1900–1981). German/British biochemist born in Germany; shared Nobel Prize 1953 for his discovery of the citric acid (Krebs) cycle.

HOLMES, F.L. DSB, 1990, **17**, 496–506.
HOLMES, F.L. Hans Krebs: Vol. 1. The formation of a scientific life 1900–1933. New York, Oxford, Oxford Univ. Press, 1991. 491 pp.
HOLMES, F.L. Hans Krebs: Vol.2. Architect of intermediary metabolism 1933–1937. Oxford, Oxford Univ. Press, 1993. 481 pp.
KORNBERG, H. & WILLIAMSON, D.H. BMFRS, 1984, **30**, 351–85.
KREBS, H. Reminiscences and reflections. Oxford, Clarendon Press, 1981. 298 pp.

KRIES, Johannes Adolf von (1853–1928). German physiologist; professor at Freiburg.

GROTE, **4**, 125–87 (CB).

KRÖNLEIN, Rudolph Ulrich (1847–1910). Swiss surgeon; professor at Zürich.

MADRITSCH, W. Der Zürcher Chirurg Ulrich Krönlein. Zürich, Juris, 1967. 53 pp.

KROGH, Schack August Steenberg (1874–1949). Danish physiologist; Nobel prizewinner 1920 for work on the physiology of capillaries.

HILL, A.V. ONFRS, 1950, **7**, 221–37.
SCHMIDT-NIELSEN, B. August and Marie Krogh: lives in science. New York, Oxford Univ. Press, 1995. 295 pp.
SNORRASON, E. DSB, 1973, **7**, 501–4.

KRONECKER, Hugo (1839–1914). German physiologist.

ROTHSCHUH, K.E. DSB, 1973, **7**, 504–5.
SCHAFER, E.A. *Proceedings of the Royal Society B*, 1917, **89**, 14–50.

KRUMBHAAR, Edward Bell (1882–1966). American physician and medical historian.

BIBLIOGRAPHY OF MEDICAL AND BIOMEDICAL BIOGRAPHY

LONG, E.R. Edward Bell Krumbhaar: physician, historian, founder of the American Association for the History of Medicine. *Bulletin of the History of Medicine*, 1957, **31**, 493–504.

KRUSEN, Frank Hammond (1898–1973). American physician.

ROBINSON, M.O. Frank H. Krusen, pioneer in physical medicine and rehabilitation. Minneapolis, Denison, 1963. 240 pp.

KÜHNE, Wilhelm Friedrich (1837–1900). German physiologist and biochemist.

ROTHSCHUH, K.E. DSB, 1973, **7**, 519–21.

KÜMMEL, Hermann (1852–1937). German surgeon, remembered for his description of traumatic spondylitis ('Kümmel's disease').

GROTE, **1**, 25–58 (CB).

KUESTER, Ernst Georg Ferdinand von (1839–1930). German surgeon.

KÖRTE, W. *Archiv für Klinische Chirurgie*, 1930, **159**, 521–6.
Archival material: Bibliothèque Royale, Copenhagen.

KÜSTNER, Otto (1849–1931). German gynaecologist and obstetrician.

GROTE, **8**, 61–163. (CB)

KUETTNER, Hermann (1870–1932). German surgeon; devised operation for tabes dorsalis.

KLEINSCHMIDT, O. *Chirurg*, 1932, **4**, 971–3.
PAYR, F. *Zentralblatt für Chirurgie*, 1932, **59**, 2866–70.

KUFFLER, Stephen W. (1913–1980). Hungarian/American neurobiologist.

KATZ, B. BMFRS, 1982, **28**, 225–59.
NICHOLLS, J.J. BMNAS, 1998, **74**, 193–208.

KUHN, Franz (1866–1929). German surgeon; introduced intrathecal insufflation anaesthetization, 1905.

GOERIG, M. Franz Kuhn (1866–1929) zum Geburtstag. Anästhesiologie, Intensivmedizin, Notfallmedizin, *Schmerztherapie*, 1991, **26**, 416–24.
SCHADEWALDT, H. Franz Kuhn: the inaugurator of peroral intubation. *Koroth*, 1985, **8**, 135–49.

KUHN, Richard (1900–1969). German chemist; Nobel Laureate (Chemistry) 1938, for his work on carotenoids and vitamins.

BURK, D. DSB, 1973, **7**, 517–8.

KUNDRAT, Hans (1845–1893). Austrian physician; first described 'Kundrat's lymphosarcoma'.

ALBERT, A. Hans Kundrat. *Wiener Klinische Wochenschrift*, 1893, **6**, 323–5.

KUNITZ, Moses (1822–1902). Russian/American biochemist.

COHEN, S.S. DSB, 1990, **17**, 515–18.

HERRIOTT, R.M. BMNAS, 1989, **58**, 305–17.

KUNKEL, Henry George (1916–1983). American physician.

BEARN, A.G. *et al.* Henry G. Kunkel (1916–1983). An appreciation of the man and of his scientific contributions and a bibliography of his research papers. *Journal of Experimental Medicine*, 1985, **161**, 869–95.

KUSSMAUL, Adolf (1822–1902). German physician; professor at Erlangen, Heidelberg, Freiburg and Strassburg.

BAST, T.H. The life and times of Adolf Kussmaul. New York, Hoeber, 1926. 131 pp.

KLUGE, F. Adolf Kussmaul, 1822–1902: Arzt und Forscher, Lehrer der Heilkunst. Freiburg im Breisgau, Rombach, *c.*2001. 543 pp.

KUSSMAUL, A. Memoirs of an old physician. Translated from German. Washington, National Library of Medicine, 1981. 352 pp. First published Stuttgart, 1899.

KUYPERS, Henricus Gerardus Maria (1925–1989). Dutch/American neuroanatomist; professor at Western Reserve University, Cleveland.

PHILLIPS, C.G. & GUILLERY, R.W. BMFRS, 1992, **38**, 187–207.

LABBÉ, Ernest Marcel (1870–1939). French physician; gave first full description of chromaffin cell tumours of the adrenal medulla, 1922.

BEZANÇON, F. *Bulletin de l'Académie Nationale de Médecine*, 1939, **121**, 804–11.

LABBÉ, Léon (1832–1916) French physician; developed pre-anaesthetic medication.

BIANCHON, H. Nos grands médecins. Paris, 1891, pp. 245–51.

LA BROSSE, Guy de (*c.*1586–1641). French physician, botanist and chemist.

GUERLAC, H. DSB, 1973, **7**, 536–41.

LACAN, Jacques (1901–1981). French psychoanalyst.

CLÉMENT, C. The lives and legends of Jacques Lacan. New York, Columbia Univ. Press, 1983. 225 pp. First published in French, 1981.

DOR, J. Thesaurus Lacan. Vol.2. Nouvelle bibliographie des travaux de Jacques Lacan. Paris, Éditions et Publications de l'Ecole Lacanienne. 1994. 278 pp.

HESNARD, A. De Freud à Lacan. Paris, Éditions ESF, 1970, 149 pp.

RODINESCO, E. Jacques Lacan. New York, Columbia Univ. Press, 1997. 574 pp.

LACASSAGNE, Antoine Marcellin (1884–1971). French oncologist; distinguished for work on radium therapy in cancer.

LATARGET, R. Notice sur la vie et les travaux de Antoine Lacassagne (1884–1971). Paris, Palais de l'Institut de France, 1973. 10 pp.

REGATO, J.A. del. Antoine Lacassagne. *International Journal of Radiation Oncology*, 1986, **12**, 2165–73.

LA COUR, Leonard Francis (1907–1984). British cytologist.

LEWIS, D. BMFRS, 1986, **32**, 357–75.

LADD-FRANKLIN, Christine (1847–1933). American psychologist; proposed 'Ladd-Franklin' theory of colour vision.

FURUMOTO, L. Christine Ladd-Franklin's color theory: strategy for claiming scientific authority? *Annals of the New York Academy of Sciences*, 1994, **727**, 91–100.

LAENNEC, René-Théophile-Hyacinthe (1781–1826). French pioneer in study of chest diseases; introduced instrumental auscultation of the chest.

BOULLE, L. *et al.* Laennec: catalogue des manuscrits scientifiques. Paris, Masson, 1982. 316 pp.
DUFFIN, J. To see with a better eye: a life of R.T.H. Laennec. Princeton, Princeton Univ. Press, 1998. 453 pp.
HEAF, F. DSB, 1973, **7**, 556–7.
KERVRAN, R. Laennec, his life and times. Translated from the French by D.C. Abrahams-Curiel. Oxford, Pergamon Press, 1960. 213 pp.
METTLING, C. Index bibliographique par ordre chronologique de l'oeuvre de René Théophile Laennec. *Presse Médicale*, 1926, 25 Dec., 1624–6.

LAGUNA [LACUNA], Andrés de (1499–1560). Spanish anatomist; physician to Charles V.

HERNANDO Y ORTEGA, T. Dos estudios históricos (vieja y nueva medicina). Espasa-Calpe, 1982. 243 pp.
LIND, L.R. Studies in pre-Vesalian anatomy. Philadelphia, American Philosophical Society, 1975. pp. 257–94.

LAIDLAW, Patrick Playfair (1881–1940). British pathologist; discovered cause and cure of dog distemper and isolated human influenza virus.

DALE, H.H. ONFRS, 1940–41, **3**, 427–47.

LAIGNEL-LAVASTINE, Maxime Paul Marie (1875–1953). French physician and medical historian.

LA VIE et l'oeuvre de Maxime Laignal-Lavastine (1875–1953). *Revue d'Histoire de la Médecine Hébraique*, 1954, **7**, 57–120.
SONDERVOORST, F.A. Maxime Laignel-Lavastine. *Histoire de Médecine*, 1953, **3**, 53–61.

LAING, Ronald David (1927–1989). British psychiatrist.

BURSTON, D. The wing of madness: the life and work of R.D. Laing. Cambridge MA, Harvard Univ. Press, 1996, 273 pp.
CLAY, J. R.D. Laing: a divided self. London, Hodder and Stoughton, 1996. 308 pp.
COLLIER, A. R.D. Laing. The philosophy and politics of psychotherapy. Hassocks, Sussex, Harvester Press, 1977. 214 pp.
HOWARTH-WILLIAMS, M. R.D. Laing; his work and its relevance for sociology. London, Routledge & Kegan Paul, 1977. 219 pp.
LAING, A. R.D. Laing: a biography. London, Peter Owen, 1994. 248 pp.
MULLAN, Bob (ed.) R.D. Laing, creative destroyer. London, Cassell, 1997. 431 pp.
MULLAN, B. R.D. Laing: a personal view. London, Duckworth, 1999. 232 pp.

LAMARCK, Jean Baptiste Pierre Antoine de Monet de (1744–1829). French biologist and zoologist.

BURKHARDT, R.W. The spirit of system. Lamarck and evolutionary biology. Cambridge, Mass., Harvard Univ. Press, 1977. 285 pp.
BURLINGAME, L.J. DSB, 1973, **7**, 584–94.
JORDANOVA, L.J. Lamarck. Oxford, Oxford Univ. Press, 1984. 118 pp.
LAURENT, G. (ed.) Jean-Baptiste Lamarck 1744–1829. Paris, Comité des Travaux Historiques et Scientifiques, 1997. 757 pp.

LA MARTINIÈRE, Pierre-Martin de (1634–1690). French physician; first to describe gonococcal arthritis.

LOUX, F. Pierre-Martin de la Martinière, un médecin au XVIIe siècle. Paris, Imago, 1988. 254 pp.

LA METTRIE, Julien Offray de (1709–1751). French physician and philosopher.

PORITZKY, J.E. Julien Offray de Lamettrie. Sein Leben und sein Werk. Geneva, Slatkine, 1971. 356 pp.
VARTANIAN, A. DSB, 1973, **7**, 605–7.
WELLMAN, K. La Mettrie: medicine, philosophy and enlightenment. Durham, N.C., Duke Univ. Press, 1992. 342 pp.

LAMY, Guillaume (*fl.* 1667–1682). French physician and philosopher.

PLANTEFOL, L. DSB, 1973, **7**, 611–13.

LANCEFIELD, Rebecca Craighill (1895–1981). American microbiologist.

McCARTY, M. BMNAS, 1987, **57**, 150–225.
McCARTY, M. DSB, 1990, **17**, 523–5.

LANCEREAUX, Étiennne (1829–1910). French physician.

POUMERY, J.J. Étiennne Lancereaux (1827–1910), sa vie, son oeuvre, ses découvertes scientifiques. *Histoire des Sciences Médicales*, 1989, **23**, 279–84.

LANCISI, Giovanni Maria (1654–1720). Italian physician.

BACCHINI, A. La vita e le opera di Giov. Maria Lancisi. Roma, Sansaini, 1920. 115 pp.
CASTELLANI, C. DSB, 1973, **7**, 613–14.

LANDIS, Eugene Markley (1901–1987). American physiologist.

PAPPENHEIMER, J.R. BMNAS, 1994, **64**, 188–206.

LANDOUZY, Louis Théophile Joseph (1845–1917). French physician.

ROGER, G.E.H. *et al.* Louis Landouzy: discours prononcés. Paris, Masson, 1923. 71 pp.

LANDRÉ-BEAUVAIS, Augustin Jacob (1772–1840). French physician; gave first accurate description of rheumatoid arthritis.

PARISH, L.C. Augustin Jacob Landré-Beauvais, 1772–1840: a neglected forerunner in the history of rheumatoid arthritis. Philadelphia, Tufts University School of Medicine, 1963. 29 pp.

LANDRY, Jean Baptiste Octave (1826–1865). French physician; described acute ascending polyneuritis ('Landry's paralysis').

BIBLIOGRAPHY OF MEDICAL AND BIOMEDICAL BIOGRAPHY

REMLINGER, P. *Presse Médicale*, 1933, **41**, 227–9.

LANDSTEINER, Karl (1868–1943). Austrian/American immunologist born in Austria; serologist and bacteriologist; Nobel Prize 1930.

HEIDELBERGER, M. BMNAS, 1967, **39**, 177–210.
ROUS, F.P. ONFRS, 1947, **5**, 295–324.
SPEISER, P. DSB, 1973, **7**, 622–5.
SPEISER, P. & SMEKAL, F.G. Karl Landsteiner: the discoverer of the blood groups and a pioneer in the field of immunology. English translation by R. Rickett. Vienna, Hollinek, 1975. 198 pp. First published in German, 1961, 2te Aufl., 1975.
Archival material: Rockefeller University; Library, American Philosophical Society, Philadelphia.

LANE, William Arbuthnot (1856–1943). British surgeon; pioneer in several aspects of surgical technique.

DALLY, A. Fantasy surgery 1880–1930, with special refererence to Sir William Arbuthnot Lane. Amsterdam, Rodopi, 1996. 359 pp.
LAYTON, T.B. Sir William Arbuthnot Lane: an enquiry into the mind and influence of a surgeon, Edinburgh, Livingstone, 1956. 128 pp.
TANNER, W.E. Sir W. Arbuthnot Lane; his life and work. 2nd ed. London, Baillière, Tindall & Cox. 1946. 192 pp.
Archival material: CMAC.

LANFRANCO of Milan (*fl.* 1290–1296). Italian surgeon.

MÜLLER, R. (ed.) Der 'Jonghe Lanfranc' (Altdeutsche Lanfranc-Übersetzungen, 1.) Thesis, Bonn, Rheinischen Friedrich-Wilhelms-Universität, 1968. 252 pp.

LANGE, Carl Georg (1834–1900). Danish physician and neurologist.

SNORRASON, E. DSB, 1973, **8**, 7–8.

LANGE, Johannes (1485–1565). German physician; gave first description of chlorosis.

MAJOR, R.H. Johannes Lange (1485–1565) of Heidelberg. *Annals of Medical History*, 1935, **7**, 133–40.

LANGENBECK, Bernhard Rudolf Konrad von (1810–1887). German surgeon; discovered *Candida albicans*; founded (*Langenbeck's) Archiv für Klinische Chirurgie*.

BERGMANN, E. von. Zur Erinnerung am Bernhard von Langenbeck. Rede bei der von der Deutschen Gesellschaft für Chirurgie. Berlin, August Hirschwald, 1888. 95 pp.
BRUNN, W. von. Die Chirurgie unter Bernhard Rudolf Konrad von Langenbeck, 1848–1882. In: Diegpen, P. & Rostock, P. (eds.) Das Universitätsklinikum in Berlin, 1810–1933. Leipzig, Barth, 1939, pp. 81–92.
GOLDWYN, R.M. Bernhard von Langenbeck, his life and legacy. *Plastic and Reconstructive Surgery*, 1969, **44**, 246–54.

LANGENBECK, Conrad Johann Martin (1776–1851). German surgeon and ophthalmologist.

BRUNN, W. Die Chirurgenfamilie Langenbeck. *Medizinische Welt*, 1936, **1**, 539–42.

LANGENBUCH, Carl Joseph August (1846–1901). German surgeon.

215

KLIMPEL, V. Das chirurgische Erbe. Carl Langenbuch – ein Pionier der modernen Gallenwegschirurgie. *Zentralblatt für Chirurgie*, 1985, **110**, 1094–107.

LANGERHANS, Paul (1847–1888). German anatomist; described 'Langerhans cells' in the pancreas.

> HAUSEN, B.M. Die Inseln des Paul Langerhans: eine Biographie in Bildern und Dokumenten. Wien, Ueberreuter Wissenschaft, 1988. 286 pp.
> MANI, N. DSB, 1973, **8**, 8–9.

LANGHANS, Theodor (1839–1915). German pathologist; first to note giant cells in lymphadenoma.

> HEDINGER, F. *Centralblatt für Allgemeine Pathologie*, 1916, **27**, 217–19.
> WEGELIN, C. *Correspondenz-Blatt für Schweizer Ärzte*, 1915, **45**, 1654–9.

LANGLEY, John Newport (1852–1925). British physiologist and histologist; professor at Cambridge University.

> GEISON, G.L. DSB, 1973, **8**, 14–19.

LANGSTAFF, George (1780–1846). British surgeon; first to report carcinoma of the prostate, 1817.

> PLARR, 679–81 (CB).

LANKESTER, Edwin (1814–1874). British physician and botanist; father of Edwin Ray Lankester.

> ENGLISH, M.P. Victorian values: the life and times of Dr Edwin Lankester, M.D., F.R.S. Bristol, Biopress Ltd., 1990. 187 pp.

LANNELONGUE. Odilon Marc (1840–1911). French surgeon; first to treat cretinism by thyroid transplantation.

> BIANCHON, H. Nos grands médecins. Paris, 1891, pp. 267–74.

LA PEYRONIE, Françoise Gigot (1678–1747). French surgeon; described plastic induration of the penis ('La Peyronie disease').

> FORGUE, E. *Montpellier Médicale*, 1937, 3 ser. **11**, 97–120.
> STEIN, J.B. *Medical Journal and Record*, 1920, **130**, 161–3.

LAPICQUE, Louis (1866–1952). French physiologist and anthropologist.

> MONNIER, A.M. DSB, 1973, **8**, 28–30.

LAQUEUR, Ludwig (1839–1909). German ophthalmologist; introduced physostigmine in treatment of glaucoma.

> SNYDER, C. Ludwig Laqueur, M.D. 1839–1909. *Archives of Ophthalmology*, 1964, **72**, 111–13.

LAROQUE, G. Paul (1876–1924). American surgeon.

> –. *Transactions of the Southern Surgical Association*, 1935, **47**, 665–7.

LARREY, Dominique Jean (1766–1842). French military surgeon; surgeon-in-chief to the Imperial armies.

BIBLIOGRAPHY OF MEDICAL AND BIOMEDICAL BIOGRAPHY

DIBLE, J.H. Napoleon's surgeon. London, Heinemann, 1970. 346 pp. [Based on the memoirs of Larrey.]

RICHARDSON, R.G. Larrey: surgeon to Napoleon's Imperial Guard. Revised ed. London, Quiller Press, 2000. 269 pp. [Based on Larrey's own letters, journals and notes.] First published 1974.

SOUBIRAN, A. Le Baron Larrey, chirurgien de Napoléon. Paris, A. Fayard, 1967. 525 pp.

Archival material: Library, American Philosophical Society, Philadelphia.

LASÈGUE, Ernest Charles (1816–1883). French physician; eponymized in Lasègue's disease (persecution mania) and Lasègue's sign (in sciatica).

POZZI, L. Charles Lasègue. *Archivio di Ortopedia*, 1968, **81**, 227–38.

STRELETSKI, C. Essai sur Ch. Lasègue, 1816–1883. Paris, G. Steinheil, 1908. 208 pp.

LASHLEY, Karl Spencer (1890–1958). American physiological psychologist.

BARTLETT, F.C. BMFRS, 1959, **5**, 107–18.

BEACH, F.A. BMNAS, 1961, **35**, 163–204.

ROOFE, P.G. DSB, 1973, **8**, 45.

WEIDMAN, N.M. Of rats and men: Karl Lashley and American psychology, 1912–1955. Ithaca, NY, [Cornell University], 1994. 339 pp.

WEIDMAN, N.M. Constructing scientific psychology: Karl Lashley's mind-brain debates. Cambridge, Cambridge University Press, 1999. 219 pp.

LATHAM, Peter Mere (1789–1875). British cardiologist; early advocate of auscultation.

BEAN, W.B. (ed.) Aphorisms from Latham. Collected and edited by William B. Bean. Iowa City, Prairie Press, 1962. 102 pp.

GREENWOOD, pp. 37–50 (CB).

LAURENTIUS, Andreas, see **DU LAURENS, André**

LAVATER, Johann Caspar (1741–1801). German anatomist and descriptive physiognomist.

DONOVAN, A. Antoine Lavoisier: science, administration and revolution. Oxford, Blackwell, 1993, 351 pp.

GRAHAM, J. Lavater's essays on physiognomy: a study in the history of ideas. Berne, Lange, 1979. 130 pp.

JATON, A.M. Johann Caspar Lavater: Philosoph, Gottesmann, Schöpfer der Physiognomik. Zürich, Schweizer Verlagshaus, 1988. 158 pp.

KUNZ, R. Johann Caspar Lavaters Physiognomielehre in Urteil von Haller, Zimmermann und anderen zeitgenössischen Ärzten. Zürich, Juris, 1970. 44 pp.

LAVATER-SLOMAN, M. Genie des Herzens: die Lebensgeschichte Johann Caspar Lavater. 5te.Aufl.. Zürich, Artemis Verlag, 1955. 478 pp.

LAVERAN, Charles Louis Alphonse (1845–1922). French parasitologist and protozoologist; Nobel Prize 1907 for contributions to malariology.

KLEIN, M. DSB, 1973, **8**, 65–6.

PHISALIX, M. Alphonse Laveran; sa vie, son oeuvre. Paris, Masson, 1923. 263 pp.

LAVOISIER, Antoine-Laurent (1743–1794). French chemist and physiologist.

DUVEEN, D.I. & GLICKSTEIN, H.S. A bibliography of the works of Antoine Laurent Lavoisier 1743–1794. London, Dawson, 1954. 491 pp. Supplement, 1965. 173 pp.

217

BIBLIOGRAPHY OF MEDICAL AND BIOMEDICAL BIOGRAPHY

GUERLAC, H. DSB, 1973, **8**, 66–91.

HOLMES, F.L. Lavoisier and the chemistry of life. Madison, Univ. of Wisconsin Press, 1985. 565 pp.

McKIE, D. Antoine Lavoisier: scientist, economist, social reformer, London, Constable, 1952. 335 pp.

POIRIER, J.P. Lavoisier: chemist, biologist, economist. Philadelphia, Univ. Pennsylvania Press, 1996. 516 pp. First published in French, 1993.

LAWRENCE, William (1793–1867). British ophthalmic surgeon.

McKINNEY, H.L. DSB, 1973, **8**, 96–8.

STINSON, D.T. The role of Sir William Lawrence in 19th century English surgery. Zürich, Juris-Verlag, 1969. 45 pp.

LAWSON, George (1831–1903). British Army surgeon; later ophthalmic surgeon.

LAWSON, G. Surgeon in the Crimea: the experiences of George Lawson recorded in letters to his family. London, Constable, 1968. 209 pp.

LAZEAR, Jesse William (1866–1900). American physician; proved causal agent in yellow fever to be transmitted by *Aedes aegypti;* died from yellow fever after mosquito bite.

CARMICHAEL, E.B. Jesse William Lazear. *Alabama Journal of Medical Sciences*, 1972, **9**, 102–14.

LEAKE, Chauncey Depew (1896–1978). American pharmacologist and medical historian; demonstrated anaesthetic properties of divinyl ether.

KEYS, T.E. Chauncey Depew Leake, 1896–1978): an appreciation. *Journal of the History of Medicine*, 1987, **33**, 428–31.

LEARMONTH, James Rognvald (1895–1967). British surgeon in Scotland.

DOUGLAS, D.M. The thoughtful surgeon, James Rognvald Learmonth. Glasgow, University of Glasgow Press, 1969. 106 pp.

LEATHES, John Beresford (1864–1965). British physiologist; professor at Sheffield.

PETERS, R. BMFRS, 1958, **4**, 185–91.

LEBEDEV, Aleksandr Nikolaevich (1881–1938). Russian biochemist.

SHAMIN, A.N. DSB, 1990, **18**, 533–4.

LEBER, Theodor (1840–1917). German ophthalmologist.

JAEGER, W. The foundation of experimental ophthalmology by Theodor Leber. *Documenta Ophthalmologica*, 1988, **68**, 71–7.

LEBERT, Hermann (1813–1878). Polish pathologist; gave first systematic account of cerebral abcess.

GOLDSCHMID, E. Über den medizinischen Aufschwung in den vierziger Jahren des 19. Jahrhunderts. Mit einem Verzeichnis der Werke von Hermann Lebert (1813–1878). *Gesnerus*, 1949, **6**, 17–33.

LE BOË, Franciscus de, *see* **SYLVIUS**

218

LE CLERC, Daniel (1652–1728). Swiss physician; wrote first comprehensive history of medicine.

RÖTHLISBERGER, P. Daniel Le Clerc (1652–1728) und seine *Histoire de la médecine Gesnerus*, 1964, **21**, 126–41.

LECONTE, Joseph (1823–1901). American biologist and physiologist.

McKINNEY, H.L. DSB, 1973, **8**, 122–3.
STEPHENS, L.D. Joseph LeConte, gentle prophet of evolution. Baton Rouge, Louisiana State Univ. Press, 1982. 340 pp.

LEDERBERG, Joshua (b.1925). American geneticist; shared Nobel Prize (Physiology or Medicine) 1958.

LEDERBERG, J. Genetic recombination in bacteria: a discovery account. *Annual Review of Genetics*, 1987, **21**, 23–46.

LEDINGHAM, John Charles Grant (1875–1944). British microbiologist; director of the Lister Institute, London.

BEDSON, S.P. ONFRS, 1945–48, **5**, 325–40.

LE DOUBLE, Anatole Félix (1848–1913). French anatomist.

HUARD, P. & IMBAULT-HUART, M.L. DSB, 1973, **8**, 125–6.

LEDUC, Stephane Armand Nicolas (1853–1939). French physician; pioneer in ionic medication and electric convulsion therapy.

COSTA, C. La electroanestesia: variaciones sobre un tema chileno. *Anales Chilenos de la Historia de la Medicina*, 1959, **1**, 77–297.
DESCLAUX, L. *Annales d'Hygiène*, 1939, **17**, 273–8.

LEEUWENHOEK, Antoni van (1632–1723). Dutch pioneer microscopist.

DOBELL, C. Antony van Leeuwenhoek and his 'little animals', being some account of the father of protozoology and bacteriology, and his multifarious discoveries in the disciplines. Collected, translated and edited from his printed works, unpublished manuscripts and contemporary records. London, Bale & Danielsson, 1932. 435 pp. Reprinted, New York, Dover, 1958.
FORD, B.J. The Leeuwenhoekiana of Clifford Dobell. *Notes and Records of the Royal Society of London*, 1986, **41**, 95–105.
FORD, B.J. The Leeuwenhoek legacy. Bristol, Biopress, 1991. 185 pp.
HENIGER, J. DSB, 1973, **8**, 126–30.
PALM, C. & SNELDERS, H.A. (eds) Antoni van Leeuwenhoek, 1632–1723. Studies on the life and work of the Delft scientist. Amsterdam, Rodopi, 1982. 209 pp.

LEGALLOIS, Julien Jean César (1770–1814). French physiologist.

KRUTA, V. DSB, 1973, **8**, 132–5.

LEGG, Arthur Thornton (1874–1939). American orthopaedic surgeon; ('Calvé-Legg-Perthes disease').

–. *New England Journal of Medicine*, 1939, **221**, 436–8.

LEGG, John Wickham (1843–1921). British physician and authority on liturgiology.

DNB.
GARROD, A.E. Obituary. *St Batholomew's Hospital Reports*, 1922, **55**, 1–6.

LEHMANN, Hermann (1910–1985). German-born chemical pathologist; professor of clinical biochemistry, Cambridge; authority on abnormal haemoglobins.

DACIE, J. BMFRS, 1988, **34**, 405–49.

LEHMANN, Karl Bernhard (1858–1961). Swiss industrial hygienist.

LEHMANN, K.B. Frohe Lebensarbeit. Erinnerungen und Bekenntnisse eines Hygienikers und Naturforschers. München, Lehmann, 1933.

LEIDY, Joseph (1823–1891). American physician, anatomist, parasitologist and protozoologist.

RITTENBUSH, P.C. DSB, 1973, **8**, 169–70.
WARREN, L. Joseph Leidy: the last man who knew everything. New Haven, Yale Univ. Press, 1998. 303 pp.
Archival material: Academy of Natural Sciences, Philadelphia; Library, American Philosophical Society, Philadelphia; College of Physicians, Philadelphia.

LEIPER, Robert Thomson (1881–1969). British helminthologist.

GARNHAM, P.C.C. BMFRS, 1970, **16**, 385–404.

LEISHMAN, William Boog (1865–1926). British physician; Director of Army Medical Services; discovered parasite of kala-azar (*Leishmania donovani*).

LEISHMAN CENTENARY NUMBER, *Journal of the Royal Army Medical Corps*, 1966, **112**, 1–26.
WILLIAM BOOG LEISHMAN, 1865–1926. *Journal of Pathology and Bacteriology*, 1925–26, **29**, 515–28.
Archival material: CMAC.

LEJEUNE, Jérôme (b.1926). French human geneticist; discovered trisomy-21, cause of Down's syndrome.

LE MÉNE, J.M. Le professeur Lejeune, fondateur de la génétique humaine. Paris, MamE, 1997. 160 pp.

LELOIR, Luis Federico (1906–1987). Argentinian biochemist, University of Buenos Aires: Nobel Prize (Chemistry) for his discovery of sugar nucleotides and their role in the biosynthesis of carbohydrates.

LELOIR, L.F. Faraway and long ago. *Annual Review of Biochemistry*, 1983, **52**, 1–15.

LEMBERG, Max Rudolf (1896–1975). German/British biochemist.

RIMINGTON, C. & GRAY, C.H. BMFRS, 1976, **22**, 257–94.

LEMERY, Louis (1677–1743). French chemist, physician and anatomist.

HANNAWAY, O. DSB, 1973, **8**, 171–2.

LEMOS, Maximiano Augusto d'Oliviera (1860–1923). Portuguese physician and medical historian.

BIBLIOGRAPHY OF MEDICAL AND BIOMEDICAL BIOGRAPHY

SAAVEDRA, A. O Professor Maximiano Lemos. Oporto, A Medicina Moderna, 1923. 39 pp. Reprinted from *Medicina Moderna*, 1923, No. 379.

LEMPERT, Julius (1890–1968). Polish/American otologist; devised operation for otosclerosis.

SHAMBAUGH, G.E. Julius Lempert 1890–1968. *Archives of Otolaryngology*, 1969, **90**, 679–705.

LENHOSSEK Mihály (1863–1937). Hungarian neuroanatomist.

GROTE, 7, 99–148 (CB).

LEONARDO DA VINCI (1452–1519). Italian artist, scientist, and founder of iconographic and physiological anatomy.

BELLONE, E. & ROSSI, P. (eds.) Leonardo e l'età della ragione. Milano, Scientia, 1982. 479 pp.
DSB, 1973, **8**, 192–245; [life, scientific methods and anatomical works, by K.D. Keele, pp. 193–206].
GUERRINI, M. Bibliotheca Leonardiana, 1493–1989. 3 vols. Milano, Ed. Bibliographica, 1990. 2216 pp.
KEELE, K.D. Leonardo da Vinci's elements of the science of man. New York, Academic Press, 1983. 385 pp.
KEMP, M. Leonardo da Vinci. The marvellous works of nature and man. London, J.M. Dent, 1981. 384 pp.
McMURRICH, J.P. Leonardo da Vinci the anatomist. Baltimore, Willimas & Wilkins, 1930. 265 pp.
O'MALLEY, C.D. & SAUNDERS, J.B. de C.M. Leonardo da Vinci on the human body; the anatomical and embryological drawings of Leonardo da Vinci, with translations, emendations and a biographical introduction. New York, Schuman, 1952. 506 pp.
RETI, L. (ed.) The unknown Leonardo. New York, McGraw Hill, 1974. 319 pp.
WHITE, M. Leonardo, the first scientist. London, Little Brown, 2000. 370 pp.

LEONICENO, Niccolò (1428–1524). Italian humanist-physician; translated Galen, corrected Pliny's botanical errors; published first scholarly treatise on syphilis.

BYLEBYL, J.J. DSB, 1973, **8**, 248–50.

LEREBOULLET, Dominique-Auguste (1804–1865). French zoologist and embryologist.

APPEL, T.A. DSB, 1973, **8**, 253–5.

LERICHE, René (1879–1955). French surgeon.

CLARKE, R. René Leriche. Paris, Sephers, 1962. 207 pp.
MONDAR, H.J.J. René Leriche, chirurgien. Paris, Ventadour, 1956. 192 pp.

LESTOURGEON, Charles, the Elder (1779–1853).

LESTOURGEON, Charles, the Younger (1808–1891).

BUSHELL, W.D. The two Charles Lestourgeons, surgeons of Cambridge. Their Huguenot ancestors and their descendants. Cambridge, Heffer, 1936. 66 pp.

LETTERER, Erich (1895–1982). German pathologist.

BIBLIOGRAPHY OF MEDICAL AND BIOMEDICAL BIOGRAPHY

HAUSER, F.E. Erich Letterers allergologische Forschung im Lichtes seines Lebenswerkes, München-Diesenhofen, Dustri-Verlag, 1990. 115 pp.

LETTSOM, John Coakley (1744–1815). British physician born in West Indies; a founder of the Medical Society of London and a notable philanthropist.

ABRAHAM, J.J. Lettsom, his life, times, friends and descendants. London, Heinemann, 1933. 498 pp.
PETTIGREW, T.J. Memoirs of the life and writings of the late John Coakley Lettsom ... with a selection from his correspondence. 3 vols. London, Longman, 1817.

LEUCKART, Karl Georg Friedrich Rudolf (1822–1898). German zoologist and parasitologist.

SCHADEWALDT, H. DSB, 1973, **8**, 269–71.
WUNDERLICH, K. Rudolf Leuckart, Weg und Werk. Jena, Fischer, 1978. 152 pp.

LEVADITI, Constantin (1874–1953). Romanian/French microbiologist and syphilologist.

WROTNOWSKA, D. DSB, 1973, **8**, 273–4.

LEVENE, Phoebus Aaron Theodor (1869–1940). Russian/American biochemist; pioneer researcher on nucleic acids.

IHDE, A.J. DSB, 1973, **8**, 275–6.
VAN SLYKE, D.D. & JACOBS, W.A. BMNAS, 1945, **23**, 75–126.
Archival material: Rockefeller Archives Center.

LEVER, Charles James (1806–1872). Irish physician and novelist.

STEVENSON, A.L. Dr Quicksilver: the life of Charles Lever. London, Chapman & Hall, 1939. 308 pp.

LEVER, John Charles Weaver (1811–1858). British obstetrician; first to report albuminous urine in puerperal convulsions, 1843.

–. Dr J.C.W. Lever. *Lancet*, 1859, **1**, 75.

LEVI, Giuseppe (1872–1965). Italian anatomist and histologist.

OLIVO, O.M. DSB, 1973, **8**, 282–3.

LEVI-MONTALCINI, Rita (b.1909). Italian/American cell biologist; Nobel prizewinner 1986 for discovery of growth factors.

LEVI-MONTALCINI, R. In praise of imperfection: my life and work. Translated by L.K. Attardi. New York, Basic Books, 1988. 220 pp.

LEVINE, Philip (1900–1987). Russian/American haematologist; discovered Rh antigen.

GIBLETT, E.R. BMNAS, 1994, **63**, 322–47.

LEVINE, Samuel Albert (1891–1966). Polish/American physician.

LEVINE, H.J. Samuel A. Levine (1891–1966). *Clinical Cardiology*, 1992, **15**, 473–6.

LEVIT, Solomon Grigorevich (1894–1938). Lithuanian medical geneticist; Director Maxim Gorky Scientific Research Institute of Medical Genetics.

ADAMS, M.B. DSB, 1990, **18**, 546–9.

LEWIN, George Richard (1820–1896). German surgeon; first to remove laryngeal growth with the aid of the laryngoscope.

LASSAR, O. *Dermatologische Zeitschrift*, 1896, **3**, 678–86.
LESSER, –. *Archiv für Dermatologie und Syphilis*, 1896, **37**, 318–20.

LEWIN, Kurt (1890–1947). German/American psychologist.

MARROW, A.J. The practical theorist: the life and work of Kurt Lewin. New York, Basic Books, 1969. 290 pp.

LEWIN, Louis (1850–1929). German pharmacologist.

HOPPE, B. Louis Lewin (1850–1929), sein Beitrag zur Entwicklung der Ethnopharmakologie, Toxikologie und der Arbeitsmedizin. Dissertation. Freie Universität Berlin, 1985. 185 pp.
MACHT, D.I. Louis Lewin, pharmacologist, toxicologist, medical historian. *Annals of Medical History*, 1931, N.S. **3**, 179–94.

LEWIS, Aubrey Julian (1900–1975). British psychiatrist.

SHEPHERD, A. A representative psychiatrist; the career, contributions and legacies of Sir Aubrey Lewis. Cambridge, University Press, 1988. 31 pp. (*Psychological Medicine*, Monograph Supplement 10.).

LEWIS, Edward B. (b. 1918). American geneticist; shared Nobel Prize (Physiology or Medicine) 1995, for discoveries concerning the genetic control of early embryonic development.

Les Prix Nobel en 1995.

LEWIS, Howard Bishop (1887–1954). American biochemist and nutritionist; professor at University of Illinois.

ROSE, W.C. & COON, M.J. BMNAS, 1974, **44**, 139–73.
Archival material: Bentley Historical Library, University of Michigan. (See Brandt, P. Survey of Sources for the History of Biochemistry and Molecular Biology, 1976, No. 4, pp. 5–7.)

LEWIS, Paul Aldin (1879–1929). American pathologist and bacteriologist.

FLEXNER, S. Paul Aldin Lewis. *Science*, 1929, **70**, 133–4.

LEWIS, Thomas (1881–1945). British physiologist, cardiologist and clinical scientist.

DRURY, A.N. & GRANT, R.T. ONFRS, 1945–48, **5**, 179–202.
HOLLMAN, A. DSB, 1973, **87**, 294–6.
HOLLMAN, A. Sir Thomas Lewis: pioneer cardiologist and clinical scientist. London, New York, Springer, 1996. 300 pp.
SNELLEN, H.A. Two pioneers of electrocardiography: the correspondence between Einthoven and Lewis from 1908 to 1926. Rotterdam, Donker, 1983. 140 pp.
Archival material: CMAC; National Institute for Medical Research, London.

LEWIS, Timothy Richards (1841–1886). British physician; tropical parasitologist.

BIBLIOGRAPHY OF MEDICAL AND BIOMEDICAL BIOGRAPHY

CLARKSON, M.J. DSB, 1973, **8**, 296–7.
Archival material: London School of Hygiene and Tropical Medicine.

LEWIS, Warren Harmon (1870–1964). American cytologist and embryologist.

CORNER, G.W. BMNAS, 1967, **39**, 323–58.
Archival material: Library, American Philosophical Society, Philadelphia.

LEWIS, William (1708–1781). British chemist and pharmacist.

EKLUND, J. DSB, 1973, **8**, 297–9.
GIBBS, F.W. William Lewis, MB, FRS (1708–1781). *Annals of Science*, 1952, **8**, 122–51.

LEXER, Erich (1867–1937). German surgeon; performed first osteoarticular joint transplant.

BÄRKLE DE LA CAMP, H. Erich Lexer: ein biographische Skizze. *Vorträge aus der praktischen Chirurgie*, Heft 78. Stuttgart, Enke, 1967. 27 pp.

LEYDEN, Ernst Victor von (1832–1910). German physician.

LEYDEN, E. von. Lebenserinnerungen...Mit einem Vorwort von Wilhelm Waldeyer. Stuttgart & Leipzig, Deutsche Verlags-Anstalt, 1910. 284 pp.

LEYDIG, Franz von (1821–1908). German comparative anatomist.

GLEES, P. DSB, 1973, **8**, 301–3.

LI, Choh Hao (1913–1987). Chinese/American endocrinologist; with co-workers isolated interstitial-cell-stimulating hormone and ACTH.

International Journal of Peptide and Protein. Memorial issue in honor of Professor Choh Hao Li. Part 1 1988, **32**, 417–598.

LIBAVIUS, Andreas [LIBAU] (*c.*1560–1616). German physician and alchemist; a founder of the Iatrochemical School; poet laureate.

HUBICKI, W. DSB, 1973, **8**, 309–12.

LICHTENBERG, Alexander von (1880–1949). Hungarian urologist; introduced opaque media in urography.

GOODWIN, W.E. *et al.* Alexander von Lichtenburg (1880–1949). Biobibliographie eines Urologen. [Thesis.] Berlin, Institut für Geschichte der Medizin der Freien Universität, 1978.

LICHTHEIM, Ludwig (1845–1928). German physician; first described subcortical sensory aphasia.

WEGELIN, C. Aus dem Memorien von Ludwig Lichtheim. *Schweizerisches Medizinische Wochenschrift*, 1956, **86**, 366–71.

LIDDELL, Edward George Tandy (1895–1981). British physiologist; Wayneflete professor, University of Oxford.

PHILLIPS, C.G. BMFRS, 1983, **29**, 333–59.
Archival material: University Laboratory of Physiology, Oxford.

LIEBEAULT, Ambroise August (1823–1904). French physician; the substitution of psychotherapy for hypnotic suggestion was due largely to his work.

LAVEYSSIÈRE, L. *Revue de l'Hypnotisme et Psychologie*, 1897–8, **12**, 289–93.

LIEBERKÜHN, Johannes Nathanael (1711–1756). German anatomist.

DAVIS, A.B. DSB, 1973, **8**, 327–8.
WATERMANN, R.A. Lieberkühn, 1711–56. Neuss, [The author], 1938. 32 pp.

LIEBERMANN, Leo von (1852–1926). Hungarian biochemist.

GROTE, 205–51. (CB)

LIEBERMEISTER, Carl von (1833–1901). Swiss physician; first chief, Basle Medical University Clinic.

BAUMBERGER, H.R. Carl Liebermeister, 1833–1901. Zürich, Juris-Verlag, 1980. 181 pp.

LIEBIG, Justus von (1803–1873). German chemist; made many contributions to biochemistry; discovered chloroform.

BLUNCK, R. Justus von Liebig. Die Lebensgeschichte eines Chemikers. Hamburg, Hammerich & Lesser, 1946. 244 pp. First published 1928.
BROCK, W.H. Justus von Liebig: the chemical gatekeeper. Cambridge, Cambridge Univ. Press, 1997. 374 pp.

LIEBREICH, Matthias Eugen Oskar (1839–1908). German physician and pharmacologist; introduced chloral hydrate into medical practice.

HOFFMANN, K.F. Der Pharmakologe Oskar Liebreich (1839–1908) *Medizinische Monatsschrift*, 1958, **12**, 475–77.

LIEBREICH, Richard (1830–1917). German ophthalmologist.

BEHRMAN, S. Richard Liebreich, 1830–1917: first iconographer of the fundus oculi. *British Journal of Ophthalmology*, 1968, **52**, 335–38.

LIEPMANN, Hugo Karl (1863–1925). German psychiatrist; gave first adequate description of apraxia.

KRAMER, F. *Monatsschrift für Psychiatrie und Neurologie*, 1935, **59**, 225–32.

LIEUTAUD, Joseph (1703–1780). French anatomist and physician.

HUARD, P. & IMBAULT-HUART, M.J. DSB, 1973, **8**, 352–4.

LIGNIÉRES, Joseph Léon Marcel (1868–1933) With J. Spatz first described the *Actinobacillus*.

BRICQ-ROUSSEU, –. *Bulletin de l'Académie de Médecine*, Paris, 1933, 3 ser., **110**, 266–8.

LILLEHEI, Clarence Walton (b.1918). American surgeon.

MILLER, G.W. King of hearts: the true story of the maverick who pioneered heart surgery. New York, Times Books, 2000. 310 pp.

LILLIE, Frank Rattray (1870–1947). Canadian/American embryologist and zoologist.

WATTERSON, R.L. DSB, 1973, **8**, 354–60.
WILLIER, B.H. BMNAS, 1957, **30**, 179–236.
Archival material: Marine Biological Laboratory, Woods Hole, Mass. (See Allen, G.E., Mendel Newsletter, 1973, No. 9, pp. 1–6); Library, University of Chicago.

LILLIE, Ralph Dougall (1896–1979). American pathologist and histologist; showed *Chlamydia psittaci* to be the causal agent in psittacosis; isolated virus of benign lymphocytic choriomeningitis.

CLARKE, A.E. Money, sex, amd legitimacy at Chicago circa 1914–1987. Lillie's center of reproductive biology. *Perspectives in Science*, 1992, **1**, 357–415, G.C. Ralph D. Lillie. *Journal of Histochemistry and Cytochemistry*, 1968, **16**, 3–16.

LIM, Robert Kho-Seng (1897–1969). Chinese physiologist; influential in medical provision, education and research in China.

DAVENPORT, H.W. BMNAS, 1980, **51**, 281–306.

LINACRE, Thomas (1460–1524). English physician; founder of the (Royal) College of Physicians, London.

JOHNSON, J.N. The life of Thomas Linacre... edited by R. Graves. London, Lumley, 1835. 363 pp.
MADDISON, F., PELLING, M. & WEBSTER, C. (eds.) Linacre studies. Essays on the life and work of Thomas Linacre. Oxford, Clarendon Press, 1977. 416 pp.
O'MALLEY, C.D. DSB, 1973, **8**, 360–61.
OSLER, W. Thomas Linacre. Linacre Lecture, 1908, St John's College, Cambridge. Cambridge, Cambridge University Press, 1908. 64 pp.

LIND, James (1716–1794). British physician; founder of naval hygiene in Britain; demonstrated antiscorbutic value of citrus fruits.

BULLOUGH, V.L. DSB, 1973, **8**, 361–3.
RODDIS, L.H. James Lind, founder of nautical medicine. New York, Schuman, 1950. 177 pp.

LINDEBOOM, Gerrit Arie (1905–1986). Dutch medical historian.

LINDEBOOM, G.A. Circa Tiliam: studia historiae medicinae Gerrit Arie Lindeboom septuagenario. Leyden, E.J. Brill, 1974. 302 pp.

LINDERSTRØM-LANG, Karl-Ulrik (1896–1959). Danish biochemist; head of Chemical Dept., Carlsberg Laboratory, Copenhagen.

EDSALL, J.T. DSB, 1990, **18**, 555–61.
HOLTER, H. K.U. Linderstrøm-Lang, 1896–1959. *Comptes Rendus des Travaux du Laboratoire Carlsberg*, 1960–62, **32**, i–xxxiii.
TISELIUS, A. BMFRS, 1960, **6**, 157–68.

LING, Per-Henrik (1776–1839). Swedish founder of modern physiotherapy.

WESTERBLAD, J. Ling, the founder of Swedish gymnastics: his life, his work and importance. Stockholm, P.A. Norstedt, 1909. 82 pp.

LINK, Karl Paul Gerhardt (1901–1978). American chemist; discovered the anticoagulant Dicumarol and synthesized Warfarin.

BALLOU, C.E. Karl Paul Gerhardt Link (1901–1978). *Advances in Carbohydrate Chemistry*, 1981, **39**, 1–12.

LINNÉ, Carl von [LINNAEUS] (1707–1778). Swedish botanist.

BLUNT, W. The compleat naturalist: a life of Linnaeus. London, Francis Lincoln, 2001. 264 pp.

FRANGSMYR, T. (ed.) Linnaeus. The man and his work. Revised ed. Canton MA, Science History Publications, 1994. 206 pp. First published Berkeley, 1983.

GOURLIE, N. The prince of botanists: Carl Linnaeus. London, Witherby, 1953. 292 pp.

HAGBERG, K.H. Carl Linnaeus. Translated from the Swedish by Alan Blair. London, J. Cape, 1952. 264 pp.

LINDROTH, S. DSB, 1973, **8**, 374–81.

LIPMANN, Fritz Albert (1899–1986). American biochemist; shared Nobel Prize with H.A. Krebs 1953.

JENCKS, W.P. & WOLFENDEN. R.B. BMFRS, 2000, **46**, 333–44.

KLEINKAUF, H., DOHREN, H. von & JAENICKE, L. (eds.). The roots of modern biochemistry: Fritz Lipmann's squiggle and its consequences. New York, De Gruyter, Hawthorne, 1988. 994 pp.

LIPMANN, F. Wanderings of a biochemist. New York, Wiley-Interscience, 1971. 229 pp.

LIPMANN, F. A long life in times of great upheaval. *Annual Review of Biochemistry*, 1984. **53**, 1–14.

LISFRANC, Jacques (1790–1847). French surgeon.

GENTY, M. Biographie Médicale, Paris, 1934, vol. 8, 357–71.

LI-SHIH-CHEN (1518–1593). Chinese physician and pharmacist.

SIVIN, N. DSB, 1973, **8**, 390–98.

LISTER, Joseph (1827–1912). British surgeon; pioneer of antiseptic surgery; surgeon to King's College Hospital, London.

DOLMAN, C.E. DSB, 1973, **8**, 399–413.

FISHER, R.B. Joseph Lister. London, Macdonald & Jane, 1977. 351 pp.

GODLEE, R.J. Lord Lister. 3rd ed., Oxford, Clarendon Press, 1924. 686 pp. First published 1917.

LeFANU, W.R. A list of the original writings of Joseph, Lord Lister, O.M. Edinburgh, London, Livingstone, 1965. 19 pp.

RAINS, A.J.H. Joseph Lister and antisepsis. Hove, Priory Press, 1977. 96 pp.

Archival material: Wellcome Institute, London; Royal College of Surgeons of England; Royal College of Surgeons of Edinburgh, Royal College of Physicians and Surgeons of Glasgow; Library, University of Edinburgh; Bodleian Library, Oxford; CMAC.

LISTER, Joseph Jackson (1786–1869); father of Lord Lister; devised the achromatic lens, leading to the modern microscope.

RAINS, A.J.H. Joseph Lister and antisepsis. Hove, Priory Press, 1977. 96 pp.

TURNER, G.L'E. DSB, 1973, **8**, 413–15.

LISTER, Martin (1638–1712). British physician, zoologist and conchologist; physician to Queen Anne.

LISTER, M. A journey to Paris in the year 1698: edited with annotations, a life of Lister and a Lister bibliography by R.P. Stearns. Urbana, Univ. of Illinois Press, 1967. 308 pp.

LISTON, Robert (1794–1847). British surgeon; professor at University College London; performed first major operation in England using anaesthesia.

COLTART, D.J. Surgery between Hunter and Lister as exemplified by the life and work of Robert Liston (1794–1847). *Proceedings of the Royal Society of Medicine*, 1972, **65**, 556–60.
FLEMING, P. Robert Liston, the first professor of clinical surgery at U.C.H. University College Hospital Magazine, 1962, **11**, 176–85.

LITTLE, Clarence Cook (1888–1971). American geneticist and immunologist; researched on histocompatibility antigens.

SNELL, G.D. BMNAS, 1975, **46**, 241–63.
SNELL, G.D. DSB, 1990, **18**, 562–4.

LITTLE, William John (1810–1894), British orthopaedic surgeon.

SCHLEICHKORN, J. 'The sometime physician': William John Little, pioneer in orthopaedic surgery, 1810–1894. Farmingdale, N.J., J. Schleichkorn, 1987. 201 pp.

LITTRÉ, Maximilien Paul Émile (1801–1881). French medical historian and lexicographer.

AQUARONE, S. The life and works of Émile Littré (1801–1881). Leyden, A.W. Sytoff, 1958. 217 pp.
EMIL LITTRÉ: Actes du Colloque, Paris, 7–9 Octobre 1981. *Revue de Synthèse*, 1982, **103**, 165–511. [18 papers on his life and work.]

LITZMANN, Carl Conrad Theodor (1815–1890). German gynaecologist and obstetrician; wrote on pelvic forms and structural anomalies.

KNOBLOCH, J. Bio- und ergographische Beiträge zur Carl Conrad Theodor Litzmann (1815–1890). Neumünster, K. Wachholtz, 1975. 180 pp.

LIVI, Carlo (1823–1877). Italian psychiatrist.

ANESCHI BOLOGNESE, S. Una luca fra le grande obre Carlo Livi per I nudi di dimente 1823–1877). Roma, Federazione Italiana Associazioni Regionali Ospedaliere, 1979. 383 pp.

LIVINGSTONE, David (1813–73). British physician; missionary in southern Africa.

BLAIKIE, W.G. The personal life of David Livingstone, chiefly from his unpublished journals and correspondence in the possession of his family. 3rd ed., London, Murray, 1882. 412 pp. First published 1880.
GELFAND, M. Livingstone, the doctor: his life and travels; a study in medical history. Oxford, Blackwell, 1957. 333 pp.
JEAL, T. Livingstone. London, Heinemann, 1973. 427 pp.
RANSFORD, O. David Livingstone: the dark interior. London, John Murray, 1978. 332 pp. [Dr Ransford considered that Livingstone suffered from cyclothymia, a manic-depressive disorder, to explain the contradictory elements in his character.]
SEAVER, G. David Livingstone: his life and letters. London, Lutterworth Press, 1957. 650 pp.

LIZARS, John (1787–1860). British surgeon in Edinburgh; professor of surgery, Royal College of Surgeons of Edinburgh.

SHEPHERD, J.A. John Lizars, a forgotten pioneer of surgery. *Journal of the Royal College of Surgeons of Edinburgh*, 1979, **24**, 49–58.

LLOYD, David P. Caradoc (1911–1985). Canadian/American neurophysiologist.

PATTON, H.D. BMNAS, 1994, **65**, 196–209.

LLOYD, John Uri (1849–1936). American chemist and pharmacist.

FLANNERY, M.A. John Uri Lloyd: the great American eclective. Carbondale and Edwardsville, Southern Illinois Univ. Press, 1998. 234 pp.
PARASCANDOLA, J. DSB, 1973, **8**, 427–8.
SIMONS, C.M. John Uri Lloyd: his life and works 1849–1936. With a history of the Lloyd Library. Cincinnati, privately printed for the author, 1972. 337 pp.

LOBSTEIN, Jean Georges Chrétien Frédéric Martin (1777–1835). French pathologist; professor at Strasbourg.

LOBSTEIN, E. Joh. Friedr. Lobstein...der Gründer des Anat. Pathol. Museums zu Strasbourg. Sein Leben und Wirken. Strasbourg, K.J. Trübner, 1878. 267 pp.

LOCKE, John (1632–1704). English philosopher, physician and scientist.

AARON, R.I. John Locke. 2nd ed. Oxford, Clarendon Press, 1955. 326 pp. Reprinted 1965.
ATTIG, J.C. (ed.) The works of John Locke: a comprehensive bibliography from the seventeenth century to the present. Westport, Conn., Greenwood Press, 1985. 185 pp.
CRANSTON, M.W. John Locke: a biography. Oxford, Oxford Univ. Press, 1985. 500 pp. First published 1957.
DEWHURST, K. John Locke, physician and philosopher. A medical biography. With an edition of the medical notes in his journals. London, Wellcome Historical Medical Library, 1963. 331 pp.
ROMANELL, P. John Locke and medicine: a new key to Locke. Buffalo, Prometheus Books, 1984. 225 pp.
YOLTON, Y.S. John Locke: a descriptive bibliography. Bristol, Thoemmes Press, 1998. 514 pp.

LOCKWOOD, Charles Barrett (1856–1914). British surgeon at St Bartholomew's Hospital, London; pioneer of aseptic surgery.

JEWESBURY, E.C.O. Life and works of Charles Barrett Lockwood (1856–1914). London, Lewis, 1916. 103 pp.

LODGE, Thomas (*c.*1558–1625). English physician, poet and dramatist.

RAY, W.D. Thomas Lodge. New York, Twayne Publishers, 1967. 128 pp.
TENNEY, E.A. Thomas Lodge. New York, Russell & Russell, 1969. 202 pp.

LOEB, Jacques (1859–1924). German/American biologist and physiologist.

FLEMING, D. DSB, 1973, **8**, 445–7.
OSTERHOUT, W.J.V. BMNAS, 1930, **13**, 318–401 (reprinted from *Journal of General Physiology*, 1928, **8**, ix–xcii).

PAULY, P.J. Controlling life: Jacques Loeb and the engineering ideal in biology. New York, Oxford Univ. Press, 1987. 252 pp.

RASMUSSEN, C. & TILMAN, R. Jacques Loeb: his science and social activities and their philosophical foundations. Philadelphia, American Philosophical Society, 1998. 176 pp.

Archival material: Rockefeller Archives Center; US Library of Congress.

LOEB, Leo (1869–1959). German/American pathologist and cancer researcher, born Germany.

GOODPASTURE, E.W. BMNAS, 1961, **35**, 205–19.

LOEB, L. Autobiographical notes. *Perspectives in Biology and Medicine*, 1958, **2**, 1–25.

PARKER, F. DSB, 1973, **8**, 447–8.

Archival material: Washington University Medical School, St Louis.

LOEB, Robert Frederick (1895–1973). American physician; professor of medicine, Columbia University; co-editor of 'Cecil & Loeb', Textbook of medicine.

BEARN, A.G. BMNAS, 1978, **49**, 149–83.

LOEFFLER, Friedrich August Johannes (1852–1915). German microbiologist.

BRIEGER, G. DSB, 1973, **8**, 448–51.

MOSCHELI, A. Friedrich Loeffler (1852–1915): ein Beitrag zur Geschichte der Bakteriologie und Virologie. Mainz, Johannes Gutenburg Universität, 1994. 88 pp. [Thesis.]

NUTTALL, G.F.H. *Parasitology*, 1924–25, **16**, 234–8.

LOEW, Oscar (1844–1941). German chemist and biochemist; prepared pyocyanase.

KLINKOWSKI, M. Oscar Loew 1844–1941. *Berichte der Deutschen Chemischen Gesellschaft*, 1941, **74A**, 115–36.

LOEWI, Otto (1873–1961). German/American physiologist and pharmacologist; shared Nobel Prize 1936 for demonstration of chemical transmission of nervous impulses.

DALE, H.H. BMFRS, 1962, **8**, 67–89.

GEISON, G.L. DSB, 1973, **8**, 451–7.

LEMBECK. F. & GIERE, W. Otto Loewi: ein Lebensbild in Dokumenten. Biographische Dokumentation und Bibliographie. Berlin, New York, Springer, 1968. 241 pp.

LOEWI, O. An autobiographic sketch. *Perspectives in Biology and Medicine*, 1960, **4**, 3–25.

Archival material: Royal Society of London.

LOMBARDINI, Luigi (1831–1898). Italian teratologist.

MANCINI, C. Luigi Lombardini, dell' Università di Pisa, fondatore della teratogenia sperimentale (1819). Pisa, Casa Editrice Giardini, 1970. 83 pp.

LOMBROSO, Cesare (1836–1909). Italian anthropologist and criminologist; professor of forensic medicine at Turin.

BULFERETTI, L. Cesare Lombroso. Torino, Utet, 1975. 605 pp.

KURELLA, H. Cesare Lombroso, a modern man of science. Translated from the German by Eden Paul. New York, Rebman, 1910. 194 pp.

PESET, J.L. & PESET, M. Lombroso y la escuela positiva italiana. Madrid, C.S. de I.C., 1975. 745 pp.

LONG, Crawford Williamson (1815–1878). American physician; first successfully to use ether vapour as anaesthetic.

RADFORD, R.L. Prelude to fame: Crawford Long's discovery of anesthesia. Los Altos, Cal., Geron-X, 1969. 175 pp.
TAYLOR, F.L. Crawford W. Long and the discovery of ether anesthesia. New York, Hoeber, 1928. 287 pp.

LONG, Cyril Norman Hugh (1901–1970). British/American biochemist and endocrinologist; professor of physiology, Yale University.

BONDY, P.K. DSB, 1990, **18**, 566–71.
SMITH, O.L. & HARDY, J.D. BMNAS, 1975, **46**, 265–309.
Archival material: American Philosophical Society, Philadelphia.

LONG, Esmond Ray (1890–1979). American pathologist and microbiologist; specialist in mycobacterial diseases.

HOWELL, P.C. BMNAS, 1987, **56**, 285–310.

LONG, John Wesley (1859–1926). American surgeon and gynaecologist.

PHILLIPS, R.L. The life and writings of John Wesley Long, M.D., 1859–1926. Greensboro, NC, Custom Graphic Impressions, 1985. 247 pp.

LONGCOPE, Warfield Theobald (1877–1953). American physician; professor at Johns Hopkins Medical School and physician-in-chief to Johns Hopkins Hospital.

TILLETT, W.S. BMNAS, 1959, **33**, 205–25.

LONGMORE, Thomas (1816–1895). British army surgeon; gave important account of heat-stroke.

Lancet, 1895, **2**, 952–4.

LOOSER, Emil (1877–1936). German pathologist; described 'Looser's syndrome'.

KIS, J. Der Knochenpathologe Emil Looser (1877–1936). Zürich, Juris, 1982. 68 pp.

LOOSS, Arthur (1861–1923). German parasitologist; elucidated life-cycle of the hookworm *Ankylostoma duodenale*, and discovered its ability to penetrate the skin.

FÜLLEBORN, F. *Archiv für Schiffs- und Tropenhygiene*, 1923, **27**, 225–8.

LORENTE DE NÓ, Rafael (1902–1990). Spanish neurologist.

WOOLSEY, T.A. BMNAS, 2001, **79**, 85–105.

LÓPEZ DE VILLALOBOS, Francisco (1473–1549). Spanish court physician; wrote important work on syphilis.

CALAMITA, C. Figuras y semblamas del imperio Francisco López de Villalobos, médico de reyes y principe de literatos. Madrid, Colección "La Nave", 1952. 309 pp.
FRIEDENWALD, H. Francisco López de Villalobos, Spanish court physician and poet. *Ibulletin of the History of Medicine*,1939, **7**, 1129–39.

LORENZ, Adolf (1854–1946). Austrian orthopaedic surgeon.

BIBLIOGRAPHY OF MEDICAL AND BIOMEDICAL BIOGRAPHY

LORENZ, A. My life and work. The search for a missing glove. London, New York, Charles Scribner, 1936. 362 pp. German translation, Leipzig, 1937.

LORENZ, Karl Zacharias (1903–1987). Austrian zoologist; shared Nobel Prize (Physiology or Medicine) 1973 for discoveries concerning organization and elicitation of individual and social behaviour patterns.

BERGER, K. Konrad Lorenz. Abbau des Göttlichen. Berneck, Schweiz, Schwengler-Verlag, 1990. 168 pp.

LORRY, Anne-Charles (1726–1783). French physician; wrote first French monograph devoted solely to dermatology.

HUARD, P. & IMBAULT-HUART, M.J. DSB, 1973, **8**, 505–7.
KISSMEYER, A. Anne-Charles Lorry et son oeuvre dermatologique. Paris, A. Legrand, 1928. 64 pp.

LOTZE, Rudolph Hermann (1817–1881). German physician and analytical psychologist.

ROTHSCHUH, K.E. DSB, 1973, **8**, 513–16.
WOODWARD, W.R. The medical realism of R. Hermann Lotze. [New Haven, Conn., c.1976.] 447 leaves. Photocopy of typescript. Ann Arbor, Mich., University Microfilms International, 1978.

LOUIS, Antoine (1723–1792). French surgeon; pioneer writer on medical jurisprudence in France.

HUARD, P. & IMBAULT-HUART, M.J. Antoine Louis. In: Huard, P.A. (ed.) Biographies médicales et scientifiques, XVIIe. siècle. Paris, R. Dacosta, 1972, pp. 33–118.
SILIE, M. Un des promoteurs de la médecine légale française, Antoine Louis (1723–1792): sa vie et son oeuvre. Lyon, Bosc Frères et Riou, 1924. 67 pp.

LOUIS, Pierre-Charles-Alexandre (1787–1872). French physician; founder of medical statistics.

GREENWOOD, pp. 733–87 (CB).
STEINER, W.R. Annals of Medical History. 1940, **2**, 451–8.
WOILLEZ, E.J. Le docteur P.-C.-A. Louis; sa vie, ses oeuvres (1787–1872). Paris, P. Dupont, 1873. 58 pp.

LOUTIT, John Freeman (1910–1992). Australian/British radiobiologist and haematologist.

LYON, M. and MOLLISON, P.L. BMFRS, 1994, **40**, 238–52.

LOVELACE, William Randolph (1907–1965). American surgeon and aerospace physician; shared in devising BLB (Boothby-Lovelace-Bulbulian) oxygen inhalation apparatus.

ELLIOTT, R.G. 'On a comet always': a biography of Dr W.Randolph Lovelace II. New Mexico Quarterly, 1966–67, **36**, 351–87.

LOWE, Peter (c.1550–1612?). Scottish surgeon; founder of the Faculty of Physicians and Surgeons of Glasgow.

FINLAYSON, J. Account of the life and works of Maister Peter Lowe, the founder of the Faculty of Physicians and Surgeons of Glasgow. Glasgow, Maclehose, 1889. 84 pp.

LOWENFELD, Henry (1900–1985). German psychoanalyst.

MÜLLER, T. Von Charlottenberg zum Central Park West. Henry Lowenfeld und die Psychoanalyse in Berlin, Prag und New York. Frankfurt am Main, Edition Déja Vu, 2000. 344 pp.

LOWENSTEIN, Otto Egon (1906–1999). German/British physiologist.

ALEXANDER, R. McN. BMFRS, 2001, **47**, 359–68.

LOWER, Richard (1631–1691). English physician and experimental physiologist; pioneer of blood transfusion.

BROWN, T.M. DSB, 1973, **8**, 523–7.
FULTON, J.F. A bibliography of two Oxford physiologists, Richard Lower (1631–1691) and John Mayow (1643–1679). Oxford, Oxford Univ. Press, 1935. 62 pp.
GOTCH, F. Two Oxford physiologists, Richard Lower 1631–1691, John Mayow 1643–1679. Oxford, Clarendon Press, 1908. 40 pp.
HOFF, E.C. the life and times of Richard Lower. *Bulletin of the InstitutIne of the History of Medicine*, 1936, **4**, 517–35.

LUBARSCH, Otto (1860–1933). German pathologist.

LUBARSCH, O. Ein bewegtes Gelehrtenleben; Erinnerungen und Erlebnisse, Kämpfe und Gedanken. Berlin, J.Springer, 1931. 606 pp.

LUC, Henri (1855–1925). French laryngologist; devised ('Caldwell-Luc') operation for abscess of maxillary sinus, 1885.

BELLINI, I. *Annales des Maladies de l'Oreille*, 1925, **44**, 1121–6.
GUISEZ, J. *Laryngoscope*, 1925, **35**, 958–60.

LUCAE, August (1835–1911). German otologist.

CLAUS, H. *Zeitschrift für Ohrenheilkunde*, 1911, **73**, iii–vi.
HEINE, B. *Archiv für Ohrenheilkunde*, 1911, **85**, i–xxii.

LUCAS, Keith (1879–1916). British physiologist; investigated properties of muscle and nerve.

FLETCHER, W. (ed.) Keith Lucas [a series of biographical sketches by various authors]. Cambridge, Heffer, 1934. 130 pp.
GEISON, G.L. DSB, 1973, **8**, 532–5.

LUCAS-CHAMPIONNIÈRE, Just Marie Marcellin (1843–1913). French surgeon; a pupil of Lister and promoter of antisepsis in France.

J. LUCAS-CHAMPIONNIÈRE [Eloges]. Clermont (Oise), Daiz and Thiron, 1916, 148 pp.

LUCATELLO, Luigi (1863–1926). Italian physician; introduced liver puncture biopsy.

PREMUDA, Luigi Lucatello a cinquant' anni dalle morte. Padova, La Garangola, 1976. 30 pp.

LUCIANI, Luigi (1840–1919). Italian physiologist; professor in Rome.

ZANOBIO, B. & PORTA, G. DSB, 1973, **8**, 535–6.

LUCK, James Murray (1899–1993). American chemist; professor at Stanford University.

LUCK, J.M. Confessions of a biochemist. *Annual Review of Biochemistry*, 1981, **50**, 1–22.

LUCKHARDT, Arno Benedict (1885–1957). American physiologist and anaesthetist; introduced ethylene as anaesthetic.

LIGHT, G.A. History, pharmacology and clinical use of ethylene. In: Eastwood, B.W. Nitrous oxide. Philadelphia, Davis, 1966, pp. 135–50.

LUDWIG, Carl Friedrich Wilhelm (1816–1895). German physiologist; professor at Marburg, Zürich, Vienna and Leipzig and one of the greatest teachers of the subject.

CRANEFIELD, P.F. (ed.) Two great scientists of the nineteenth century. Correspondence of Emil Du Bois-Reymond and Carl Ludwig. Baltimore, Johns Hopkins Univ. Press, 1982. 183 pp.
ROSEN, G. DSB, 1973, **8**, 540–42. 1982. 183 pp.
SCHRÖER, H. Carl Ludwig, Begründer der messenden Experimentalphysiologie, 1816–1895. Stuttgart, Wissenschaftliche Verlagsgesellschaft, 1967. 340 pp.

LUDWIG, Wilhelm Friedrich von (1790–1865). German physician; described 'Ludwig's angina'.

BURKE, J. Angina Ludovici [1836]. A translation, together with a biography. *Bulletin of the History of Medicine*, 1939, **7**, 1115–26.

LUKENS, Francis Dring Wetherill (1899–1978). American physician; produced experimental diabetes by artificially induced hyperglycaemia.

WINEGRAD, A.I. In memoriam Francis D.W. Lukens. *Endocrinology*, 1979, **105**, 574.

LUNDY, John Silas (b. 1894). American anaesthesiologist; introduced thiopentone sodium as anaesthetic, 1935.

FORSBURGH, L.C. From this point in time: some memories of my part in the history of anesthesia – John S. Lundy, MD. *AANA Journal*, 1997, **65**, 323–8.

LURIA, Salvador Edward (1912–1991). Italian/American microbiologist and geneticist; shared Nobel Prize 1969.

LURIA, S.E. A slot machine, a broken test tube. An autobiography. New York, Harper & Row, 1984. 228 pp.

LUSCHKA, Hubert von (1820–1875). German anatomist.

DVORAK, J. & SANDLER, A. Hubert von Luschka: a pioneer of clinical anatomy. *Spine*, 1994, **19**, 2478–82.

LUSK, Graham (1866–1932). American nutritionist and clinical investigator.

C., E.P. ONFRS, 1932–35, **1**, 143–6.
DUBOIS, E.F. BMNAS. 1941, **21**, 95–142.
ROSENBERG, C.E. DSB, 1973, **8**, 555–6.

LUTZ, Adolpho (1855–1940). Brazilian scientist; described South American blastomycosis (Lutz's disease).

ADOLPHO LUTZ (1855–1940). *Memorias do Instituto Oswaldo Cruz*, 1956, **54**, 447–89.

ADOLPHO LUTZ (1855–1940): vida e obra do grande cientista brasileiro. Rio de Janeiro, " Commissão do Centenario de Adolpho Lutz", Conselho Nacional de Pesquisas, 1956. 55 pp.

LUZZI, Mondino de', see MONDINO DE' LUZZI

LWOFF, André M. (1902–1994). French microbiologist. Member of Institut Pasteur; shared Nobel Prize (Physiology or Medicine) 1965, for work on genetic control of enzyme and virus synthesis.

LWOFF, A. From protozoa to bacteria and viruses; fifty years with microbes. *Annual Review of Microbiology*, 1971, **25**, 1–26.
JACOB, F.Y. & GIRARD, M. BMFRS, 1998, **44**, 255–63.

LYNEN, Feodor Felix Konrad (1911–1979). German biochemist: shared Nobel Prize (Physiology or Medicine) 1964 for work on mechanism and regulation of cholesterol and fatty acid metabolism.

FEODOR LYNEN, 6 April 1911 bis 6 August 1979. *Naturwissenschaftliche Rundschau*, 1980, **33**, 213–32.
KREBS, H. & DECKER, K. BMFRS, 1982, **28**, 261–317.
LYNEN, F. Life, luck and logic in biochemical research. *Perspectives in Biology and Medicine*, 1969, **12**, 200–208 [P.161].

LYON, Elias Potter (1867–1937). American physician; Dean, Minnesota Medical School.

WANGENSTEEN, O.H. Elias Potter Lyon, Minnesota's leader in medical education. St Louis, Missouri, Warren H. Green, 1981. 292 pp.

LYSENKO, Trofim Denisovich (1898–1976). Russian geneticist.

JORAVSKY, D. The Lysenko affair. Cambridge, Harvard Univ. Press, 1970. 459 pp.
LECOURT, D. Lyssenko: histoire réele d'une "science prolétarienne". Paris, Maspero, 1976. 257 pp.
MEDVEDEV, Z.A. The rise and fall of T.D. Lysenko. Translated by I.M. Lerner. New York, Columbia Univ. Press, 1969. 284 pp.

MacALISTER, Donald (1854–1934). British physician; considerable linguist.

MacALISTER, E.F.B. Sir Donald MacAlister of Tarbert. With chapters by Sir Robert Rait and Sir Norman Walker. London, Macmillan, 1935. 392 pp.

MacALISTER, John Young Walker (1856–1925). British librarian and later secretary of the Royal Society of Medicine, for the formation of which he was responsible by the amalgamation of 15 existing London medical societies.

GODBOLT, S. & MUNFORD, W.A. The incomparable Mac. A biographical study of Sir John Young Walker MacAlister. London, Library Association, 1983. 142 pp.
Sir J.Y.W. MacALISTER : a memorial for his family and friends. London, Hazel, Watson & Viney, 1926. 88 pp.

MACALLUM, Archibald Byron (1858–1934). Canadian physiologist and biochemist.

L., J.B., ONFRS, 1932–35, **1**, 287–91.

McRAE, S.F. A.B. Macallum and physiology at the University of Toronto. In: Geison, G. (ed.) Physiology in the American context, 1850–1940. Baltimore, American Physiological Society, 1987. pp. 97–114.
RICHARDSON, R.A. DSB, 1973, **8**, 583–4.
Archival material: Public Archives of Canada, Ontario.

MacCALLUM, John Bruce (1876–1906). American histologist and physiologist.

MALLOCH A. Short years: the life and letters of John Bruce MacCallum 1876–1906. Chicago, Normandie House, 1938. 343 pp.

MacCALLUM, William George (1874–1944). Canadian-born pathologist; professor at Johns Hopkins University.

LONGCOPE, W.T. BMNAS, 1943–46, **27**, 339–64.

McCANCE, Robert Alexander (1898–1993). British physician; professor of experimental medicine, Cambridge; published (with E.M.Widdowson) tables on *The composition of foods*.

ASHWELL, M. (ed.) McCance & Widdowson: a scientific partnership of 60 years. London, British Nutrition Foundation, 1993. 264 pp.
WIDDOWSON, E.M. BMFRS, 1995, **41**, 262–80.
Archival material: CMAC.

McCARRISON, Robert (1878–1960). British physician and nutritionist.

GOPALAN, C. Food and health: McCarrison's prescription endures. *Nutrition and Health*, 1992, **8**, 1–16.

MACARTNEY, James (1770–1843). Irish physician; professor of anatomy and chirurgery, University of Dublin.

MACALISTER, A. James Macartney. London, Hodder & Stoughton, 1900. 293 pp.

McCARTY, Maclyn (b.1911). American biochemist; with O.T. Avery and C.M. Macleod identified DNA as genetic material.

McCARTY, M. Reminiscences of the early days of transformation. *Annual Review of Genetics*, 1980, **14**, 1–15.
McCARTY, M. The transforming principle: discovering that genes are made of DNA. New York, London, W. W. Norton, 1985. 252 pp.

McCLINTOCK, Barbara (1902–1992). American geneticist; Nobel prizewinner, 1983.

COMFORT, N.C. The tangled field: Barbara McClintock's search for the patterns of genetic control. Cambridge, MA, Harvard Univ. Press, 2001. 357pp.
FEDOROFF, N.V. BMFRS, 1994, **40**, 266–80.
FEDOROFF, N.V. BMNAS, 1995, **67**, 211–35.
FEDOROFF, N. & BOTSTEIN, D. (eds.). The dynamic genome: Barbara McClintock's ideas in the century of genetics. Cold Spring Harbor, Cold Spring Harbor Laboratory Press, 1992. 422 pp.
FINE, Barbara McClintock: Nobel Prize geneticist. Springfield NJ, Enslow Publications, 11988. 1288 pp.
KELLER, E. A feeling for the organism; the life and work of Barbara McClintock. San Francisco, Freeman, 1983. 235 pp.

McCLUNG, Clarence Erwin (1870–1946). American cytologist.

ALLEN, G.E. DSB, 1973, **8**, 586–90.
WENRICH, D.H. *Journal of Morphology*, 1940, **66**, 635–88.

McCOLLUM, Elmer Verner (1879–1967). American nutritional scientist; discovered vitamins A and D.

BECKER, S.L., DSB, 1973, **8**, 590–91.
CHICK, H. & PETERS, R.A. BMFRS, 1969, **15**, 159–71.
DAY, H.G. BMNAS, 1974, **45**, 263–335.
McCOLLUM, E.V. From Kansas farm boy to scientist: the autobiography of Elmer Verner McCollum. Lawrence, Univ. of Kansas Press, 1964. 253 pp.
Archival material: Johns Hopkins University; University of Kansas.

McCORMAC, William (1836–1901). British surgeon at St Thomas' Hospital.

PAVY, F.W. *Medico-chirurgical Transactions*, 1902, **85**, 113–21.

McCRAE, John (1872–1918). Canadian physician and poet; author of poem "In Flanders field the poppies blow".

GRAVES, D. A crown of life: the world of John McCrae. Staplehurst, Spellmount, 1997. 300 pp.
PRESCOTT, J.F. In Flanders fields. The story of John McCrae. Erin, Ont., Boston Mills Press, 1985. 144 pp.

McDERMOTT, Walsh (1909–1981). American physician; tuberculologist; prominent in foundation of Institute of Medicine, National Academy of Sciences.

BEESON, P.B. BMNAS, 1990, **59**, 282–307.

MACDONALD, Greville (1856–1944). British physician; consulting physician and emeritus professor, King's College Hospital, London.

MACDONALD, G. Reminiscences of a specialist. London, Allen & Unwin, 1932. 422 pp.

MacDONALD, John Smyth (1867–1941). British physiologist; professor at Liverpool University.

RAPER, H.S. ONFRS, 1940–41, **3**, 853–66.

McDONALD, Sydney Fancourt (1885–1967). Australian paediatrician.

PEARN, J.H. The highest traditions: a biography of Sydney Fancourt McDonald. Brisbane, Dept. of Child Health, Royal Children's Hospital, 1985. 261 pp.

McDOUGALL, William (1871–1938). British psychologist; professor at Harvard University.

GREENWOOD, M. & SMITH, M. ONFRS, 1940–41, **3**, 39–62.
ROBINSON, A.L. William McDougall, 1871–1938, MB, DSc, FRS, a bibliography, together with an outline of his life. Durham, Duke Univ. Press, 1943. 54 pp.
SMITH, J.G. The mystery of the mind: a biography of William McDougall. Dissertation, University of Texas, Austin, 1980. Dissertation Abstracts International, 1980, 41, 3163-A. University Microfilms Order No. 91-00964. 322 pp.

McDOWELL, Ephraim (1771–1830). American pioneer in abdominal surgery.

GRAY, L.A. The life and times of Ephraim McDowell. Louisville, Kentucky, V.G.Read and Sons, Printers, 1907. 157 pp.

RIDENBAUGH, M.Y. The biography of Ephraim McDowell, MD, 'the Father of ovariotomy'. New York, Charles C. Webster & Co. 1890. 588 pp. Revised editions 1894 and 1897.

SCHACHNER, A. Ephraim McDowell, 'father of ovariotomy', the founder of abdominal surgery. With an appendix on Jane Todd Crawford. Philadelphia, London, Lippincott, 1921. 331 pp.

MACEWEN, William (1848–1924). British surgeon; Regius Professor of Surgery, University of Glasgow.

BOWMAN, A.K. The life and teaching of Sir William Macewen: a chapter in the history of surgery. London, Hodge, 1942. 425 pp.

MACFARLANE, Robert Gwyn (1907–1987). British haematologist; professor of clinical pathology, Oxford.

BORN, G.V.R. & WEATHERALL, D.J. BMFRS, 1990, **35**, 209–45.

ROBB-SMITH, A. Life and achievements of Professor Robert Gwyn Macfarlane, FRS, pioneer in the care of haemophiliacs. London, Royal Society of Medicine, 1993. 104 pp.
Archival material: CMAC.

McGOVERN, John Phillip (b.1921). American allergologist.

APPRECIATIONS, reminiscences and tributes honoring John P. McGovern. Houston, Health Sciences Institute, 1980. 680 pp.

MacGREGOR, William (1814–1919). British physician and colonial administrator; contributed to advances in tropical medicine.

JOYCE, J.B. Sir William MacGregor. Oxford, Oxford Univ.Press, 1971. 484 pp.

McGRIGOR, James (1771–1858). Director-General of the British Army Medical Department from 1815; 'Father' of the Royal Army Medical Corps.

BLANCO, R.L. Wellington's Surgeon General: Sir James McGrigor. Durham, N.C., Duke University Press, 1974. 235 pp.

McGRIGOR, M. (ed.) Sir James McGrigor: the scalpel and the sword The autobiography of the Father of army medicine. Dalkeith, Scottish Cultural Press, 2000. 320 pp. A revised and annotated edition of McGrigor, J. *The autobiography and services of the late Sir James McGrigor.* London, Longman, 1861.

McGUIRE, Hunter Holmes (1835–1900). American physician and surgeon; medical director, Confederate Army of Shenandoah and of Northern Virginia.

McGUIRE, S. Hunter Holmes McGuire, M.D., LL.D. *Annals of Medical History*, 1938, **10**, 1–14, 136–51.

SHAW, M.F. Stonewall Jackson's surgeon Hunter Holmes McGuire: a biography. Lynchburg, VA, H.E. Howard, 1993. 114 pp.

SCHILDT, J.W. Hunter Holmes McGuire, doctor in gray. Chewsville, Md., J.W. Schildt, 1986. 135 pp.

MACHEBOEUF, Michel (1900–1953). French biochemist.

MONNIER, M. DSB, 1973, **8**, 607–8.

MACHT, David Israel (1882–1961). Russian/American pharmacologist; demonstrated antispasmodic action of theophylline.

WILK, D. David Israel Macht (1882–1961). *Koroth*, 1983, **8**, 305–17.

McINDOE, Archibald Hector (1900–1960). New Zealand surgeon; plastic surgeon to St Bartholomew's Hospital, London, and to Queen Victoria Hospital, East Grinstead; consultant plastic surgeon to Royal Air Force.

McLEAVE, H. McIndoe: plastic surgeon. London, Muller, 1961. 231 pp.
MOSLEY, L. Faces from the fire. The biography of Sir Archibald McIndoe. London, Weidenfeld & Nicolson, 1962. 268 pp.

MacINTOSH, Frank Campbell (1909–1992). Canadian physiologist and pharmacologist.

BURGEN, A. BMFRS, 1994, **40**, 254–64.

McINTOSH, James (1882–1948). British pathologist and bacteriologist; isolated *Lactobacillus odontolyticus*.

FILDES, P. James McIntosh 1882–1948. *Journal of Pathology and Bacteriology*, 1949, **61**, 285–89.

MACINTOSH, Robert Reynolds (1897–1989). New Zealand anaesthetist.

SMITH, W.D.A. & PATERSON, G.M.C. (eds.) A tribute to Sir Robert Macintosh on his 90th birthday. London, Royal Society of Medicine, 1988. 45 pp.
Archival material: CMAC.

MACINTYRE, John (1857–1928). British radiologist in Scotland; introduced X-ray cinematography; photographed renal calculi.

GOODALL, A.L. John Macintyre, pioneer radiologist. *Surgo* (Glasgow), 1958, **24**, 119–26.
WALTHER, K. John Macintyre 1857–1928. *Röntgenblätter*, 1959, **12**, 337–41.

MACKENRODT, Alwin Karl (1859–1925). German gynaecologist.

MACKENRODT, H. *Zentralblatt für Gynäkologie*, 1926, **50**, 1042–50.

MACKENZIE, James (1853–1925). British general practitioner for 28 years, during which he made pioner studies of the cardiovascular system, then became a consultant cardiologist.

MAIR, A. SIR James Mackenzie, M.D., 1853–1925, general practitioner. Edinburgh, Livingstone, 1973. 366 pp. (Photoreproduction, with new chapter, London, Royal College of General Practitioners, 1986. 374 pp.)
MONTEITH, W.B.R. Bibliography with synopsis of the original papers of the writings of Sir James Mackenzie, London, Humphrey Milford, 1930. 99 pp.
WILSON, R.M. The beloved physician: Sir James Mackenzie. A biography. London, John Murray, 1926. 316 pp. Reprinted 1946.

MACKENZIE, Morell (1837–92). British laryngologist.

HAWEIS, H.R. Sir Morell Mackenzie, physician and operator. A memoir compiled and edited from private papers and personal reminiscences. London, Allen, 1893. 376 pp.

STEVENSON, R.S. Morell Mackenzie: the story of a Victorian tragedy. London, Heinemann, 1946. 194 pp.

McKENZIE, Robert Tait (1867–1938). Canadian physician; director of physical training at McGill University and later University of Pennsylvania; sculptor of athletes.

KOZAR, A.J. R. Tait McKenzie, the sculptor of athletes. Knoxville, Univ. of Tennessee Press, 1975. 118 pp.

McGILL, J. The joy of effort. A biography of R. Tait McKenzie. Bewdley, Ont., Clay Publishing Co., 1980. 241 pp.

MACKENZIE, William (1791–1868). British ophthalmologist; founder of the Glasgow Eye Infirmary.

THOMSON, A.M.W. The life and times of Dr William Mackenzie, founder of the Glasgow Eye Infirmary. Glasgow, University Press, 1973. 132 pp.

MacLEOD, Colin Munro (1909–1972). American biochemist; with O.T. Avery and M. McCarty demonstrated role of DNA in genetic transmission.

McCARTY, M. The transforming principle: discovering that genes are made of DNA. New York, London, W. W. Norton, 1985. 250 pp.

McDERMOTT, W. BMNAS, 1983, **54**, 183–219.

OLBY, R. DSB, 1990, **18**, 587–9.

McKESSON, Elmer Isaac (1881–1935). American anaesthetist; introduced intermittent gas-oxygen machine, 1911.

MORRIS, L.E. A perspective: E.I. McKesson (1881–1930). In: International Symposium on the History of Anaesthesia (2nd). London, History of Anaesthesia, 1988, pp. 575–8.

MACLAGAN, Thomas John (1838–1903). British physician; introduced salicylates in rheumatism.

STEWART, W.K. & FLEMING, L.W. Perthshire pioneer of anti-inflammatory agents (Thomas John Maclagan). *Scottish Medical Journal*, 1987, **32**, 141–6.

McLEAN, Jay (1890–1957). American surgeon; extracted heparin from dog liver.

COUCH, N.P. About heparin or ... whatever happened to Jay McLean? *Journal of Vascular Surgery*, 1989, **10**, 1–8.

LAM, C.R. The strange story of Jay McLean, the discoverer of heparin. *Henry Ford Hospital Journal*, 1985, **33**, 18–23.

MARCUM, J. William Henry Howell and Jay McLean: the experimental context for the discovery of heparin. *Perspectives in Biology and Medicine*, 1990, **33**, 214–30.

McLEOD, James Walter (1887–1978). British bacteriologist.

WILSON, G. & ZINNEMAN, K.S. BMFRS, 1979, **25**, 421–44.

MACLEOD, John James Rickard (1876–1935). British physiologist; professor in Toronto and Aberdeen; shared Nobel Prize 1923 for work on insulin.

CATHCART, E.P. ONFRS, 1932–35, **1**, 585–9.
WILLIAMS, M.J. J.J.R. Macleod: the co-discoverer of insulin. *Proceedings of the Royal College of Physicans of Edinburgh*, 1993, **23**, No. 3, Suppl. 1. 125 pp.

McMASTER, Philip Duryée (1891–1973). American pathologist and immunologist.

GOEBEL, W.F. BMNAS, 1979, **50**, 287–308.
Archival material: Rockefeller Archives Center.

McMICHAEL, John (1904–1993). British physician and clinical scientist; Professor of Medicine, Royal Postgraduate Medical School, London. First, with E.P. Sharpey-Schafer, to perform cardiac catheterization in man.

DOLLERY, C. BMFRS, 1995, **41**, 282–96.
Archival material: CMAC.

MACMICHAEL, William (1784–1839). British physician; physician to William IV; author of *The gold-headed cane*.

HUNT, T.C. William Macmichael, MD, FRS, author of "The gold-headed cane". *Journal of the Royal College of Physicians of London*, 1968, **2**, 372–80.
[The edition of *The gold-headed cane* published in New York, Froben Press, 1932 includes an essay on William Macmichael, his life, his works and his editors, by H.S. Robinson, originally published in *Medical Life*, 1937, **34**, 467–88.]

MACNAMARA, Jean (1899–1968). Australian medical scientist; worked on treatment of poliomyelitis.

ZWAR, D. The Dame; the life and times of Dame Jean Macnamara, medical pioneer. South Melbourne, McMillan, 1984. 168 pp.

MacNIDER, William de Berniere (1881–1951). American pharmacologist; professor at the University of North Carolina.

RICHARDS, A.N. BMNAS, 1959, **33**, 238–72.

McQUARRIE, Irvine (1891–1961). American paediatrician.

McQUARRIE, I. Autobiographic sketch. *Perspectives in Biology and Medicine*, 1962, **6**, 61–74.

McSWINEY, Bryan Austin (1894–1947). British gastric physiologist; professor at St Thomas's Hospital Medical School, London.

BROWN, G.L. ONFRS, 1948–49, **6**, 147–60.

MACVICAR, Neil (1871–1949). South African physician and teacher of medicine.

SHEPHERD, R.H.W. A South African medical pioneer: the life of Neil Macvicar. Lovedale, SA, Lovedale Press, 1952. 249 pp.

MacWILLIAM, John Alexander (1857–1937). British physiologist; professor, University of Aberdeen.

KEITH, A., ONFRS, 1936–39, **2**, 335–8.

BIBLIOGRAPHY OF MEDICAL AND BIOMEDICAL BIOGRAPHY

MADELUNG, Otto Wilhelm (1846–1926). German surgeon.

PERTHES, G. *Archiv für Klinische Chirurgie*, 1926, **143**, i–vii.

MADSEN, Thorvald Johannes Marius (1870–1957). Danish bacteriologist; Director, Statens Serum Institut, Copenhagen.

PARNAS, J. Thorvald Madsen, 1870–1957, leader in international public health. Copenhagen, for the author, 1980. 16 pp.

MAFFUCCI, Angelo (1845–1903). Italian physician.

COSTA, A. Ricordo di Angelo Maffucci (1845–1903) primo cattedratico de anatomia patologica nello Studio Pisano. *Archivio de Vecchi per l'Anatomia Patologica*, 1965, **45**, i–xix.

MAGATI, Cesare (1579–1647). Italian surgeon; followed Paré in his method of treating gunshot wounds.

CASTELLANI, C. L'attività clinico-medico di Cesare Magati con sedici consulta inedita. Milano, Edizioni Stedar, 1959. 115 pp.
MÜNSTER, L. & ROMAGNOLI, G. Cesare Magati (1579–1647) lettore de chirurgia nello studio ferrarese. Ferrara, Università degli Studi, 1968. 77 pp.
PREMUDA, L. DSB, 1974, **9**, 4–5.

MAGENDIE, François (1783–1855). French physiologist; pioneer of experimental physiology and pharmacology.

DELOYERS, L. François Magendie 1783–1855: precursor de la médecine expérimentale. Bruxelles, Presses Universitaires de Bruxelles, 1970. 279 pp.
GRMEK, M.D. DSB, 1974, **9**, 6–11.
OLMSTED, J.M.D. François Magendie, pioneer in experimental physiology and scientific medicine in XIX century France. New York, Schuman's, 1944. 290 pp.

MAGGI, Bartolomeo (1477–1552). Italian surgeon; notable for his soothing treatment of gunshot wounds.

GENTILI, G. La vita e l'opera di Bartolomeo Maggi .. Bologna, Tipografia Vighi e Rizzoli, 1967, 91 pp.
PREMUDA, L. DSB, 1974, **9**, 11–12.

MAGILL, Ivan Whiteside (1888–1986). British anaesthetist.

THOMAS, K.B. Sir Ivan Whiteside Magill. A review of his publications and other references to his life and work. *Anaesthesia*, 1978, **33**, 628–34.

MAGNAN, Jacques Joseph Valentin (1835–1916). French psychiatrist.

SERIEUX, P.V. Magnan: sa vie et son oeuvre (1835–1916). Paris, Masson, 1921. 174 pp.

MAGNUS, Rudolf (1873–1927). German physiologist and pharmacologist.

MAGNUS, O. Rudolf Magnus, physiologist and pharmacologist. Dordrecht, Kluwer Academic Publishers, 2003.
SWAZEY, J.P. DSB, 1974, **9**, 19–21.

242

BIBLIOGRAPHY OF MEDICAL AND BIOMEDICAL BIOGRAPHY

MAGNUS-LEVY, Adolf (1869–1955). German/American physiologist with particular interest in metabolism.

IHDE, A.J. DSB, 1990, **18**, 589–92.

MAGRATH, George Burgess (1870–1933). American pathologist.

CANAVAN, M.M. *Archives of Pathology*, 1939, **27**, 620–23.
CHRISTIAN, N.A. *Harvard Alumni Bulletin*, 1938–39, **13**, 59–61.

MAHFOUZ, Naguib (1882–1974). Egyptian gynaecological surgeon.

MAHFOUZ, N. The life of an Egyptian doctor. Edinburgh, Livingstone, 1966. 191 pp.

MAHONEY, John (1889–1957). American physician; introduced penicillin in treatment of syphilis.

PARASCANDOLA, J. John Mahoney and the introduction of penicllin to treat syphilis. Pharmacy in History, 2001, **43**, 3–13.

MAIER, Michael (1568–1622).

CRAVEN, J.B. Count Michael Maier, doctor of philosophy and of medicine, alchemist, Rosicrucian, mystic, 1568–1622. Kirkwall, William Peace, 1910. 165 pp. Reprinted, London, Dawsons, 1968.

MAIMONIDES [MOSES BEN MAIMON] (1135 or 1138–1204). Jewish theologian, physician and scientist, born Córdoba.

HAYOUN, M.R. Maimonides: Arzt und Philosoph im Mittelalter. Eine Biographie. München, C.H. Beck, 1999. 336 pp. First published in French, 1994.
HESCHEL, A.J. Maimonides. A biography. New York, Farrar, Strauss & Giroux, 1982. 284 pp.
LEAMAN, O. Moses Maimonides. London, Routledge, 1990. 190 pp.
ORMSBY, E.L. (ed.) Maimonides and his time. Washington DC, Catholic University of America Press. 1999. 180 pp.
PINES, S. DSB, 1974, **9**, 27–32.
ROSNER, F. The medical legacy of Moses Maimonides. Hoboken, NJ, KTAV Publishing House, 1998. 308 pp.

MAISONNEUVE, Jacques Gilles Thomas (1809–1882) French urologist; introduced hair catheter, 1895.

DOYEN, E. *Archives Provinciales de Chirurgie*, 1897, **6**, 388–92.

MAITLAND, Huth Bethune (1895–1972). Canadian/British bacteriologist; introduced medium for cultivation of vaccinia virus.

DOWNIE, A.W. Hugh Bethune Maitland. *Journal of Medical Microbiology*, 1973, **6**, 253–8.

MAIZELS, Montague (1899–1976). British haematologist.

MACFARLANE, R.G. BMFRS, 1977, **23**, 345–66.

MAJOCCHI, Domenico (1848–1929). Italian dermatologist; first to describe purpura annularis telangiectodes.

243

BIBLIOGRAPHY OF MEDICAL AND BIOMEDICAL BIOGRAPHY

DIASIO, F.A. Domenico Majocchi: a biographical appreciation. *Medical Life*, 1932, **39**, 597–601.

MAJOR, Johann Daniel (1634–1693). German physician; first successfully to inject drugs into the human body, 1662.

REINBACHER, W.R. Leben, Arbeit und Umwelt des Arztes Johann Daniel Major (1634–1693). Eine Biographie aus dem 17. Jahrhundert; mit neuen Erkenntnissen. Linsengericht, Verlag M. Kroeber, 1998. 240 pp.

MAJOR, Ralph Hermon (1884–1970). American physician and medical historian.

MIDDLETON, W.S. Intellectual crossroads – an appreciation of Ralph H. Major. *Perspectives in Biology and Medicine*, 1971, **14**, 651–8.

al-MAJUSI, ABU'L HASAN ALI IBN 'ABBAS, *see* **HALY ABBAS**

MALASSEZ, Louis-Charles (1842–1909). French haematologist; designed haemocytometer.

JOLLY, J. Notice sur la vie et les travaux de Louis Malassez (1842–1909). *Comptes Rendus de la Société de Biologie*, 1910, **68**, 1–18.

MALCOLM, Andrew (1818–1856). British physician.

CALWELL, H.G. Andrew Malcolm of Belfast 1818–1856: physician and historian. Belfast, Brough Cox and Dunn, 1977. 134, 139, xxxii pp. With facsimile reproduction of History of the Belfast General Hospital and the Principal Medical Institutions of the Town.

MALGAIGNE, Joseph François (1806–1865). French surgeon; 'the greatest surgical historian and critic whom the world has yet seen' – Billings.

PILASTRE, E. Malgaigne. Étude sur la vie et ses idées d'après ses écrits, des papiéres de famille et des souvenirs particuliers. Paris, F.Alcan, 1905. 246 pp.

MALL, Franklin Paine (1862–1917). American anatomist and embryologist; professor at Johns Hopkins University.

CORNER, G.W. DSB, 1974, **9**, 55–8.
SABIN, F.R. Franklin Paine Mall: the story of a mind. Baltimore, Johns Hopkins Press, 1934. 342 pp.
SABIN, F.R. BMNAS, 1931, **16**, 65–122.
Archival material: Welch Medical Library, Johns Hopkins University.

MALMSTEN, Pehr Henrik (1811–1883). Swedish physician; discovered the parasitic protozoon *Balantidium coli*.

LENNMALM, C.G.F. Pehr Henrik Malmsten, minminnestecking. Stockholm, 1911. 99 pp. (Vol. 39 of *Svenska Läkaresällskapets Handlingar*.)

MALPIGHI, Marcello (1628–1694). Italian physician, 'founder of histology' and pioneer microscopist.

ADELMANN, H.B. Marcello Malpighi and the evolution of embryology. Vol 1. Ithaca, NY, Cornell Univ. Press, 1966. 750 pp.
ADELMANN, H.B. (ed.) The correspondence of Marcello Malpighi. 5 vols., New York, Cornell Univ. Press, 1975. 2227 pp.

BELLONI, L. DSB, 1974, **9**, 62–6.
MELI, D.B. (ed.) Marcello Malpighi: anatomist and physician. Firenze, Olschki, 1997. 325 pp.

MALTHUS, Thomas Robert (1766–1834). British cleric; professor of history and political economy at East India College, Haileybury.

JAMES, P. Population Malthus, his life and times. London, Routledge & Kegan Paul, 1979. 524 pp.
PETERSEN, W. Malthus. Cambridge, Mass., Harvard Univ. Press, 1979. 302 pp.
SIMPKINS, D.M. DSB, 1974, **9**, 67–71.

MANARDO, Giovanni (1462–1536). Italian physician, botanist, translator and editor.

ATTI del Convegno Internazionale per la Celebrazione del V Centenarion della Nascità di Giovanni Manardo 1462–1536, Ferrara, 8–9 dicembre 1962. Ferrara, Università degli Studi, 1963. 298 pp.
COTTON, J.H. DSB, 1974, **9**, 74–5.

MANDL, Felix (1892–1957). Austrian physician.

DOLEV, E. [Prof. Felix Mandl (1892–1957): a forgotten genius. *Harefuah*, 1993, **124**, 375–8.

MANN, Frank Charles (1887–1962). American physiologist, specializing in digestive system; at Mayo Clinic for 38 years.

VISSCHER, M.B. BMNAS, 1965, **38**, 161–204.

MANN, Ida (1893–1983). British ophthalmologist.

BUCKLEY, E.I. & POTTER, D.U. Ida and the eye; a woman in British ophthalmology. Speldhurst, Kent, Parapress, 1998. 322 pp.

MANSON, Patrick (1844–1922). British physician; 'Father of tropical medicine'.

CLARKSON, M.J. DSB, 1974, **9**, 81–3.
HAYNES, D.N. Imperial medicine: Patrick Manson and the control of tropical medicine. Philadelphia, Univ. of Philadelphia Press, *c*. 2001. 229 pp.
MANSON-BAHR, P.H. Patrick Manson, the father of tropical medicine. London, Nelson, 1962. 192 pp.
MANSON-BAHR, P.H. & ALCOCK, A. The life and works of Sir Patrick Manson. London, Cassell, 1927. 273 pp.

MANSON-BAHR, Philip Heinrich (1881–1966). British physician; specialist in tropical diseases.

DUGGAN, A.J. A bibliography of Sir Philip Manson-Bahr, 1881–1966. London, Wellcome Foundation, 1970, 50 pp.
Archival material: CMAC.

MANTEGAZZA, Paolo (1831–1910). Italian physician.

PASINI, W. Paolo Mantegazza: ovvero l'elogio dell'eclettismo. Rimini, Panozzo Editore, 1999. 254 pp.

BIBLIOGRAPHY OF MEDICAL AND BIOMEDICAL BIOGRAPHY

MANTELL, Gideon Algernon (1790–1852). British surgeon and geologist; followed hobby of geology throughout his practice; his collection of fossils was sold to the British Museum.

CHALLINOR, J. DSB, 1974, **9**, 86–8.

DEAN, D.R. Gideon Mantell and the discovery of dinosaurs. Cambridge, Cambridge Univ. Press, 1999. 290 pp.

SPOKES, S. Gideon Algernon Mantell, surgeon and geologist. London, Bale and Danielsson, 1927. 263 pp.

MANZONI, Alessandro (1746–1819). Italian surgeon.

DACCO, G. & ROSSETTO, M. (eds.) Alessandro Manzoni: società, storia, medicina. Milano, Leonardo Arte, 2000. 153 pp.

MARAÑÓN Y POSADILLO, Gregorio (1887–1960). Spanish pathologist; professor in Madrid.

BOTELLA LIUSIA, J. Gregorio Marañón: el hombre, la vida, la obra. Toledo Centro Universitario de Toledo, 1972. 70 pp.

GOMEZ-SANTOS, M. Vida de Gregorio Marañón. Esplugas de Llobregat, Plaza y Janés, 1977. 478 pp.

IZQUIERDO HERNANDEZ, M. Marañón. Madrid, Ed.Cid, 1965. 275 pp.

MARAT, Jean Paul (1743–1793). French physician and revolutionary.

CONNER, C.D. Jean Paul Marat: scientist and revolutionary. [Atlantic Highlands], Humanities Press, 1997. 285 pp.

MASSIN, J. Marat. Aix-en-Provence, Alinea, 1988. 253 pp.

MICHEL, A. Jean Paul Marat: his career in England and France before the revolution. London, Methuen, 1924. 144 pp.

MARCET, Alexander John Gaspard (1770–1822). Swiss physician; first to describe alkaptonuria.

COLEY, N.G. Alexander Marcet (1770–1822), physician and animal chemist. *Medical History*, 1968, **12**, 394–402.

MARCHAND, Felix (1846–1928). German pathologist; wrote memorable works on inflammation and wound healing.

GROTE, 1923, **1**, 59–104.

MARCHI, Vittorio (1851–1908). Italian pathologist.

BELLONI, L. DSB, 1974, **9**, 93–4.

MARCHIAFAVA, Ettore (1847–1935). Italian pathologist.

CASTELLANI, C. DSB, 1974, **9**, 94–5.

MARCHIAFAVA, G. Ettore Marchiafava. *Scientia Medica Italica*, 1958, **7** (1), 157–98.

MARCHIONINI, Alfred (1899–1965). German dermatologist.

POMER, A. Alfred Marchionini (1899–1965): Leben und Werk. Inauguraldissertation zur Erlangung des Doktorgrades der Medizin der Johannes Gutenberg-Universität Mainz. Mainz, 1990. 300 pp.

BIBLIOGRAPHY OF MEDICAL AND BIOMEDICAL BIOGRAPHY

MARCHOUX, Émile (1862–1943). French pathologist.

RAMON, G.L. *et al.* Émile Marchoux, 1862–1943. Laval, 1944, 57 pp.

MARCI OF KRONLAND, Johannes Marcus (1595–1667). Bohemian physician.

NOVÝ, L. DSB, 1974, **9**, 96–8.

MARCINKOWSKI, Karol (1800–1848). Polish physician.

WRZOSEK, A. Karol Marcinowski. 2 vols. Warzawa, Panstwowy Zaklad Wydawnictw Lekarskich, 1960–61. 329+228pp.

MARESCHAL, Georges (1658–1736). French surgeon; surgeon to Louis XIV and XV; instrumental in improvement of social status of surgeons in France.

DOOLIN, W. Georges Mareschal (1658–1736). liberator of surgery. *Annals of the Royal College of Surgeons of England*, 1952, **10**, 78–95.
KUFFERATH, G. Sire, votre chirurgien. [Paris?], Rouff, 1974. 229 pp.
MARESCHAL DE BIÈVRE, G. Georges Mareschal, Seigneur de Bièvre: chirurgien et confident de Louis XIV (1658–1736). Paris, Librarie Plon, 1906. 600 pp.

MAREY, Etienne-Jules (1830–1904). French physiologist; introduced graphical recording and cinematography in the study of his subject.

BRAUN, M. Picturing time: the work of Etienne-Jules Marey (1830–1904). Chicago, Univ. Chicago Press, 1992. 450 pp.
DRAGONET, F. Etienne-Jules Marey. Paris, Hazan, 1987. 139 pp.
GROSS, M. DSB, 1974, **9**, 101–3.
NOTICE sur les titres et travaux scientifiques de Dr Marey. Paris, Typographie Lahure, 1876. 83 pp.

MAREY, Pierre (1853–1940). French neurologist; professor of clinical neurology at La Salpétrière.

CRITCHLEY, M. Pierre Marie (1853–1940).In his: *The black hole and other essays*, London, Pitman, 1964, pp. 146–54.
SWAZEY, J.P. DSB, 1974, **9**, 108–9.

MARFAN, Bernard Jean Antonin (1858–1942). French paediatrician; described the skeletal deformities in the'Marfan syndrome'.

RENAULT, –. *Bulletin de l'Academie de Médecine*, Paris. 1942, **126**, 129–31.
SNELLEN, H.A. E.J. Marey and cardiology. Rotterdam, Kooyker, 1980. 264 pp.

MARINE, David (1880–1976). American scientist; made important contributions to knowledge of the thyroid gland.

MATOVINOVIC, J. David Marine (1880–1976): Nestor of thyroidology. *Perspectives in Biology and Medicine*, 1978, **21**, 565–89.

MARINESCU, Georghe (1863–1938). Romanian neurologist; professor of clinical neurology, Bucharest.

MARINESCU, M. Georghe Marinescu (1863–1938). *Leopoldina*, 1965, Ser.3, **11**, 226–40.
MARINESCU, G.M. & BRATESCU, G. Gheorghe Marinescu. Bucuresti, Editura Stiintifica, 1968. 207 pp.

247

MARKHAM, Roy (1916–1979). British virologist; Director, John Innes Institute.

ELSDEN, S.R. BMFRS, 1982, **28**, 319–45.
Archival material: CMAC.

MARRACK, John Richardson (1886–1976). British immunologist.

MAZUMDAR, P.M.H. DSB, 1990, **18**, 592–5.

MARRIAN, Guy Frederic (1904–1981). British biochemist; isolated hormones.

GRANT, J.K. BMFRS, 1982, **28**, 347–78.
Archival material: CMAC.

MARRIOTT, Williams McKim (1885–1936). American paediatrician; authority on child nutrition.

MEMORIAL address at a meeting held in the Auditorium of the Washington University School of Medicine, St Louis, 3 January 1937. St Louis, privately printed, Washington University, 1937. 46 pp.
VERDER, B. Pediatric profiles: Williams McKim Marriott (1885–1935). *Journal of Pediatrics*, 1955, **47**, 791–801.

MARSDEN, William (1796–1867). British surgeon; founder of the Royal Free and the Royal Marsden hospitals, London.

SANDWITH, F. Surgeon compassionate: the story of Dr William Marsden, founder of the Royal Free and Royal Marsden hospitals, by his great-granddaughter. London, Davies, 1960. 235 pp.

MARSH, Frederick Howard (1839–1915). British surgeon; professor at University of Cambridge; surgeon, St Bartholomew's Hospital, London.

MARSH, V.S. A memoir of Howard Marsh, London, Murray, 1921. 86 pp.

MARSH, James (1794–1846). British chemist; devised test for detection of arsenic.

CAMPBELL, W.A. Some landmarks in the history of arsenic testing. *Chemistry in Britain*, 1965, **1**, 198–202.

MARSHALL, Eli Kennerley (1889–1966). American pharmacologist; professor at Johns Hopkins University Medical School.

MAREN, T.H. BMNAS, 1987, **56**, 313–52.

MARSHALL, Francis Hugh Adam (1878–1949). British physiologist specializing in reproductive physiology.

PARKES, A.S. ONFRS, 1950–51, **7**, 239–51.

MARSTON, Hedley Ralph (1900–1965). Australian biochemist.

SYNGE, R.L. BMFRS, 1967, **13**, 267–93.

MARTEN, Benjamin (*fl*. 1720). British physician; suggested a parasitic micro-organism as cause of tuberculosis.

DOETSCH, R.N. Benjamin Marten and his 'New Theory of Consumptions'. *Microbiological Reviews*, 1978, **42**, 521–8.

MARTIN, Archer John Porter (1910–2002). British chemist; shared Nobel Prize (Chemistry) 1952 for invention of partition chromatography.

STAHL, G.A. *Journal of Chemical Education*, 1997, **54**, 80–83.

MARTIN, August Eduard (1847–1933). German gynaecologist.

MARTIN, A. Werden und Wirken eines deutschen Frauenarztes. Berlin, S.Karger, 1924. 370 pp.

MARTIN, Charles James (1866–1955). British biochemist; director, Lister Institute of Preventive Medicine.

CHICK, H., BMFRS, 1956, **2**, 173–208.
FYE, W.B. DSB, 1990, **18**, 597–9.
Archival material: CMAC; Basser Library, Australian Academy of Sciences, Canberra.

MARTIN, Edward (1859–1938). American surgeon.

JEQUIER, A.M. Edward Martin (1859–1938). The founding father of clinical andrology. *International Journal of Andrology*, 1991, **14**, 1–10.

MARTIN, Franklin Henry (1857–1935). American gynaecologist.

THE JOY of living: an autobiography of Dr Franklin H. Martin. 2 vols. Garden City, NY, Doubleday, Doran, 1933. 491+526 pp.

MARTIN, Henry Newell (1846–1896). British physiologist; professor at Johns Hopkins University, Baltimore.

FYE, W.B. H. Newell Martin – a remarkable career destroyed by neurasthenia and alcoholism. *Journal of the History of Medicine*, 1985, **40**, 133–66.

MARTIN, James Ranald (1796–1874). British surgeon to the East India Company.

FAYRER, J. Inspector-General Sir James Ranald Martin, F.R.S. London, Innes, 1897. 203 pp.

MARTINDALE, Louisa (1872–1966). British surgeon and gynaecologist.

MARTINDALE, L. A woman surgeon. London, Gollancz, 1951. 253 pp.
Archival material: CMAC.

MARTIN-LEAKE, (Arthur) (1874–1953). British medical officer; first to be awarded double Victoria Cross (in Boer War, 1902 and Great War, 1914).

CLAYTON, A. Martin-Leake, double V.C. London, L. Cooper, 1994. 250 pp.

MASCAGNI, Paolo (1755–1815). Italian anatomist; professor at Siena.

ALLODI, F. DSB, 1974, **9**, 153–4.
VANNOZI, F. (ed.) La scienza illuminata: Paolo Mascagni nel suo tempo (1755–1815). Siena, Nuova Immagine Editrice, 1996. 254 pp.

MASIUS, Jean Baptiste Nicolas Voltaire (1836–1912). Belgian physician.

ZUNZ, –. *Bulletin de l'Académie Royale de Médecine de Bellgique*, 1926, 5 sér., **6**, 286–301.

MASLOW, Abraham (1908–1970). American psychologist.

HOFFMAN, E. The right to be human: a biography of Abraham Maslow. Los Angeles, Crucible, 1988. 382 pp.

MASSA, Niccolo (1485–1569). Italian anatomist and syphilologist.

LIND, L.R. Studies in pre-Vesalian anatomy. Philadelphia, American Philosophical Society, 1975, pp. 167–253.
O'MALLEY, C.D. DSB, 1974, **9**, 165–6.

MASSERMAN, Jules Hymen (b. 1905). Polish/American psychiatrist.

MASSERMAN, J. H. A psychiatric odyssey. New York, Science House, 1971. 624 pp.

MATAS, Rudolph (1860–1957). American surgeon; professor at Tulane University, Louisiana.

COHN, I. & DEUTSCH, H.B. Rudolph Matas: a biography of one of the great pioneers in surgery. New York, Doubleday, 1960. 431 pp.

MATHER, Cotton (1663–1728). American Congregational minister and author; pioneer of inoculation against smallpox.

BEALL, O.T. & SHRYOCK, R.H. Cotton Mather, first significant figure in American medicine. Baltimore, Johns Hopkins Press, 1954. 241 pp.
COHEN, B. (ed.) Cotton Mather and American science and medicine. 2 vols. New York, Arno, 1980.
HOLMES, T.J. Cotton Mather: a bibliography of his works. 3 vols.. Cambridge, Mass., Harvard Univ. Press, 1940. 1395 pp.
LEVIN, D. Cotton Mather, the young life of the Lord's remembrancer, 1663–1703. Cambridge MA, Harvard Univ. Press, 1978. 360 pp.
SILVERMAN, K. The life and times of Cotton Mather. Philadelphia, Harper & Row, 1984. 479 pp.

MATHER, Kenneth (1911–1990). British geneticist; professor at Birmingham University.

LEWIS, D. BMFRS, 1992, **38**, 249–66.

MATHIJSEN, Antonius (1805–1878). Danish surgeon; introduced plaster-of-Paris bandage.

SPOELSTRA, D. Dr Antonius Mathijsen; uitvinder van het gipsverband, 1805–1878. Assen, Van Gorcum, 1970. 475 pp.

MATTEUCCI, Carlo (1811–1868). Italian physiologist and physicist.

BERNABEO, R. Carlo Matteucci (1811–1868): profilo della vita e dell'opera. Ferrara, Università degli Studi, 1972. 37 pp.
GUERRA, G. del & SCARPINI, A. Carlo Matteucci, 1811–1868. Con documenti relativi alla sua dimora Pisana. Pisa, Giardini, 1966. 47 pp.
MORUZZI, G. DSB, 1974, **9**, 167.

MATTHEWS, Bryan Harold Cabot (1906–1986). British neurophysiologist; professor of physiology, Cambridge.

GRAY, J. BMFRS, 1990, **35**, 263–79.

MATTIOLI, Pietro Andrea (1500–1577). Italian physician; considered mercury a specific in treatment of syphilis.

FERRI, S. Pietro Andrea Mattioli, Siena 1501 – Trento 1580. La vita, le opere. Perugia, Quattroemme, 1997. 405 pp.

LECLERC, H. Un naturaliste irascible: P.A. Mattiole de Siena *Janus*, 1927. **31**, 336–45.

MATY, Matthew (1718–1776). Huguenot physician; Principal Librarian, British Museum, 1772–76.

GUNTHER, A.E. Matthew Maty, MD, FRS (1718–1776) and science at the foundation of the British Museum, 1759–80. London, British Museum (Natural History), *c*.1987. 58 pp. (*Bulletin of the British Museum* (*Natural History*).)

MAUDSLEY, Henry (1835–1918). British psychiatrist; instrumental in the foundation of the Maudsley Hospital, London.

COLLIE, M. Henry Maudsley: Victorian psychiatrist. A bibliographical study. Winchester, St Paul's Bibliographies, 1988. 205 pp.

TURNER, T. Henry Maudsley – psychiatrist, philosopher and entrepreneur. *Psychological Medicine*, 1988, **18**, 551–74.

MAUGHAM, William Somerset (1874–1965). British physician, novelist and playwright.

CALDER, R.W. Willie: the life of W. Somerset Maugham. London, Heinemann, 1989. 429 pp.

CONNON, B. Somerset Maugham and the Maugham dynasty. London, Sinclair-Stevenson, 1997. 207 pp.

RAPHAEL, F. W.Somerset Maugham and his world. London, Thames & Hudson, 1976. 128 pp.

MAURICEAU, François (1637–1709). French obstetrician.

BUESS, H. Die Anfange wissenschaftlicher Geburtshilfe im Jahre 1668, François Mauriceau (1637–1709) und sein Handgriff. *Gynaekologische Rundschau*, 1968, **5**, 77–80.

MAVOR, Osborne Henry (1868–1951). British physician and dramatist; practised medicine until 1938 and was professor of medicine, Anderson College, Scotland; wrote under pseudonym 'James Bridie'.

LUYBEN, H.L. James Bridie, clown and philosopher. Philadelphia, Univ. of Pennsylvania Press, 1965. 180 pp.

MAVOR, O.H. One way of living. London, Constable, 1939. 299 pp.

MAVOR, R. Dr Mavor and Mr Bridie. Edinburgh, Canongate Publishing, 1988. 150 pp.

MAXCY, Kenneth Fuller (1889–1966). American epidemiologist; described murine typhus ('Maxcy's disease').

WOOD, W.B. & WOOD, M.L. BMNAS, 1971, **42**, 161–73.

MAYDL, Karel (1853–1913). Bohemian surgeon; performed first successful colostomy; devised operation of uretero-intestinal anastomosis.

–. *Casopis Lékaru Ceských*, 1913, **52**, 1383–7.

MAYER, Adolf (1843–1942). German phytopathologist; described and named tobacco virus disease.

–. *Chemische Zeitung*, 1943, **67**, 47.
MAYER, A. *Naturwissenschaften*, 1924, **12**, 905–11.

MAYER, Ernst (1883–1952). German physician.

RAAFLAUB, W. Ernst Mayer, 1883–1952. Leben und Werk. Bern, Hans Huber, 1986. 127 pp.

MAYER, Julius Robert (1814–1878). German physician and scientist; demonstrated principle of conservation of energy as far as physiological processes are concerned.

DÜHRING, E. Robert Mayer, der Galilei des neunzehnten Jahrhunderts. 2 Bde. Leipzig, C.G. Naumann, 1895–1904.
SCHMOLZ, H. & WECKBACH, H. Robert Mayer: sein Leben und Werk in Dokumenten. Weissenhorn, A.H.Konrad, 1964. 186 pp.

MAYER, Manfred Martin (1916–1984). German/American microbiologist and immunologist; professor of microbiology, Johns Hopkins University.

AUSTEN, K.F. BMNAS, 1990, **59**, 256–80.

MAYER-GROSS, Wilhelm (1881–1961). German/British psychiatrist.

HAAS, K.F. Wilhelm Mayer-Gross; Leben und Werk. Dissertation, Mainz, 1976. 69 pp.

MAYERNE, Theodor Turquet de, *see* **TURQUET DE MAYERNE, Theodore**

MAYNARD, Leonard Amby (1887–1972). American nutritionist.

ROE, D.A. BMNAS, 1998, **62**, 296–309.

MAYNEORD, William Valentine (1902–1988). Britiish radiologist; professor of physics as applied to medicine, University of London.

SPIERS, F.W. BMFRS, 1991, **37**, 341–64.
Archival material: CMAC.

MAYO FAMILY. Charles Horace (1869–1939); **Charles William** (1898–1968); **William James** (1861–1939); **William Worrall** (1819–1911).

CLAPESATTLE, H.B. The doctors Mayo. Minneapolis, Univ. Minnesota Press, 1941. 822 pp.; 2nd ed., 1967. 426 pp.
MAYO, C.W. The story of my family and my career. Garden City, NY, Doubleday, 1968. 351 pp.
NAGEL, G.W. The Mayo legacy. Springfield, C.C. Thomas, 1966. 157 pp.

MAYO, Herbert (1796–1852). British neuroanatomist; founded Middlesex Hospital Medical School, London.

LeFANU, W. DSB, 1974, **9**, 241–2.

MAYOR, Mathias (1775–1847). Swiss surgeon.

GERSTER, H. Mathias Mayor 1775–1847, Chef Chirurg des Kantonspitals Lausanne. Zürich, Juris, 1968. 30 pp.

BIBLIOGRAPHY OF MEDICAL AND BIOMEDICAL BIOGRAPHY

MAYOW, John (1643–1679). British chemist and physiologist.

BROWN, T.M. DSB, 1974, **9**, 242–6.

FULTON, J.F. A bibliography of two Oxford physiologists, Richard Lower (1631–1691) and John Mayow (1643–1679). Oxford, Oxford Univ. Press, 1935. 62 pp.

GOTCH, F. Two Oxford physiologists, Richard Lower 1631–1691, John Mayow 1643–1679. Oxford, Clarendon Press, 1908. 40 pp.

MEAD, Richard (1673–1754). British physician; supported the practice of inoculation against smallpox.

FERGUSON, V.A. A bibliography of the works of Richard Mead. Submitted in part requirement for the University of London Diploma in Librarianship, June 1959. 52 pp.

MACMICHAEL, W. The gold-headed cane. (CB).

MEADE, R.H. In the sunshine of life. A biography of Dr Richard Mead. Philadelphia, Dorrance & Co., 1974. 196 pp.

ZUCKERMAN, A. Dr Richard Mead (1673–1754); a biographical study. Thesis (PhD.). University of Illinois, 1965. Ann Arbor, University Microfilms International, 1980. 288 leaves.

MEANS, James Howard (1885–1967). American physician; authority on Graves' disease.

MEANS, J.H. Experiences and opinions of a full-time teacher. *Perspectives in Biology and Medicine*, 1959, **2**, 127–62.

MEARS, James Ewing (1838–1919). American surgeon.

HARTE, H.R. Transactions of the College of Physiciansa of Philadelphia, 1921, 3 ser., **43**, lxi–lxvii.

W., J.C. *Boston Medical and Surgical Journal*, 1919, **181**, 552–4.

MECHNIKOV [METCHNIKOFF], Ilya Ilyich (1845–1916). Russian-born bacteriologist and immunologist at Institut Pasteur, Paris; shared Nobel Prize 1908.

BESREDKA, A. The story of an idea: E. Metchnikoff's work, embryogenesis – inflammation – immunity – aging – pathology – philosophy. Bend, Oregon, Maverick Publications, 1979. 89 pp. First published in French, 1921.

BRIEGER, G.H. DSB, 1974, **9**, 331–5.

METCHNIKOFF, O. Life of Élie Metchnikoff 1845–1916, trans. E. Ray Lankester. London, Constable, 1921. 297 pp. Published in French 1920.

TAUBER, A.I. & CHERNYAK, L. Metchnikoff and the origins of immunity. New York, Oxford Univ. Press, 1991. 247 pp.

MECKEL, Johann Friedrich, *the Younger* (1781–1833). German anatomist and pathologist; professor at Halle.

BENEKE, R.C.A.C. Johann Friedrich Meckel der jungere. Halle (Saale), M. Niemeyer, 1934. 159 pp.

MEADER, R.G. The Meckel dynasty and medical education. *Yale Journal of Biology and Medicine*, 1937, **10**, 1–29.

RISSE, G.B. DSB, 1974, **9**, 252–3.

MEDAWAR, Peter Brian (1915–1987). British biologist and immunologist; professor of zoology, Birmingham and London; Director, National Institute for Medical Research; shared Nobel Prize 1960, for work on immunological tolerance.

MEDAWAR, J. A very decided preference. Life with Peter Medawar. Oxford, Oxford University Press, 1990. 256 pp.
MEDAWAR, P.B. Memoir of a thinking radish: an autobiography. Oxford Univ. Press, 1986. 209 pp.
MITCHISON, N.A. BMFRS, 1990, **35**, 281–301.
Archival material: CMAC.

MEDIN, Oskar (1847–1927). Swedish physician; first noted the epidemic character of poliomyelitis ('Heine-Medin disease').

ERNBERG, H. *Hygiea*, Stockholm, 1928, **90**, 49–66.

MEDUNA, Ladislaus Joseph (1896–1965). Hungarian neuropsychiatrist; introduced cardiazol convulsion therapy in schizophrenia.

FINK, M. Meduna and the origins of convulsive therapy. *American Journal of Psychiatry*, 1984, **141**, 1034–41.

MEEK, Walter Joseph (1878–1963). American cardiac physiologist; professor, University of Wisconsin.

BROOKS, C.M. BMNAS, 1983, **54**, 251–68.

MEEKEREN, Job Janszoon van (1611–1666). Dutch physician; first to record a bone graft, which was later removed by order of the Church.

BAUMANN, E.D. Jan van Meekeren. *Bijdragen tot de Geschiedenis der Geneeskunde*, 1923, **3**, 25–46.

MEISSNER, Georg (1829–1905). German anatomist and physiologist; professor of physiology at Göttingen.

ROTHSCHUH, K.E. DSB, 1974, **9**, 258–60.

MÉLIER, François (1798–1888). French physician.

BERGERON, E.J. Éloge de M.Mélier. Paris, G.Masson. 1889. 38 pp.

MELLANBY, Edward (1884–1955). British physiologist; Secretary, Medical Research Council.

DALE, H.H. BMFRS, 1955, **1**, 193–222.
GEISON, G.L. DSB, 1978, **15**, 417–20.
PLATT, B.S. Sir Edward Mellanby: the man, research worker, and statesman. *Annual Review of Biochemistry*, 1956, **25**, 1–28.
Archival material: CMAC; Medical Research Council, London: National Institute for Medical Research, London.

MELLANBY, John (1878–1939). British physiologist; professor at St Thomas's Hospital Medical School and later at University of Oxford.

LEATHES, J.B. ONFRS, 1940–41, **3**, 173–95.

MELTZER, Samuel James (1851–1920). Russian/American physiologist and pharmacologist, born in Russia.

HOWELL, W.H. BMNAS, 1926, **21**, 1–23.

PARASCANDOLA, J. DSB, 1974, **9**, 265–6.

MENDEL, Bruno (1897–1959). German-born physiologist; professor at Toronto and Amsterdam; distinguished for his work on cholinesterases.

FELDBERG, W.S. BMFRS, 1960, **6**, 191–9.

MENDEL, Johann Gregor (1822–1884). Austrian geneticist.

ILTIS, H. Life of Mendel, trans. by E. & C. Paul. London, Allen & Unwin, 1966. 336 pp. First published in German, 1924.
JAKUBICEK, M. & KUBICEK, J. Bibliographia mendeliana. Brno, Universtni knihovna v Brne, 1965. 74 pp. Supplementum 1965–69, 1970. 77 pp. Supplementum 1970–1974, 1976. 56 pp.
KRUTA, V. & OREL, V. DSB, 1974, **9**, 277–83.
MENDEL, J.G. (IN).
OREL, V. Gregor Mendel, the first geneticist. London, Oxford Univ. Press, 1994. 363 pp. First published in Czech.
Archival material: Mendelianum, Brno, Czechoslovakia.

MENDEL, Lafayette Benedict (1872–1935). American physiological chemist and nutritionist; professor at Yale University.

CHITTENDEN, R.H. BMNAS, 1937, **18**, 123–55.
VICKERY, H.B. DSB, 1974, **9**, 284–90.
Archival material: Medical Library, Yale University.

MENEGHETTI, Egidio (1892–1961). Italian pharmacologist.

FRANCESCHINI, P. DSB, 1974, **9**, 295–6.

MENETRIER, Pierre Émile (1859–1935). French physician and medical historian.

LAIGNEL-LAVASTINE, M. M. le professeur Menetrier (1859–1935). *Bulletin de la Société Française de l'Histoire de Médecine*, 1935, **29**, 245–7.
LAIGNEL-LAVASTINE, M. L'oeuvre historique de Pierre Menetrier. *Archeion*, 1936, **16**, 23–8.

MENGELE, Joseph (1911–1979). German physician; experimented on humans at Auschwitz extermination camp.

ASTOR, G.A. The 'last' Nazi: the life and times of Dr Joseph Mengele. London, Weidenfeld & Nicolson, 1985. 305 pp.
POSNER, G.L. & WARE, J. Mengele: the complete story. New York, Cooper Press, 2000. 364 pp. First published New York, 1987.

MENGHINI, Vincenzo Antonio (1704–1759). Italian physician; demonstrated presence of iron in blood.

BELLONI, L. DSB, 1974, **9**, 302–3.
BRIGHETTI, A. Il Menghini e la scoperta del ferro nel sangue. *Pagine di Storia della Medicina*, 1968, **12**, 63–80.

MENIÈRE, Prosper (1799–1862). French physician; described 'Menière's disease'.

MENIÈRE, P. Journal du docteur Prosper Menière; publié par son fils le dr. E. Menière. précédé d'une biographie par le dr. Fiessinger. Paris, Plon-Nourrit et Cie., 1903. 466 pp.

BIBLIOGRAPHY OF MEDICAL AND BIOMEDICAL BIOGRAPHY

MENIÈRE, P. Menière's original papers, reprinted with an English translation...and biographical sketch by Miles Atkinson. *Acta Otolaryngologica*, 1991, Suppl. **162**, 77 pp.

MENNINGER FAMILY (20th century).

FRIEDMAN, L.J. Menninger: the family and the clinic. New York, Knopf, 1990. 472 pp.

MENNINGER, Charles Frederick (1862–1953). American physician; established Menninger Diagnostic Clinic.

WINSLOW, W. The Menninger story [biography of Dr Charles Frederick Menninger and the story of the clinic he founded]. Garden City, N.Y., Doubleday Co., 1956. 350 pp.

MENNINGER, William Claire (1899–1966). American psychiatrist; professor at Menninger School of Psychiatry, Topeka, Kansas.

MENNINGER, W.C. A psychiatrist for a troubled world: selected papers of William C. Menninger...with biographical sketch by Henry W. Brosin. New York, Viking Press, 1967. 871 pp.

MERCER, Hugh (*c*.1726–1777). Scottish soldier who emigrated to America after Culloden; practised medicine there and was killed at the Battle of Princeton.

WATERMAN, J.M. With sword and lancet. The life of General Hugh Mercer. Richmond, Va., Garrett & Massie, 1941. 177 pp.

MERCK, Georg Franz (1825–1873). German chemist.

HOFMANN, A.W. Georg Merck. *Berichte der Deutschen Chemischen Gesellschaft*, 1873, **6**, 1583–5.

MERCURIALE, Geronimo (1530–1606). Italian physician; professor of medicine, Bologna and Pisa; wrote authoritatively on skin diseases, children's diseases and medical gymnastics.

PAOLETTI, H. Girolamo Mercuriale e il suo tempo. Bologna, Università degli Studi, 1963. 64 pp.
SIMILI, A. Gerolamo Mercuriale, lettore e medico a Bologna. Bologna, Azzogoidi, 1966. 86 pp.

MERING, Joseph von (1849–1908). German physician; with co-workers produced experimental diabetes.

DITTRICH, H.M. & HAHN VON DROSCHE, H. Die Entdeckung des Pankreasdiabetes durch von Mering und Minkowski. *Anatomischer Anzeiger*, 1978, **143**, 509–17.
WINTERNITZ, H. & ZUNZ, N. Josef v. Mering. *Münchener Medizinische Wochenscbrift*, 1908, **55**, 400–402.

MERRILL, John Putnam (1917–1986). American surgeon; first successfully to transplant a kidney.

FRIEDMAN, E.A. John P. Merrill. The father of nephrology as a craft. *Nephron*, 1978, **22**, 6–8, 265–80.

MERRITT, Hiram Houston (1902–1979). American neurologist; professor at College of Physicians and Surgeons, New York; editor, *Archives of Neurology*.

YAHR, M.D. (ed.) H.Houston Merritt: memorial volume. New York, Raven Press, 1983. 221 pp.

MÉRY, Jean (1645–1722). French surgeon and anatomist.

WILLIAMS, W.C. DSB, 1974, **9**, 322–3.

MERYON, Edward (1807–1880). British physician; gave first description of Duchennes's muscular atrophy, 1852.

EMERY, M.L. & EMERY, A.E.H. Edward Meryon (1807–1880) his life and Huguenot background. *Journal of Medical Biography*, 1998, **6**, 21–7.

MERZBACHER, Ludwig (1875–1942). German physician; recorded familial centrolobar sclerosis ('Pelizaeus-Merbacher disease') 1908.

PEIFFER, J. & GEHRMANN, J. Ludwig Merzbacher (1875–1942): the man behind the disease. *Brain Pathology*, 1995, **5**, 311–18.

MESMER, Franz Anton (1734–1815). German physician; introduced treatment by hypnosis (mesmerism).

BURANELLI, V. The wizard from Vienna. New York, Coward, McCann & Geoghegan, 1975. 256 pp.
DARNTON, R. DSB, 1974, **9**, 325–8.
FLOREY, E. Ars magnetica: Franz Anton Mesmer, 1734–1815, Magier vom Bodensee. Konstanz, Univ.-Verlag, *c*.1995. 286 pp.
PATTIE, F.A. Mesmer and animal magnetism. Hamilton, NY, Edmonston Publ., 1993. 274 pp.
WALMSLEY, D.M. Anton Mesmer. London, Hale, 1967, 192 pp.

MESNIL, Félix (1868–1938). French biologist and parasitologist.

TÉTRY, A. DSB, 1974, **9**, 328–9.

MESTREZAT, William (1883–1928). French physician; gave first accurate description of the chemical constitution of the cerebrospinal fluid.

DERRIEN, E. *Bulletin de la Société de Chimie Biologique*, 1929, **11**, 1067–83.

METCHNIKOFF, Élie, *see* **MECHNIKOV, Ilya Ilyich**

METTAUER, John Peter (1787–1875). American gynaecologist; successfully treated vesico-vaginal fistula, 1840.

RUCKER, P. Dr John Peter Mettauer (1787–1875), an early southern gynecologist. *Annals of Medical History*, 1938, **10**, 36–46.

METTENHEIMER, Carl Friedrich Christian (1824–1898). German physician.

METTENHEIMER, H. von. Carl von Mettenheimer (1824–1898). Werden, Wollen und Wirken eine alten Arztes in Briefen und Niederschriften. Frankfurt am Main, Verlag Waldemar Kramer, 1985. 560 pp.

MEYER, Adolf (1866–1950). Swiss-born American psychiatrist; professor at Johns Hopkins University and Director, Henry Phipps Psychiatric Clinic, Baltimore.

FEIERSTEIN, S. Adolf Meyer: life and work. Dissertation, Zurich, 1965. 23 pp.

MEYER, A. The commonsense psychiatry of Adolf Meyer. Fifty-two selected papers edited with biographical narrative by A.Lief. New York, Arno Press, 1973. 677 pp.

MEYER, Hans Horst (1853–1939). German pharmacologist; professor at Dorpat, Marburg and Vienna.

GROTE, **2**, 139–68 (CB).

MEYER, Karl Friedrich (1884–1974). Swiss/American microbiologist and epidemiologist; professor, University of California.

SABIN, A.D. BMNAS, 1980, 52, **2**, 269–332.
Archival material: University of California, Berkeley.

MEYER, Robert (1864–1947). German gynaecologist, emigrated to USA.

MEYER, R. Autobiography of Dr Robert Meyer (1864–1947). A short abstract of a long life. With a memoir by Emil Novak. New York, Henry Schuman, 1949. 126 pp.

MEYERHOF, Otto Fritz (1884–1951). German/American biochemist; shared Nobel prize 1922 for work on physiology of muscle.

FRUTON, J.S. DSB, 1974, **9**, 359.
NACHMANSOHN, D. *et al.* BMNAS, 1960, **35**, 153–82.
PETERS, R.A. ONFRS, 1954, **9**, 175–200.

MEYNERT, Theodor Hermann (1833–1892). Professor of neurology and psychiatry, Vienna.

STOCKERT-MEYNERT, D. Theodor Meynert und seine Zeit. Wien, Österreichischer Bundesverlag, 1930. 297 pp.

MIBELLI, Vittorio (1860–1910). Italian dermatologist.

ALLEGRA, F. Vittorio Mibelli and the tale of "porokeratosis". *American Journal of Dermatopathology*, 1986, **8**, 169–72.

MICHAEL SCOT (*c*.1175–1235). Scottish scholar, physician, traveller and translator of Aristotle and Arabic writers.

BROWN, J.W. An enquiry into the life and legend of Michael Scot. Edinburgh, Daniel Douglas, 1897. 281 pp.
FERGUSON, J. A short biography and bibliography of Michael Scotus. Glasgow, Bibliographical Society, 1931. 27 pp.
THORNDIKE, L. Michael Scot. London, Nelson, 1965. 143 pp.

MICHAELIS, Gustav Adolf (1798–1848). German obstetrician; wrote classic work on pelvic deformities.

SEMM, K. Gustav Adolf Michaelis (1798–1848), *Zentralblatt für Gynäkologie*, 1988, **110**, 1234–42.

MICHAELIS, Leonor (1875–1949). German biochemist; member of the Rockefeller Institute for Medical Research.

FRUTON, J.S. DSB, 1990, **18**, 620–25.
MacINNES, D.A. & GRANICK, S. BMNAS, 1958, **31**, 282–321.

MIDDELDORPF, Albrecht Theodor (1824–1868). German surgeon.

SUDERMANN, H. Albrecht Theodor Middeldorpf und die Anfänge der Elektrochirurgie (Galvanokaustik). *Würzburger Medizinhistorische Mitteilungen*, 2000, **19**, 59–110.

MIESCHER, Guido (1887–1961). Swiss dermatologist.

DEVIGUS, A. Der Dermatologe Guido Miescher, 1887–1961. Zürich, Juris, 1990. 92 pp.

MIESCHER, Johann Friedrich II (1844–1895). Swiss physiologist and biochemist; discovered nucleoprotein (nuclein) later shown to be the hereditary genetic material, DNA.

MIRSKY, A.E. The discovery of DNA. *Scientific American*, 1968, **218** (6), 78–88.
OLBY, R. DSB, 1974, **9**, 380–81.

MIKULICZ-RADECKI, Johann von (1850–1905). Polish surgeon; professor at Cracow and Wroclaw; improved antiseptic methods in surgery.

NEUGEBAUER, J. Weltruhm deutscher Chirurgie. Johann von Mikulicz. Ulm/Donau, Haug, 1965. 316 pp.

MILES, Arnold Ashley (1904–1988). British pathologist; professor of experimenatal pathology, London; Director of Lister Institute, London.

NEUBERGER, A. BMFRS, 1990, **35**, 303–26.

MILES, Walter Richard (1885–1978). American psychiatrist; professor at Yale University.

HILGARD, E.R. BMNAS, 1985, **55**, 411–32.

MILES, William Ernest (1869–1947). British surgeon; devised the operation of abdomino-perineal resection.

GILBERTSEN, V.A. Contributions of William Ernest Miles to surgery of the rectum for cancer. *Diseases of the Colon and Rectum*, 1964, **7**, 375–80.

MILLER, Frederick Robert (1881–1967). Canadian physiologist.

STAVRAKY, G.W. BMFRS, 1969, **15**, 173–84.

MILLER, Henry George (1913–1976). Professor of neurology and dean of medicine, University of Newcastle upon Tyne.

LOCK, S. & WINDLE, H. (eds.) Remembering Henry. London, British Medical Association, 1977. 166 pp.

MILLER, Willoughby Dayton (1843–1907). American dentist and stomatologist; made important contributions to the knowledge of dental bacteriology.

DERRICK, D. Willoughby Dayton Miller: ein Pionier der Zahnheilkunde. *Phillip Journal für Restaurative Zahnmedizin*, 1992, **9**, 462–4.
HIRSCHFELD, J.J. W.D. Miller and the "chemoparasitic" theory of dental caries. *Bulletin of the History of Dentistry*, 1978–79, 26–7, 11–20.

MILLS, Charles Karsner (1845–1931). American neurologist; first to describe uilateral descending paralysis, 1906.

LLOYD, J.H. *Archives of Neurology aand Psychiatry*, 1924, **4**, 198–201.

MILNE, Malcolm Davenport (1915–1991). British physician: professor of medicine, University of London.

PEART, S. BMFRS, 1995, **41**, 298–307.
Archival material: CMAC.

MILSTEIN, César (1927–2002). Argentinian molecular biologist; shared Nobel Prize (Physiology or Medicine) 1984 for discovery of the principle of production of mononuclear antibodies.

Les Prix Nobel 1984.

MINKOWSKI, Oskar (1858–1931). Lithuanian pathologist; studied role of the pancreas in diabetes.

ZELLER, S. & BLISS, M. DSB, 1990, **18**, 626–33.

MINOR, William Chester (1834–1920). American surgeon, homicidal maniac and lexicographer; while confined in Broadmoor Asylum, England was a prolific and valued contributor to the compilation of the *Oxford English Dictionary*.

WINCHESTER, S. The surgeon of Crowthorne: a tale of murder, madness and the love of words. London, Viking, 1998. 207 pp. Republished Penguin, 1999.

MINOT, Charles Sedgwick (1852–1914). American anatomist and embryologist.

CORNER, G.W. DSB, 1974, **9**, 416.
MORSE, E.S. BMNAS, 1920, **9**, 263–85.
Archival material: Library, Harvard University Medical School.

MINOT, George Richards (1885–1950). American physician; shared Nobel Prize 1934 for his work on pernicious anaemia.

CASTLE, W.B. DSB, 1974, **9**, 416–17.
CASTLE, W.B. BMNAS, 1974, **45**, 337–83.
RACKEMANN, F.M. The inquisitive physician: the life and times of George Richards Minot. Cambridge, Mass., Harvard Univ. Press, 1956. 288 pp.
Archival material: Harvard University Medical School.

MIRAULT, Germanicus (1796–1879). French surgeon; modified Malgaine's operation for cleft palate.

DUPOIREAUX, L. & PENNEAU, M. La contribution de Germanicus Mirault à la chirurgie maxillofaciale en générale et à celle du bec-de-lièvre en particulier. *Revue de Stomatologie et de Chirurgie Maxillo-faciale*, 1996, **97**, 195–201.

MIRSKY, Alfred Ezra (1900–1974). American biochemist and cytologist; member of the Rockefeller Institute.

COHEN, S.S. DSB, 1990, **18**, 633–6.
COHEN, S. BMNAS, 1998, **73**, 323–32.

MITCHELL, John Kearsley (1793–1858). American physician.

BURR, A.R. Weir Mitchell: his life and letters. New York, Duffield, 1929. 424 pp. [Contains much information on Weir's father John Kearsley Mitchell.]

MITCHELL Joseph Stanley (1909–1987). British radiologist; professor of radiotherapeutics and Regius Professor of Physic, Cambridge.

MARRIAN, D.H. BMFRS, 1988, **34**, 581–607.
Archival material: Cambridge University.

MITCHELL, Peter Dennis (1920–1972). British biochemist; Nobel Prize (Chemistry), 1978, for his contribution to the understanding of biological energy transfer through the formulation of the chemiosmotic theory.

HAYWARD, A. *et al.* Catalogue of the papers and correspondence of Peter Dennis Mitchell FRS (1920–1992) deposited in Cambridge University Library. 4 vols. Bath, University of Bath, 1997.
SLATER, E.C. BMFRS, 1994, **40**, 282–305.

MITCHELL, Silas Weir (1829–1914). American neurologist, novelist and poet.

BAILEY, P. BMNAS, 1958, **32**, 334–53.
BURR, A.R. Weir Mitchell: his life and letters. New York, Duffield, 1929. 424 pp.
BYNUM, W.F. DSB, 1974, **9**, 422–3.
EARNEST, E. S. Weir Mitchell, novelist and physician. Philadelphia, Univ. Pennsylvania Press, 1950. 279 pp.
LOVERING, J.P. S. Weir Mitchell. New York, Twayne, 1971. 176 pp.
WALTER, R.D. S. Weir Mitchell, neurologist: a medical biography. Springfield, Ill., C.C. Thomas, 1970. 232 pp.
Archival material: College of Physicians of Philadelphia.

MITSUDA, Kensuke (1876–1964). Japanese leprologist .

FASAL, P. Histopathology of leprosy: a tribute to Kensuke Mitsuda. *Cutis*, 1976, **18**, 66–72.

MIXTER, William Jason (1880–1958). American surgeon; with J.S. Barr demonstrated the causal role of invertebral disc herniation in sciatica.

PARISIEN, R.C. & BALL, P.A. William Jason Mixter (1880–1958); ushering the dynasty of the disc. *Spine*, 1998, **23**, 2363–6.

MÖBIUS, Paul Julius (1853–1907). German neurologist and psychiatrist.

SCHILLER, F. A Möbius strip. Fin-de-siècle neuropsychiatry and Paul Möbius. Berkeley, Univ. California Press, 1982. 134 pp.
WALDECK-SEMADENI, E.K. Paul Julius Möbius, 1853–1907. Leben und Werk. Dissertation, Bern, Medizinisches Institut der Universität, 1980. 222 pp.

MÖLLENDORF, Wilhelm von (1887–1944). German anatomist; professor at Hamburg, Kiel and Frieburg; edited important *Handbuch* on microscopic anatomy.

ZURGILGEN, B.M. Der Anatom Wilhelm von Möllendorf (1887–1944). Zürich, Juris Druck, 1991. 116 pp.

MÖRCH, Ernst (b. 1908). Danihsh/American surgeon.

BIBLIOGRAPHY OF MEDICAL AND BIOMEDICAL BIOGRAPHY

ROSENBERG, H. & AXELROD, J.K. Ernst Mörch: inventor, medical pioneer, heroic freedom fighter. *Anesthesia and Analgesia*, 2000, **90**, 218–21.

MOIR, John Chassar (1900–1977). British obstetrician and gynaecologist; with H.W. Dudley isolated ergotamine, 1935.

DUNN, P.M. John Chassar Moir (1900–1977) and the discovery of ergotamine. *Archives of Disease in Childhood, Fetal and Neonatal ed.*, 2002, **87**, F152–F154.

MOLESCHOTT, Jacob (1822–1893). Dutch physician and physiologist.

GIEST-HOFMAN, A.M. DSB, 1974, **9**, 456–7.
HAGELHANS, U. Jacob Moleschott als Physiologe. Frankfurt a.M., New York, Peter Lang, 1985. 156 pp.
LAAGE, R.J.V.C. Jacques Moleschott, een markante personlijkheid in negentiende eeuwse fysiologie? Zeist. Gregoriushuis, [1980?].
MOSER, W. Der Physiologe Jakob Moleschott, 1822–1893, und seine Philosophie. Zürich, Juris-Verlag, 1967. 55 pp.

MOLL, Albert (1862–1939). German psychiatrist; made study of sexual problems.

SCHULTZ, J.H. Albert Molls ärztliche Ethik. Zürich, Juris, 1986. 100 pp.

MOLLESON, Ivan Ivanovich (1842–1920). Pioneer Russian public health officer.

RÜMPEL, E. Ivan Ivanovich Molleson (1842–1920) der erste Hygienearzt in der Zemstvo-Medizin. Berlin, Osteuropa-Institut, 1968, 92 pp.

MOLONEY, William Curry (b. 1907). American haematologist.

MOLONEY, W.C. & JOHNSON, S. Pioneering hematology: the research and treatment of malignant disorders – reflections on a life's work. Boston, Francis A. Countway Library of Medicine, 1997. 196 pp.

MONAKOW, Constantin von (1853–1930). Russian neuroanatomist; professor in Zürich.

MONAKOW, C. von. Vita mea – mein Leben. Hrsg. von A.W. Gubser und E.H. Ackerknecht. Bern, Huber, 1970. 133 pp.
WASER, M. Begegnung am Abend; ein Vermächtnis. Stuttgart, Deutsche Verlagsanstalt, 1933. 415 pp.

MONARDES, Nicolás Bautista (1493–1588). Spanish botanist and physician; wrote first treatise on Central American drugs.

BOXER, C.R. Two pioneers of tropical medicine: Garcia d'Orta and Nicolas Monardes. London, Wellcome Historical Medical Library, 1963. 36 pp.
GUERRA, F. Nicolás Bautista Monardes, su vida e su obra. Monterrey, S.A., Mexico, Compãiná Fundidora de Hiero Acero de Monterrey, 1961. 226 pp.
GUERRA, F. DSB, 1974, **9**, 466.

MONDEVILLE, Henri de (*c.*1260–*c.*1320). French surgeon.

BULLOUGH, V.L. DSB, 1972, **6**, 276–7.
MONDEVILLE, H. de. Chirurgie...biographie...par E. Nicaise. Paris, Alcan, 1893, pp. v–lxxxii.

262

MONDINO DE' LUZZI (*c*.1275–1325). Italian anatomist; wrote first modern book on anatomy based on human dissection.

BULLOUGH, V.L. DSB, 1974, **9**, 467–9.

MONDOR, Henri (1885–1962). French surgeon; professor of surgical pathology at La Saltpêtrière.

BINET, J.P. Les vies multiples de Henri Mondor. Paris, Masson, 1933. 178 pp.
COMMÉMORATION du centenaire de Henri Mondor. *Chirurgie*, 1985, **111**, 387–426, i–xxii.
NUMÉRO SPÉCIAL consacré au Professeur Henri Mondor à l'occasion du centenaire de sa naissance (1885–1985). *Journal de Chirurgie*, 1985, **122**, 611–54.

MONIZ, *see* **EGAS MONIZ**

MONKS, George Howard (1853–1933). American plastic surgeon.

GOLDWYN, R.M. George H. Monks, M.D.: a neglected innovator. *Plastic and Reconstructive Surgery*, 1971, **48**, 478–84.

MONOD, Jacques Lucien (1910–1976). French molecular biologist; shared Nobel Prize 1965.

DEBRÉ, P. Jacques Monod. Paris, Flammarion, 1996. 366 pp.
LWOFF, A. BMFRS, 1977, **23**, 385–412.
LWOFF, A. & ULLMAN, A. Origins of molecular biology: a tribute to Jacques Monod. New York, Academic Press, 1979. 246 pp.
SOULIER, J.P. Jacques Monod: le choix de l'objectivité. Paris, Frisons-Roche, 1997. 225 pp.

MONRO FAMILY. John (1670–1740); **Alexander** *primus* (1697–1767); **Alexander** *secundus* (1733–1817); **Alexander** *tertius* (1773–1859); **David** (1813–1871). Edinburgh anatomists and surgeons.

WRIGHT-St. CLAIR, R.E. Doctors Monro: a medical saga. London, Wellcome Historical Medical Library, 1964. 190 pp.

MONRO, Alexander *primus* (1697–1767). British anatomist; professor, University of Edinburgh.

FINLAYSON, C.P. DSB, 1974, **9**, 479–82.
WRIGHT-ST CLAIR, R.E. Doctors Monro: a medical saga. London, Wellcome Historical Medical Library, 1964. 190 pp.
Archival material: Library, University of Edinburgh; Medical Library, University of Otago, New Zealand; Royal College of Surgeons of England.

MONRO, Alexander *secundus* (1733–1817). British anatomist; professor, University of Edinburgh.

FINLAYSON, C.P. DSB, 1974, **9**, 482–4.
WRIGHT-ST CLAIR R.E. Doctors Monro: a medical saga. London, Wellcome Historical Medical Library, 1964. 190 pp.
Archival material: Library, University of Edinburgh; Medical Library, University of Otago, New Zealand; Royal College of Surgeons of England.

MONTE [MONTANUS], Giambattista da (1498–1552). Italian physician and medical philologist; introduced bedside teaching in Padua, *c*.1543.

CERVETTO, G. Giambattista da Monte e dello medicina italiana del secolo XVI. Verona, G. Antonelli, 1839. 121 pp.

MONTESSORI, Maria (1870–1952). Italian psychiatrist and educationist; first woman to graduate MD at Rome University.

KRAMER, R. Maria Montessori: a biography. Oxford, Blackwell, 1978. 410 pp.

MONTGOMERY, William Fetherston (1797–1859). Irish obstetrician.

FLEMING, J.B. Montgomery and follicles of the areola as a sign of pregnancy (1837). *Irish Journal of Medical* Science, 1966, 169–82.

MOON, William (1818–1894). British inventor, blind at age 22, devised the Moon type, a simplified form of roman letters, endorsed on paper, for use by blind readers.

SAKULA, A. That the blind may read. *Medical History*, 1998, **6**, 21–7.

MOORE, Carl Richard (1892–1955). American zoologist and endocrinologist.

PRICE, D. BMNAS, 1974, **45**, 385–412.

MOORE, Carl Vernon (1908–1972). American physician.

LOWRY, O.H. BMNAS, 1994, **64**, 278–303.

MOORE, Charles Hewitt (1821–1870). British surgeon; established basic principles for the surgical treatment of cancer.

–. Charles Hewitt Moore, F.R.C.S. *British Medical Journal*, 1870, **1**, 641–2. Plarr, 2, 63–4.

MOORE, Francis (1657–*c*.1715). English astrologer and physician; obtained licence to practise medicine 1698; published *Vox stellarum*, an alamanac based on astrology which still appears annually as *Old Moore's Almanac*.

DNB.

MOORE, Francis Daniels (b.1913). American surgeon; performed experimental liver transplantation, 1959.

MOORE, F.D. A miracle and a privilege; recounting a half century of surgical advance. Washington, DC, Joseph Henry Press, 1995. 450 pp.

MOORE, John (1729–1802). Scottish physician and novelist; father of General Sir John Moore (1761–1809).

DNB.

MOORE, Norman (1847–1922). British physician and medical historian.

LINNETT, M. The life and works of Sir Norman Moore. London, 1947, 16 pp. [Reprinted from *St Bartholomew's Hospital Journal*, vol. 51, No. 6–9.]

MOORE, Stanford (1913–1982). American biochemist; professor at Rockefeller University, New York; Nobel Prize (chemistry), jointly, 1972, for work on ribonuclease.

SMITH, E.L. & HIRS, C.H.W. BMNAS, 1987, **56**, 354–85.

MOOREN, Albert (1828–1899). German ophthalmologist; described 'Mooren's ulcer'.

HOSS, J. Albert Mooren, ein Augenarzt von 19. Jahrhundert. Düsseldorf, Triltsch, 1980. 183 pp.

MOOSER, Hermann (1891–1971). Swiss bacteriologist; differentiated murine from epidemic typhus, 1928.

MARTI, R. Hermann Mooser (1891–1971). Der Entdecker des murinen Fleckfiebers. Zürich, Juris, 1978. 66 pp.

MORAN, Lord, *see* **WILSON, Charles McMoran**

MORAT, Jean-Pierre (1846–1920). French physiologist.

GROSS, M. DSB, 1974, **9**, 505–7.

MORAX, Victor (1866–1935). Swiss ophthalmologist.

BAILLART. L'oeuvre de Victor Morax (à l'occasion du centenaire de sa naissance). *Annales d'Oculistique*, 1966, **199**, 225–37.

MOREAU, Jacqued Joseph (1804–1884). French psychiatrist.

SCARCELLA, M. Un pionnier de la pédopsychiatrie moderne: Moreau de Tours. *Annales Médicopsychologiques*, 1964, **122**, 717–29.
WEBER, M. J.J. Moreau de Tours (1804–1884) und die experimentelle und therapeutische Verwendung von Haschich in der Psychiatrie. Zürich, Juris, 1971. 27 pp.

MORESTIN, Hippolyte (1869–1919). French plastic surgeon.

ROGERS, B.O. Hippolyte Morestin (1869–1919). Part 1. A brief biography. *Aesthetic and Plastic Surgery*, 1982, **6**, 141–7.

MORGAGNI, Giovanni Battista (1682–1771). Italian physician; founder of pathological anatomy.

BELLONI, L. DSB, 1974, **9**, 510–12.
CAMERON, G.R. The life and times of Giambattista Morgagni (1682–1771). *Notes and Records of the Royal Society of London*, 1952, **9**, 217–43.
GIORDANI, D. Giambattista Morgagni. Torino, Unioni tipografico-editrice torinese, 1941. 268 pp.

MORGAN, John (1735–1789). American physician; professor of the theory and practice of medicine, College of Philadelphia; established first school of medicine in North America.

BELL, W.J. John Morgan: continental doctor, Philadelphia, Univ. of Pennsylvania Press, 1965. 301 pp.
Archival material: Library, American Philosophical Society, Philadelphia.

MORGAN, Thomas Hunt (1866–1945). American zoologist and geneticist; Nobel Prize 1933.

ALLEN, G.E. DSB, 1974, **9**, 515–26.
ALLEN, G.E. Thomas Hunt Morgan: the man and his science. Princeton, NJ, Princeton Univ. Press, 1978. 447 pp.

DE BEER, G. ONFRS, 1945–48, **5**, 451–66.

SHINE, I. & WROBEL, S. Thomas Hunt Morgan: pioneer of genetics. Lexington, Univ. Press of Kentucky, 1976. 160 pp.

STURTEVANT, A.H. BMNAS, 1959, **33**, 283–325.

Archival material: Library, American Philosophical Society, Philadelphia; California Institute of Technology, Pasadena.

MORITZ, Friedrich (1861–1938). German physician; introduced cardiac orthodiagraphy.

CORNELY, M.E. Friedrich Moritz, Arzt und Lehrer: der Nachlass in Halle. Köln, Institut für Geschichte der Medizin, 1995. 262 pp.

MORO, Ernst (1874–1951). Austrian paediatrician.

HOEFNAGEL, D. & LUEDERS, D. Ernst Moro. *Pediatrics*, 1962, **29**, 643–5.

MORQUIO, Luis (1867–1935). Uruguayan physician; described osteochondrodystrophy ("Morquio's disease").

SOTO, J.A. Radiología clínica en la obra cientifica de Morquio. *Archivos de Pediatria de Uruguay*, 1965, **36**, 441–54.

MORRIS, William Richard, 1st Viscount Nuffield (1877–1963). Automobile manufacturer; medical benefactor.

BINNS, F.J. (ed.) Wealth well-given: the enterprise and benevolence of Lord Nuffield. Stroud, Alan Sutton, 1994. 318 pp.

HOLDER, D.W. BMFRS, 1966, **12**, 387–404.

BENEFACTIONS of Lord Nuffield [a chronological list]. *British Medical Journal*, 1963, **2**, 553.

MORRISON, George Ernest (1862–1920). Australian physician, explorer and statesman.

PEARL, C. Morrison of Peking. Sydney, Angus & Robertson, 1967. 431 pp.

MORTON, Richard (1637–1698). English phthisiologist.

TRAIL, R.R. Richard Mofton (1637–1698) *Medical History*, 1970, **14**, 166–74.

MORTON, Richard Alan (1899–1977). British biochemist; professor, Liverpool University.

GLOVER, J., PENNOCK, J.F., PITT, G.A.J. & GOODWIN, T.W. BMFRS, 1978, **24**, 409–42.

Archival material: University of Liverpool.

MORTON, Samuel George (1799–1851). American physician and anthropologist.

BELL, W.J. DSB, 1974, **9**, 540–41.

MEIGS, C.D. A memoir of Samuel George Morton, M.D., Philadelphia, Collins, 1851. 48 pp.

Archival material: Library, American Philosophical Society, Philadelphia.

MORTON, Thomas George (1835–1903). American surgeon; gave first complete description of anterior ('Morton's') metatarsalgia.

LEWIS, M.J. *Transactions of the College of Physicians of Philadelphia*, 1914, lxviii–lxxii.

MORTON, William Thomas Green (1819–1868). American dentist; pioneer in use of ether as anaesthetic.

MacQUILTY, B. Victory over pain; Morton's discovery of anaesthesia. New York, Taplinger, 1971. 208 pp. Previously published as *Battle for oblivion*.
WOLF, R.J. Tarnished idol: William Thomas Green Morton and the introduction of surgical anesthesia. San Anselmo CA, Jeremy Norman, 2001. 672 pp.
WOODWARD, G.S. The man who conquered pain: a biography of William Thomas Green Morton. Boston, Beacon Press, 1962. 175 pp.

MORVAN, Augustin Marie (1819–1917). French physician.

SCHOTT, B. *et al.* From Morvan's fibrillary chorea to the 'mal des ardents'. *Journal of the History of the* Neurosciences, 1996, **5**, 265–73.

MOSETIG-MOORHOF, Albert von (1838–1907). Austrian surgeon; introduced iodoform dressing in surgery.

–. *Medizinische Blätter*, Wien, 1907, **30**, 206–8.

MOSHER, Harris Peyton (1867–1954). American surgeon.

HARRIS PEYTON MOSHER, *Eye, Ear, Nose and Throat Monthly*, 1971, **50**, 399–402.

MOSS, William Lorenzo (1876–1957). American pathologist; classified blood groups.

DAVIS, A.B. DSB, 1974, **9**, 545–6.
Archival material: University of Georgia School of Medicine.

MOSSO, Angelo (1846–1910). Italian physiologist; professor at Turin.

ANGELO MOSSO, la sua vito e la sue opera: in memoriam novembre 1912. Milano, 1912. 244 pp.
CASTELLANI, C. DSB, 1974, **9**, 546–7.

MOTT, Frederick Walter (1853–1926). British neurologist and psychiatrist.

LORD, J.R. (ed.) Contributions in psychiatry, neurology and serology dedicated to the late Sir Frederick Mott. London, H.K. Lewis, 1929. [Biography by C. von Monakow, pp. 383–9; bibliography of Mott, pp. 391–401.]
MEYER, A. Frederick Mott, founder of the Maudsley laboratories. *British Journal of Psychiatry*, 1973, **122**, 497–516.

MOTT, Valentine (1785–1865). American surgeon; pioneer in vascular surgery; professor of surgery, University of the City of New York.

BISHOP, L.F. Brief sketch of Valentine Mott (1785–1865). *Medical Life*, 1928, **35** 577–82.
GROSS, S.D. Memoir of Valentine Mott. New York, Appleton; Philadelphia, Linsay & Blakiston, 1868. 96 pp.

MOUAT, Frederick John (1816–1897). British physician.

MOUAT, T.B. An eminent physician: Frederick John Mouat, M.D., F.R.C.S. (1816–1897). *Report of Proceedings, Scottish Society for the History of Medicine*, 1960–61, 33–5.

MOURANT, Arthur Ernest (1904–1994). British geneticist.

BIBLIOGRAPHY OF MEDICAL AND BIOMEDICAL BIOGRAPHY

Catalogue of the papers of A.E. Mourant. National Cataloguing Unit, Archives of Contemporary Scientists. Bath, 1999. 235 pp.
MISSON, G.P., BISHOP, A.C. & WATKINS, W.M. BMFRS, 1999, **45**, 331–48.
Archival material: CMAC.

MOYNIHAN, Berkeley George Andrew (1865–1936). British surgeon; professor of clinical surgery, University of Leeds.

BATEMAN, D.S. Berkeley Moynihan, surgeon. London, Macmillan, 1940. 355 pp.
FRANKLIN, A.W. Lord Moynihan: a short biography. In: Selected writings of Lord Moynihan. London, Pitman Medical Publishing, 1967, pp. 1–9.
KEYNES, G.L. Moynihan of Leeds. *Annals of the Royal College of Surgeons of England*, 1966, **38**, 1–21.

MUCH, Hans Christian (1880–1932). German immunologist, novelist and poet; director, tuberculosis and immunity institutes, Hamburg.

GROTE, 1925, **4**, 189–226.
MUCH, H. Vermächtnis. Bekentnisse von einem Arzt und Menschen. Dresden, Carl Meissner, 1933. 259 pp.
WIRTZ, R. Leben und Werk des Hamburger Arztes, Forschers und Schriftstellers Hans Much (1880–1932) unter besonderer Berücksichtigung seiner medizintheoretischen Schriften. Herzogenrath, Verlag Murke-Altrugge, 1991. 185 pp.

MUDD, Stuart (1893–1975). American microbiologist; professor at University of Pennsylvania.

MUDD, S. Sequences in medical microbiology; some observations over fifty years. *Annual Review of Microbiology*, 1969, **23**, 1–28.

MÜLLER, Friedrich von (1858–1941). German professor of medicine at Marburg, Basel and Munich; demonstrated increased metabolism in exophthalmic goitre.

MÜLLER, F. Lebenserinnerungen. München, J.F.Lehmann, 1951. 264 pp.
PFARRWALLER-STIEVE, A. Friedrich von Müller (1858–1941) und seine Stoffwechseluntersuchungen. Zürich, Juris Druck, 1983. 58 pp.

MÜLLER, Heinrich (1820–1864). German anatomist; discovered rhodopsin (visual purple).

KÖLLIKER, A. *Würzberger Naturwissenschaftlicher Zeitschrift*, 1864, **5**, xxix–xlvi.

MÜLLER, Johannes Peter (1801–1858). German anatomist and physiologist.

DU BOIS-REYMOND, E. Gedächtnisrede auf Johannes Müller. Berlin, K. Akademie der Wissenschaften, 1860. 191 pp.
HABERLING, W. Johannes Müller (1801–1858). Das Leben des Rheinischen Naturforschers. Leipzig, Akademische Verlagsgesellschaft, 1924, 501 pp.
KOLLER, G. Das Leben des Biologen Johannes Müller, 1801–1858. Stuttgart, Wissenschaftliche Verlagsgesellschaft, 1958. 267 pp.
RATHER, L.J., RATHER, P. & FRERICHS, J.B. Johannes Mueller and the nineteenth-century origins of tumour cell theory. Canton, MA, Science History Publications, 1986. 193 pp.
STEUDEL, J. DSB, 1974, **9**, 567–74.

MÜLLER, Paul Hermann (1899–1965). Swiss chemist; awarded Nobel Prize (Physiology or Medicine) 1948 for introduction of DDT insecticide.

– . Dr Paul Müller. *Nature*, 1965, **208**, 1043–4.

BIBLIOGRAPHY OF MEDICAL AND BIOMEDICAL BIOGRAPHY

MÜLLER-EBERHARD, Hans Joachim (1927–1998). German/American immunologist.

BEARN, A.G. BMNAS, 2001, **79**, 247–60.

MUELLER, John Howard (1891–1954). American microbiologist.

PAPPENHEIMER, A.M. BMNAS, 1987, **57**, 307–21.

MUELLER, Otto Friedrich (1730–1784). Danish biologist; first to attempt systematic classification of bacteria (1786).

ANKER, J. Otto Friedrich Mueller, et bidrag til den biologisk forsknings historie. 2 vols. Kjøbenhaven, E. Munksgaard, 1943– . *Acta Historica Scientiarum et Medicinalium*, Vol. 2.

MÜNSTERBERG, Hugo (1863–1916). German/American psychiatrist; pioneer in applied psychology.

HALE, M. Human science and social disorder: Hugo Münsterberg and the origins of applied psychology. Philadelphia, Temple Univ. Press, 1980. 239 pp.
MÜNSTERBERG, M.A.A. Hugo Münsterberg: his life and work. New York, Appleton, 1922. 449 pp.

MUIR, Robert (1864–1959). British pathologist; professor at Glasgow University.

CAMERON, R. BMFRS, 1959, **5**, 149–73.

MULLER, Hermann Joseph (1890–1967). American geneticist; Nobel Prize 1946 for X-ray mutation of gene.

CARLSON, E.A. Genes, radiation and society: the life and work of J.H. Muller. Ithaca, NY, Cornell Univ Press., 1981. 457 pp.
CARLSON, E.A. DSB, 1974, **9**, 564–5.
PONTECORVO, G. BMFRS, 1968, **14**, 349–89.
Archival material: Lilly Library, Indiana University, Bloomington, Ind., USA.

MULLIKEN, Robert Sanderson (1896–1986). American molecular biologist.

BERRY, R.S. BMNAS, 2000, **78**, 263–79.

MUMEY, Nolie (1891–1984). American surgeon.

MUMEY, N.L. Nolie Mumey, M.D. 1891–1984; surgeon, aviator, author, philosopher and humanitarian. Boulder, Col., Johnson Publishing Co., 1987. 274 pp.

MUNDINUS, *see* **MONDINO DE' LUZZI**

MUNTHE, Axel Martin Fredrik (1857–1949). Swedish physician and author.

MUNTHE, G. & UEXKÜLL G. The story of Axel Munthe. Translated from the Swedish by Malcolm Munthe, and from the German by Lord Sudley. New York, Dutton, 1953. 217 pp.

MURAD, Ferid (b. 1936). American physician; shared Nobel Prize (Physiology or Medicine), 1998, for discoveries concerning nitrous oxide as a signalling molecule in the cardiovascular system.

Les Prix Nobel en 1998.

MURALT, Johannes von (1645–1733). Swiss surgeon, anatomist and medical teacher; in 1686 opened college in Zürich with teaching in vernacular.

BOSCHUNG, U. Johannes von Muralt, 1645–1733: Arzt, Chirurg, Anatom, Naturforscher, Philosoph. Zürich, Rohr, 1983. 105 pp.
WOLF, J.H. DSB, 1974, **9**, 581–2.

MURLEY, Reginald Sydney (1916–1997). British surgeon; President, Royal College of Surgeons of England 1977–1980.

MURLEY, R. Surgical roots and branches. London, Memoir Club/British Medical Journal, 1990. 341 pp.

MURPHY, James Bumgardner (1884–1950). American biologist; researcher in cancer.

DAVIS, A.B. DSB, 1974, **9**, 586–7.
LITTLE, C.C. BMNAS, 1960, **34**, 183–203.
Archival material: Library, American Philosophical Society, Philadelphia.

MURPHY, John Benjamin (1857–1916). American surgeon; professor of clinical surgery. Chicago Postgraduate Medical School.

DAVIS, L.E. Surgeon extraordinary: the life of J.B. Murphy. London, Harrap, 1938. 286 pp. Published in New York as *J.B. Murphy, stormy petrel of surgery*, 1938. 311 pp.
SCHMITZ, R.I. & OH, T.T. The remarkable surgical practice of John Benjamin Murphy. Urbana, Univ. of Illinois Press, 1993. 207 pp.

MURPHY, William Parry (1892–1987). American physician; shared Nobel Prize (Physiology or Medicine) 1934, for introducing raw liver diet in the treatment of pernicious anaemia.

Les Prix Nobel en 1934.

MURRAY, Everitt George Dunne (1890–1964). Canadian bacteriologist.

STEVENSON, J.W. & COWAN, S.T. E.G.D. Murray (1890–1964). *Journal of General Microbiology*, 1967, **46**, 1–21.

MURRAY, James Alexander (1873–1950). British pathologist; Director, Imperial Cancer Research Fund.

FINDLAY, G.M. ONFRS, 1950–51, **7**, 445–52.

MURRAY, Joseph E. (b.1919). American surgeon; collaborated in the first successful kidney transplantation, 1956; shared Nobel Prize (Physiology or Medicine) 1990, for discoveries concerning cell transplantation in treatment of human disease.

MURRAY, J.E. Surgery of the soul: reflections on a curious career. Canton, MA, Science History Publications, 2001. 255 pp.

MURRELL, Christine (1874–1933). British physician; first woman elected to Council of British Medical Association; member of General Medical Council.

ST JOHN, C. Christine Murrell, M.D.: her life and work. London, Williams and Norgate, 1935. 133 pp.
Archival material: CMAC.

MURRELL, William (1853–1912). British physician; introduced Trinitrin (nitroglycerine) in angina.

SMITH, E. & HART, F.D. William Murrell, physician and practical therapist. *British Medical Journal*, 1971, **3**, 632–3.

MURRI, Augusto (1841–1932). Italian physician; professor of medicine in Bologna.

BUSACCHI, V. Augusto Murri. *Scientia Medica Italica*, 1959, 2 ser. **8**, 153–73.

MUSTARD, William Thornton (1914–1987). Canadian surgeon.

DUNLOP, M. Bill Mustard, surgical pioneer. Toronto, Oxford, Hannah Institute & Dundurn Press. 1989. 108 pp.

MUYBRIDGE, Edweard James (1830–1904). British photographer; investigated mechanics of motion by means of serial photographs of consecutive movements of animals and man.

HAAS, R.B. Muybridge man in motion. Berkeley, Cal., Univ. of California Press, 1976. 207 pp.
HENDRICKS, G. Edweard Muybridge: the father of the motion picture. London, Secker & Warburg, 1975. 271 pp.

MYERS, Charles Samuel (1873–1946). British psychologist; established the National Institute of Industrial Psychology.

BARTLETT, F.C. ONFRS, 1945–48, **5**, 767–77.

NABARRO, David Nunes (1874–1958). British clinical pathologist; discovered cause and mechanism of trypanosomiasis.

SIGNY, A.G. David Nunes Nabarro 27th February 1874–3rd October 1958. *Journal of Pathology and Bacteriology*, 1960, **79**, 429–35.

NACHMANSOHN, David (1899–1983). Russian/American biochemist; professor at Columbia University.

NACHMANSOHN, D. Biochemistry as part of my life. *Annual Review of Biochemistry*, 1972, **41**, 1–28.
OCHOA, S. BMNAS, 1989, **58**, 357–404.

NAEGELE, Franz Carl (1778–1851). German obstetrician; first described the obliquely-contracted ('Naegele') pelvis.

ROHLFS, H. Medizinische Klassiker Deutschlands, Stuttgart, 1880, **2**, 499–566.

NAEGELI, Otto (1871–1938). Swiss haematologist; professor at Tübingen and Zürich.

DUFEK, W.M. Der Internist Otto Naegeli 1871–1938. Zürich, Juris, 1983. 76 pp.

NAFIS, ibn al- (1210–1288). Syrian physician; described lesser circulation.

BITTAR, E.E. A study of Ibn Nafis. *Bulletin of the History of Medicine*, 1955, **29**, 353–68, 429–47.
CHÉHADE, A.K. Ibn al-Nafis et la découverte de la circulation pulmonaire. Damascus, Institut Français de Damas, 1955. 54 pp.

COPPOLA, E.D. The discovery of the pulmonary circulation: a new approach. *Bulletin of the History of Medicine*, 1957, **31**, 44–77.
ISKANDAR, A.Z. DSB, 1974, **9**, 602–5.

NAGAI, Nagajosi (1848–1986). Japanese pharmacologist; isolated ephedrine.

TSUZUKI, Y. *Journal of Chemical Education*, 1970, **47**, 495–6.

NANSEN, Fridtjof (1861–1930). Norwegian neurologist and Arctic explorer; Nobel Peace Prize, 1922.

GIHR, M. & PERMI, J. Fridjof Nansens zoologische und neuroanatomische Arbeiten; seine Beziehung zu Camillo Golgi und seine Bedeutung als Neuroanat. Ostermundigen, Schweiz, Hirnanatomische Institut, Psychiatrische Universitätsklinik, 1979. 60 pp.
NOCKHER, L. Fridtjof Nansen: Polarforscher und Helfer der Menschenheit. Stuttgart, Wissenschaftliche Verlagsgesellschaft, 1955. 236 pp.
WYKE, B. Fridtjof Nansen, GCVO, DSc, DCL, PhD. 1861–1930. A note on his contribution to neurology on the occasion of the centenary of his birth. *Annals of the Royal College of Surgeons of England*, 1962, **30**, 243–52.

NASSE, Christian Friedrich (1778–1851). German physician; propounded 'Nasse's law' of female immunity in haemophilia.

NOORDEN, W. von Christian Friedrich Nasse; ein Vorkämpfer und Wegbereiter des deutschen Arzttums. Berlin, Eberling, 1936. 424 pp.

NATHANS, Daniel (1918–1999). American microbial geneticist; shared Nobel Prize (Physiology or Medicine) 1978 for discovery of restriction enzymes and their application to problems of molecular genetics.

DIMAIO, D. BMNAS, 2001, **79**, 263–79.

NAUDE, Gabriel (1600–1653). Librarian of the Vatican; Librarian to Mazarin; he compiled a medical dictionary, 1632–1647.

CLARKE, J.A. Gabriel Naude, 1600–1653. Hamden, Conn., Archon Books, 1970. 183 pp.

NAUNYN, Bernard (1839–1925). German professor of medicine; published classical work on gallstones.

NAUNYN, B. Memories, thoughts and convictions. Edited with an introduction by D.L. Cowen. Canton MA, Science history Publications, 1994. 371 pp. First published in German, 1925.

NEEDHAM, Dorothy Mary Moyle (1896–1987). British biochemist.

WITKOWSKI, J.A. Optimistic analysis – chemical embryology in Cambridge. *Medical History*, 1987, **31**, 247–68.
Archival material: Girton College, Cambridge.

NEEDHAM, Joseph (1900–1995). British biochemist and historian of science and medicine.

GOLDSMITH, M. Joseph Needham: 20th-century Renaissance man. Paris, UNESCO, 1995. 170 pp.
GURDON, J.B. & RODBARD, B. BMFRS, 2000, **46**, 365–76.

BIBLIOGRAPHY OF MEDICAL AND BIOMEDICAL BIOGRAPHY

MUKHERJEE, S.K. & GHOSH, A. (eds.) The life and works of Joseph Needham. Calcutta, The Asiatic Society, 1997. 204 pp.

POWELL, T.E. Catalogue of papers and correspondence of Joseph Needham, C.H., F.R.S. (1900–1995) material relating to chemical and biological warfare. Bath, Univ. of Bath, 1995. 32 leaves.

POWELL, T.E. Supplementary catalogue of the papers and correspondence of Joseph Needham, CH, FRS, (1905–1995). 2 vols. Bath, National Cataloguing Unit for the Archives of Contemporary Scientists, 1999. NCUACS Catalogue No. 81/2/99.

SAID, H.M. (ed.) Essays on science: felicitation volume in honour of Joseph Needham. Karachi, Hamdard Foundation Pakistan, 1990.

NEEL, James van Gundia (b. 1915). American geneticist.

SCHULL, W.J. Scientist, journalist, orchidist – will the real James G. Neel please stand up. *Progess in Clinical and Biological Research*, 1986, **218**, 1–9.

SCHULL, W.J. BMNAS, 2002, **81**, 214–33.

NEEL, Louis Boyd (1905–1981). British physician and conductor; founder of Boyd Neel Orchestra.

DNB, 1981–85.

NEERGAARD, Kurt von (1887–1947). Swedish physician; professor in Zürich.

SCHMID, H. Kurt von Neergaard (1887–1947), Professor für physikalische Therapie. Zürich, Juris, 1986. 75 pp.

NEGRI, Adelchi (1876–1917). Italian pathologist.

ZANOBIO, B. DSB, 1974, **10**, 15–16.

NEHER, Erwin (b. 1944). German biohysical chemist; shared Nobel Prize (Physiology or Medicine) 1991, for discoveries concerning the function of single ion channels in cells.

Les Prix Nobel en 1991.

NEILD, James Edward (1824–1906). English/Australian forensic pathologist, medical editor and drama critic.

LOVE, H. James Edward Neild, Victorian virtuoso. Melbourne, Melbourne Univ. Press, 1989. 372 pp.

NEISSER, Albert Ludwig Sigesmund (1855–1916). German dermatologist and bacteriologist; discovered gonococcus and confirmed Hansen's work on the leprosy bacillus.

SCHADEWALDT, H. DSB, 1974, **10**, 17–19.

SCHMITZ, S. Albert Neisser: Leben und Werk auf Grund neuer, unveröffentlicher Quellen. Düsseldorf, W.Triltsch, 1968. 85 pp.

NÉLATON, Auguste (1807–1873). French surgeon; distinguished teacher at Hôpital St Louis, Paris.

OUGOUAG, A.T. Thèse pour le doctorat de médecine, Paris. Biobiblographie de Nélaton August (1807–1873). Paris, Dupuytren-copy, 1975. 119 leaves.

NEMESIUS (*fl.* 390–400). Bishop of Emesa [Homs], Syria; wrote *The nature of man*, influential text of the Middle Ages, remarkable for its sound physiological content.

O'MALLEY, C.D. DSB, 1974, **10**, 20–21.

TELFER, W. (ed.) Cyril of Jerusalem and Nemesius of Emesa. London; Westminster Press, 1955, 466 pp.

NENCKI, Marceli (1847–1901). Polish biochemist.

BICKEL, M.M. Marceli Nencki, 1847–1901. Bern, Huber, 1972. 102 pp.
NIEMIERKI, W. DSB, 1974, **10**, 22–3.

NEPVEU, Gustav (1841–1903). German pathologist; first to see trypanosomes in human blood, 1901.

–. *Marseille Médicale*, 1903, **40**, 289–94.

NESFIELD, Vincent Blumhardt (1879–1972). Indian-born British ophthalmic surgeon.

STRONG, P. Doctor anonymous. London, Covenant Publishing Co., 1967. 222 pp.

NETTLESHIP, Edward (1845–1913). British ophthalmic surgeon; his description of urticaria pigmentosa led to this being named 'Nettleship's disease'.

LAWFORD, J.B. *British Journal of Ophthalmology*, 1923, **7**, 1–9.

NEUBAUER, Otto (1874–1957). German/British biochemist.

BECHTEL, W. DSB, 1990, **18**, 663–4.
SCHEPARTZ, B. Otto Neubauer, a neglected biomedical scientist. *Transactions and Studies of the College of Physicians of Philadelphia*, 1984, **6**, 139–54.
Archival material: CMAC.

NEUBERGER, Albert (1908–1996). British biochemist.

ALLEN, A.K. & MUIR, H.M. BMFRS, 2001, **47**, 371–82.

NEUBURGER, Max (1868–1955). Austrian teacher and writer on the history of medicine.

BERGHOFF, E. Max Neuburger: Werden und Wirken eines österreichischen Gelehrten. Wien, W. Maudrich, 1948. 144 pp.
KAGAN, S.R. Professor Max Neuburger. A biography and bibliography, *etc. Bulletin of the History of Medicine*, 1943, **14**, 423–48.

NEUFELD, Fred (1861–1945). German pathologist; jointly named and described bacteriotropins; introduced antipneumococcus serum.

KLEINE, F.K. *Zeitschrift für Hygiene*, 1947, **127**, 185–6.

NEUMANN, Ernst (1834–1918). German haematologist.

NEUMANN-REDLIN v. MEDING, E. Der Pathologe Ernst Neumann (1834–1918) und sein Beitrag zur Begründung der Hämatologie in 19. Jahrhundert. Gräfeling, Demeter Verlag, 1987. 239 pp.

NEWLAND, Henry Simpson (1873–1969). Australian surgeon; founder of Royal Australasian College of Surgeons and Australian Medical Association.

HUGHES, J.E. Henry Simpson Newland. A biography by J. Estcourt Hughes and South Australian Fellows of the Royal Australasian College of Surgeons. Adelaide, Griffin Press, 1972. 120 pp.

NEWMAN, Horatio Hackett (1875–1957). American zoologist.

BOGIN, M.M. DSB, 1990, **18**, 665–8.

NEWSHOLME, Arthur (1857–1943). British physician and public health administrator; Chief Medical Officer, England and Wales.

EYLER, J.M. Sir Arthur Newsholme and state medicine. Cambridge, Cambridge Univ. Press, 1997. 422 pp.

NEWSHOLME, A. The last thirty years in public health: resollections and reflections on my official and post-official life. London, Allen & Unwin, 1936. 410 pp.

NICAISE, Jules Édouard (1838–1896). French surgeon and historian of surgery.

SEGOND, P. Éloge de Édouard Nicaise prononcé à la Société de Chirurgie dans la Séance annuelle du février 1903. Paris, Masson, 1903. 39 pp.

NICHOLAS, John Spangler (1895–1963). American embryologist; professor of biology at Johns Hopkins University.

OPPENHEIMER, J.M. BMNAS, 1969, **40**, 240–89.

NICHOLLS, Frank (1699–1778). British anatomist and physiologist; physician to George II; first to describe dissecting aneurysm of aorta, 1761.

MUNK, **2**, 123–6.

ROSS, J.K. The death of King George II, with a biographical note on Dr Frank Nicholls, physician to the King. *Journal of Medical Biography*, 1999, **7**, 228–33.

NICOLL, Henry Maurice Dunlop (1884–1953). British; exponent of psychological medicine.

POGSON, B. Maurice Nicoll: a portrait, London, Stuart, 1961. 288 pp.

NICOLLE, Charles Jules Henri (1866–1936). French bacteriologist; Nobel prizewinner 1928 for his proof of the role of lice in the transmission of typhus.

GRMEK, M.D. DSB, 1978, **15**, 453–5.

HUET, M. Le pommier et l'olivier. Charles Nicolle; une biographie (1866–1936) Paris, Sauramps Médical, 1995. 243 pp.

LOT, F. Charles Nicolle, un grand biologiste. Paris, Éditions de la Liberté, 1946. 111 pp.

PELIS, K.A. Pasteur's imperial missionary: Charles Nicolle (1866–1936) and the Pasteur Institute of Tunis. Dissertation submitted to Johns Hopkins University ... for the degree of Doctor of Philosophy, 1995. Baltimore, 1995, University Microfilms, 1997.

NICOLLE, Maurice (1862–1932). French parasitologist; demonstrated passive anaphylaxis.

MABROU, J. L'oeuvre scientifique de Maurice Nicolle. Paris, Masson, [1934]. 150 pp.

NICOLLE, J. Maurice Nicolle: un homme de la Renaissance et à notre époque. Paris, Editiones du Vieux Colombier, 1957. 174 pp.

NIEHANS, Paul (1882–1971). German physician; introduced 'cellular therapy', the injection of fresh living animal cells for regeneration of the elderly.

FISCHER, K.J. Niehans, Arzt des Papstes. München, W. Andermann, 1957. 318 pp.

HANNON, L.F. The second chance: the life and work of Dr Paul Niehans. London, New York, Allen, 1972. 242 pp.

BIBLIOGRAPHY OF MEDICAL AND BIOMEDICAL BIOGRAPHY

NIEMANN, Albert (1834–1861). German physician; isolated cocaine.

DIEPGEN, pp. 96–99 (CB).

NIGHTINGALE, Florence (1820–1910). British reformer of hospital nursing, born in Florence.

BISHOP, W.J. & GOLDIE, S. A biobibliography of Florence Nightingale. London, Dawsons, 1962. 160 pp.
COOK, E. The life of Florence Nightingale. 2 vols., London, Macmillan, 1913. 507 + 510 pp.
DOSSEY, B.M. Florence Nightingale: mystic, visionary, healer. Springhouse, PA, Springhouse Corporation, 1999. 440 pp.
SMITH, B. Florence Nightingale: reputation and power. London, St Martins Press, 1982. 216 pp.
WOODHAM-SMITH, C. Florence Nightingale, 1820–1910. London, Constable, 1950. 615 pp. Reprinted 1992.
Archival material: CMAC.

NIRENBERG, Marshall Warren (b. 1927). American biochemist; shared Nobel Prize (Physiology or Medicine), 1968, for work on DNA.

Les Prix Nobel en 1968.

NISSEN, Henry Wieghorst (1901–1958). American biologist and psychologist; director, Yerkes Laboratories of Primate Biology, Florida.

CARMICHAEL, L. BMNAS, 1965, **38**, 205–22.

NISSEN, Rudolf (1896–1981). German surgeon: performed first total pneumonectomy, 1931.

NISSEN, R. Helle Blätter – dunkel Blätter. Erinnerungen eines Chirurgen. Stuttgart, Deutsche Verlags-Anstalt, [*c*.1969]. 398 pp.

NISSL, Franz (1860–1919). German psychiatrist and neuropathologist.

GLEES, P. DSB, 1974, **10**, 130–31.

NITZE, Max (1848–1906). German urologist; devised electrically-illuminated cystoscope.

HAUSMANN, H. Das urologische Erbe. Maximilian Nitze (1848–1906). Seine Bedeutung für die Entwicklung der Urologie. *Zeitschrift für Urologie und Nephrologie*, 1987, **80**, 539–43.

NOBEL, Alfred Bernhard (1833–1896). Swedish engineer and chemist; founded Nobel prizes. [For information on lists of Nobel prizewinners see Collective Biographies section.]

FANT, K. Alfred Nobel: a biography. Translated from the Swedish edition (Stockholm, 1991) by M. Ruuth. New York, Arcade, 1993. 342 pp.
LJUNGGREN, B. & BRUYN, G.W. The Nobel Prize in Medicine and the Karolinska Institute: the story of Axel Munthe and Alfred Nobel. Basel, Karger, 2001. 232 pp.
ODELBERG, W. (ed.) Alfred Nobel: the man and his prizes. 3rd ed. New York, American Elsevier, 1972. 659 pp.
SOHLMAN, R. The legacy of Alfred Nobel: the story behind the Nobel prizes; translated from the Swedish by E.H. Schubert. London, Bodley Head, 1983. 144 pp.

NOCARD, Edmond (1850–1903). French veterinarian; described *Nocardia.*

NUTTALL, G.H.F. In memoriam: Edmond Nocard. *Journal of Hygiene*, 1903, **3**, 517–19.

NOCHT, Bernard Albrecht Eduard (1857–1927). Director, Hamburg Institute of Naval and Tropical Medicine.

MARTINI, E.C.W. Bernard Nocht; ein Lebensbild. Hamburg, Th.Dingwart & Sohn, 1957. 175 pp.

NOEGGERATH, Emil (1827–1895). American gynaecologist.

DIEGPEN, P. Emil Noeggerath und die Gynäkologie in der zweiten Hälfte des 19. Jahrhunderts. *Archiv für Geschichte der Medizin*, 1928, **20**, 198–232.

NOGUCHI, Hideyo (1876–1928). Japanese bacteriologist and pathologist; studied aetiology of syphilis, trachoma, and yellow fever, and immunity against snake venoms.

DOLMAN, C.E. DSB, 1974, **10**, 141–5.
ECKSTEIN, G. Noguchi. London, Harper, 1931. 419 pp.
PLESSET, I.R. Noguchi and his patrons. Rutherford, NJ, Fairleigh Dickinson Univ. Press [*c.*1980.] 314 pp.
Archival material: CMAC.

NOLAND, Lloyd (1880–1949). American surgeon and hospital administrator.

SMITH, A. Lloyd Nolan, MD, the boss. Fairfield, Ala., Lloyd Nolan Foundation, 1986. 548 pp.

NOORDEN, Carl Hermann Johannes von (1858–1944). German physician; professor in Berlin, Frankfurt, Vienna; made important investigations on metabolic disorders.

HAUK, J. Carl Harko Hermann von Noorden (1858–1944). [Thesis]. Mainz, Universität, Medizinisches Institut, 1980.

NORRIS, Charles (1867–1935). American physician; isolated spirochaete found to be causal organism of American relapsing fever.

EWING, J. Charles Norris, M.D., 1867–1935. *Bulletin of the New York Academy of Medicine*, 1935, **11**, 633–6.

NORTH, Elisha (1771–1843). American physician in New England; pioneer of Jennerian vaccination; published treatise on cerebrospinal meningitis.

ITALIA, S.R. Elisha North: experimentalist, epidemiologist, physician, 1771–1843. *Bulletin of the History of Medicine*, 1957, **31**, 505–36.

NORTHROP, John Howard (1891–1987). American biochemist and virologist; crystallized pepsin; shared Nobel Prize (Chemistry) 1946.

HERRIOTT, R.M. BMNAS, 1994, **63**, 422–50.

NOTHNAGEL, Carl Wilhelm Hermann (1841–1905). German physician; professor of medicine at Freiburg, Jena and Vienna.

NEUBERGER, M. Hermann Nothnagel. Leben und Wirken eines Deutschen Klinikers. Wien, Rikola, 1922. 469 pp.

NOTT, Josiah Clark (1804–1873). American physician.

HORSMAN, R. Josiah Nott of Mobile: southerner, physician and racial theorist. Baton Rouge, Louisiana State Univ. Press, 1987. 348 pp.

BIBLIOGRAPHY OF MEDICAL AND BIOMEDICAL BIOGRAPHY

NOVINSKY, Mstislav Alexandrovich (1841–1914). Russian onclogist.

SHABAD, L.M. M.A.Novinsky: rodonachalnik eksperimentalnoy onkologii. [M.A. Novinsky: father of experimental oncology.] Moskva, Izd. Akademii Nauk SSSR, 1950. 258 pp.
SHIMKIN, M.B. M.A.Novinsky: a note on the history of transplantation of tumors. *Cancer*, 1955, **8**, 653–5.

NOVY, Frederick George (1864–1957). American microbiologist; professor of bacteriology, University of Michigan.

LONG, E. BMNAS, 1959, **33**, 326–50.
PARASCANDOLA, J. DSB, 1974, **10**, 154–5.

NÜSSLEIN-VOLHARD, Christiane (b.1942). German geneticist; shared Nobel Prize (Physiology or Medicine) 1995, for discoveries concerning the genetic control of early embryonic development.

Les Prix Nobel en 1995.

NUFFIELD, VISCOUNT, *see* **MORRIS, William Richard**

NUMA, Shosaku (1929–1992). Japanese neurobiologist.

IMURA, H. BMFRS, 1995, **41**, 310–14.

NURSE, Paul Maxime (b.1949). British oncologist; shared Nobel Prize (Physiology or Medicine) 2001, for discovery of key regulators of the cell cycle.

Les Prix Nobel en 2001.

NUTTALL, George Henry Falkiner (1862–1937). British pathologist and parasitologist; professor of biology, University of Cambridge; founder of *Journal of Hygiene* and *Parasitology*.

GRAHAM-SMITH, G.S. & KEILIN, D. ONFRS, 1936–39, **2**, 493–9.
GRAHAM-SMITH, G.S. & KEILIN, D. George Henry Falkiner Nuttall (1862–1937). *Parasitology*, 1930, **30**, 403–18.

NUTTING, Mary Adelaide (1858–1948). American nursing pioneer.

MARSHALL, H.E. Mary Adelaide Nutting, pioneer of modern nursing. Baltimore, Johns Hopkins Press, 1972. 396 pp.

OAKLEY, Cyril Leslie (1907–1975). British immunologist.

EVANS, D.G. BMFRS, 1976, **22**, 295–305.

OBERMEIER, Otto Hugo Franz (1843–1873). German bacteriologist; discovered causal agent in relapsing fever.

ZEISS, H. Otto Hugo Franz Obermeier (1843–1873). *Archiv für Geschichte der Medizin*, 1923, **15**, 161–4.

OCHOA, Severo (1905–1993). Spanish biochemist; professor at New York University; shared Nobel Prize (Physiology or Medicine) 1959.

GRUNBERG-MANAGO, M. BMFRS, 1997, **43**, 349–65.
KORNBERG, A. *et al.* (eds.) Reflections in biochemistry in honour of Severo Ochoa. Oxford, Pergamon Press, 1976. 465 pp. [Biography of Ochoa on pp. 1–17.]
OCHOA, S. The pursuit of a hobby. *Annual Review of Biochemistry*, 1980, **49**, 1–30.
SALGADO GOMEZ, E. Severo Ochoa. Barcelona, Dopesa, 1971. 78 pp.

OCHSNER, Edward William Alton (1896–1981). American surgeon, professor, University of Illinois.

WILDS, J. & HARKEY, I. Alton Ochsner, surgeon of the South. Baton Rouge, Louisiana State University, 1990. 268 pp.

ODDI, Ruggero (1864–1913). Italian physiologist; 'sphincter of Oddi' named after him.

BELLONI, L. DSB, 1974, **10**, 175–6.

ODRIOZOLA, Ernesto (1862–1921). Peruvian physician.

VALDIZÁN, H. *Anales de la Facultad de Medicina, Lima*, 1921–1922, 31–41.

O'DWYER, Joseph P. (1841–1898). American physician; first to intubate in diphtheria.

GELFAND, C. Diphtheria: Dr Joseph O'Dwyer and his intubation tubes. *Caduceus*, 1987, **3**, 1–34.

OGDEN, Henry Vining (1857–1931). American/Canadian physician; friend of Osler, Cushing and other eminents.

WEISTROP, H.V. The life and letters of Dr Henry Vining Ogden, 1857–1931. Milwaukee, Milwaukee Academy of Medicine Press, 1986. 338 pp.

OGSTON, Alexander (1844–1929). British surgeon; professor in Aberdeen.

OGSTON, W.H. Alexander Ogston, KCVO: memories and tributes...with some autobiographical writings. Aberdeen, Aberdeen Univ. Press, 1943. 198 pp.

OGSTON, Alexander George (1911–1996). British biochemist.

SMITHIES, O. BMFRS, 1999, **45**, 351–64.

OHNO, Susumu (1928–2000). Japanese geneticist.

BEUTLER, E. BMNAS, 2002, **81**, 234–5.

OKEN [OKENFUSS], Lorenz (1779–1851). German biologist.

ECKER, A. Lorenz Oken. A biographical sketch. Translated into English by A. Tulk. London, Kegan Paul, 1883. 183 p. First published Stuttgart, 1880.
KLEIN, M. DSB, 1974, **10**, 194–6.

OLBERS, Heinrich Wilhelm Matthaus (1758–1840). German physician and astronomer.

MULTHAUF, L.S. DSB, 1974, **10**, 197–9.
Archival material: Staatsbibliothek, Bremen.

OLDS, James (1922–1976). American neuroscientist.

THOMPSON, R.F. BMNAS, 1999, **77**, 246–63.

OLITSKY, Peter Kosciusko (1886–1964). American pathologist and immunologist; isolated and propagated poliomyelitis virus in pure culture.

GOEBEL, W.F. Peter K. Olitsky. *Rockefeller Institute Review*, 1964, **2**, 9–11.

OLIVER, George (1841–1915). British physiologist; elucidated function of endocrine glands.

FRENCH, R.D. DSB, 1974, **10**, 204–6.
Archival material: Royal College of Physicians of London; Wellcome Institute for the History of Medicine.

OLLIER, Louis Xavier Edouard Leopold (1830–1900). French surgeon: pioneered bone allografting.

VINCENT, E. Le professeur Ollier. Paris, [n.p.]. 1901. 199 pp.

OLSHAUSEN, Robert von (1835–1915). German gynaecologist.

HENKEL, –. *Monatschrift für Geburtshulfe und Gynäkologie*, 1915, **41**, 285–91.
KÜSTNER, –. *Zentralblatt für Gynäkologie*, 1915, **39**, 177–85.
WINTER, G. *Zeitschrift für Geburtshulfe und Gynäkologie*, 1915, **58**, 656–734.

O'MALLEY, Charles Donald (1907–1970). American medical historian.

BURKE, J.G. Eloge. Charles Donald O'Malley. *Isis*, 1970, **61**, 371–8.
DEBUS, A.G. (ed.) Medicine in seventeenth-century England, a symposium held at UCLA in honor of C.D. O'Malley. Berkeley, University of California Press, [c.1974]. 485 pp. [Biographical contributions by L. White and M.T. Gnudi.]
POYNTER, F.N.L. Charles Donald O'Malley (1 April 1907–6 April 1970). *Clio Medica*, 1970, **5**, 197–201.

OPARIN, Aleksandr Ivanovich (1894–1980). Russian biochemist and evolutionist.

ADAMS, M.B. DSB, 1990, **18**, 695–700.

OPIE, Eugene Lindsay (1873–1971). American pathologist.

LONG, E.R. BMNAS, 1975, **47**, 293–320.
Archival material: Library, American Philosophical Society, Philadelphia.

OPPENHEIM, Hermann (1858–1919). German neurologist; first to describe amyotonia congenita ('Oppenheim's disease'), 1900.

CASSIRER, R. *Berliner Klinische Wochenschrift*, 1919, **56**, 669–71.
ERB, W. Deutsche *Zeitschrift für Nervenheilkunde*, 1918, **58**, v–viii.
LIEPMANN, H. *Zeitshchrift für die Gesamte Neurologie und Psychiatrie*, 1919, **52**, 1–6.
NONNE, M. *Neurologisches Zentralblatt*, 191, **38**, 386–90.

ORBELI, Leon Abgarovich (1882–1958). Russian physiologist.

GRIGORIAN, N.A. DSB, 1974, **10**, 220–21.
GRIGORIAN, N.A. L.A.Orbelli i razvitie sovetskoi fiziologii. Moskva, Meditsina, 1985. 102 pp.
LEIBSON, L.B. Leon Abgarovich Orbeli. Leningrad, Nauka, 1973. 449 pp.

ORD, William Miller (1834–1902). British physician; introduced the term 'myxoedema'.

–. William Miller Ord, M.D., F.R.C.P. Lond. Consulting Physician to St Thomas's Hospital. *Lancet*, 1902, **1**, 1494–7.

ORFILA, Mathieu Joseph Bonaventure (1787–1853). Toxicologist born in Minorca; a founder of the Académie de Médecine, Paris.

FAYOL, A. La vie et l'oeuvre d'Orfila. Paris, Albin Michel, 1930. 315 pp.
HUERTES GARCIA-ALEJO, R. Orfila: saber y poder médico. Madrid, Consejo Superior de Investigaciones Médicas, 1988. 456 pp.

ORIBASIUS (AD 325–403). Greek physician and medical compiler.

BALDWIN, B. The career of Oribasius. *Acta Classica*, 1975, **18**, 85–97.
KUDLIEN, F. DSB, 1974, **10**, 230–31.

ORTA, Garcia da, *see* **GARCIA DA ORTA**

ORTH, Johannes (1847–1923). German pathologist; first to describe kernicterus, 1895.

–. *Berliner Klinische Wochenschrift*, 1917, **54**, 101–3.
FICK, R. *Sitzungsberichte der Preussischen Akademie der Wissenschaften*, 1924, xcii–xcvii.

OSBORNE, Thomas Burr (1859–1929). American chemist.

VICKERY, H.B. BMNAS, 1931, **14**, 261–304.
VICKERY, H.B. DSB, 1974, **10**, 241–4.

OSLER, William (1849–1919). Canadian physician and humanist scholar, bibliophile; physician to Montreal General Hospital and Johns Hopkins Hospital; Regius Professor of Medicine, Oxford.

BLISS, M. William Osler: a life in medicine. Oxford, Oxford Univ. Press, 1999. 581 pp.
CUSHING, H.W. The life of Sir William Osler. 2 vols. Oxford, Clarendon Press, 1925. 685 + 728 pp. Second edition, London, Oxford Univ. Press, 1940. 1417 pp. Reprinted 1947.
GOLDEN, R.L. & ROLAND, C.G. (eds.) Sir William Osler: an annotated bibliography with illustrations. San Francisco, Norman Publishing, 1988. 214 pp. [Based on M.E. Abbott's classified and annotated bibliography of Sir William Osler's publications, 2nd ed., 1939. Corrigenda et addenda. 5 pp. San Francisco, Norman, 1990]; Addenda to Sir William Osler: an annotated bibliography. Huntington NY, The Author, 1997. 67 pp.
GOLDEN, R.L., NATION, E.F., ROLAND, C.G. & McGOVERN, J.P. An annotated checklist of Osleriana. Kent, Ohio, Kent State Univ. Press, 1976. 289 pp.
NATION, E.F. An up-dated checklist of Osleriana, supplement 1977–1985. Dr E.F. Nation, 112 N. Madison Ave, Pasadena, Cal. 91101, USA.
SAKULA, A. The portraiture of Sir William Osler. London, Royal Society of Medicine Services, 1991. 91 pp.
Archival material: Osler Library, McGill University, Montreal.

OSTERHOUT, Winthrop John Vanleuven (1871–1964). American botanist and physiologist.

BLINKS, L.R. BMNAS, 1974, **44**, 213–49.
PAULY, P.J. DSB, 1990, **18**, 702–4.

OTT, Isaac (1847–1916). American physiologist; discovered heat-regulating centre in the brain.

RADBILL, S.X. DSB, 1974, **10**, 252–3.

OTTESEN-JENSEN, Elise (1866–1973). Swedish sex educationist and social reformer.

LINDER, D.H. Crusader for sex education: Elise Ottesen-Jensen (1866–1973) in Scandinavia and the international scene. Lanham MD, University Press of America, 1996. 319 pp.

OTTO, Bodo (1711–1787). German settler in Pennsylvania; senior surgeon, active in the War of American Independence.

GIBSON, J.E. Dr Bodo Otto and the medical background of the American Revolution. Springfield, Ill., C.C.Thomas, 1937. 345 pp.

OTTO, John Conrad (1774–1844). American physician; described haemophilia, 1803.

KRUMBHAAR, E.B. John Conrad Otto (1774–1844) and the recognition of hemophilia. *Bulletin of the Johns Hopkins Hospital*, 1930, **6**, 123–40.

OTTO, Richard (1872–1952). German bacteriologist and immunologist.

BLUMENTHAL, G. R. Otto zum Gedächtnis zugleich eine historische Darstellung seiner Verdienste um die Anaphylaxieforschung. *Zentralblatt für Bakteriologie*, 1952, **159**, 5–10.

OUDOT, Jacques (*fl*. 1981). French surgeon.

NATALI, J. Jacques Oudot and his contribution to surgery of the aortic bifurcation. *Annals of Vascular Surgery*, 1992, **6**, 185–92.
NATALI, J. Hommage à Jacques Oudot pour le 50e. anniversaire de la premier graffe de bifurcation aortique. *Chirurgie*, 1999, **124**, 454.

OVERTON, Charles Ernest (1865–1933). British cell physiologist and pharmacologist.

COLLANDER, R. DSB, 1974, **10**, 256–7.

OWEN, Richard (1804–1892). British comparative anatomist, physiologist and palaeontologist.

OWEN, R. The life of Sir Richard Owen. By his grandson. 2nd ed. 2 vols. London, J. Murray, 1895. 409 + 393 pp.
RUPKE, N.A. Richard Owen: Victorian naturalist. New Haven, Yale Univ.Press, 1994. 462 pp.
WILLIAMS, W.C. DSB, 1974, **10**, 260–63.
Archival material: British Museum (Natural History), London; Royal College of Surgeons of England, London; Library, American Philosophical Society, Philadelphia.

PACCHIONI, Antonio (1665–1726). Italian anatomist and histologist.

FRANCESCHINI, P. DSB, 1974, **10**, 266.

PACINI, Filippo (1812–1883). Italian physician; professor of medicine, Florence; described *Vibrio cholerae*, causal organism of cholera, 1854.

BARBENSI, G. Il pensiero scientifico di Filippo Pacini, medico e matematico. *Rivista di Storia delle Scienze Mediche e Naturali*, 1940, **31**, 101–18, 139–59.
CASTALDI, L. Filippo Pacini. *Rivista di Storia delle Scienze Mediche e Naturali*, 1923, **14**, 182–212.
HOWARD-JONES, N. Choleranomalies: the unhistory of medicine as exemplified by cholera. *Perspectives in Biology and Medicine*, 1972, **15**, 422–33.

PADGETT, Earl Calvin (1893–1946). American plastic surgeon; introduced dermatome.

STEPHENSON, K.L. As I remember Earl C. Padgett. *Annals of Plastic Surgery*, 1981, **6**, 142–57.

PAGE, Earl (1880–1961). Australian surgeon and politician.

PAGE, E. Truant surgeon. The inside story of forty years of Australian political life. Sydney, Angus & Robertson, 1963. 421 pp.

PAGE, Irving Heinly (1901–1991). American physician; first described use of sodium amytal as an anaesthetic; with O.M. Helmer isolated angiotensin.

DUSTAN, H.P. BMNAS, 1995, **68**, 237–50.

PAGEL, Julius Leopold (1852–1912). German medical historian.

GROMER, J. Julius Leopold Pagel (1851–1912), Medizinhistoriker und Arzt. Köln, 1985. 239 pp. (*Arbeiten der Forschungsstelle des Instituts für Geschichte der Medizin der Universität zu Köln*, Bd. 38.)
PAGEL, W. Julius Leopold Pagel. Victor Robinson Memorial Volume. New York, Froben, 1948, pp. 273–97.
Archival material: CMAC.

PAGEL, Walter Traugott Ulrich (1898–1983). German/British pathologist and medical historian.

BUESS, H. Walter Pagel – 12. November 1898 – 25.März 1983 – zum Gedenken. *Clio Medica*, 1983, **18**, 233–9.
HERTEL, F. Klinische Monatsblätter für Augenheilkunde, 1919, **62**, 822–4.
WINDER, M. A bibliography of the writings of Walter Pagel. In: Debus, A. (ed.) Science, medicine and history in the Renaissance. New York, Neale Watson Academic Publications Inc., 1972, vol. 2, pp. 273–97.
Archival material: Hunter, I. The papers of Walter Pagel in the Contemporary Medical Archives Centre. *Medical History*, 1998, **42**, 89–96.

PAGENSTECHER, Ernst Hermann (1844–1918). German ophthalmologist.

HERTEL, F. Klinische *Monatsblätter für Augenheilkunde*, 1919, **62**, 822–4.
PAGENSTECHER, E.H. Aus den Lebenserinnerungen. 3 vols. Leipzig, R. Voigtländer, 1913.

PAGET, James (1814–1899). British surgeon; surgeon to St Bartholomew's Hospital, London; Sergeant-Surgeon to Queen Victoria.

PAGET, J. Memoirs and letters ... Edited by Stephen Paget. 3rd ed. London, New York, Longmans Green, 1903. 465 pp. First published 1901.
ROBERTS, S. Sir James Paget: the rise of clinical surgery. London, Royal Society of Medicine Services, 1989. 223 pp.
Archival material: Bodleian Library, Oxford; American Philosophical Society, Philadelphia; Royal College of Physicians of London; Royal College of Surgeons of England.

PAINTER, Theophilus Shickel (1889–1969). American zoologist and geneticist.

ALLEN, G.E. DSB, 1974, **10**, 276–7.
GLASS, B. BMNAS, 1990, **59**, 308–37.

BIBLIOGRAPHY OF MEDICAL AND BIOMEDICAL BIOGRAPHY

PALADE, George Emil (b.1912). Romanian cell biologist; shared Nobel Prize (Physiology or Medicine) 1974 for discoveries concerning structural and functional organization of the cell.

Les Prix Nobel, 1974.

PALFIJN [PALFYN], Jan (1650–1730). Belgian surgeon and obstetrician; re-invented or re-introduced obstetric forceps.

BROECKART, A. Der Bader von Kortryk. Das Leben des Jan Palfyn, 1650–1730. Leipzig, L. Staackmann, 1944. 333 pp. Originally published in Flemish.

PALLADIN, Aleksandr Vladimirovich (1885–1972). Russian biochemist.

SHAMIN, A.N. DSB, 1990, **18**, 704–5.
UTEVSKI, A.M. Alexander Vladimirovich Palladin. Kiev, Naukova Dumka, 1981. 128 pp. [in English].

PALMER, Daniel David (1845–1913). American founder of chiropractic.

GIELOW, V. Old dad chiro: biography of D.D. Palmer, founder of chiropractic. Davenport, Iowa, Bawden Bros., 1981. 184 pp.
PALMER, D. The Palmers: a pictorial life story; memoirs of D.D.Palmer. Davenport, Iowa, Bawden Bros., [1977]. 313 pp.

PANAS, Photinas (1832–1903). Greek ophthalmologist; professor at l'Hôtel-Dieu, Paris.

LAPERSONNE, F. de. Le professeur Panas. Paris, G. Steinheil, 1903. 24 pp.

PANCOAST, Joseph (1805–1882). American surgeon; first successfully to operate for extroversion of bladder.

RADBILL, S.X. Joseph Pancoast (1805–1882), Jefferson anatomist and surgeon and his world. *Transactions and Studies of the College of Physicians of Philadelphia*, 1986, 5 ser., **8**, 233–45.

PANDER, Christian Heinrich (1794–1865). Russian embryologist and comparative anatomist; developed germ-layer theory.

RAIKOV, B.E. Christian Heinrich Pander; ein bedeutenden Biologe und Evolutionist; an important biologist and evolutionist, 1794–1865. Frankfurt am Main, Verlag Waldemar Kramer, 1984. 144 pp. First published in Russian, 1954.

PANTRIDGE, James Francis (b.1916). British cardiologist; professor at Queen's University, Belfast.

PANTRIDGE, J.F. An unquiet life: memoirs of a physician and cardiologist. Belfast, Royal Victoria Hospital, 1989. 122 pp.

PANUM, Peter Ludvig (1820–1885). Danish physiologist, pathologist and epidemiologist.

CARSTENSEN, J. Peter Ludvig Panum, Professor der Physiologie in Kiel, 1853–1864. Neumünster, K.Wachholtz, 1967. 163 pp.
GAFAFER, W.M. Bibliographical biography of Peter Ludvig Panum (1820–1885), epidemiologist and physiologist. *Bulletin of the Institute of the History of Medicine*, 1934, **2**, 259–80.

284

BIBLIOGRAPHY OF MEDICAL AND BIOMEDICAL BIOGRAPHY

PAPANICOLAOU, George Nicholas (1883–1962).Greek/ American anatomist and cytologist.

CARMICHAEL, D.E. The Pap smear: life of George N. Papanicolaou. Springfield, Ill., C.C. Thomas, 1973. 122 pp.

SPEERT, H. DSB, 1974, **10**, 291–2.

PAPEZ, James Wensceslas (1882–1958). Austrian neuroanatomist.

LIVINGSTONE, K.E. & HORNYKIEWICZ, O. (eds.) Limbic mechanisms. The continuing evolution of the limbic system. New York, Plenum, 1976. 542 pp. [Tribute to James W. Papez: biographical comtributions by P.D. MacLean, J.B. Angevine, P.I. Yakovker.]

PAPPENHEIMER, Alwyn Max (1908–1995). American immunologist; isolated diphtheria toxin.

LAWRENCE, H.S. BMNAS, 1999, **77**, 264–80.

PARACELSUS [BOMBASTUS AB HOHENHEIM], Aureolus Philippus Theophrastus] (*c*.1493–1541). Swiss physician, surgeon and alchemist.

BECHTEL, G. Paracelse, ou la naissance de la médecine alchimique. Paris, Culture, Art, Loisirs, 1970. 286 pp.

DOPSCH, H., GOLDHAMER, K. & KRAMML, P.F. (eds.) Paracelsus (1493–1541) "Keine andern Knecht". Salzburg, Anton Puslet, 1993. 396 pp.

GRELL, O.P. (ed.) Paracelsus: the man and his reputation. Leiden, Brill, 1998. 351 pp.

LETTER, P. Paracelsus: Leben und Werk. Krummwich, Königsfurt, 2000, 400 pp.

PAGEL, W. Paracelsus: an introduction to philosophical medicine in the era of the Renaissance. 2nd ed., Basel, Karger, 1982, 399 pp.

PAGEL, W. DSB, 1974, **10**, 304–13.

RUEB, F. Mythos Paracelsus. Werk und Leben von Philippus Aureolus Theophrastus Bombastus von Hohenheim. München, Quintessenz Verlag, 1995. 350 pp.

SUDHOFF, K. Bibliographia Paracelsica. Besprechung der unter Hohenheims Namen 1527–1893 erschienenen Drucksschrift. Graz, Akademische Druck- und Verlagsanatalt, 1958. 722 pp. First published 1894. Nachweise zur Paracelsus-Literatur, *Acta Paracelsica*, 1932, Suppl., 68 pp. Continued by WEIMANN, K.M. Paracelsus-Bibliographie, 1932–1960. Wiesbaden, Steiner, 1963. 100 pp.

PARÉ, Ambroise (1510–1590). French military surgeon.

DOE, J. A bibliography, 1545–1940, of the works of Ambroise Paré, 1510–1590, premier chirurgien & conseiller du Roi. Descriptive and annotated bibliography of the original books, collected works, extracts and translations. With an introduction, appendixes and an index. Amsterdam, Gérard Th. van Heusden, 1976. 275 pp. First published Chicago, 1937. [The above incorporates the Addenda and Errata, 1940, and further addenda by the author, 1970.]

DUMAITRE, P. Ambroise Paré: chirurgien de quatre rois de France. Paris, Perrin, Fondation Singer-Polignac, 1986. 409 pp.

HAMBY, W.B. Ambroise Paré, surgeon of the Renaissance. St Louis, W.H. Green, 1967. 251 pp.

HAMBY, W.B. DSB, 1974, **10**, 315–17.

PACKARD, F.R. Life and times of Ambroise Paré (1510–1590), with a new translation of his Apology, and a new account of his journey in divers places. New York, Hoeber, 1921. 297 pp.

PAGET, S. Ambroise Paré and his times 1510–1590. New York, G.P. Putnam, 1897. 309pp.

PEDRON, F. Histoire d'Ambroise, chirurgien du roi. Paris, Orban, 1980. 468 pp.

PARIS, John Ayrton (1785–1856). British physician; first to describe cancer due to arsenic ingestion.

MUNK, W. A memoir of the life and writings of John Ayrton Paris. London, Bell and Daldy, 1857. 30 pp.

PARK, Henry (1744–1831). British surgeon.

LOWNDES, F.W. Henry Park of Liverpool. *Liverpool Medico-chirurgical Journal*, 1916, **36**, 122–6.

PARK, Mungo (1771–1806). British physician and explorer of the Nile.

DUFFILL, M. Mungo Park: West African explorer. Edinburgh, NMS Publishing, 1999. 144 pp.

PARK, William Hallock (1863–1939). American bacteriologist and hygienist: founder of the William Hallock Park Laboratory.

OLIVER, W.W. The man who lived for tomorrow. A biography of William Hallock Park, M.D. New York, Dutton, 1941. 507 pp.

PARKER, George Howard (1864–1955). American zoologist and physiologist; professor at Harvard University; worked on sensory physiology.

ROWE, A.S. BMNAS, 1967, **39**, 359–90.

PARKER, Peter (1804–1888). American physician; first Protestant medical missionary in China; established hospitals, lay clinics and medical schools.

GULICK, E.V. Peter Parker and the opening of China. Cambridge, Mass., Harvard Univ. Press, 1973. 282 pp.
STEVENS, G.B. & MARKWICK, W.F. The life, letters, and journals of the Rev. and Hon. Peter Parker, M.D., missionary, physician, and diplomatist; founder of the Ophthalmic Hospital in Canton. Boston, Congregational Sunday School and Pub. Soc., 1896. 362 pp. Reprinted Wilmington, Del., Scholarly Resources, 1972. 362 pp.

PARKER, Willard (1800–1884). American surgeon; first in USA to operate for appendicitis; professor of surgery, College of Physicians and Surgeons, New York.

RUHRÄH, J. Willard Parker. *Annals of Medical History*, 1933, N.S. **5**, 205–14, 376–89, 458–83.

PARKER, William Kitchen (1823–90). General practitioner in London; comparative anatomist.

PARKER, T.J. William Kitchen Parker, F.R.S., sometime Hunterian Professor of Anatomy and Physiology in the Royal College of Surgeons of England: a biographical study. London, New York, Macmillan, 1893. 145 pp.

PARKES, Alan Sterling (1900–1990). British endocrinologist.

PARKES, A.S. Off-beat biologist: the autobiography of Alan S. Parkes. Cambridge, Galton Foundation, 1985. 485 pp.
PARKES, A.S. Biologist at large, the background to *Offbeat biologist*. Cambridge, A.S. Parkes, 1988. 465 pp.
Archival material: CMAC.

PARKINSON, James (1755–1824). British physician and palaeontologist; described paralysis agitans.

GARDNER-THORP, C. James Parkinson 1755–1824, and a reprint of *The shaking palsy*. 2nd ed. Exeter, Dept. of Neurology, Royal Devon and Exeter Hospital, 1988. 82 pp + 66 pp.

GERSTNER, P.A. DSB, 1974, **10**, 321–3.

MORRIS, A.D. James Parkinson: his life and times. Edited by F.C. Rose. Basle, Birkhäuser, 1989. 207 pp.

ROBERTS, S. James Parkinson 1755–1824: from apothecary to general practitioner. London, Royal Society of Medicine, 1997. 136 pp.

PARKINSON, John (1885–1976). British cardiologist.

McDONALD, L. Sir John Parkinson – a leader in cardiology. *Clinical Cardiology*, 1980, **12**, 546–8.

PARMENTIER, Antoine-Augustin (1737–1813). French nutritionist and pharmacist.

MURATORI-PHILIP, A. Parmentier. Paris, Plon, 1994. 398 pp.

PARNAS, Jakub Karol (1884–1949). Polish biochemist.

KORZYBSKI, T.A. DSB, 1974, **10**, 326–7.

PARRY, Caleb Hillier (1755–1822). British general practitioner in Bath; gave classical early account of exophthalmic goitre.

GLASER, S. The spirit of enquiry: Caleb Hillier Parry, MD, FRS. Stroud, Gloucestershire, Alan Sutton, 1995. 177 pp.

ROLLESTON, H.D. Caleb Hillier Parry, MD, FRS. *Annals of Medical History*, 1925, N.S. **7**, 205–15; N.S. **9**, 1–12.

PARRY, John S. (1843–1876). American pathologist; wrote first authoritative work on ectopic pregnancy, 1876.

INGRAM, J.V. Memoir of John S. Parry, M.D., late President of the Obstetrical Society of Philadelphia, etc. *Transactions of the College of Physicians of Philadelphia*, 1876, 3 ser., **2**, xlv-lviii.

PARSONS, John Herbert (1868–1957). British ophthalmologist.

DUKE-ELDER, S. BMFRS, 1958, **4**, 204–14.

PARSONS, Leonard Gregory (1879–1950). British paediatrician; professor, University of Birmingham.

CAMERON, G.R. ONFRS, 1950–51, **7**, 453–68.

PARSONS, Usher (1788–1868). American surgeon; a founder of the American Medical Association.

GODOWSKY, S.J. Yankee surgeon: the life and times of Usher Parsons (1788–1868). Boston MA, Francis A. Countway Library, 1988. 450 pp.

PASSOW, Adolf (1859–1926). German otologist.

BIBLIOGRAPHY OF MEDICAL AND BIOMEDICAL BIOGRAPHY

HAAKE, K. Zum 125. Geburtstag von Adolf Passow. *Charité-Annalen*, 1984, N.F. **4**, 312–17.

PASTEUR, Louis (1822–1895). French chemist, microbiologist and immunologist.

DARMON, P. Pasteur. Paris, Fayard, 1995. 411 pp.
DEBRÉ, P. Louis Pasteur. Translation by E. Forster. Baltimore, Johns Hopkins Univ, Press, 1998. First published in French, 1994.
DUBOS, R.J Pasteur and modern science, ed. T. Brock. London, Berlin, Springer, 1988. 168 pp.
GEISON, G.L. DSB, 1974, **10**, 350–416.
GEISON, G.L. The private science of Louis Pasteur. Princeton NJ, Princeton Univ. Press, 1995. 378 pp.
VALLERY-RADOT, R. The life of Louis Pasteur. Translated from the French by R.L. Devonshire. 12th ed. London, Constable, 1923. 484 pp. First English edition 1911.
Archival material: Bibliothèque Nationale, Paris.

PATERSON, Herbert John (1868–1940). British surgeon; pioneer of postgraduate teaching in London.

PATERSON, H.J. A surgeon looks back: the memoirs of the late Herbert John Paterson. London, Epworth Press, 1941. 253 pp.

PATIN, Guy (1601–1672). French physician; professor at the Collège de France and founder of the École de Médecine.

PACKARD, F.R. Guy Patin and the medical profession in Paris in the XVIIth century. New York, Hoeber, 1925. 334 pp.
PIC, P. Guy Patin. Paris, Steinheil, 1911. 300 pp.

PATON, William Drummond MacDonald (1917–1993). British pharmacologist; professor at Oxford University.

BECKETT, C. A coherent collection: the personal papers of Sir William Paton. *Friends of the Wellcome Library...Newsletter*, 2003, **28**, 7–11.
PATON, W.D. On becoming and being a pharmacologist. *Annual Review of Pharmacology and Toxicology*, 1986, **26**, 1–22.
RANG, H.P. & PERRY, W.. BMFRS, 1996, **42**, 290–314.
Archival material: CMAC.

PATTERSON, John Thomas (1873–1960). American embryologist and geneticist.

WAGNER, R.P. DSB, 1990, **18**, 707–9.

PATTISON, Granville Sharp (1791–1851). British anatomist in Edinburgh and U.S.A.

PATTISON, F.L.M. Granville Sharp Pattison, anatomist and antagonist 1791–1851. Edinburgh, Canongate, 1987. 284 pp.

PAUL OF AEGINA (AD 625–690). Greek physician.

THOMAS, P.D. DSB, 1974, **10**, 417–19.

PAUL, John Rodman (1893–1971). American epidemiologist; devised Paul-Bunnell test for glandular fever.

HORSTMANN, D.M. & BEESON, P.B. BMNAS, 1975, **47**, 323–68.

PAULESCU, Nicolas Constantin (1869–1931). Romanian physiologist; preceded Banting and Best in the isolation of insulin.

MURRAY, I. Paulesco and the isolation of insulin. *Journal of the History of Medicine*, 1971, **26**, 150–57.
PAVEL, I. The priority of N.C.Paulescu in the discovery of insulin. Bucuresti, Edit. Academei Republicii Socialiste Romania, 1976. 251 pp.

PAULING, Linus Carl (1901–1994). American chemist and molecular biologist; Nobel prizewinner for chemistry (1954) and peace (1962).

DUNITZ, J,D, BMFRS, 1996, **42**, 316–38.
DUNITZ, J.D. BMNAS, 1997, **71**, 221–61.
GOERTZEL, B. & GOERTZEL, G. Linus Pauling: a life in science and politics. New York, Basic Books, 1995. 300 pp.
HAGER, T. Force of nature: the life of Linus Pauling. New York, Simon & Schuster, 1995. 791 pp.
HAGER, T. Linus Pauling and the chemistry of life. New York, Oxford Univ. Press, *c.*1998. 142 pp.
MEAD, C. & HAGER, T. (eds.) Linus Pauling: scientist and peacemaker. Corvallis, Oregon State Univ. Press, 2001. 272 pp.
SERAFINI, A. Linus Pauling: a man and his science. New York, Paragon House, 1989. 310 pp.
Archival material: Kerr Library, Oregon State University; Wellcome Library, London.

PAULLI, Simon (1603–1680). Danish anatomist and botanist.

FIGALA, K. DSB, 1974, **10**, 426–7.

PAVLOV, Ivan Petrovich (1849–1936). Russian physiologist; notable for his work on conditioned reflexes and digestion; Nobel Prize 1904.

ANREP, G.V. ONFRS, 1936–39, **2**, 1–18.
BABKIN, B.P. Pavlov: a biography. London, Gollancz, 1951. 365 pp.
CUNY, H. Ivan Pavlov: the man and his theories. Transl. by Patrick Evans. London, Souvenir Press, 1964. 174 pp.
GRAY, J.A. Ivan Pavlov. New York, Viking, 1980. 153 pp.
GRIGORIAN, N.A. DSB, 1974, **10**, 431–6.

PAVLOVSKI, Evgenii Nikanorovich (1884–1965). Russian medical parasitologist; professor at Military Medical Academy, Leningrad.

BORCHERT, A. E.N. Pawlowski: Leben und Werk. Berlin, Deutscher Verlag der Wissenschaften, 1959. 177 pp.
HOARE, C.A. E.N. Pavlovsky, MD, DSc. *British Medical Journal*, 1965, **1**, 1677.

PAVY, Frederick William (1829–1885). British physician.

TAYLOR, F. Frederick William Pavy M.D., F.R.S., F.R.C.P., *Guy's Hospital Reports*. 1912, **66**, 1–23.

PAWLIK, Karel (1849–1914). Bohemian surgeon.

–. *Monatschrift für Geburtshulfe und Gynäkologie*, 1914, **40**, 684–6.

PAYNE, Fernandez (1881–1977). American geneticist.

CARLSON, E.A. DSB, 1990, **18**, 709–10.

PAYNE, Rose Marise (1909–1999). American hamatologist and immunologist; introduced leucocyte typing, 1957.

–. Rose Payne 1909–1999. *Tissue Antigens*, 1999, **54**, 102–5.

PAYR, Erwin (1871–1946). Austrian surgeon; transplanted thyroid.

BECKER, T. Das chirurgische Erbe. Erwin Payr zum Gedenken. *Zentralblatt für Chirurgie*, 1981, **106**, 1563–8.
GROTE, 1924, **3**, 121–64. (CB).
PAYR, E. Am Wege. Erinnerungen und Betrachtungen eines Chirurgien. aus dem Nachlass herausgegeben von Joachim Krebs. Leipzig, J.A. Barth, 1994. 232 pp.

PEABODY, Francis Weld (1881–1927). American physician; Harvard's professor of medicine at Boston City Hospital; director of Thorndike Memorial Laboratory.

PAUL, O. The caring physician: the life of Dr Francis W. Peabody. Boston, Francis H. Countway Library of Medicine, 1991. 220 pp.

PEACHEY, *see* PEACHIE, PECHIE

PEACHIE, John (*c.*1632–1692). English physician.

PEACHEY, G.G. The two John Peacheys, seventeenth century physicians, their lives and times. *Janus*, Leiden, 1918, **23**, 121–58.

PEACOCK, Thomas Bevill (1812–1882). British physician at St Thomas' Hospital.

PORTER, I. The nineteeth-century physician and cardiologist Thomas Bevill Peacock. *Medical History*, 1962, **6**, 240–54.

PÉAN, Jules Émile (1830–1898). French surgeon and gynaecologist.

DIDIER, R. Péan. Paris, Maloine, 1948. 245 pp.

PEARCE, Louise (1885–1959). American physician; intoduced tryparsamide in treatment of trypanosomiasis.

FAY, M. Louise Pearce 5th March 1885–9th August 1959. *Journal of Pathology and Bacteriology*, 1961, **82**, 542–45.

PEARL, Raymond (1879–1940). American biologist, geneticist and statistician.

JENNINGS, H.S. BMNAS, 1943, **22**, 295–347.
PARKER, F. DSB, 1974, **10**, 444–5.
Archival material: Library, American Philosophical Society, Philadelphia.

PEARSON, Egon Sharpe (1895–1980). British biometrician.

BARTLETT, MS. BMFRS, 1981, **27**, 425–43.

PEARSON, George (1751–1828). British chemist and physician; advocate of smallpox vaccination.

SCOTT, E.L. DSB, 1974, **10**, 445–7.

PEARSON, Karl (1857–1936). British statistician; founder of the science of biometrics.

EISENHART, C. DSB, 1974, **10**, 447–73.

MORANT, G.M. A bibliography of the statistical and other writings of Karl Pearson. Compiled by G.M. Morant with the assistance of B.L. Welch. Cambridge, Cambridge Univ. Press, 1939. 119 pp.

PEARSON, E.S. Karl Pearson; an appreciation of some aspects of his life and work. Cambridge, Cambridge Univ. Press, 1938. 170 pp.

YULE, G.U. ONFRS, 1936, **2**, 73–104.

Archival material: University College London [catalogue by M. Merrington *et al.*, London, University College, 1983. 164 pp.]; Royal Society of London; CMAC.

PEAT, Stanley (1902–1969). British biochemist.

HIRST, E.L. & TURVEY, J.R. BMFRS, 1970, **16**, 441–62.

PECHEY-PHIPSON, Mary Edith (1845–1908). Pioneer British woman physician and surgeon, MD Berne 1877; worked in India and founded Pechey-Phipson sanatorium near Bombay.

LUTZKER, E. Edith Pechey-Phipson. M.D. The story of England's foremost pioneering woman doctor. New York, Exposition, 1973. 259 pp.

PECHIE, John (1654–1718). English physician.

PEACHEY, G.G. The two John Peacheys, seventeenth century physicians, their lives and times. *Janus*, Leiden, 1918, **23**, 121–58.

PECQUET, Jean (1622–1674). French anatomist; discovered thoracic duct.

BENTATA, Y. Un médecin anatomiste du XVIIme siècle, Jean Pecquet. Paris, 1932. 47 pp.

HUARD, P. DSB, 1974, **10**, 476–8.

PEKELHARING, Cornelis Adrianus (1848–1922). Dutch physician and physiological chemist; anticipated discovery of vitamins.

BAART DE LA FAILLE, J.M., WESTENBRINK, H.G.K. & NIEWENHUIJSE, P. Leven en werken van Cornelius Adrianus Pekelharing, 1848–1922. Utrecht, A.Oosthoek, 1984. 217 pp.

ERDMAN, A.M. DSB, 1974, **10**, 492–3.

PEL, Pieter Klazes (1852–1919). Dutch physician.

DELPRAT, C.C. *Nederlandsch Tijdschrift für Geneeskunde*, 191, **1**, 569–72.

PELLETIER, Pierre-Joseph (1788–1842). French chemist and pharmacist; isolated plant alkaloids.

BERMAN, A. DSB, 1974, **10**, 497–9.

PELLIER de QUENGSBY, Guillaume (?1750–1835). French ophthalmologist; wrote first book on ophthalmic surgery.

CHIRILA, T.V. & HICKS, C.R. The origins of the artificial cornea: Pellier de Quengsby and his contribution to the modern concept of keratoprosthesis. *Gesnerus*, 1999, **56**, 96–106.

BIBLIOGRAPHY OF MEDICAL AND BIOMEDICAL BIOGRAPHY

TERSON, A. Pellier de Quengsby. Paris, G. Steinheil, 1895. 29 pp.

PEMBREY, Marcus Seymour (1868–1934). British physiologist; professor at Guy's Hospital Medical School.

DOUGLAS, C.R. ONFRS, 1932–35, **1**, 563–7.

PENFIELD, Wilder Graves (1891–1976). American/Canadian neurosurgeon.

ECCLES, J. & FEINDEL, W. BMFRS, 1978, **24**, 473–513.
LEWIS, J. Something hidden. A biography of Wilder Penfield. Toronto, Doubleday Canada Ltd., 1981. 311 pp.
PENFIELD, W.G. No man alone: a neurosurgeon's life. Boston, Little, Brown, 1977. 398 pp.

PENNELL, Theodore Leighton (1867–1912). British medical missionary on the North-West Frontier of India.

PENNELL, A.M. Pennell of the Afghan frontier. The life of Theodore Leighton Pennell, 2nd ed., London, Seeley, Service, 1914. 464 pp.

PENROSE, Lionel Sharples (1898–1972). British geneticist and medical statistician.

HARRIS, H. BMFRS, 1973, **19**, 521–61.
LAXOVA, R. Lionel Sharples Penrose, 1898–1972: a personal memoir in celebration of the centenary of his birth. *Genetics*, 1998, **50**, 1333–40.
MERRINGTON, M. *et al.* (comp.) A list of the papers and correspondence of Lionel Sharples Penrose (1898–1972) held in the Manuscripts Room, University College London Library. London, University College, 1979. 90 pp.
SMITH, M. Lionel Sharples Penrose: a biography. Colchester, Michael Smith, 1999. 56 pp.

PENZOLDT, Franz (1849–1927). German pharmacologist; professor in Erlangen.

GROTE, 1923, **2**, 169–86 (CB).

PEPPER, William (1843–1898). American physician.

THORPE, F.N. William Pepper, M.D., LL.D. (1843–1898), Provost of the University of Pennsylvania. Philadelphia, London, Lippincott, 1904. 555 pp.

PEPPERDENE, Frank Simpson (1863–1932). British radiologist.

CAMERON, M.D. F.S. Pepperdene: pioneer radiologist and X-ray martyr. Windsor, Ont., Electra Press, 1994. 93 pp.

PERCIVAL, Thomas (1740–1804). British physician; early writer on medical ethics.

LEAKE, C.D. (ed.) Percival's *Medical ethics*. Reprint of 1927 edition with new historical introduction...by C.R. Burns. Huntington, N.Y., Krieger, 1975. 299 pp.

PERCY, Pierre François (1754–1825). French military surgeon; one of Napoleon's leading surgeons.

LONGIN, E. Journal des campagnes du Baron Percy, chirurgien en chef de la Grande Armée (1754–1825). 2nd ed. Paris, Librairie Plon, 1904. 537 pp.

PEREIRA, Helio George (1918–1994). Brazilian virologist.

BIBLIOGRAPHY OF MEDICAL AND BIOMEDICAL BIOGRAPHY

SKEHEL, J.J. & TYRRELL, D.A.J. BMFRS, 1999, **45**, 381–96.

PERLS, Friedrich Salomon (1893–1970). German-born psychoanalyst; creator of Gestalt therapy.

CLARKSON, P. & MACKEWN, J.. Fritz Perls. London, Sage Publications, 1993. 205 pp.
GAINES, J. Fritz Perls, here and now. Millbrae CA, Celestial Arts, 1979. 440 pp.
SHEPARD, M. Fritz (an intimate portrait of Fritz Perls and Gestalt therapy). New York, E.P. Dutton & Co., 1975. 235 pp.

PERRONCITO, Edoardo (1847–1936). Italian parasitologist and bacteriologist; diagnosed hookworm disease in St Gotthard miners.

CASTELLANI, C. DSB, 1974, **10**, 527.

PERTHES, Georg Clemens (1869–1927). German surgeon; besides being a pioneer in radiotherapy, he described juvenile osteochondritis deformans (Calvé-Legg-Perthes' disease).

GOERIG, M. *et al.* Georg Perthes (1869–1927) a pioneer of local anaesthesia by blocking peripheral nerves with the use of electrical stimulation. Second International Symposium on the History of Anaesthesia, London, 1987. London, International Congress & Symposium Series, No. 134, 1988, pp. 551–8.

PERUTZ, Max Ferdinand (1914–2002). Austrian/British molecular biologist; shared Nobel Prize (Chemistry) 1962 for work on globular proteins.

EISENBERG, D. Max Perutz's achievements: how did he do it? *Protein Science*, 1994, **3**, 1625–8.

PESCHIER, Jacques (1769–1832). Swiss pharmacist and chemist.

REICHENBACH, K.R. Jacques Peschier (1779–1832): ein Genfer Apotheker und Chemiker: sein Lebensweg und sein besonderer Berucksichtigung bisher unveroffentlicher Dokumente. Stuttgart, Wissenschaftliche Verlagsgesellschaft, 2001. 498 pp.

PETER of Abano, *see* **PIETRO D'ABANO**

PETERS, John Punnett (1887–1955). American physician and biochemist; professor of internal medicine, Yale University School of Medicine.

PAUL, J.R. & LONG, C.N. H. BMNAS, 1958, **31**, 347–76.

PETERS, Rudolph Albert (1889–1982). British biochemist.

PETERS, R. Forty-five years of biochemistry. *Annual Review of Biochemistry*, 1957, **26**, 1–16.
THOMPSON, R.H.S. & OGSTON, A.G. BMFRS, 1983, **29**, 495–523.
Archival material: Bodleian Library, Oxford; CMAC.

PETIT, Jean-Louis (1674–1750). French surgeon; published important work on bone diseases; gave first acount of osteomalacia.

LAUGIER, –. Éloge de J.-L. Petit. *Union Médicale*, 1865, 2 sér. **28**, 228–39.

PETREN, Karl (1866–1927). Swedish physician; professor of internal medicine, Uppsala and Lund.

BIBLIOGRAPHY OF MEDICAL AND BIOMEDICAL BIOGRAPHY

GROTE, 1924, **3**, 165–99.

PETRI, Richard (1852–1921). German bacteriologist; introduced Petri dish.

RICHARD PETRI, or how to immortalize one's name in a laboratory. *Medical Sciences Historical Society Newsletter*, 1999, No. 21.

PETRIE, George Ford (1874–1955). British bacteriologist and immunologist.

McCLEAN, D. George Ford Petrie 10th March 1874 – 17th March 1955. *Journal of Pathology and Bacteriology*, 1955, **70**, 559–62.

PETROV, Nikolai Nikolaevich (1876–1964). Russian oncologist.

MEL'NIKOV, R.A. [N.N.Petrov – founder of the experimental study of carcinogenesis in primates.] [In Russian.] *Voprosy Onkologii*, 1976, **22** (11), 34–40.

PETRUS HISPANUS, Pope John XXI (*c*.1210–1277). Spanish physician: wrote *Thesaurus pauperum*, a popular medieval formulary (published 1485).

SCHIPPERGES, H. Arzt im Purpur. Grundzüge eine Krankheitslehre bei Petrus Hispanus (*ca*.1210–1277) Berlin, Springer-Verlag, 1994. 133 pp.

PETTENKOFER, Max Josef von (1818–1901). German experimental hygienist and epidemiologist.

BREYER, H. Max von Pettenkofer: Arzt im Vorfeld der Krankheit. Leipzig, Hirzel, 1980. 234 pp.
DOLMAN, C.E. DSB, 1974, **10**, 556–63.
HUME, E.E. Max von Pettenkofer. New York, Hoeber, 1927, 142 pp.
WEININGER, K. Max von Pettenkofer: Das Leben eines Wohltäters. München, Hugendubel, 1987. 217 pp.

PETTY, William (1623–1687). Statistician; professor of anatomy, Oxford.

DALE, P.G. Sir W.P. of Romsey. Romsey, Hants: LTVAS Group (3 Linden Rd, Romsey) for the author, 1987. 60 pp.
KEYNES, G. A bibliography of Sir William Petty FRS, Oxford, Clarendon Press, 1971. 103 pp.
STRAUSS, E. Sir William Petty: portrait of a genius. London, Bodley Head, 1954. 260 pp.

PEYER, Johann Conrad (1653–1712). Swiss physiologist.

HUARD, P. & IMBAULT-HUART, M.-J. DSB, 1974, **10**, 567–8.
LANG, R. Johann Conrad Peyer. *Janus*, 1914, **19**, 61–83.

PEZARD, Albert (1875–1927). French endocrinologist.

MONNIER, A.M. DSB, 1974, **10**, 569–71.

PFAFF, Philipp (1716–1780). German dentist; published first important German dental manual, in which he described taking impressions and casting moulds for false teeth.

WITT, D. Beiträge zur Leben und Werk von Philipp Pfaff. Berlin, Institut für Geschichte der Medizin der Freien Universität, 1969. 86 pp.

PFAFFMAN, Carl (1913–1994). American neurologist.

RIGGS, L.A. BMNAS, 1997, **71**, 263–79.

PFANNENSTIEL, Hermann Johannes (1862–1909). German gynaecologist; gave first detailed description of familial icterus gravis neonatorum.

CAESAR, T. Hermann Johannes Pfannenstiel (1862–1929): Vertreter der Geburtshilfe und Frauenheilkunde. Giessen, W. Schmitz, 1994. 124 pp.

PFEIFFER, Emil (1846–1921). German physician; accredited with first description of infectious mononucleosis.

LEHNDORFF, W. *Archives of Pediatrics*, 1946, **63**, 218–23.

PFEIFFER, Richard Friedrich Johannes (1858–1945). Polish bacteriologist, professor in Berlin; discoverer of bacteriolysis.

FILDES, P. BMFRS, 1956, **2**, 237–47.

PFLÜGER, Eduard Friedrich Wilhelm (1829–1910). German physiologist.

NUSSBAUM, M. E.F.W. Pflüger als Naturforscher. Bonn, M.Hager, 1909. 40 pp.
ROTHSCHUH, K.E. DSB, 1974, **10**, 578–81.

PHAER, Thomas (1510–1560). English paediatrician; compiled first work on diseases of children to be written by an Englishman. 1545.

BOWERS, R. Thomas Phaer and the Boke of Chyldren (1544). Tempe. Arizona Center for Medieval and Renaissance Studies, 1999. 100 pp.
RUHRÄH, J. Thomas Phaer. *Annals of Medical History*, 1919, **2**, 334–47.

PHILIP, Robert William (1857–1939). British geneticist, tuberculologist in Edinburgh.

SIR ROBERT W. PHILIP, 1857–1939. Memories of his friends and pupils one hundred years after his birth. London, Edinburgh, NAPT, 1957. 96 pp.

PHILLIPS, Charles Garrett (1916–1994). British neurophysiologist.

PORTER, R. BMFRS, 1996, **42**, 340–62.

PHILLIPS, David Chilton (1924–1999). British molecular physicist.

JOHNSON, L.N. BMFRS, 2000, **46**, 377–401.

PHYSICK, Philip Syng (1768–1837). American surgeon, the 'Father of American surgery'; studied under John Hunter; professor at University of Pennsylvania.

CRELLIN, J.K. (ed.) The surgical teaching of Philip Syng Physick. Durham, NC, Duke Univ. Medical Center, 1981. 101 pp.
MIDDLETON, W.S. Philip Syng Physick, father of American surgery. *Annals of Medical History*, 1929, N.S. **1**, 562–82.
RANDOLPH, J. A memoir on the life and character of Philip Syng Physick. Philadelphia, Collins, 1839. 114 pp.

PIAGET, Jean (1896–1980). Swiss psychologist; specialist in child psychology; professor of general psychology, Lausanne.

BODEN, M.A. Jean Piaget. New York, Viking, 1979. 176 pp.

CATALOGUE of the Jean Piaget archives, University of Geneva. Boston, G.K. Hall, 1975.
MODGIL, S. & MODGIL, C. (eds.) Jean Piaget. Consensus and controversy. New York, Praeger, 1982. 446 pp.

PICK, Arnold (1851–1924). Czech neurologist.

PICK, A. Aphasia. Translated...by J.W. Brown. Springfield, Ill., Thomas [c.1973]. 148 pp. [Includes biographical sketch by F. Stieglmayr.]

PICK, Ludwig (1868–1935). German pathologist; associated with 'Niemann–Pick disease'.

GRUBER, G.G. In memoriam Ludwig Pick (31.8.1868 bis 3.2.1944). *Verhandlungen der Deutschen Gesellschaft für Pathologie*, 1968, **52**, 574–80.

PICKERING, George White (1904–1980). British physician; Regius Professor of Medicine, Oxford; authority on hypertension.

McMICHAEL, J. & PEART, W.S. BMFRS, 1982, **28**, 431–49.
Archival material: CMAC.

PICKLES, William Norman (1885–1969). British general practitioner and epidemiologist.

PEMBERTON, J. Will Pickles of Wensleydale. 2nd ed. London, Geoffrey Bles, 1972, Reprinted, London, Royal College of General Practitioners, 1984. 224 pp.

PIETRO D'ABANO (1250–1316). Italian physician.

FERRARI, S. Per la biografia e per gli scritti di Pietro d'Abano. *Memorie della R.Accademia dei Lincei, Cl. di Scienze Morale*, 1918, Ser.5., **15**, 629–727.
NORPOTH, L. Zur Bio-, Bibliographie und Wissenschaftslehre des Pietro d'Abano, Mediziners, Philosophen und Astronomen in Padua. *Kyklos*, 1930, **3**, 292–353.
PASCHETTO, E. Pietro d'Abano, medico e filosofo. Firenze, Nuovedizione, E. Vallechi, 1984. 358 pp.
PERMUDA, L. DSB, 1970, **1**, 4–5.

PILCHER, Lewis Stephen (1845–1934). American surgeon and medical editor.

PILCHER, L.S. A surgical pilgrim's progress: reminiscences of Lewis Stephen Pilcher. Philadelphia, Lippincott, 1925. 451 pp.

PILLSBURY, Walter Bowers (1872–1960). American experimental psychologist; professor at University of Michigan.

MILES, W.R. BMNAS, 1964, **37**, 267–91.

PINCUS, Gregory Goodwin (1903–1967). American endocrinologist; developed contraceptive pill.

CHANG, M.C. DSB, 1974, **10**, 610–11.
INGLE, D.J. *et al.* Gregory Pincus, 1903–1967.*Perspectives in Biology and Medicine*, 1968, **11**, 337–521. [Special issue in honour of Pincus.]
INGLE, D.J. BMNAS, 1971, **42**, 229–70.
Archival material: US Library of Congress.

PINEL, Philippe (1745–1826). French physician, notable for humane treatment of the insane.

CHABBERT, P., DSB, 1974, **10**, 611–14.

BIBLIOGRAPHY OF MEDICAL AND BIOMEDICAL BIOGRAPHY

GARRABÉ, J. Philippe Pinel. Sous la direction de Jean Garrabé. La Plessis-Robinson, Synthelabo, 1994. 156 pp.

HACKLER, B. Philippe Pinel; unchainer of the insane. New York, Watts, 1968. 118 pp.

SÉMELAIGNE, R. Alienistes et philanthropes: les Pinel et les Tuke. Paris, Steinheil, 1912, 548 pp.

SEMELAIGNE, R. Philippe Pinel et son oeuvre point de vue de la médecine mentale. New York, Arno Press, 1975. 173 pp. First published in Paris, 1888.

PINKUS, Felix (1868–1947). German dermatologist; first to describe lichen nitidus.

MEHREGAN, A.M. Felix Pinkus, M.D. (1868–1947). *Journal of the American Academy of Dermatology*, 1988, **18**, 1158–64.

PIORRY, Pierre Adolphe (1794–1879). French physician.

SAKULA, A. Pierre Adolphe Piorry (1794–1879): pioneer of percussion and pleximetry. *Thorax*, 1979, **34**, 575–81.

PIROGOV, Nikolai Ivanovich (1810–1881). Russian military surgeon.

MIKULINSKY, S.R. DSB, 1974, **10**, 619–21.

PETROVSKOGO, B.V. Pirogov. Moskva, Zk. Viksm., 1965. 301 pp. [In Russian.]

PIRQUET VON CESENATICO, Clemens von (1874–1929). Austrian paediatrician; professor of pediatrics at Johns Hopkins University, Baltimore, and at universities of Breslau and Vienna.

WAGNER, R. Clemens von Pirquet: his life and work. Baltimore, Johns Hopkins Press, 1968. 214 pp.

PISO, Willem (*c*.1611–1678). Dutch physician, worked in Dutch Settlement in Brazil (1636–44).

PIES, E. Willem Piso (1611–1678). Begründer der kolonialen Medizin und Leibarzt des Grafen Johann Moritz von Nassau-Singen in Brasilien: eine Biographie. Düsseldorf, Interna-Orb Verlagsgruppe, 1981. 227 pp.

VAN DEN PAS, P. W. DSB. 1974, **10**, 621–2.

PITCAIRN, David (1749–1809). British physician; physician to St Bartholomew's Hospital.

MACMICHAEL, W. The gold-headed cane. (CB).

PITCAIRN, William (1711–1791). British physician and botanist; president of the Royal College of Physicians of London, 1775–85.

MACMICHAEL, W. The gold-headed cane (CB).

PITCAIRNE, Archibald (1652–1713). Scottish physician; originator of the Edinburgh Medical School.

BROWN, T.M. DSB, 1975, **11**, 1–3.

JOLLY, L. Archibald Pitcairne. *Edinburgh Medical Journal*, 1953, **60**, 39–51.

Archival material: British Library.

PITRES, Jean Albert (1843–1928). French physician; published important accounts of agraphia and paraphrasia.

FLEURY, M. de. *Revue de Médecine*, 1928, **44**, 1275–82.

PITT-RIVERS, Rosalind Venetia (1907–1990). British endocrinologist.

TATA, J.R. BMFRS, 1994, **39**, 326–48.

PITTS, Robert Franklin (1908–1977). American renal physiologist.

BERLINER, R.W. & GIEBISCH, G.H. BMNAS, 1987, **57**, 322–44.

PLÁCIDO DA COSTA, Antonio (1840–1916). Portuguese ophthalmologist; introduced keratoscope.

LEVENE, J.R. The true inventors of the keratoscope and photo-keratoscope. *British Journal of the History of Science*, 1965, **2**, 324–42.

PLATER, Felix, *see* **PLATTER, Felix**

PLATT, Robert (1900–1978). British physician; professor of medicine, Manchester University; President, Royal College of Physicians of London.

PLATT, R. Private and controversial. London, Cassell, 1972. 180 pp.

PLATTER, Felix (1536–1614). Swiss physician; city physician, Basle; provided first modern classification of mental diseases.

PILET, P.E. DSB, 1975, **11**, 33.
PLATTER, F. Beloved son Felix. The journal of Felix Platter, a medical student in Montpellier in the sixteenth century. Translated and introduced by S. Jennett. London, Muller, 1961. 157 pp.
TAGEBUCH (Lebensbeschreibung) 1536–1567)...herausgegeben von V. Lötscher. Basel, Schwabe & Co. Verlag, 1976. 579 pp.
TRÖHLER, U. (ed.) Felix Platter (1536–1614) in seiner Zeit. Basel, Schwabe, 1991. 86 pp.

PLAUT, Hugo Carl (1858–1928). German physician; gave first comprehensive description of 'Plaut's angina', 1894.

LIEBESCHUETZ, E.A.R. Hugo Carl Plaut. 2 vols. [in 1]. Balham, London, Printique, 1989. 212 pp.

PLENČIČ [Plenciz], Marcus Antonius (1705–1786). Austrian physician; early proponent of the germ theory.

KRUTA, V. DSB, 1975, **11**, 37–8.

PLENCK, Joseph Jacob von (1738–1807). German physician.

HOLUBAR, K. & FRANKL, J. Joseph Plenck A forerunner of modern dermatology. *Journal of the American Academy of Dermatology*, 1984, **10**, 326–32.

PLOTZ, Harry (1890–1947). American bacteriologist; Head of Laboratory at Institut Pasteur 1921–1939.

–. Harry Plotz. *Annales de l'Institut Pasteur*, 1947, **73**, 584–6.

PLOUCQUET, Wilhelm Gottfried (1744–1814). German physician; compiled first important classified bibliography of medical literature (1793–97).

DUISBERG, R.A. Der Gerichtsmediziner Wilhelm Gottfried Ploucquet (1744–1814). Dissertation. Tübingen, 1974. 170 pp.

PLUMMER, Henry Stanley (1874–1936). American physician; professor of medicine at Mayo Foundation for Medical Education and Research.

WILLIUS, F.A. Henry Stanley Plummer: a diversified genius. Springfield, Ill., Thomas, 1960. 71 pp.

POISEULLE, Jean Leonard Marie (1797–1869). French physiologist.

PEDERSEN, K.M. DSB, 1975, **11**, 62–4.
Archival material: Centre des Recherches Alexandre Koyré, Paris.

POL, Nicolaus (*c*.1470–1532). Austrian physician and theologian who collected a remarkable library.

FISCH, M. Nicolaus Pol doctor, 1494. With a critical text of his Guaiac tract. Edited with a translation by D.M. Schullian. New York, Reichner, for Cleveland Medical Library Association, 1947. 246 pp. [Includes list of books known to have been collected by Pol.]

POLLENDER, Franz Aloys Anton (1800–1879). German physician, discovered anthrax bacillus.

SCHADEWALDT, H. DSB, 1975, **11**, 68–71.

POLLITZER, Sigmund (1859–1937). American dermatologist.

–. *Archives of Dermatology*, 1938, **37**, 499–503.

POLLOCK, Martin Rivers (1914–1999). British molecular biologist.

AMBLER, R.P. & MURRAY, R. BMFRS, 2002, **48**, 359–73.
Archival material: Edinburgh University; University of Bradford Department of Peace Studies.

PÓLYA, Eugen [Jenö] Alexander (1876–1944?). Hungarian surgeon; modified Billroth II operation for pyloric cancer.

PETRI, G. Das chirurgische Erbe. Tragische Chirurgenschicksale. Eugen Alexander Pólya. *Zentralblatt für Chirurgie*, 1985, **110**, 46–52.

POMMER, Christoph Friedrich von (1787–1841). Swiss physician.

ROMINGER, S. Christoph Friedrich von Pommer 1787–1841, Internist in Zürich. Zürich, Juris Verlag, 1992. 96 pp.

PONCE DE LEON, Pedro (1510–1584). Spanish monk; instituted method of instruction of deaf-mutes.

PÉREZ DE URBEL, J. Fray Pedro Ponce de Leon y el origen del arte de enseñar a habler a los mudos. Madrid, Editorial Obras Selectas, 1973. 295 pp.

PONCET, Antonin (1849–1913). French physician.

FISCHER, L. Antonin Poncet. Conférences d'Histoire de la Médecine, 1996. Lyon, Fondation Marcel Mérieux, 1996, pp. 105–38.

PONFICK, Emil (1844–1913). German pathologist; contributed to knowledge on human actinomycosis.

EMIL PONFICK. *Verhandlungen der Deutschen Pathologischen Gesellschaft*, 1914, **17**, 598–600.

PONTECORVO, Guido (1907–1999). Italian/British geneticist.

SIDDIQI, O. BMFRS, 2002, **48**, 377–90.

POPJÁK, George Joseph (1914–1998). Hungarian/British/American biochemist.

AKHTAR, M. BMFRS, 2000, **46**, 403–24.

POPPER, Hans (1903–1988). Austrian/American hepatologist; professor at Columbia and Mount Sinai School of Medicine.

BERK, P.D. *et al*. Hans Popper: a tribute. New York, Raven Press, 1992. 179 pp.
SCHMIDT, R. BMNAS, 1994, **65**, 290–309.
THALER, H. & SHERLOCK, S. Hans Popper (1903–1988): Leben und Werk. Freiburg i.Br., Falk Foundation, 1997. 63 pp.

PORGES, Otto (1879–1968). Austrian physician; introduced gastrophotography.

NOVAK, J. Zur Erinnerung an Professor Dr Otto Porges. *Wiener Klinische Wochenschrift*, 1968, **80**, 559–60.

PORRO, Edoardo (1842–1902). Italian gynaecologist; introduced 'Porro's operation'.

CAFFARATTO, T.M. Ricordo di Edoardo Porro nel centenarion dell sua operazioni. *Minerva Ginecologica*, 1976, **28**, 1033–40.

PORTA, Giambattista della (1535–1615). Italian scientist; initiated the study of physiognomy.

MURARO VAIANI, L. Giambattista della Porta: mago e scienziato. Milani, Feltrinelli, 1978. 221 pp.
RIENSTA, M.H. DSB, 1975, **11**, 95–8.

PORTA, Luigi (1800–1875). Italian surgeon.

ZANOBIO, B. DSB, 1975, **11**, 98–9.

PORTAL, Antoine Portier Paul (1866–1962). French physiologist; described anaphylaxis.

MONNIER, A.M. DSB, 1975, **11**, 101–2.

PORTEN, Ernst von der (1884–1940). German anaesthesiologist.

TSCHOP, M. Ernst von der Porten 1884–1940 in der Geschichte der Anästhesiologie. Berlin, Springer, 1996. 96 pp.

PORTER, Rodney Robert (1917–1985). British biochemist and immunologist; shared Nobel Prize (Physiology or Medicine) 1972.

PERRY, S.V. BMFRS, 1987, **33**, 445–89.
SPECIAL ISSUE in memory of Rodney Porter. *Bioscience Reports*, 1985, **5**, 809–1014.
Archival material: Bodleian Library, Oxford.

PORTIER, Paul Jules (1866–1962). French physiologist; with C.R. Richet published first full description of anaphylaxis.

MAY, C.D. The ancestry of allergy; being an account of the original induction of hypersensitivity recognizing the contribution of Paul Portier. *Journal of Allergy and Clinical Immunology*, 1985, **75**, 485–95.

POSADAS, Alejandro (1870–1902). Argentinian surgeon.

CRANWELL, D.J. Nuestros grandes cirujanos. Buenos Aires, El Ateneo, 1939, pp. 128–48.

POSNER, Carl (1854–1928). German urologist.

GROTE, 1928, **7**, 151–84. (CB).

POST, Wright (1766–1828). American surgeon; professor at Columbia (College of Physicians and Surgeons).

MOTT, V. A biographical memoir of Wright Post M.D. New York, E. Conrad, 32 pp.

POTAIN, Pierre Carl Édouard (1825–1901). French physician.

TEISSIER, T.J. Pierre Carl Potain. Paris, [1902?]. 52 pp.

POTT, Percivall (1714–1788). British surgeon; surgeon to St Bartholomew's Hospital, London.

DOBSON, J. Percivall Pott. *Annals of the Royal College of Surgeons of England*, 1972, **50**, 54–65.
EARLE, J. Life. *In*: Pott's *Chirurgical works*, London, 1790, vol. 1, i–xlv.
Archival material: Royal College of Surgeons of England; Royal College of Physicians of London.

POTTER, Jared (1742–1810). American physician.

THOMS, H. The doctors Jared of Connecticut. Hamden, Conn., Shoe String Press, 1958. 76 pp. [Deals with Jared Potter, Jared Potter Kirtland, and Jared Eliot.]

POTTS, Jonathan (1745–1781). American physician.

BLANCO, R.L. Physician of the American Revolution: Jonathan Potts. New York, London, Garland STPM Press, 1979. 276 pp.

POUPART, François (1661–1708). French surgeon; described the inguinal ('Poupart's') ligament.

DELAUNAY, P. François Poupart. *France Médicale*, Paris, 1904, **51**, 357–61.

POURFOUR DU PETIT, François (1664–1741). French physiologist and surgeon.

BEST, A.E. DSB, 1975, **11**, 111–13.
ZEHNDER, E. François Pourfour du Petit (1664–1741) und sein experimentelle Forschung über das Nervensystems. Zürich, Juris, 1968. 39 pp.

POWELL, Thomas Philip Stroud (1923–1996). British neuroanatomist.

GUILLERY, R.W. BMFRS, 1997, **43**, 411–27.

POWER, D'Arcy (1855–1941). British surgeon; medical historian, biographer and bibliographer; surgeon to St Bartholomew's Hospital, London.

CHARLTON, C.A.C. Sir D'Arcy Power: surgeon and historian. *Journal of Medical Biography*, 1994, **2**, 137–145.
LE FANU, W.R. Sir D'Arcy Power. PLARR, 1930–1951, 638–45 (CB).

POWER, Henry (1829–1911). British surgeon; ophthalmic surgeon to St Bartholomew's Hospital.

POWER, H. A brief sketch of my life: a fragment. Stratford-upon-Avon, privately printed, 1912. 95 pp.

POYNTON, Frederick John (1869–1943). British rheumatologist.

HOLT, L.W.J. The memoirs of F. John Poynton. Dissertation, Wellcome Institute for the History of Medicine, 1993. London, 185 leaves.

PRATT, Frederick Haven (1873–1958). American physiologist.

ROBINSON, G. DSB, 1975, **11**, 125–6.

PRAUSNITZ(-GILES), Otto Carl Willy (1876–1963). German/British immunologist; devised Prausnitz-Kuestner hypersensitivity test.

DOWNIE, A.W. Carl Prausnitz (Giles). *Journal of Pathology and Bacteriology*, 1966, **92**, 241–52.
Archival material: CMAC.

PRAVAZ, Charles Gabriel (1791–1853). French physician; introduced a hypodermic syringe and invented the modern galvanocautery.

CHAMPEAU, D. Un novateur: Charles Gabriel Pravaz, 1791–1853. Paris, Thèse, 1931. 52 pp.
MONNIER–KUHN, A. Un novateur d'orthopédie au XIXe siècle Charles-Gabriel Pravaz. *Revue Lyonnaise de Médecine*, 1966, **15**, 937–44.
RHEIMS, A. Pravaz et la découverte de la méthode parénterale d'administration des médicaments. Lyon, Bose et Rion, 1944. 48 pp.

PRAXAGORAS OF COS (b.c.340 BC). Greek anatomist and physiologist.

LONGRIGG, J. DSB, 1975, **11**, 127–8.
STECKERL, F. The fragments of Praxagoras of Cos and his school. Leiden, Brill, 1958. 132 pp.

PRESTON, Ann (1813–1872). American physician; early female medical graduate in USA.

FOSTER, P.P. Ann Preston M.D. (1813–1872): a biography. The struggle to obtain training and acceptance for women physicians in mid-nineteenth century America. Ph.D. dissertation, University of Pennsylvania, Philadelphia, 1984. 470 pp.

PRÉVOST, Jean-Louis (1790–1850). Swiss physiologist and embryologist.

BRUNNER, V. Der Genfer Arzt Jean-Louis Prévost, 1790–1850, und sein Beitrag zur Entwicklungsgeschichte und Physiologie. Zürich, Juris, 1966. 34 pp.
PILET, P.E. DSB, 1975, **11**, 132–3.

PREYER, Thierry William (1841–1897). British physiologist and psychologist; professor of physiology at Jena.

GEUS, A. DSB, 1975, **11**, 135–6.

PRICHARD, James Cowles (1786–1848). British physician and anthropologist.

AUGSTEIN, H.F. James Cowles Prichard's anthropology: remaking the science of man in early nineteenth-century Britain. Amsterdam, Rodopi, 1999. 294 pp.
ODOM, H.M. DSB, 1975, **11**, 136–8.
SYMONDS, J.A. Some account of the life, writings, and character of the late James Cowles Prichard. Bristol, Evans & Abbott, 1849. 54 pp.

PRIESSNITZ, Vincenz (1799–1851). German hydrotherapist.

HELD-RITT, E. von. Priessnitz und Gräfenberg, oder treue Darstellung seiner Heilverfahrens mit kalten Wasser. Würzburg, Bergstädt-Verlag Wilhelm Gottlieb Korn, 1988.
METCALFE, R. Life of Vincent Priessnitz, founder of hydrotherapy. London, Simpkin, Marshall, Hamilton, Kent & Co., 1898. 210 pp.
SELINGER, J.E.M. Vincenz Priessnitz: einer Lebenbeschreibung. Wien, Gerold & Sohn, 1852. 208 pp.

PRIESTLEY, John Gillies (1879–1941). British physiologist.

FRENCH, R.D. DSB, 1975, **11**, 138–9.

PRIESTLEY, Joseph (1733–1804). British chemist and clergyman; isolated oxygen.

GIBBS, F.W. Joseph Priestley: adventurer in science and champion of truth. London, Nelson, 1965. 258 pp.
SCHOFIELD, R.E. DSB, 1975, **11**, 139–47.

PRIESTLEY, William Overend (1829–1900). British gynaecologist.

–. *Medico-chirurgical Transactions*, 1901, **84**, 102–5.
–. *Transactions of the Obstetrical Society of London*, 1902, **43**, 50–59.

PRINGLE, John (1707–1782). British physician; founder of modern military medicine.

RADBILL, S.X. DSB, 1975, **11**, 147–8.
SELWYN, S. Sir John Pringle: hospital reformer, moral philosopher and pioneer of antiseptics. *Medical History*, 1966, **10**, 266–74.
SINGER, D.W. Sir John Pringle and his circle. *Annals of Science*, 1949–50, **6**, 127–80. 229–61.
Archival material: Royal College of Physicians of Edinburgh.

PROCHÁSKA, Georg (1749–1820). Czech physician; professor of anatomy and physiology at Prague.

KRUTA, V. DSB, 1975, **11**, 158–60.

PROETZ, Arthur Walter (1888–1966). American otolaryngologist; introduced displacement method of treatment for nasal sinusitis.

HILL, F.T. Arthur Walter Proetz, M.D. *Annals of Otology, Rhinology and Laryngology*, 1966, **75**, 605–15.
SCHENK, A.P. Arthur Walter Proetz: an appreciation. *Annals of Otology, Rhinology and Laryngology*, 1966, **75**, 612–21.

PROFETA, Guiseppe (1840–1911). Italian venereologist; ('Profeta's law' – a non-syphilitic child of syphilitic parents is immune).

SPRECHER, –. *Bollettino della Real Academia di Medicina di Genova*, 1911, **36**, 52–8.

PROUT, William (1785–1850). British physician and biochemist.

BROCK, W.H. The life and work of William Prout. *Medical History*, 1965, **9**, 101–26.
BROCK, W.H. DSB, 1975, **11**, 172–4.
BROCK. W.H. From protyle to proton: William Prout and the nature of matter, 1875–1985. Bristol, H. Hilger, 1985. 252 pp.
Archival material: Royal Society of London; Royal College of Physicians of London; Wellcome Institute for the History of Medicine.

PROWAZEK, Stanislaus (1875–1915). Bohemian protozoologist.

KRUTA, V. DSB, 1975, **11**, 174–5.

PRUDDEN, Theophil Mitchell (1849–1924). American pathologist and bacteriologist.

BIOGRAPHICAL SKETCHES and letters of T. Mitchell Prudden. New Haven, Yale Univ. Press, 1927. 311 pp.
HEKTOEN, L. BMNAS, 1929, **12**, 73–98.
ROBINSON, G. DSB., 1975, **11**, 175–7.
Archival material: Yale University Library.

PRUNER-BEY, Franz Ignace (1808–1882). German physician and anthropologist; his important public health work in Egypt gained him the title 'Bey'.

SCHÄFER, A. Leben und Wirken des Arztes Franz Pruner-Bey. Leiden, E.J. Brill, 1931. 98 pp. (Also *Janus*, 1931–32, **35–36**.).
WROTNOVSKA, D. DSB, 1975, **11**, 177–9.

PRUSINER, Stanley B. (b. 1942). American physician; awarded Nobel Prize (Physiology or Medicine) 1997, for discovery of prions – a new biological principle of infection.

Les Prix Nobel en 1997.

PUNNETT, Reginald Crundall (1875–1967). British geneticist and morphologist.

CREW, F.A.E. BMFRS, 1967, **13**, 323–6.
CREW, F.A.E. DSB, 1975, **11**, 211–12.

PUPIN, Michael Idvorsky (1858–1935). Bohemian/American physicist; introduced intensifying screen into radiology.

PUPIN, M.I. From immigrant to inventor. New York, Scribners, 1923. 396 pp. German translation 1929.

PURKYNE, Jan Evangelista [PURKINJE] (1787–1869). Bohemian physiologist, histologist and embryologist; professor of physiology, Prague.

JAN EVANGELISTA PURKYNE, 1787–1869. Prague, State Medical Publishing House, 1962. 143 pp.
JOHN, H. J. Evangelista Purkyne: Czech student and patriot. Philadelphia, American Philosophical Society, 1959. 94 pp.

KRUTA, V. J.E. Purkyne (1787–1869), physiologist: a short account of his contributions to the progress of physiology. With a bibliography of his works. Prague, Academia, 1969. 137 pp.

KRUTA, V. DSB, 1975, **11**, 213–17.

WADE, N. & BROZECK, J. Purkynje's vision: the dawning of neuroscience. Marwah, N.J., Lawrence Erlbaum Associates, 2001. 159 pp.

PURTSCHER, Otto (1852–1927). Austrian ophthalmologist; first described traumatic angiopathy of retina ('Purtscher's disease') 1912.

KRAMER, R.R. *Klinische Monatsblätter für Augenheilkunde*, 1928, **80**, 99–101.

PURVES-STEWART, James (1869–1949). British neurologist; consultant physician to Westminster Hospital, London.

PURVES-STEWART, J. Sands of time; recollections of a physician in peace and war. London, Hutchinson, [1940]. 356 pp.

PUSCHMANN, Theodor (1844–1899). German physician and medical historian.

NEUBERGER, M. *Wiener Medizinische Presse*, 1899, **40**, 1688–91.

PAGEL, J.A. *Janus*, 1899, **4**, 567–9.

PUSEY, William Allen (1865–1940). American dermatologist and historian of dermatology.

WILLIAM ALLEN PUSEY. An appreciation by friends and co-workers. Special issue of *Archives of Dermatology and Syphilology*, 1937, **35**, No.1. 71 pp.

PUTNAM, James Jackson (1846–1918). American neurologist.

VASILE, R.G. James Jackson Putnam: from neurology to psychoanalysis. Oceanside, NY, Dabor Science, 1977. 116 pp.

WILKERSON, S.W. Mind over body. Durham, NC, [*c*.1978]. 377 leaves. Ann Arbor, Mich., University Microfilms International, 1979.

PUTTI, Vittorio (1880–1940). Italian orthopaedic surgeon; developed and improved kineplastic surgery.

BONOLA, A. Il contributo di Vittorio Putti alla storia della medicina. *Chirurgia degli Organi di Movimento*, 1952, **38**, 8–43.

GUI, L. & PANTALEONI, M. La raccolta Vittorio Putti di quadri disegni incisioni ritratti di medici e uomino illustrati. Bologna, Gaggi, 1966. 217 pp.

PYBUS, Frederick Charles (1882–1975). British surgeon.

EMMERSON, J. F.C. Pybus: the man and his books. In: Newcastle School of Medicine 1834–1984; sesquicentennial celebrations. Newcastle upon Tyne, Faculty of Medicine, University of Newcastle upon Tyne, 1985, pp. 41–7.

PYLARINO, Giacomo (1659–1710). Italian physician; first to practise variolation for smallpox.

ALVISATOS, C.N. The first immunologist, James Pylarino (1659–1718), and the introduction of variolation. *Proceedings of the Royal Society of Medicine*, 1934, **27**, 25–30.

PYMAN, Frank Lee (1882–1944). British chemist; Director of Research, Boots Pure Drug Co., Nottingham.

KING, H. ONFRS, 1942–44, **4**, 681–97.

QUASTEL, Juda Hirsch (1899–1987). British biochemist; professor at McGill University, Montreal.

MACINTOSH, F.C. & SOURKES, T.L. BMFRS, 1990, **36**, 379–418.

QUERCETANUS, Josephus, *see* **DUCHESNE, Joseph**

QUERVAIN, Johann Friedrich de (1868–1940). Swiss surgeon.

TROEHLER, U. Der Schweizer Chirurg J.F. de Quervain (1868–1940). Aarau, Sauerländer, 1973. 137 pp.

QUETELET, Lambert-Adolphe-Jacques (1796–1874). Belgian astronomer and statistician; founder of biostatistics; he devised a technique for the application of statistics to the data of biology and the social sciences.

FREUDENTHAL, H. DSB, 1975, **11**, 236–8.
HANKINS, F.H. Quetelet as statistician. New York, London, P.S. King 1908. 134 pp.

QUFF, ibn al- (1233–1286). Arabian physician and surgeon.

HAMARNEH, S.K. The physician, therapist and surgeon, Ibn al-Quff (1233–1286). An introductory survey of his time, life and works. Cairo, Atlas Press, 1974. 199 pp.

QUICK, Armand James (1894–1977). American haematologist; devised tests for liver function and prothrombin clotting time.

EBEL, E. The Quick test: the life and work of Dr Armand J. Quick, Blacksburg, VA, Pocahontas Press, 1996. 454 pp.

QUIN, Henry (1718–1791). Irish physician.

KIRKPATRICK, T.P.C. Henry Quin, M.D., President and Fellow of the King and Queen's College of Physicians in Ireland, and King's Professor of the Practice of Physick, 1718–1791. Dublin, University Press, 1919. 66 pp.

QUINCKE, Heinrich Irenaeus (1842–1922). German neurologist; introduced lumbar puncture independently of W.E. Wynter, 1891; reported aneurysm of hepatic artery.

BETHE, H. Heinrich Quincke, 1842–1922; sein Leben und Werk unter Berücksichtigung der Kieler Fakultätsgeschichte. Neumünster, Wacholtz, 1968. 139 pp.
FREDERIKS, J.A.M. & KOEHLER, P.J. The first lumbar puncture. *Journal of the History of the Neurosciences*, 1997, **6**, 147–53.

QUINCY, John (d.1722). English physician and medical lexicographer.

HOWARD-JONES, N. John Quincy, M.D. (died 1722), apothecary and iatrophysical writer: a study of his works, etc. *Journal of the History of Medicine.* 1951, **6**, 149–75.

QUITTENBAUM, Carl Friedrich (1793–1852). German surgeon; established splenectomy as a surgical procedure.

WISCHHUSEN, H.G. Anatomie und Chirurgie an der Universität Rostock in der ersten Hälfte des XIX. Jahrhunderts. NTM. *Zeitschrift für Geschichte der Naturwissenschaften, Technik und Medizin*, 1969, 6, 87–101, 1970, **7**, 76–84.

BIBLIOGRAPHY OF MEDICAL AND BIOMEDICAL BIOGRAPHY

RABANUS MAURUS (*c.* 776–856). Archbishop of Mainz; his *De sermonum proprietate*, 1467?, is the earliest known printed book to include a section on medicine.

PAXTON, F.S. Curing bodies – curing souls. Hrabanus Maurus, medical education, and the clergy in the ninth century. *Journal of the History of Medicine*, 1995, **50**, 230–52.

RABELAIS, François (*c.*1495–1553). French physician, satirist and priest.

ANTONIOLI, R. Rabelais et la médecine. Genève, Librairie Droz, 1976. 394 pp.
HENRY, G. Rabelais. Paris, Perrin, 1988. 311 pp.
HUARD, P. & IMBAULT-HUART, M.-J. DSB, 1975, **11**, 253–4.
POWYS, J.C. Rabelais, his life: the story told by him. Selections here newly translated, and an interpretation of his genius and his religion. London, Bodley Head, 1948. 424 pp.
SCREECH, M.J. Rabelais. London, Duckworth, 1979. 494 pp.

RABL FAMILY. Includes **Carl** (1787–1850), surgeon and obstetrician; **Carl** (1819–1889), physician; **Carl** (1853–1917), anatomist, embryologist and cytologist.

RABL, R. Die oberösterreichische Arztefamilie Rabl. 1620–1970. Wels, Verlag Welsmühl, 1971. 118 pp.

RABL, Carl (1853–1917). Austrian anatomist, embryologist and cytologist.

ROBINSON, G. DSB, 1975, **11**, 254–6.

RACE, Robert Russell (1907–1984). British haematologist.

CLARKE, C. BMFRS, 1985, **31**, 455–92.

RACKER, Ephraim (1913–1991). Austrian/American biochemist.

SCHATZ, G. BMNAS, 1996, **90**, 321–46.

RADCLIFFE, John (1652–1714). British physician to William III, Queen Mary, and Queen Anne; still acts as a benefactor through the Radcliffe Fund.

HONE, C.R. The life of Dr John Radcliffe, 1652–1714, benefactor of the University of Oxford. London, Faber, 1950. 149 pp.
MACMICHAEL, W. The gold-headed cane. (CB).
NIAS, J.B. Dr John Radcliffe: a sketch of his life with an account of his fellows and foundations. Oxford, Clarendon Press, 1918.

RADEMACHER, Johann Gottfried (1772–1850). German physician; introduced system of treatment similar to homoeopathy.

KRACK, N. Dr Gottfried Rademacher: sein Leben, sein Lehre, seine Heilmittel und wir. Heidelberg, Haug, 1984. 181 pp.

RAE, John (1813–1893). British physician and Arctic explorer.

RICHARDS, R.I. Dr John Rae. Whitby, Caedmon, 1985. 231 pp.

RAFFEL, Sidney (b.1911). American immunologist; professor at Stanford University.

RAFFEL, S. Fifty years of immunology. *Annual Review of Microbiology*, 1982, **36**, 1–26.

RAHN, Hermann (1912–1990). German/American physiologist.

307

PAPPENHEIMER, J. BMNAS, 1996, **69**, 243–67.

RAISTRICK, Harold (1890–1971). British biochemist.

BIRKINSHAW, J.H. BMFRS, 1972, **18**, 489–509.

RAJCHMAN, Ludwik (1881–1965). Polish bacteriologist; Director of League of Nations Health Office; launched United Nations Children's Emergency Fund.

BALÍNSKI, M. Rajchman, medical statesman. Budapest, Central European Univ.Press, 1998. 293 pp. First published in French, 1995.

RAMAZZINI, Bernardino (1633–1714). Italian physician; published first systematic work on occupational diseases, 1700.

KOELSCH, F. Bernardino Ramazzini, der Vater der Geweberhygiene (1633–1714), sein Leben und seine Werke. Stuttgart, F. Enke, 1912. 35 pp.
PIETRO, P. di. Bibliografia di Bernardino Ramazzini. Roma, Istituto Italiano di Medicina Sociale, 1977. 142 pp.
RAMAZZINI, B. Diseases of tradesmen ... together with biographical notes, translated from the French of Claude Mayer (1928). New York, Medical Lay Press, 1933. 95 pp.

RAMMELKAMP, Charles H. Jr (1911–1981). American physician.

ROBBINS, F.C. BMNAS, 1994, **64**, 354–66.

RAMON, Gaston Léon (1886–1963). French immunologist.

DELAUNAY, A. DSB, 1975, **11**, 271–2.
WELSCH, M. Notice sur la vie et l'oeuvre de Gaston Ramon. *Bulletin de l'Académie de Médecine de Belgique*, 1963, Ser.7, **3**, 740–58.

RAMÓN Y CAJAL, Santiago (1852–1934). Spanish histologist and neurologist; awarded Nobel Prize in 1906.

ALBARRACIN, A. Santiago Ramón y Cajal o la pasion de España. Madrid, Labor, 1982. 319 pp.
CANNON, D.F. Explorer of the human brain: the life of Santiago Ramón y Cajal (1852–1934). With a memoir of Dr Cajal by Sir Charles Sherrington. New York, Schuman, 1949. 303 pp.
CRAIGIE, E.H. & GIBSON, W.C. The world of Ramón y Cajal. With a selection from his non-scientific writings. Springfield, Ill., C.C. Thomas, 1968. 295 pp.
LOPEZ-PIÑERO, J.M. *et al.* Bibliografia cajaliana: ediciones de los escritos de Santiago Ramón y Cajal y estudios sobre su vida y su obra. Valenca, Albatros, 2000. 377 pp.
MUÑOS, G. & BURON, F.A. Ramón y Cajal. Tomo 1. Vida e obra. Zaragoza, Institución "Fernando el Catolico", 1960. 544 pp.
RAMÓN Y CAJAL, S. Recollections of my life. New ed. Caambridge MA, M.I.T. Press, 1989. 638 pp. First published in Spanish, 2 vols., Madrid, 1901–17.
SHERRINGTON, C.S. ONFRS, 1935, **1**, 425–41.
TAYLOR, D.W. DSB, 1975, **11**, 273–6.

RAMSTEDT, Wilhelm Conrad (1867–1963). German surgeon; devised operation for relief of hypertrophic pyloric stenosis.

BORGWARDT, G. Conrad Ramstedt – an appreciation. *Zeitschrift für Kinderchirurgie*, 1986, **41**, 195–200.

RANDALL, John Turton (1905–1984). British biophysicist and molecular biologist.

WILKINS, M.H.F. BMFRS, 1987, **33**, 493–535.

RANK, Otto [born ROSENFELD] (1884–1939). Austrian psychoanalyst; worked in USA.

LIEBERMANN, E.J. Acts of will. The life and work of Otto Rank. New York, Free Press, 1985. 485 pp.

RANVIER, Louis-Antoine (1835–1922). French histologist.

APPEL, T.A. DSB, 1975, **11**, 295–7.
JOLLY, J. Louis Ranvier (1835–1922): notice biographique. *Archives d'Anatomie Microscopique*,1923, **9**, i–lxxii.

RAPER, Henry Stanley (1882–1951). British physiologist; professor, University of Manchester.

HARTLEY, P. ONFRS, 1952–53, **8**, 567–82.

RAPER, Kenneth Bryan (1908–1987). American microbiologist and botanist; professor of bacteriology, Wisconsin.

BURRIS, R.H. & NEWCOMB, E.H. BMNAS, 1991, **60**, 250–65.

RASPAIL, François-Vincent (1794–1878). French biologist; pioneer cytochemist and cytopathologist.

KLEIN, M. DSB, 1975, **11**, 300–302.
TOTIRIGATELLI, L. François-Vincent Raspail: una vita per la medicina e la rivoluzione. Padova, Isonomia. 1997. 161 pp.
WEINER, D.B. Raspail, scientist and reformer. New York, Columbia Univ. Press, 1968. 336 pp.

RATHKE, Martin Heinrich (1793–1860). German embryologist.

BULLOUGH, V.L. DSB, 1975, **11**, 307–8.
MENZE, H. Martin Heinrich Rathke (1793–1860): ein Embryologe des 19. Jahrhunderts. Marburg, Basilisk-en-Presse, 2000. 280 pp.

RATNER, Sarah (b.1903). American biochemist; head of department at Public Health Research Institute, New York.

RATNER, S. A long view of nitrogen metabolism. *Annual Review of Biochemistry*, 1977, **46**, 1–24.

RAUWOLF, Leonhard (*c*.1537–1596). German physician and botanist.

DANNENFELDT, K.H. Leonhard Rauwolf, sixteenth-century physician, botanist and traveller. Cambridge, Mass., Harvard Univ. Press, 1968. 321 pp.
DANNENFELDT, K.H. DSB, 1975, **11**, 311–12.

RAVITSCH, Mark Mitchell (1910–1989). American surgeon; professor at University of Chicago.

RAVITSCH, M. Second thoughts of a surgical curmudgeon. Chicago, Year Book, 1987. 240 pp.

RAY, Isaac (1807–1881). American physician; wrote first treatise on medico-legal aspects of insanity (1838).

HUGHES, J.S. In the law's darkness: Isaac Ray and the medical jurisprudence of insanity in nineteenth-century America. New York, Oceana Publications, 1986. 206 pp.

RAYER, Pierre François Olive (1793–1867). French physician; authority on dermatology and kidney diseases.

CAVERIBERT, R. La vie et l'oeuvre de Rayer (1793–1867). Paris, M. Vigné, 1931. 61 pp.
THÉODORIDES, J. Un demi-siècle de Pierre Rayer (1793–1867) médecine française. Paris, Editiones Louis Pariente, 1997. 266 pp.

RAYNAUD, Maurice (1834–1881). French physician; described 'Raynaud' disease' (1862).

MULLOOLY, J. P. *et al.* Maurice Raynaud (1834–1881) : a commemorative tribute. *Linacre Quarterly*, 1981, **48**, 115–92.

AL-RAZI, ABU BAKR MUHAMMAD IBN ZAKARIYYA, *see* **RHAZES**

READ, Grantly Dick, *see* **DICK-READ, Grantly**

RÉCAMIER, Joseph Claude Anthelm (1774–1852). French surgeon and gynaecologist.

SAUVÉ, L. Le docteur Récamier, 1774–1852; sa famille, ses amies. Paris, Spes, 1938. 253 pp.
TRIAIRE, P. Récamier et ses contemporains. Paris, J.B. Baillière, 1899. 471 pp.

RECKLINGHAUSEN, Friedrich Daniel von (1833–1910). German pathologist; described neurofibromatosis, 'Recklinghausen's disease'.

CHIARI, H. Friedrich Daniel von Recklinghausen. *Verhandlungen der Deutschen Pathologischen Gesellschaft*, 1912, **15**, 478–88.

RECLUS, Paul (1847–1914). French surgeon; gave classic description of chronic cystic mastitis.

–. *Bulletin de l'Académie de Médecine*, Paris, 1914, 3 sér, **72**, 113–16.
LEJARS, F. *Bulletin et Mémoires de la Société de Chirurgie de Paris*, 1919, **45**, 148–62.

REDI, Francesco (1626–1697). Italian toxicologist and parasitologist.

BELLONI, L. DSB, 1975, **11**, 341–2.
BERNARDI, I. & GUERRINI, L. Francesco Redi, un protagonista della scienza moderna. Documenti, esperimenti, immagine. Firenze, Leo SD Olschki, 1999. 388 pp.
COLE, R. Francesco Redi (1626–97), physician, naturalist, poet. *Annals of Medical History*, 1926, **8**, 347–59.
VIVIANI, U. Vita, opera, iconografia, bibliografia...di Francesco Redi. 3 pts in 2 vols. Arezzo, U. Viviani, 1924–31.

REED, Walter (1851–1902). US Army physician; proved mosquito transmission of yellow fever.

BEAN, W.B. DSB, 1975, **11**, 345–7.
BEAN, W.B. Walter Reed: a biography. Charlottesville, Univ. Press of Virginia, 1982. 190 pp.
KELLY, H.A. Walter Reed and yellow fever. 3rd ed., Baltimore, Norman Remington Co. 1923. 355 pp.

WOOD, L. Walter Reed, doctor in uniform, New York, Messner, 1943. 277 pp.

REES, George Owen (1813–1889). British physician.

COLEY, N.G. George Owen Rees, MD, FRS (1813–89): pioneer of medical chemistry. *Medical History*, 1986, **30**, 173–90.

REFSUM, Sigvald (1907–1991). Norwegian psychiatrist.

COHEN, M.M. & VIDAVER, D. The man behind the syndrome: Sigvald Refsum. *Journal of the History of the Neurosciences*, 1992, **1**, 277–84.

REGAUD, Claudius (1870–1940). French radiologist.

REGAUD, J. Claudius Regaud (1870–1940) pionnier de la cancérologie, créateur de la Fondation Curie: chronique de sa vie et son oeuvre. Paris, Maloine, 1982. 233 pp.

REHN, Ludwig Mettler (1849–1909). German surgeon; performed first successful thyroidectomy for exophthalmic goitre (1880) and first successful heart suture (1896).

ABSOLON, K. & NAFICY, N.A. First successful cardiac operation on a human, 1896. A documentation: the life and times and the work of Ludwig Rehn (1849–1909). Rockville, MD, 2002. 246 pp.
SCHMIEDEN, D. Ludwig Rehn, geboren 13.IV.1849, Gestorben 29.V.1980. *Deutsche Medizinische Wochenschrift*, 1930, **56**, 1185–87.

REICH, Ludwig (1849–1930). German surgeon; performed first thyroidectomy for exophthalmic goitre, 1880; successfully sutured heart, 1896.

GROTE, 1924, **3**, 201–44. (CB).

REICH, Wilhelm (1897–1957). Austrian psychoanalyst; worked in Denmark, Norway and USA; advanced theory of 'orgone energy'.

CATTIER, M. The life and work of Wilhelm Reich. Translated from the French by G. Boulanger. New York, Horizon Press, 1971. 224 pp. First published in French, 1969.
MANN, W.E. & HOFFMANN, E. Wilhelm Reich, the man who dreamed of tomorrow. The life and thought of Wilhelm Reich. Wellingborough, Crucible, 1990. 295 pp.
SHARAF, M. Fury on earth. A biography of Wilhelm Reich. London, Andrew Deutsch. 1983. 550 pp.

REICHERT, Karl Bogislaus (1811–1883). German embryologist and histologist.

KRUTA, V. DSB, 1975, **11**, 360–61.

REICHSTEIN, Tadeus (1897–1996). Polish biochemist; shared Nobel Prize (Physiology or Medicine) 1950 for isolation of cortisone; with co-workers synthesized vitamin C.

BOETTCHER, C. Das Vitaminbuch: die Geschichte der Vitaminforschung. Köln, Kiepenheur & Witsch, 1865, pp. 215–17.
ROTHSCHILD. M. BMFRS, 1999, **45**, 451–67.

REID, Edward Waymouth (1862–1948). British physiologist; professor at University College, Dundee, for 46 years.

CATHCART, E.P. & GARRY, R.C. ONFRS, 1948–9, **6**, 213–18.

REID, James (1849–1923). British physician; Physician-in-Ordinary to Queen Victoria, Edward VII and George V.

REID, M. Ask Sir James. London, Hodder & Stoughton, 1987. 315 pp.

REIL, Johann Christian (1759–1813). German physician and physiologist.

KAISER, E. & VÖLCKER, A. (eds.) Johann Christian Reil (1759–1813) und seine Zeit. Halle, Universität, 1989. 242 pp.
RISSE, G.B. DSB, 1975, **11**, 363–5.
ZAUNICK, R. *et al.* Johann Christian Reil, 1759–1813. Leipzig, J.A. Barth, 1960. 159 pp.

REITER, Hans (1881–1969). German bacteriologist and public hygienist; described 'Reiter's disease'.

FREIHERR, D. Hans Reiter (1881–1969) Hygiene, Sozialhygiene und Eugenik. Mainz, Johannes Gutenberg-Universität, Mainz. Thesis, 1992.

REMAK, Robert (1815–1865). German histologist, embryologist and neurologist.

HINTZSCHE, E. DSB, 1975, **11**, 367–70.
SCHMIEDEBACH, H.P. Robert Remkak (1815–1865); ein jüdischer Arzt im Spannungsfeld von Wissenschaft und politik. Stuttgart, Gustav Fischer, 1995. 374 pp.

REMMELIN, Johann (b.1583). German anatomist; compiled the *Catoptron microcosmicum* (1613), one of the earliest anatomical atlases with superimposed plates.

RUSSELL, K.F. A bibliography of Johann Remmelin the anatomist. St Kilda, Victoria, Russell, 1991. 124 pp.

RENAUDOT, Théophraste (1586–1633). French physician; introduced pawnshops, intelligence offices and newspapers in France.

GREEN, E. The worlds of Dr Renaudot. London, Minerva Press, 1997. 288 pp.
SOLOMON, H.M. Public welfare, science, and propaganda in seventeenth-century France. The innovations of Théophraste Renaudot. Princeton, N.J., Princeton Univ. Press, 1972. 290 pp.

RENAUT, Joseph-Louis (1844–1917). French anatomist and histologist.

WEINER, D.B. DSB, 1975, **11**, 373–4.

RENDU, Henri Jules Louis (1844–1902). French physician.

–. *Bulletin Médicale*, Paris, 1902, **16**, 1107–13.
–. *Bulletin et Mémoires de la Sociéte des Hôpitaux, Paris*, 1902, 3 sér., **19**, 1118–24.
SERRATRICE, G. Henri Rendu: a leading French clinician of the nineteenth century. *Journal of Medical Biography*, 1993, **1**, 199–201.

RENUCCI, Simon François (*fl.*1834–47). Rediscovered the itch-mite, *Sarcoptes scabiei.*

FRIEDMAN, R. The story of scabies. Written for the centenary of Renucci's rediscovery of the *Acarus scabiei. Medical Life*, 1934, **41**, 381–424, 426–76; 1935, **42**, 218–68, 551–64.

RENZI, Salvatore de (1800–1872). Italian physician and medical historian.

GARZA, A. Lettere e scritti vari de Salvatore de Renzi. Napoli, Giaminini, 1999. 76 pp.

RETZIUS, Anders Adolf (1796–1860). Swedish comparative anatomist, histologist and anthropologist.

KRUTA, V. DSB, 1975, **11**, 379–81.
LARSELL, O. Anders Adolf Retzius. *Annals of Medical History*, 1924, **6**, 16–24.

RETZIUS, Magnus Gustav (1842–1919). Swedish anatomist, histologist and anthropologist.

RUDOLPH, G. DSB, 1975, **11**, 381–3.

REVERDIN, Jaques Louis (1842–1929). Swiss surgeon; among first to operate on thyroid gland.

REVERDIN, H. Jaques Louis Reverdin 1842–1929, un chirurgien à l'aube d'une ère nouvelle. Aarau, Sauerländer, 1971. 226 pp.
SAUDAN, G. Jaques-Louis Reverdin (1842–1929) and his cousin Auguste (1848–1908) of Geneva; or how surgical clinical pracrice prevailed over experimental physiology. *Journal of Medical Biography*, 1993, **1**, 144–50, 207–14.

REYBARD, Jean François (1795–1863). French surgeon.

SABIONCELLO, B. Reybard, 1795–1863: sa vie et son oeuvre. Paris, Thèse, 1930. 74 pp.

REYNOLDS, James Henry (1844–1932). British physician; surgeon, Royal Army Medical Corps.

STEVENSON, L. The Rorke's Drift doctor: James Henry Reynolds, V.C. and the defence of Rorke's Drift, 22nd–23rd January, 1879. Brighton, Lee Stevenson, 2001. 284 pp.

RHAZES (854–925). Persian physician and philosopher.

FIROUZABADI, H. Bibliographie der medizinischen Werke Rhazes, bu Bakar Muhammad Ibn Zakaryya. Düsseldorf, Medizinischen Fakultät der Universität Düsseldorf, 1969. 42 pp.
PINES, S. DSB, 1975, **11**, 323–6.
RANKING, G.S.A. The life and work of Rhazes. *Proceedings of XVII International Congress of Medicine 1913*, London, 1914, Sect.**23**, pp. 237–68.

RHINE, Joseph Banks (1895–1980). American biologist and parapsychologist.

BRIAN, D. The enchanted voyager; the life of J.B. Rhine. Englewood Cliffs, Prentice-Hall Inc., 1976. 367 pp.

RHOADS, Cornelius Packard (1898–1959). American pathologist; introduced triethylene melamine in treatment of Hodgkin's disease.

–. C.P. Rhoads. *British Medical Journal*, 1959, **2**, 309–10.

RHOADS, Jonathan Evans (1907–2002). American surgeon.

ROMBEAU, J.L. Jonathan E. Rhoads, M.D.: Quaker sense and sensibility in the world of surgery. Philadelphia, Hanley & Belfus, 1997. 308 pp.

RIBBERT, Moritz Wilhelm Hugo (1855–1920). German pathologist; modern protagonist of the theory of the embryonal origin of cancer.

METTLER, M. Der Pathologe Hugo Ribbert (1855–1920). Zürich, Juris, Druck, 1991. 144 pp.

RIBEIRO SANCHES [SANCHEZ], Antonio Nuñes (1699–1783). Portuguese physician; worked in Russia as public health official; an authority on venereal disease.

WILLEMSE, D. Antoine Nuñes Ribeiro – et son importance pour la Russie. Leiden, Brill, 1966. 188 pp.

RICH, Arnold Rice (1893–1968). American pathologist and immunologist; professor at Johns Hopkins University.

OPPENHEIMER, E.H. BMNAS, 1979, **50**, 331–50.

RICHARDS, Alfred Newton (1876–1966). American physiologist and pharmacologist.

SCHMIDT, C.F. BMFRS, 1967, **13**, 327–42.
SCHMIDT, C.F. BMNAS, 1971, **42**, 271–318.
STARR, I. ed. Dr Alfred Newton Richards, scientist and man. *Annals of Internal Medicine*, 1969, Suppl. 8. 89 pp.
SWANN, J.P. DSB, 1990, **18**, 734–6.
Archival material: University of Pennsylvania.

RICHARDS, Dickinson Woodruff (1895–1973). American physician; professor of medicine, College of Physicians and Surgeons, New York; shared Nobel Prize 1956 for investigations on cardiac physiology.

COURNAND, A. BMNAS, 1989, **58**, 459–87.

RICHARDSON, Benjamin Ward (1828–1896). British physician and pharmacologist; medical reformer.

BYNUM, W.F. DSB, 1975, **11**, 418–19.
MacNALTY, A.S. A biography of Benjamin Ward Richardson. London, Harvey & Blythe, 1950. 92 pp.
RICHARDSON, B.W. Disciples of Aesculapius. With a life of the author by his daughter, Mrs George Martin London, Hutchison & Co., 1900. 827 pp.
RICHARDSON, B.W. Vita medica. Chapters of medical life and work. London, Longmans, Green, 1897. 495 pp.
Archival material: Royal College of Physicians of London; University College London; Library, Wellcome Institute, London.

RICHARDSON, Edward Henderson (b. 1877). American gynaecologist; taught at Johns Hopkins University School of Medicine.

RICHARDSON, E.H. A doctor remembers. New York, Vantage Press, 1959. 252 pp.

RICHARDSON, John (1787–1865). British surgeon, Arctic explorer and naturalist.

JOHNSON, R.E. Sir John Richardson, Arctic explorer, natural historian, naval surgeon. London, Taylor & Francis, 1976. 209 pp.

RICHET, Charles Robert (1850–1935). French physiologist; described anaphylaxis; Nobel Prize 1913.

HOLMES, F.L. DSB, 1975, **11**, 425–32.
JURI, M. Charles Richet, physiologiste (1850–1935). Zürich, Juris, 1965. 51 pp.
RICHET, C. Souvenirs d'un physiologiste. Paris, J.Peyronnet, 1933. 156 pp.

WOLF, S. Brain, mind and medicine: Charles Richet and the origins of physiological psychology. New Brunswick, Transaction Publishers, 1993.

RICHTER, August Gottlieb (1742–1812). German surgeon; lecturer at Göttingen.

ROHLFS, H. Geschichte der deutschen Medizin. Leipzig, 1883, **3**, 33–172.

RICHTER, Derek (b.1907). British psychiatrist; director MRC Neuropsychiatry Unit.

RICHTER, D. Life in research. Kingswood, Surrey, Stuart Phillips, 1989. 170 pp.

RICKETTS, Howard Taylor (1871–1910). American pathologist; contributed to the knowledge of rickettsial infections.

MULLEN, P.C. DSB, 1975, **11**, 442–3.
Archival material: Rocky Mountain Laboratory (National Institute of Allergy and Infectious Diseases), Hamilton, Montana; Harper Memorial Library, University of Chicago.

RICORD, Philippe (1800–1889). French venereologist.

EGINER, C.A.E. Philippe Ricord: 10 decembre 1800–20 octobre 1889: sa vie, son oeuvre. Paris, Librairie Le François, 1939. 122 pp.
ORIEL, J.D. Eminent venereologists. 3. Philippe Ricord. *Genitourinary Medicine*, 1989, **65**, 388–93.

RIDDLE, Oscar (1877–1968). American physiologist and endocrinologist.

MAIENSCHEIN, J. DSB, 1990, **18**, 736–8.

RIDLON, John (1852–1936). American orthopaedic surgeon.

ORR, H.W. On the contributions of Hugh Owen Thomas of Liverpool, Sir Robert Jones of Liverpool and London, John Ridlon, M.D., of New York and Chicago, to modern orthopedic surgery. With a supplement on Ridlon and his share in moulding orthopedic surgery by A. Steindler. Springfield, C.C. Thomas, 1949. 253 pp.

RIGBY, Edward (1747–1821). British physician; first to differentiate between the two main causes of antepartum haemorrhage.

BASKETT, T.F. Edward Rigby (1747–1821) of Norwich and his Essay on the uterine haemorrhage. *Journal of the Royal Society of Medicine* , 2002, **95**, 618–22.
CROS, J.A. A memoir of the life of Edward Rigby, MD. Norwich, Burks & Kinebrook, 1821.

RIGGS, John M. (1810–1885). American dentist.

EPSTEIN, H.D. John Riggs and his disease [pyorrhoea alveolaris]. *Bulletin of the History of Dentistry*. 1969, **17**, 1–6.

RILLIET, Frédéric (1814–1861). Swiss paediatrician.

DUVAL, A.J. Le docteur Rilliet, sa vie et ses oeuvres. Genève, Rambos & Schuchart, 1861. 72 pp.

RIMINGTON, Claude (1902–1993). British chemical pathologist.

NEUBERGER, A. & GOLDBERG, A. BMFRS, 1996, **42**, 363–78.

BIBLIOGRAPHY OF MEDICAL AND BIOMEDICAL BIOGRAPHY

RINGER, Sydney (1835–1910). British physician and physiologist.

BYNUM, W.F. DSB, 1975, **11**, 462–3.
MOORE, B. In memory of Sidney Ringer. Some account of the fundamental discoveries of the great pioneer of the bio-chemistry of crystallo-colloids in living cells. *Biochemical Journal*, 1910–11, **5**, i–xix.

RINGLEB, Otto (1875–1946). German urologist; professor in Berlin.

KLUG, M.H. Otto Ringleb (17.5.1875–8.11.1946). Biobibliographie eines Urologen. [Thesis.] Berlin, Frei Universität, 1983, 205 pp.

RIO-HORTEGA, Pio del (1882–1945). Argentinian neurohistologist.

CANO DIAZ, P. Una contribución histológica: la obra de Don Pio del Rio-Hortega. Madrid, Consejo Superior de Investigaciones Científicas, 1985. 219 pp.
GLICK, T.F. DSB, 1975, **11**, 465–6.
RIERA, J. Pío del Río-Hortega y la ciencia de su tiempo. Valladolid, Univ. Vallalodid, 1994. 107 pp.

RIOLAN, Jean, Jr (1580–1657). French anatomist and physician.

BYLEBYL, J.J. DSB, 1975, **11**, 466–8.
MANI, N. Jean Riolan II and medical research. *Bulletin of the History of Medicine*, 1968, **42**, 121–44.

RITTENBERG, David (1906–1990). American biochemist.

SHEMIN, D. & BENTLEY, R. BMNAS, 2001, **80**, 257–74.

RIVA-ROCCI, Scipione (1863–1937). Italian physician; introduced mercury sphygmomanometer.

BELLONI, L. DSB, 1975, **11**, 481–2.
PONTICACCIA, L. La vita e l'opera di Scipione Riva-Rocci. *Giornale di Clinica Medica*, 1959, **40**, 631–47.

RIVERS, Thomas Milton (1888–1962). American virologist; member of Rockefeller Institute.

HORSFALL, F.L. BMNAS, 1965, **38**, 263–94.
RIVERS, T.M. Tom Rivers: reflections on a life in medicine and science. An oral history memoir prepared by Saul Benison. Cambridge, Mass., M.I.T. Press, 1967. 682 pp.
Archival material: Library, American Philosophical Society, Philadelphia.

RIVERS, William Halse Rivers (1864–1922). British neurologist and anthropologist.

SLOBODIN, R. W.H.R. Rivers; pioneer anthropologist, psychiatrist of the Ghost Road. Stroud, Sutton Publishing, 1997, 299 pp. First published 1978.

RIVIÈRE, Lazare (1589–1655). French physician; first to note aortic stenosis.

DULIEU, L. Lazare Rivière. *Revue d'Histoire de la Pharmacie*, 1966, **18**, 205–11.

ROBB, George Douglas (1899–1974). New Zealand surgeon; president BMA, 1961–62.

ROBB, G.D. Medical odyssey. Auckland, London, Collins, 1967. 201 pp.

ROBBINS, Frederick Chapman (b. 1916). American microbiologist; shared Nobel Prize (Physiology or Medicine), 1954, for work on the poliomyelitis virus leading to vaccine production.

316

Les Prix Nobel en 1954.

ROBERTS, John Alexander Fraser (1899–1987). British medical genticist.

POLANI, P. BMFRS, 1992, **38**, 307–22.

ROBERTS, Richard Brooke (1910–1980). American biophysicist.

BRITTEN, R.J. BMNAS, 1993, **62**, 326–48.

ROBERTS, Richard John (b. 1943). British molecular biologist; shared Nobel Prize (Physiology or medicine) 1993, for discovery of split genes.

Les Prix Nobel en 1993.

ROBERTSON, Douglas Argyll (1837–1909). British ophthalmic surgeon; described 'Argyll Robertson pupil'.

BENNETT, D.C. Argyll Robertson: a breadth of vision. *Journal of Medical Biography*, 1993, **1**, 186–8.
HÖRNSTEN, G. The man behind the syndrome: Argyll Robertson, Scotland's ambassador extraordinary in the realm of ophthalmology. *Journal of the History of the Neurosciences*, 1994, **3**, 61–5.
RAVIN, J.G. Argyll Robertson: 'twas better to be his pupil than have his pupil'. *Ophthalmology*, 1998, **105**, 867–70.

ROBERTSON, Muriel (1883–1973). British microbiologist and immunologist.

BISHOP, A. & MILES, A. BMFRS, 1974, **20**, 317–47.

ROBERTSON, Oswald Hope (1886–1966). British/American microbiologist and haematologist.

COGGESHALL, L.T. BMNAS, 1971, **42**, 319–38.
Archival material: Library, American Philosophical Society, Philadelphia.

ROBIN, Charles-Philippe (1821–1885). French biologist and histologist.

GENTY, V. Un grand biologiste, Charles Robin (1821–1885); sa vie, ses amités philosophiques et littéraires. Lyon, A. Ray, 1931. 141 pp.
GRMEK, M.D. DSB, 1975, **11**, 491–2.

ROBINSON, James (1813–1862). British dentist; first to administer anaesthesia in Britain (19 Dec.1846); wrote first textbook on surgical anaesthesia, 1847.

ELLIS, R.H. The life of James Robinson, Britain's first anaesthetist. *Dental History*, 1985, Oct. (10), 18–20.

ROBINSON, Robert (1886–1975). British chemist; Wayneflete professor, Oxford; Nobel Prize (Chemistry) 1947.

ROBINSON, R. Memories of a minor prophet; seventy years of organic chemistry. Amsterdam, Elsevier, 1976. 252 pp.
TODD, Lord, & CORNFORTH, J.W. BMFRS, 1976, **22**, 415–527.
WILLIAMS, T.I. Robert Robinson: chemist extraordinary. London, Oxford University Press, 1990. 201 pp.
Archival material: Royal Society of London.

BIBLIOGRAPHY OF MEDICAL AND BIOMEDICAL BIOGRAPHY

ROBIQUET, Pierre Jean (1780–1840). French chemist; isolated codeine, 1832.

BUSSY, A. *Journal de Pharmacie*, 1840, **26**, 443–52.

ROBISON, Robert (1883–1941). British biochemist; professor at Lister Institute.

HARINGTON, C.R. ONFRS, 1940–41, **3**, 929–39.

ROCHA E SILVA, Mauricio Oscar da (1910–1987). Brazilian pharmacologist; discovered bradykinin.

BERALDO, W.T. Mauricio Rocha e Silva and the discovery of bradykinin. *Agents and Actions*, Supplements, 1992, **36**, 134–9.
HAWGOOD, B.J. Mauricio Rocha e Silva, M.D.: snake venom, bradykinin and the rise of autopharmacology. *Toxicon*, 1997, **35**, 1569–1580.
HENRIQUES, S.B. The scientific thinking of Professor Mauricio Rocha e Silva. *Agents and Actions*, Supplements, 1992, **36**, 7–13.
MOUSSATCHE, H. Mauricio: a friendship at all times. *Agents and Actions*, Supplements, 1992, **36**, 14–17.
ROTHSCHILD, A.M. Mauricio Rocha e Silva: man and work. *Agents and Actions*, Supplements, 1992, **36**, 1–6.

ROCHA-LIMA, Henrique da (1879–1956). Brazilian physician: isolated *Rickettsia prowazecki*.

CERQUEIRA FALCÃO, E. de. A vida cientifica de Henrique Da Rocha-Lima. *Revista Brasileira de Malariologia*, 1967, **19**, 353–8.

ROCK, John (1890–1984). American gynaecologist; performed first in vitro fertilization of human ovum; worked on contraceptive pill.

McLAUGHLIN, L. The pill, John Rock and the church: the biography of a revolution. Boston, Little Brown, *c.*1982. 243 pp.

ROCKEFELLER, John Davison (1839–1937). American businessman; made large gifts for the promotion of medical and scientific research and public health.

HARR, J.E. & JOHNSON, P.J. The Rockefeller century. New York, Scribner, 1988. 621 pp.
HAWKE, D.F. John D. The founding father of the Rockefellers. New York, Harper & Row, 1980. 260 pp.

ROCKEFELLER, John Davison Jr. (1874–1960). American industrialist and philanthropist; continued the work of his father in the promotion of medical research.

CORNER, G.W. BMFRS, 1961, **6**, 247–57.

RODBELL, Martin (b. 1925). American biochemist; shared Nobel Prize (Physiology or Medicine), 1994, for discovery of G-proteins and their role in signal transduction in cells.

Les Prix Nobel en 1994.

RODDICK, Thomas George (1846–1923). Canadian surgeon; dean of the medical faculty, McGill University.

MACDERMOT, H.E. Sir Thomas Roddick: his work in medicine and public life. Toronto, Macmillan, 1938. 160 pp.

318

RODGERS, John Kearny (1793–1851). American surgeon; successfully wired ununited fracture of humerus, 1827.

DELAFIELD, F. Biographical sketch of J.Kearny Rodgers. (Read before the New York Academy of Medcine). New York, G.A.C Van Beuren, 1851. 28 pp.
HOSACK, A.F. A history of the case of the late J. Kearny Rodgers. New York, S.W. Benedict, 1851. 47 pp.

RODRIGUES, João, *see* AMATUS LUSITANUS

ROEDERER, Johann Georg (1726–1763). German physician and obstetrician; professsor at Göttingen; published exhaustive study of typhoid, 1762.

RAMSAUER, L. Johann Georg Roederer, der Begründer der wissenschaftlichen Geburtshilfe. Göttingen, Vanderhoeck & Ruprecht, 1936. 58 pp.

ROEHL, Wilhelm (1881–1929). German physician; introduced pamaquine in treatment of malaria.

HORLWEIN, H. *Therapeutische Berichte*, 1929, **6**, 259–63.

ROELANTS, Cornelis (d.1525). Flemish clinician; early writer on paediatrics.

SUDHOFF, K. Nochmals Dr Cornelis Roelants von Mechelen, etc. *Janus*, 1915, **20**, 443–58.

RÖNTGEN, Wilhelm Conrad (1845–1923). German physicist, discoverer of X rays; first Nobel prizewinner for Physics, 1901.

GLASSER, O. Dr W.C. Röntgen. 2nd ed., Springfield, C.C.Thomas, 1958. 169 pp. [Includes translation of 'Ueber eine neue Art von Strahlen'.] Reprinted 1972. First published 1945.
GLASSER, O. Wilhelm Conrad Röntgen and the early history of the Roentgen rays ... With a chapter: Personal reminiscences of W.C. Röntgen by M. Boveri. London, Bale & Danielsson, 1933. 494 pp. Reprinted, San Francisco, Norman Publishing, 1991. 494 pp.
LEICHT, H. Wilhelm Conrad Röntgen: Biographie. München, Ehrenwurth, 1994. 193 pp.
NITSKE, W.R. The life of Wilhelm Conrad Röntgen, discoverer of the X ray. Tucson, Univ. Arizona Press, 1971. 375 pp.
TURNER, G. L'E. DSB, 1975, **11**, 529–31.
Archival material: Deutsches Röntgen Museum, Remscheid-Lennep, Germany.

RÖSSLIN, Eucharius (d.1526). German physician; wrote earliest textbook for midwives (1513).

BOAS, K. Eucharius Rösslins Lebensgang. *Archiv für Geschichte der Medizin*, 1908, **1**, 429–41.
POWER, D'A. The birth of mankind ... A bibliographical study. *Library*, 1927, 4 ser., **8**, 1–37; subsequently published in book form.

ROGERS, Carl (1902–1997). American psychotherapist.

COHEN, D. Carl Rogers: a critical biography. London, Constable, 1997. 253 pp.

ROGERS, Leonard (1868–1962). British physician; pioneer in study of leprosy and tropical diseases; founder of British Empire Leprosy Relief Association and Calcutta School of Tropical Medicine.

BOYD, J.S.K. BMFRS, 1963, **9**, 261–85.

POWER, H.J. Sir Leonard Rogers FRS (1869–1962): tropical medicine in the Indian Medical Service. Thesis submitted to the University of London for the degree of Doctor of Philosophy, 1993. 307 pp.

ROGERS, L. Happy toil: fifty-five years of tropical medicine. London, Muller, 1950. 271 pp.

Archival material: CMAC.

ROGET, Peter Mark (1779–1869). Swiss/British physician; professor of physiology, Royal Institution, London; compiled a thesaurus of synonyms and antonyms.

EMBLEN, D.L. Peter Mark Roget: the word and the man. London, Longman, 1970. 368 pp.

SHELL, W.E. Peter Mark Roget, F.R.C.P. (1799–1869). *Journal of the Royal College of Physicians of London*, 1974, **8**, 276–82.

ROGOFF, Jules Moses (1884–1966). Latvian/American endocrinologist; used adrenal cortical extract in treatment of Addison's disease.

FISHER, B.M. Julius M. Rogoff. *Pharmacologist*, 1967, **9**, 23–4.

ROKITANSKY, Carl (1804–1878). Bohemian pathologist.

LESKY, E. (ed.). Carl von Rokitansky. Selbstbiographie und Antrittsrede. Wien, H. Bohlaus Nachf., 1960. 111 pp. (*S.B. Öst. Akad. Wiss.*, 1960, **234**, 3 Abhandlung.)

MICIOTTO, R.J. Carl Rokitansky: nineteenth-century pathologist and leader of the new Vienna school. Ph.D. Dissertation, Johns Hopkins University, 1979. 301 leaves. Dissertation Abstracts International, 1979, **40**, 2847-A. University Microfilms Order No. 79-24638.

NEUBERGER, M. *Medical Life*, 1934, **41**, 542–56.

ROLANDO, Luigi (1773–1831). Italian anatomist and physiologist; professor of anatomy at Turin.

CASTELLANI, C. DSB, 1975, **11**, 510–11.

ROLLET, Joseph-Pierre-Martin (1824–1894). French venereologist.

WEINER, D.B. DSB, 1975, **11**, 515–16.

ROLLIER, Auguste (1874–1954). Swedish physician; introduced heliotherapy in treatment of tuberculosis.

HELLER, G. Leysin et son passé médical, *Gesnerus*, 1990, **47**, 329–44.

ROMANES, George John (1848–1894). British physiologist, comparative psychologist and evolutionist.

LESCH, J.E. DSB, 1975, **11**, 516–20.

ROMANES, E. The life and letters of George John Romanes. Written and edited by his wife. New edition, London, Longmans, 1896. 391 pp.

ROMBERG, Moritz Heinrich (1795–1873). German neurologist; professor in Berlin.

JACOBY, E. Der Neurologe Moritz Heinrich Romberg, 1795–1873. Zürich, Medizinhistorisches Institut, 1965. 24 pp.

ROMERO, Francisco (*fl*.1795–1815). Spanish cardiovascular surgeon: performed first successful pericardiocentesis.

RODRIGUEZ, D. Francisco Romero: padre de la cirugía cardiaca. Barcelona, Centro de Documentación de Historia de la Medicina de J. Uriach, 1985. 26 pp.

ROONHUYSE, Hendrick van (1622–1672). Dutch gynaecologist; published first 'modern' work on gynaecology, 1663.

BAUMANN, E.D. Hendrick van Roonhuyse (c.1622–1672). *Bijdragen tot de Geschiedenis der Geneeskunde*, 1922, **2**, 63–81.
BAUMANN, E.D. *Nederlandsch Tijdschrift voor Geneeskunde*, 1922, **66**, I, 856–74.
THIÉRY, M. Hendrick van Roonhuyse (c.1622–1672): surgeon, obstetrician and gynaecologist. In: Obstetrics and gynaecology of the Low Countries. Zeist, Medical Forum International, 1997, pp. 7–17.

RORSCHACH, Hermann (1884–1922). Swiss psychiatrist; devised 'Rorschach' test of personality traits.

ELLENBERGER, H. The life and work of Hermann Rorschach (1884–1922). *Bulletin of the Menninger Clinic*, 1954, **18**, 173–219.

ROSE, Francis Leslie (1909–1988). Pharmaceutical chemist, Research Division, ICI, Nottingham.

SUCKLING, C.W. & LANGLEY, B.W. BMFRS, 1990, **36**, 489–524.

ROSE, Mary Swartz (1874–1941). American nutritionist.

EAGLES, J.A., *et al*. Mary Swartz Rose, 1874–1941: pioneer in nutrition. New York, Teachers College, Columbia Univ., 1979. 172 pp.

ROSE, William (1847–1910). British surgeon; first performed Gasserian ganglionectomy for trigeminal neuralgia, 1890.

–. *Lancet*, 1910, **1**, 1654–7.

ROSE, William Cumming (1887–1985). American biochemist.

CARTER, H.E. & COON, M.J. BMNAS, 1995, **68**, 253–271.

ROSEN, George (1910–1977). American hygienist.

GEORGE ROSEN: in memoriam. With contributions by [eight authors]. *Journal of the History of Medicine*, 1987, **33**, 244–31.

ROSÉN VON ROSENSTEIN, Nils (1706–1773). Swedish paediatrician; considered the founder of modern paediatrics; wrote important text on the subject (1764).

KRISTALY-KANYA, S. Nils Rosen von Rosenstein und seine "Anweisung zur Kenntnis und Kur der Kinderkrankheiten (1764). Inaugural-Dissertation...Universität Zürich, Juris-Verlag, 1965. 49 pp.
VAHLQUIST, B. & WALLGREN, A. (eds.) Nils von Rosenstein and his textbook on paediatrics. *Acta Paediatrica*, 1964, Suppl.**156**. [Biography by A.Wallgren, pp. 11–26; bibliography by A. Dintler, pp. 40–46.]

ROSENBACH, Julius Friedrich (1842–1923). German surgeon; professor in Göttingen; differentiated between staphylococcus and streptococcus.

GROTE, 1923, **2**, 187–92. (CB)

BIBLIOGRAPHY OF MEDICAL AND BIOMEDICAL BIOGRAPHY

ROSENBACH, Ottomar (1851–1907). German physician.

ENGEL, J. Ottomar Rosenbach. Inaugural-Dissertation...Universität Zürich, Juris-Verlag, 1965. 57 pp.

ROSENBLUETH, Arturo (1900–1970). Mexican neurophysiologist.

MONNIER, A.M. DSB, 1975, **11**, 545–7.

ROSENHEIM, Max Leonard (1908–1972). British physician; President, Royal College of Physicians of London.

PICKERING, G. BMFRS, 1974, **20**, 349–58.
Archival material: Royal College of Physicians of London.

ROSENHEIM, Sigmund Otto (1871–1955). German-born biochemist; worked at King's College and National Institute for Medical Research, London.

KING, H. BMFRS, 1956, **2**, 257–67.

ROSENTHAL, Wolfgang (1884–1971). German surgeon; professor in Berlin.

AUGNER, P.M. Wolfgang Rosenthal.Leipzig, Teubner, 1991. 79 pp.

ROSS, Donald Nixon (b. 1922). British cardiac surgeon.

GUNNING, A.J. Ross's first homograph replacement of the aortic valve. *Annals of Thoracic Surgery*, 1992, **54**, 809–10.

ROSS, Ian Clunies (1899–1959). Australian parasitologist.

HUMPHREYS, L.R. Clunies Ross: Australian visionary. Melbourne, Miegunyah Press, 1998. 250 pp.
ROSS, I.C. Ian Clunies Ross: memoirs and papers, with some fragments of autobiography. Melbourne, Oxford Univ. Press, 1961. 240 pp.

ROSS, Ronald (1857–1932). British physician and poet; proved the transmission of malaria by mosquitoes; Nobel Prize 1902.

CRELLIN, J.K., DSB, 1975, **11**, 555–7.
INDIAN JOURNAL OF MALARIOLOGY. Issue dedicated to Sir Ronald Ross, Vol. 34, No 2, pp. 47–116, 1997. (Includes bibliography of the published works of Surgeon-Major Sir Ronald Ross, pp. 111–16.)
KAMM, J. Malaria Ross. London, Methuen, 1963. 179 pp.
NUTTALL, G.H.F. ONFRS, 1932–35, **1**, 108–14.
NYE, E.R. & GIBSON, M.E. Ronald Ross: malariologist and polymath: a biography. London, Macmillan, 1997. 316 pp.
REES, M.J. [Complete bibliography in unpublished dissertation for Diploma of Librarianship and Archives, University of London, 1966.]
ROSS, R. Memoirs, with a full account of the great malaria problem and its solution London, Murray, 1923. 547 pp.
ROSS, R. The great malaria problem and its solution. From the Memoirs of Ronald Ross, with an introduction by L.J. Bruce-Chwatt. London, British Medical Association, Keynes Press, 1988. 236 pp. First published as Part II of Memoirs, 1923 [see above].
ROWLAND, J. The mosquito man: the story of Sir Ronald Ross. London, Lutterworth Press, 1958. 150 pp.

BIBLIOGRAPHY OF MEDICAL AND BIOMEDICAL BIOGRAPHY

Archival material: London School of Hygiene and Tropical Medicine; Bodleian Library, Oxford, University of Liverpool; Royal College of Physicians and Surgeons of Glasgow; CMAC.

ROSSITER, Roger James (1913–1976). Australian biochemist and neurochemist.

USSELMAN, M.C. DSB, 1990, **18**, 753–5.

ROSTAN, Léon Louis (1790–1866). French physician, professor of medicine at Hôtel-Dieu, Paris.

GRMEK, M.D. DSB, 1975, **11**, 559–60.

ROTHBERGER, Carl Julius (1871–1945). Austrian cardiologist.

WYKLICKY, H. Wiener Experimentalkardiologie am Beginn des 20. Jahrhunderts: Rothbergers Untersuchungen zur Klärung d. Herzstromkurve. Wien, Hollinek, *c*.1974. 120 pp.

ROTHERA, Arthur Cecil Hamel (1880–1915). British biochemist; introduced test for acetone.

HOPKINS, F.G. Arthur Cecil Hamel Rothera. *Biochemical Journal*, 1916, **10**, 11–13.

ROTHMUND, August (1830–1906). German ophthalmologist; described poikiloderma congenitale, 1868.

EVERBUSCH, O. *Münchener Medizinische Wochenschrift*, 1900, **47**, 1082, 1084.

ROTHSCHILD, Nathaniel Mayer Victor (1910–1990). British biologist and zoologist.

REEVE, S. BMFRS, 1994, **39**, 364–80.

ROTHSCHUH, Karl Eduard (1908–1984). German medical historian; professor of medical history, Münster University.

MÜNSTER Universität, Medizinische Fakultät. Karl Eduard Rothschuh 1908–1984. Münster, Vereinigung der Freunde der Medizinische Fakultät, 1985. 65 pp.
TOELLNER, R. (ed.) Karl Eduard Rothschuh: Bibliographie, 1935–1983. 2te Aufl. Tecklenberg, Burgverlag, 1983. 54 pp.

ROUGET, Charles Marie Benjamin (1824–1904). French physiologist and histologist.

BEST, A.E. DSB, 1975, **11**, 565–7.

ROUGHTON, Francis John Worsley (1899–1972). British physiologist and biochemist.

EDSALL, J.T. DSB, 1990, **18**, 759–62.
GIBSON, Q.H. BMFRS, 1973, **19**, 563–82.
Archival material: Library, American Philosophical Society, Philadelphia; Cambridge University Library.

ROUS, Francis Peyton (1879–1970). American pathologist; shared Nobel Prize 1966 with C.B. Huggins for work on experimental cancer.

ANDREWES, C.H. BMFRS, 1971, **17**, 643–62.
DULBECCO, R. BMNAS, 1976, **48**, 275–306.

BIBLIOGRAPHY OF MEDICAL AND BIOMEDICAL BIOGRAPHY

A NOTABLE career in finding out: Peyton Rous (1879–1970). [Tributes from J.S. Henderson, P.D. McMaster, J.G. Kidd & C. Huggins]. New York, Rockefeller Univ. Press, 1970. 47 pp.

Archival material: Library, American Philosophical Society, Philadelphia; Rockefeller Archives Center, New York.

ROUX, César (1857–1934). Swiss surgeon.

CLÉMENT, G. César Roux: l'homme et le chirurgien. Lausanne, 1935. 38 pp.
SAEGESSER, F. César Roux: son époque et le nôtre. Lausanne, Editions de l'Aire, 1989. 195 pp.

ROUX, Philibert Joseph (1780–1854). French surgeon.

STEPHENSON, J. Repair of cleft palate by Philibert Roux in 1819. Stephenson's "De velosynthesi" translated by W.W. Francis; introductory note by Lloyd G. Stephenson. *Journal of the History of Medicine*, 1962, **18**, 209–19.

ROUX, Pierre Paul Émile (1853–1933). French bacteriologist; member of staff, Institut Pasteur, Paris.

DELAUNAY, A. DSB, 1975, **11**, 569.
LAGRANGE, É. Monsieur Roux. Bruxelles, Ad. Goemaere, 1955. 251 pp.
ONFRS, 1932–35, **1**, 197–204.

ROUX, Wilhelm (1850–1924). German anatomist and embryologist; founder of developmental mechanics ('Entwicklungsmechanik').

CHURCHILL, F.B. DSB, 1975, **11**, 570–75.
GROTE, 1923, **1**, 141–206. (CB)
MOCEK, R. Wilhelm Roux, Hans Driesch. Zur Geschichte der Entwicklungsphysiologie der Tiere. Jena, G. Fischer, 1974. 229 pp.

ROVSING, Niels Thorkild (1862–1927). Danish surgeon.
SCHNOHR, E. Thorkild Rovsing. København, Gaardbroder-Sanafund, 1948. 174 pp.

ROWNTREE, Leonard George (1883–1959). American physician; Director of Philadelphia Institute of Medical Research.

ROWNTREE, L.G. Amid masters of twentieth-century medicine. Springfield, Ill., C.C. Thomas, [c.1958]. 684 pp.

RUBNER, Max (1854–1932). German physiologist and hygienist; professor of hygiene, Berlin.

ROTHSCHUH, K.E. DSB, 1975, **11**, 585–6.

RUDBECK, Olof (1630–1702). Swedish physician and anatomist.

LARSELL, O. Olof Rudbeck the elder. *Annals of Medical History*, 1928, **10**, 301–13.
LINDROTH, S. DSB, 1975, **11**, 586–8.

RUDOLPHI, Karl Asmund (1771–1832). Swedish anatomist, physiologist and helminthologist.

KRUTA, V. DSB, 1975, **11**, 592–3.

324

RUEFF, Jacob (1500–1558). Swiss obstetrician; his improved version of Rösslin's Swangern frawen, 1554, contains the first accurate anatomical pictures in a gynaecological work.

GRÜNWALS, L. Jakob Rueff und die Anfänge der Teratologie. Janus, 1926, **30**, 27–33.
NAEGELI-AKERBLOM, H. Jacobus Rueff, chirurgicus turicensis. *Gynecologia Helvetica*, 1910, **10**, 214–20.

RUETE, Christian Georg Theodor (1810–1867). German ophthalmologist; produced the first practical ophthalmoscope.

ZEHENDER, W. Christian Georg Theodor Ruete. *Klinische Monatsblätter für Augenheilkunde*, 1867, **5**, 187–209.

RUFFER, Marc Armand (1859–1917). Franco-British palaeopathologist.

SANDISON, A.T. DSB, 1975, **11**, 595–6.

RUFFINI, Angelo (1864–1929). Italian histologist and embryologist.

FRANCESCHINI, P. DSB, 1975, **11**, 596–8.
LAMBERTINI, G. Il pensiero e i trovati di Angelo Ruffini nell sua opera 'Physiogeny'. [Italian and English]. Padova, Piccin, 1980. 111 pp.

RUFUS of Ephesus (*fl.* AD 98–117). Greek physician and surgeon.

ILBERG, J. Rufus von Ephesus: ein griechischer Arzt in Trajanischer Zeit. Leipzig, S. Hirzel, 1930. 53 pp.
KUDLIEN, F. DSB, 1975, **11**, 601–3.

RUSH, Benjamin (1745–1813). American physician and chemist; signatory of the Declaration of Independence.

BINGER, C.A.L. Revolutionary doctor: Benjamin Rush, 1745–1813. New York, Norton, 1966. 326 pp.
CARLSON, E.T. DSB, 1975, **11**, 616–18.
FOX, C.G., MILLER, G.J. & MILLER, J.C. (comp.). Benjamin Rush, M.D.: a bibliographic guide. Westport, Greenwood Press, 1996. 216 pp.
HAWKE, D.F. Benjamin Rush, revolutionary gadfly. Indianapolis, Bobbs-Merrill, 1971. 490 pp.
RUSH, B. The autobiography of Benjamin Rush: his 'Travels through life', together with his Commonplace book for 1789–1813. Edited with introduction and notes by G.W. Corner. Princeton, NJ, Princeton Univ. Press, 1948. 399 pp.
Archival material: Library, American Philosophical Society, Philadelphia.

RUSHD, ibn, *see* **AVERROËS**

RUSHTON, William (1901–1982). British physiologist (vision).

BARLOW, H.B. BMFRS, 1986, **32**, 423–59.

RUSK, Howard (b.1901). American physician; pioneer in the use of rehabilitation therapy.

RUSK, H. A world to care for. [The autobiography of Howard Rusk.] New York, Random House, 1972. 307 pp.

RUSSELL, Elizabeth S. (1913–2001). American mammalian geneticist.

BARKER, J.C. & SILVERS, W.K. BMNAS, 2002, **81**, 258–77.

RUSSELL, James Burn (1837–1904). British physician; first full-time medical officer of health, Glasgow.

ROBERTSON, E. Glasgow's doctor: James Burn Russell, MOH, 1837–1904. East Linton, Tuckwell Press, 1998. 248 pp.

RUST, Johann Nepomuk Philip (1755–1840). German surgeon; first described tuberculous spondylitis of cervical vertebrae, 1834.

WACHHOLZ, L. Johann Nep. Philip Rust. Beitrag zur Geschichte der medizinische Fakultät in Krakau. *Archiv für Geschichte der Medizin*, 1938, **31**, 40–51.

RUTTY, John (1698–1775). Irish physician; gave first clear description of relapsing fever, 1770.

SHARPLESS, W.T.S. Dr John Rutty (1698–1775) of Dublin and his 'spiritual diary and soliloquies'. *Annals of Medical History*, 1928, **10**, 249–57.

RUYSCH, Frederik (1638–1731). Dutch physician and anatomist; professor of anatomy, Leiden and Amsterdam.

LINDEBOOM, G.A. DSB, 1975, **12**, 39–42.

RUŽICKA, Leopold (1887–1976). Croatian organic chemist; professor at Eidgennössische Technische Hochschule, Zürich; shared Nobel Prize 1939 for synthesis of androsterone.

BORELL, M. DSB, 1990, **12**, 764–5.
RUŽICKA. L. In the borderland between bioorganic chemistry and biochemistry. *Annual Review of Biochemistry*, 1973, **42**, 1–20.

RYDYGIER, Ludwik (1850–1920). Polish surgeon; first to extirpate carcinomatous polyp.

MOLL, J. Das chirurgische Erbe. Ludwik Rydygier (1850–1920). *Zentralblatt für Chirurgie*, 1983, **108**, 1532–5.
RUDOWSKI, W. Special comment: Ludwik Rydygier. In: Nyhus, L.M. & Wastell, C. (eds.). Surgery of the stomach and duodenum. 4th ed. Boston. Little, Brown, 1986, pp. 41–5.

RYFF, Walther Hermann (d.1548). Strassburg physician and surgeon; wrote first monograph on dentistry.

WAGNER, I.V. Walther Hermann Ryff: seine Verdienste um die deutsche Zahnheilkunde. *Deutsche Stomatologie*, 1970, **20**, 314–18.

RYLE, John Alfred (1889–1950). British physician.

PORTER, D. John Ryle: doctor of revolution. In: *Doctors, politics and society*. Amsterdam, Clio Medica, 1993, pp. 247–74.

RYMSDYK (Rijmsdyk), Jan van (*c*.1730–*c*.1789). Medical artist, born in Holland, worked in England.

THORNTON, J.L. Jan van Rymsdyk: medical artist of the eighteenth century. Cambridge, Oleander Press, 1982. 111 pp.

SABIN, Albert Bruce (1906–1999). Russian/American microbiologist; made important contributions to virology, including a live attenuated poliovirus vaccine.

BENISON, S. International medical cooperation . Dr Albert Sabin, live poliovirus vaccine and the Soviets. *Bulletin of the History of Medicine*, 1982, **56**, 460–83.

SABIN, Florence Rena (1871–1953). American anatomist; professor at Johns Hopkins University; first woman member of the National Academy of Sciences.

BLUEMEL, E. Florence Sabin, Colorado woman of the century. Boulder, Colo., Univ. Colorado Press, 1959, 238 pp.
BRIEGER, G.H. DSB, 1975, **12**, 48–9.
KRONSTADT, J. Florence Sabin, medical researcher. New York, Chelsea House, 1990. 110 pp.
McMASTER, P.D. & HEIDELBERGER, M. BMNAS, 1960, **34**, 271–305.
PHELAN, M.K. Probing the unknown. The story of Dr Florence Sabin. New York, Thomas Y. Crowell, 1969. 176 pp.
Archival material: Smith College, Northampton, Mass.; Rockefeller University; Johns Hopkins University; Library, American Philosophical Society, Philadelphia.

SABOURAUD, Raimond Jacques Adrien (1864–1938). French dermatologist.

JUBILÉ scientifique du Docteur Sabouraud, 28 juillet 19929. Paris, Masson, 1930. 103 pp.

SACHS, Bernard (1858–1944). American neurologist; described cerebral changes in 'Tay-Sachs' disease. *Proceedings of the Charaka Club*, 1947, **11**, 231–8.

SACHS, Ernest (1879–1958). American neurosurgeon.

SACHS, E. Fifty years of neurosurgery: a personal story. New York, Vantage Press, 1958. 186 pp.

SACHS, Hans (1877–1945). German immunologist; "Sachs-Georgi" reaction for diagnosis of syphilis.

BROWNING, C.H. Prof. Hans Sachs. *Nature*, 1945, **155**, 600.

SAEMISCH, Edwin Theodor (1833–1909). German ophthalmologist.

HAMMAN, K. Edwin Theodor Saemisch: eine historische Studie. Dissertation, Bonn, 1969. 65 pp.

SÄNGER, Max (1853–1903). German gynaecologist; wrote classics on caesarean section and on chorionic tumors.

ROSTHORN, A. *Monatsschrift für Geburtshülfe und Gynäkologie*, 1903, **17**, 131–5.

SAGER, Ruth (1918–1998). American geneticist and oncologist.

PARDEE, A.R. BMNAS, 2001, **80**, 277–89.

SAHLI, Hermann (1856–1933). Swiss physician; director, University Clinic, Berne.

GROTE, **5**, 177–235 (CB).

SAINT-YVES, Charles de (1667–1736). French ophthalmologist; removed cataract from living patient (reported 1722).

MÜNCHOW, W. Charles de Saint-Yves (1667–1738). *Annales d'Oculistique*, 1969, **2**, 751–62.

SAJOUS, Charles Euchariste de Medicis (1852–1929). French/American endocrinologist.

ROBINSON, V. Charles Euchariste de Medicis Sajous. *Medical Life*, 1925, **32**, 3–21.

SAKHAROV, Vladimir Vladimirovich (1902–1969). Russian geneticist.

BABKOFF, V. DSB, 1975, **12**, 76–7.
MALINOVSKY, A.A. & ANDREEV, V.S. Vladimir Vladimirovich Sakharov. *Genetika* (Moscow), 1969. **5** (2), 177–82.

SAKEL, Manfred Joshua (1900–1957). Austrian/American physician; introduced shock therapy in schizophrenia, 1934.

BARBIER, D. Le cure de Sakel, *Soins Psychiatrique*, 1992, **144**, 37–41.

SAKMANN, Bert (b. 1942). German physiologist; shared Nobel prize (Physiology or Medicine) 1991 for discoveries concerning function of single ion channels in cells.

Les Prix Nobel en 1991.

SALA, Angelus (1576–1637). Italian chemist and physician.

GANTENBEIN, U.L. Die Chemiater Angelus Sala 1576–1637: ein Arzt in Selbstzeugnissen und Krankengeschichten. Zürich, Juris Druck, 1992. 276 pp.

SALAMAN, Redcliffe Nathan (1874–1955). British pathologist and virologist.

SMITH, K.M. BMFRS, 1955, **1**, 239–45.
Archival material: Cambridge University.

SALAZAR, Tomas (1830–1917). Peruvian physician; gave the name 'verruga peruana' to bartellonosis (Carrión's disease').

–. *Chronica Médica*, Lima, 1917, **34**, 117–19, 269–73.

SALIMBENI, Alessandro (1867–1942). Italian immunologist and microbiologist.

MARTIN, L. A.-T. Salimbeni. *Annales de l'Institut Pasteur*, 1942, **68**, 369–72.

SALK, Jonas Edward (b.1914). American virologist; developed an anti-poliomyelitis vaccine.

CARTER, R, Breakthrough: the saga of Jonas Salk. New York, Trident Press, 1966. 435 pp.
ROWLAND, J. The polio man. The story of Dr Jonas Salk. London, Lutterworth Press, 1960. 128 pp.
SHERROW, V. Jonas Salk. New York, Facts on File, *c.*1993, 134 pp.

SALKOWSKI, Ernst Leopold (1844–1923). German physiologist and chemical pathologist; described pentosuria.

Neuberg, C. Ernst Salkowski. *Biochemische Zeitschrift*, 1923, **138**, 1–4.

SALMON, Daniel Elmer (1850–1914). American veterinary pathologist, after whom the Salmonellae tribe of bacteria is named.

SACKMAN, W. Daniel Elmer Salmon (1850–1914) *Historiae Medicinae Veterinariae*, 2000, **25**, 39–41.

SALMON, Thomas William (1876–1927). American psychiatrist; professor at Columbia University; co-founder of National Committee for Mental Hygiene.

BOND, E.D. & KOMORA, P.O. Thomas W. Salmon, psychiatrist. New York, Norton, 1950. 237 pp. Reprinted, New York, Arno Press, 1980.

SALOMONSEN, Carl Julius (1847–1924). Danish bacteriologist.

SNORRASON, E. DSB, 1975, **12**, 87–9.

SALTER, Henry Hyde (1823–1871). British physician; wrote important work on asthma, 1860.

BEALE, A.F. Some thoughts and experiments on respiration and on asthma, with special reference to Henry Hyde Salter. *Medical History*, 1963, **7**, 247–57.
COHEN, S.G. Asthma among the famous: Henry Hyde Salter (1823–1871) *Allergy and Asthma Proceedings*, 1997, **18**, 256–8.

SAMOILOV, Aleksandr Filippovich (1867–1930). Russian electrophysiologist and electrocardiographer; wrote first book on electrocardiogram, 1909.

GRIGORIAN, N.A. DSB, 1975, **12**, 95.
KRIKLER, D.M. The search for Samoiloff, a Russian physiologist in times of change. *British Medical Journal*, 1987, **295**, 1624–7.

SAMOILOVICH, Danilov (1744–1805). Ukrainian epidemiologist.

GROMBAKH, S.M. Danilov Samoilowitz: an eighteenth-century Ukrainian epidemiologist and his role in the Moscow plague (1770–72). *Journal of the History of Medicine*, 1972, **27**, 434–46.

SAMUELSSON, Bengt Ingemar (b. 1934). Swedish biochemist; shared Nobel Prize (Physiology or Medicine), 1982, for discoveries concerning the prostaglandins and related biologically active substances.

Les Prix Nobel en 1982.

SANARELLI, Giuseppe (1864–1940). Italian physician and bacteriologist.

AMBROSIONI, P. DSB, 1975, **12**, 96–7.

SANCTORIUS, *see* **SANTORIO**

SANDERSON, John Scott Burdon (1828–1905). British physiologist and pathologist; Waynflete Professor of Physiology (1882) and Regius Professor of Medicine (1895–1903), Oxford.

ROMANO, T.M. Making medicine scientific: John Burdon Sanderson and the culture of Victorian science. Baltimore, Johns Hopkins Univ. Press, *c*.2002. 255 pp.
SANDERSON, G.H. Sir John Burdon Sanderson: a memoir. Oxford, Clarendon Press, 1911. 315 pp.
Archival material: University College London.

BIBLIOGRAPHY OF MEDICAL AND BIOMEDICAL BIOGRAPHY

SANDSTRÖM, Ivar Victor (1852–1899). Swedish anatomist; first to describe the parathyroid glands.

ASK-UPMARK, E. *et al.* Ivar Sandström and the parathyroid glands. Stockholm, Almqvist & Wiksell, 1967. 13 pp.

SANGER, Frederick (b.1918). British biochemist; Nobel Prize (Chemistry) 1958, for elucidating structure of insulin; shared Nobel Prize (chemistry) 1980, for work in DNA sequencing.

SANGER, F. Sequences, sequences and sequences. *Annual Review of Biochemistry*, 1988, **57**, 1–28.

SANGER, Margaret Higgins (1879–1966). American pioneer of birth control.

CHESSLER, E. Woman of valor: Margaret Sanger and the birth control movement in America. New York, Simon and Schuster, 1992. 639 pp.
DOUGLAS, E.T. Margaret Sanger, pioneer of the future. New York, Holt, Rinehart & Winston, 1970. 274 pp.
GEISON, G.L. DSB, 1973, **2**, 598–9.
GRAY. M. Margaret Sanger: a biography of the champion of birth control. New York, Richard Marek, 1979. 494 pp.
KENNEDY, D.M. Birth control in America. The career of Margaret Sanger. New Haven, Yale Univ. Press, 1970. 370 pp.
LADER, L. & MELTZER, M. Margaret Sanger, pioneer of birth control. New York, Thomas Y. Crowell, 1969. 174 pp.

SANTORINI, Giovanni Domenico (1681–1737). Italian physician and anatomist.

LEAKE, C.D. DSB, 1975, **12**, 100–101.

SANTORIO [SANCTORIUS], Santorio (1561–1636). Italian physician; professor at Padua and inventor of measuring instruments.

CASTIGLIONI, A. La vita e l'opera di Santorio Santorio Capodistriano (1561–1636). Bologna, L.Cappelli, 1920. Translation in *Medical Life*, 1931, **38**, 729–85.
ETTARI, L.S & PROCOPIO, M. Santorio Santorio: la vita e le opera. Roma, Istituto Nazionale della Nutrizione, Città Universitaria, 1968. 165 pp.
GRMEK, M.D. DSB, 1975, **12**, 101–4.

SANTOS, Reynaldo dos (1880–1970). Portuguese surgeon; pioneer in arteriography and thorotrast aortography.
CARDOSO, E.M. Elogio historico de Reynaldo dos Santos. *Memorias da Academia das Ciências de Lisboa*, 1971, **15**, 123–36.
JORNAL da Sociedad das Ciencias Medicas de Lisboa,1981, **145**, No.4 [Memorial issue].

SARETT, Lewis Hastings (1917–1999). American chemist; synthesized cortisone.

PATCHETT, A.A. BMNAS, 2002, **81**, 278–92.

SARGANT, William Walters (1907–1988). British psychiatrist.

SARGANT, W.W. The unquiet mind; the autobiography of a physician in psychological medicine. London, Heinemann, 1967. 240 pp.
Archival material: CMAC.

BIBLIOGRAPHY OF MEDICAL AND BIOMEDICAL BIOGRAPHY

SARTON, George Alfred Leon (1884–1956). Belgian/American historian of science.

OYE, P. van. George Sarton: de mens en zijn werk breven an vrienden an kennissen. Brussel, Paleis der Academiën, 1965. 166 pp.

ISIS, 1957, **48**, 283–350. George Sarton memorial issue, includes memorial essays and (pp. 337–46) a bibliography of the publications (pp. 336–50) by K. Strelsky.

SARTON, M. A world of light: portraits and celebrations. New York, Norton , 1976. 254 pp. Includes biography of Sarton.

THACKRAY, A. & MERTON, R.K. DSB, 1975, **12**, 107–14.

Archival material: Houghton Library, Harvard University; Carnegie Institution, Washington; see also DSB above.

SAUCEROTTE, Nicolas (1741–1814). French surgeon.

RICKLIN, J. Contribution à l'étude de la chirurgie en Lorraine. Nicolas Saucerotte (1741–1814). Nancy, Société d'Imp. Typ., 1924. 64 pp.

SAUERBRUCH, Ernst Ferdinand (1875–1951). German surgeon.

GENSCHOREK, W. Ferdinand Sauerbruch, ein Leben für die Chirurgie. 7te Aufl. Leipzig, Hirzel, 1987. 234 pp.

SAUERBRUCH, F. A surgeon's life. Translated by F.G. Renier and A. Cliff. London, Deutsch, 1953. 297 pp. First published in German. Bad Worishofen, 1951. New ed. 1976.

THORWALD, J. The dismissal: the last days of Ferdinand Sauerbruch, surgeon. English translation by R. and C. Winston. London, Thames & Hudson, 1961. 256 pp.

SAUNDERS, Cicely Mary Strode (b.1918). British physician.

CLARK, D. (ed.) Cicely Saunders – founder of the hospice movement. Selected letters 1959–1999. Oxford, Oxford Univ. Press, 2002. 397 pp.

Du BOULAY, S. Cicely Saunders. London, Hodder & Stoughton, 1993. 276 pp. First published 1984.

SAWYER, Wilbur Augustus (1879–1951). American physician and immunologist; contributed to yellow fever prophylaxis.

FINDLAY, G.M. Wilbur A. Sawyer, M.D. *British Medical Journal*, 1951, **2**, 1347–8.

SAYRE, Lewis Albert (1820–1900). American orthopaedic surgeon.

SHANDS, A.R. Lewis Albert Sayre, the first professor of orthopaedic surgery in the United States (1820–1900). *Current Practice in Orthopaedic Surgery*, 1969, **4**, 22–42.

SAWYER, Wilbur Augustus (1879–1951). American physician and immunologist; contributed to yellow fever prophylaxis.

FINDLAY, G.M. Wilbur A. Sawyer, M.D. *British Medical Journal*, 1951, **2**, 1347–8.

SCARLETT, Earle Parkhill (1887–1982). Canadian physician.

MUSSELWHITE, F.W. Earle P. Scarlett: a study in scarlet. Toronto, Hannah Institute, 1991. 185 pp.

SCARPA, Antonio (1752–1832). Italian anatomist and surgeon; professor at Modena and Pavia.

FRANCESCHINI, P. DSB, 1975, **12**, 136–8.

GAMBACORTA, G. Antonio Scarpa: anatomico, chirurgo, oculisto. Milano, Asclepio, 2000. 87 pp.

331

BIBLIOGRAPHY OF MEDICAL AND BIOMEDICAL BIOGRAPHY

MONTI, A. Antonio Scarpa in scientific history and his role in the fortunes of the University of Pavia ... Translation by F.L. Loria. New York, Vigo Press, 1957. 125 pp.

SCHADE, Heinrich (1876–1935). German molecular pathologist.

HADJAMU, J. Professor Dr. med. Heinrich Schade, Begründer der Molekularpathologie 1876–1935. Leben und Werk. Düsseldorf, Triltsch, 1971. 121 pp.

SCHÄFER, Edward Albert, *see* **SHARPEY-SCHAFER, Edward Albert**

SCHAFFER, Károly (1864–1939). Hungarian neuropathologist.

CAROLI SCHAFFER. In memoriam. *Acta Medica Academia Scientifica Hungarica*, 1965, **21**, 363–448 [10 papers and a bibliography].

SCHALLY, Andrew Victor (b.1926). Polish/American endocrinologist; shared Nobel Prize 1977 for work on isolation of hypothalamic hormones.

WADE, N. The Nobel duel. Two scientists' 21-year race to win the world's most coveted research prize. Garden City, NY, Anchor Press, Doubleday, 1981. 321 pp.

SCHAMBERG, Jay Frank (1870–1934). American dermatologist.

SHELLEY, W.B. & BIERMAN, H. Jay Frank Schamberg (1870–1934). *American Journal of Dermatopathology*, 1984, **6**, 441–4.

SCHARLIEB, Mary Ann Dacomb (1845–1930). British surgeon and gynaecologist.

SCHARLIEB, M. Reminiscences. London, Williams & Norgate, 1925. 239 pp.

SCHARRER, Berta Vogel (1906–1995). German/American neuroendocrinologist.

PURPURA, D.P. BMNAS, 1998, **74**, 289–307.

SCHATZ, Albert Israel (b.1920). American microbiologist; co-discoverer of streptomycin.

WAINWRIGHT, M. Streptomycin: discovery and resultant controversy. *History and Philosophy of the Life Sciences*, 1991, **13**, 97–124.

SCHAUDINN, Fritz Richard (1871–1906). German protozoologist; discovered causal organism of syphilis.

HESSE, P. & HOHMANN, J.S. Friedrich Schaudinn (1871–1906), Sein Leben und Werk als Mikrobiologe. Eine Biographie. Frankfurt am Main, Lang, 1995, 252 pp.
RISSE, G.B. DSB, 1975, **12**, 141–3.

SCHEELE, Carl Wilhelm (1742–1786). Swedish chemist and pharmacist; discovered oxygen and uric acid.

BOKLUND, U. DSB, 1975, **12**, 143–50.
CASSEBAUM, H. Carl Wilhelm Scheele. Leipzig, B.G. Teubner, *c*.1982. 108 pp.
URDANG, G. The apothecary chemist Carl Wilhelm Scheele. A pictorial biography. 2nd ed. Madison, Wisconsin, 1958. 66 pp.
ZEKERT, O. Carl Wilhelm Scheele: Apotheker, Chemiker, Entdecker. Stuttgart, Wissenschaftliche Verlagsgesellschaft, 1963. 149 pp.

332

SCHEFFLER, Johannes [ANGELUS SILENSIS] (1624–1677). German philosopher and poet; practised medicine but in 1653 joined Roman Catholic Church and wrote religious works and hymns; ordained priest 1661.

ELLINGER, G. (ed.) Angelus Silensis: heilige Seelenlust oder geistlicher Hirtenlieder. Halle, 1901. 312 pp.

SCHICK, Bela (1877–1967). Hungarian/American allergologist and immunologist; developed Schick test of susceptibility to diphtheria.

GRONOWICZ, A. Bela Schick and the world of children. New York, Abelard-Schuman, 1954. 216 pp.

SCHIFF, Moritz (1823–1896). German physiologist; professor at Geneva.

RIEDO, P. Der Physiologe Moritz Schiff (1823–1896) und der Innervation des Herzens. Zürich, Juris, 1970. 37 pp.
RISSE, G.B. DSB, 1975, **12**, 164–5.

SCHILD, Heinz Otto (1906–1984). Italian/British pharmacologist.

BLACK, J. BMFRS, 1994, **39**, 382–415.

SCHILDER, Paul Ferdinand (1886–1940). Austrian/American neurologist and psychiatrist; described encephalitis periaxialis diffusa ('Schilder's disease').

LANGER, D. Dr Paul Ferdinand Schilder. Leben und Werk, Dissertation. Mainz, 1979. 259 pp.
SHASKAN, D.A. & ROLLER, W.D. (eds.). Paul Schilder: mind explorer. New York, Human Sciences Press, 1985. 316 pp.

SCHILLER, Johann Christoph Friedrich von (1759–1805). German poet, philosopher and physician.

DEWHURST, K. & REEVES, N. Friedrich Schiller's medicine, psychology and literature, with the first English edition of his complete medical and psychological writings. Oxford, Sandford Publications, 1978. 413 pp.
THEOPOLD, W. Schiller: sein Leben und die Medizin im 18. Jahrhundert. Stuttgart, G. Fischer, 1964. 251 pp.

SCHILLER, Walter (1887–1960). Austrian/American pathologist: introduced test for cervical carcinoma, 1933.

GRUHN, J.G. & ROTH, L.M. History of gynecological pathology: Dr Walter Schiller. *International Journal of Gynecological Pathology*, 1998, **17**, 380–86.

SCHILLING, Robert Selwyn Francis (1911–1997). British physician.

SCHILLING,, R. A challenging life: sixty years in occupational health. London, Canning, 1998. 207 pp.

SCHIMMELBUSCH, Curt (1860–1895). German surgeon; described mammary cystadenoma ('Schimmelbusch's disease').

DALTON, M. & GROZINGER, K.H. Curt Schimmelbusch and Schimmelbusch's disease. *Surgery*, 1968, **63**, 859–61.

BIBLIOGRAPHY OF MEDICAL AND BIOMEDICAL BIOGRAPHY

SCHINDLER, Rudolf (1888–1968). German gastroenterologist.

DAVIS, A.B. Rudolf Schindler's role in the development of gastroscopy. *Bulletin of the History of Medicine*, 1972, **46**, 150–70.
GORDON, M.E. & KIRSNER, J.B. Rudolf Schindler, pioneer endoscopist. Glimpses of the man and his work. *Gastroenterology*, 1979, **77**, 354–61.

SCHLATTER, Carl Bernhard (1864–1934). Swiss surgeon; performed first successful total gastrectomy.

ZIROJEVIC, D. Der Unfallchirurg Carl Schlatter (1864–1934). Zürich, Juris Druck, 1990. 52 pp.

SCHLEICH, Carl Ludwig (1859–1922). German physician; introduced infiltration anaesthesia.

SCHLEICH, C.L. Those were good days. Translated by B. Miall. New York, W.W. Norton & Co., 1936. 280 pp. First published in German, Berlin, 1925.
HALBLÜTZEL, N. Carl Ludwig Schleich und seine Gedanken über die Neuroglia. Zürich, Juris-Verlag, 1966. 59 pp.

SCHLEIDEN, Matthias Jacob (1804–1881). German botanist.

FRANKE, W.W. Matthias Jacob Schleiden and the definition of the cell theory. *European Journal of Cell Biology*, 1988, **47**, 145–56.
KLEIN, M. DSB, 1975, **12**, 173–6.

SCHLINK, Herbert Henry (1883–1962). Australian gynaecologist.

MADDOX, K. Schlink of Prince Alfred. A biography of Sir Herbert Schlink. Sydney, Royal Prince Alfred Hospital, 1978. 283 pp.

SCHLOFFER, Hermann (1868–1937). Austrian surgeon.

MÄNNL, F. Hermann Schloffer (1868–1937) ein Pionier auf dem chirurgischen Lehrstuhl der Deutschen Universität in Prag. *Wüzburger medizinhistorische Mitteilungen*, 2002, **21**, 287–318.

SCHLOSSBERGER, Julius Eugen (1819–1860). German biochemist.

HESSE, F. Julius Eugen Schlossberger (1818–1860). Begründer der physiologischen Chemie in Tübingen – Leben und Werk. Düsseldorf, Triltsch, 1976. 172 pp.

SCHLOSSMANN, Arthur (1867–1932). German paediatrician.

HABERLING, W.G.M. Arthur Schlossmann: sein Leben und sein Werk. Düsseldorf, L. Schwann, 1927. 42 pp.
WUNDERLICH, P. & RENNER, K. Arthur Schlossmann und die Düsseldorfer Kinderklinik. Festschrift zur Feier des 100. Geburtstag am 16 Dezember, 1967. Düsseldorf, M. Triltsch, 1967. 123 pp.

SCHMIDT, Carl Frederic (1893–1988). American pharmacologist.

KOELLE, G.B. BMNAS, 1995, **68**, 273–88.

SCHMIDT, Gerhardt (1901–1981). German-born biochemist; professor at Tufts University Medical School, Boston.

334

KALCKAR, H.M. BMNAS, 1987, **57**, 398–429.

SCHMIDT-NIELSEN, Knut (b. 1915). Norwegian/American physiologist.

SCHMIDT-NIELSEN, K. The camel's nose: memoirs of a curious scientist. Washington, Island Press, 1998. 339 pp.

SCHMIEDEBERG, Johann Ernst Oswald (1838–1921). German pharmacologist; a founder of *Archiv für experimentelle Pathologie und Pharmakologie.*

IHDE, A.J. DSB, 1990, **18**, 789–91.

SCHMIEDEN, Victor (1874–1945). German surgeon.

LOTZ, G.W. Der Chirurg Victor Schmieden (1874–1945). [Thesis.] Frankfurt a.M., Senckenbergisches Institut für Geschichte der Medizin der Universität, 1978. 183 pp.

SCHMITT, Francis Otto (1903–1995). American neurologist.

ADELMANN, G. & SMITH, B. BMNAS, 1998, **75**, 343–54.

SCHMUZIGER, Pierre (1894–1971). Swiss oral surgeon; professor in Zürich.

SCHMID, S. Der Kieferchirurg Pierre Schmuziger (1894–1971). Zürich, Juris, 1990. 80 pp.

SCHNEIDER, Friedrich Anton (1831–1890). German zoologist and cytologist; described cell division.

ROBINSON, G. DSB, 1975, **12**, 192–4.

SCHNITZLER, Arthur (1862–1931). Austrian physician, poet, playwright and novelist.

NESBIT, L. Arthur Schnitzler (1862–1931). *Medical Life*, 1935, **42**, 511–64.
SCHEIBLE, H. Arthur Schnitzler in Selbstzeugnissen und Bilddokumeten. Reinbek bei Hamburg, Rowohlt, 1976. 156 pp.
SCHNITZLER, H., BRANSTATTER, C. & URBACH, R. Arthur Schnitzler: sein Leben sein Werk, seine Zeit. Frankfurt am Main, J. Fischer, 1981, 368 pp.

SCHOENHEIMER, Rudolph (1898–1941). German/American biochemist.

BERTHOLD, H.K. Rudolf Schönheimer (1898–1941): Leben und Werk. Bonn, H.K. Berthold, 1998. 90 pp.
FRUTON, J.S. DSB, 1990, **18**, 791–5.
PEYER, R. Rudolf Schönheimer (1898–1941) und die Beginn der Tracer-Technik bei Stoffwechseluntersuchungen. Zürich, Juris-Verlag, 1972. 37 pp.

SCHÖNLEIN, Johann Lucas (1793–1864). German physician; first to attribute a fungus, *Achorion schönleinii*, as the cause of a disease (favus).

ROBINSON, G. DSB, 1975, **12**, 202–3.
SCHEMMEL, B. & STEBER, A. Johann Lukas Schönlein: Arzt und Mäzen. Ausstellung der Staatsbibliothek Bamberg. Bamberg, Staatsbibliothek, 1993. 160 pp.

SCHOTTMÜLLER, Hugo Adolf Georg (1867–1936). German physician; isolated *Streptococcus viridans* amd *S. muris ratti.*

BUDELMANN, G. Hugo Schottmüller, 1867–1936. Das Problem der Sepsis. *Internist* (Berlin), 1969, **10**, 92–101.

BIBLIOGRAPHY OF MEDICAL AND BIOMEDICAL BIOGRAPHY

SCHRADER, Franz (1891–1962). German/American cytologist.

COOPER, K.W. BMNAS, 1993, **62**, 368–80.

SCHROEDER VAN DER KOLK, Jacobus Ludovicus Conradus (1797–1862). Dutch neurologist and psychiatrist.

ESCH, P. van der. Jacobus Ludovicus Conradus Schroeder van der Kolk, 1797–1862: leven en werken. Academisch proefschrift...Amsterdam, 1954. 196 pp.
SNELDERS, H.A.M. DSB, 1975, **12**, 223–5.

SCHRÖDINGER, Erwin (1887–1961). Austrian physicist; his book *What is life?* had considerable influence on the development of molecular biology; Nobel Prize (Physics) 1933.

HEITLER, W. BMFRS, 1961, **7**, 221–8.
KILMISTER, C.W. (ed.) Schrödinger: centenary celebration of a polymath. Cambridge, Cambridge Univ. Press, 1987. 253 pp.
MOORE, W. Schrödinger: life and thought. Cambridge, Cambridge University Press, 1989. 513 pp.
MOORE, W. A life of Erwin Schrödinger. Cambridge, Cambridge University Press, 1994, 349 pp.

SCHÜLLER, Arthur (1874–1958). Austrian neurologist.

SCHINDLER, E. Arthur Schüller: pioneer of neuroradiology. *American Journal of Neuroradiology*, 1997, **18**, 1297–302.
SCHINDLER, E. Arthur Schüller: Vater der Neuroradiologie oder: ein österreichisches Wissenschaftler-Schicksal. *Wiener Klinische Wochenschrift*, 1998, **110**, 162–6.

SCHULTES, Johann, *see* **SCULTETUS, Johannes**

SCHULTHESS, Wilhelm (1855–1917). Swiss orthopaedic surgeon.

RÜTTIMAN, B. Wilhelm Schulthess (1855–1917) und die Schweizer Orthopädie seiner Zeit. Zürich, Schulthess Polygraphischer Verlag, 1983. 272 pp.

SCHULTZ, Jack (1904–1971). American geneticist at Institute for Cancer Research, Philadelphia.

ANDERSON, T.F. BMNAS, 1975, **47**, 393–422.
Archival material: American Philosophical Society, Philadelphia.

SCHULTZE, Friedrich (1848–1934). German neurologist.

GROTE, **2**, 193–215 (CB).

SCHULTZE, Max Johann Sigismund (1825–1874). German anatomist, microscopist and cytologist.

GEISON, G.L. DSB, 1975, **12**, 230–32.
HAST, T.H. Max Johann Sigismund Schultze (1825–1874). *Annals of Medical History*, 1931, **3**, 166–78.
LUECKER, R.R. Max J.S. Schultze (1825–1874) und die Zellenlehre des 19. Jahrhunderts. [Thesis.] Bonn, Friedrich-Wilhelms-Universität, 1977, 222 pp.

BIBLIOGRAPHY OF MEDICAL AND BIOMEDICAL BIOGRAPHY

SCHULZE, Johann Heinrich (1687–1744). German physician and medical historian.

EDER, J.M. Johann Heinrich Schulze: der Lebenslauf des Erfinders des ersten photographischen Verfahrens und des Begrunders der Geschichte der Medizin. Wien, K.K. Graphische Lehr- und Versuchsanstalt, 1917. 79 pp.

SCHWABACH, Dagobert (1846–1920). German otologist; devised the Schwabach hearing test.

–. *Archiv für Ohren-, Nasen- und Kehlkopfheilkunde*, 1920, **106**, Heft 2–3, pp. i–iii.

SCHWANN, Theodor Ambrose Hubert (1810–1882). German anatomist and physiologist; professor of anatomy and physiology, Liège; advanced cell theory of development.

CAUSEY, G. The cell of Schwann. Edinburgh, E. & S. Livingstone, 1960. 120 pp.
FLORKIN, M. DSB, 1975, **12**, 240–45.
WATERMANN, R. Theodor Schwann; Leben und Werk. Düsseldorf, L. Schwann, 1960. 364 pp.

SCHWABACH, Dagobert (1846–1920). German otologist; devised the Schwabach hearing test.

–. *Archiv für Ohren-, Nasen- und Kehlkopfheilkunde*, 1920, **106**, Heft 2–3, pp. i–iii.

SCHWARTZE, Heinrich Hermann Rudolf (1837–1910). German otologist; with A. Eysell modernized the mastoid operation.

AMBERG, E. *Detroit Medical Journal*, 1911, **11**, 922–8.
KRETSCHMANN, F. *Archiv für Ohrenheilkunde*, 1910, **10**, 83–i–xvi.

SCHWEITZER, Albert Louis Philipp (1875–1965). French physician, philosopher, musician and theologian; founder of the mission and hospital at Lambaréné, West Africa.

BRABAZON, J. Albert Schweitzer: a biography. 2nd ed. Syracuse, NY, Syracuse Univ. Press, 2000. 555 pp. First published 1976.
MARSHALL, G. & POLING, D. Schweitzer. London, Geoffrey Bles, 1971. 342 pp.
SEAVER, G. Albert Schweitzer: the man and his mind. 6th ed., London, Black, 1969. 365 pp. First published 1947.

SCOT, Michael, *see* **MICHAEL SCOT**

SCOTT, David Aylmer (1892–1971). Canadian biochemist; worked on insulin preparations.

BEST, C.H. & FISCHER, A.M. BMFRS, 1972, **18**, 511–24.

SCOTT, Walter (1787–1854). Scottish surgeon; worked in Scotland and Australia.

PEARN, J. In the capacity of a surgeon: a biography of Walter Scott, surgeon, colonist and first civilian of Queensland. Brisbane, Amphion Press (Dept. of Child Health, Royal Children's Hospital), 1988. 226 pp.

SCOTT, William Wallace (b. 1913). American urologist; jointly introduced adrenalectomy for prostate cancer.

ENGEL, R.M. William Wallace Scott, Snr. *Journal of Urology*, 1998, **160**, 2370–74.
SCOTT, W.W. Autobiography of William Wallace Scott. Middletown NJ, Till & Till Inc., 1996.

BIBLIOGRAPHY OF MEDICAL AND BIOMEDICAL BIOGRAPHY

SCRIMGER, Francis Alexander Carron (1880–1937). Canadian physician; awarded Victoria Cross while serving in Canadian Army Medical Corps during First World War.

KINGSMILL, S. Francis Scrimger: beyond the call of duty. Toronto, Hannah Institute and Dundurn Press, 1991. 112 pp.

SCUDDER, Ida Sophia (1870–1960). American physician; medical missionary in India for fifty years.

WILSON, D.C. Dr Ida: the story of Dr Ida Scudder of Vellore. London, Hodder & Stoughton, 1960; reprinted 1964. French edition, Geneva, 1971.

SCULTETUS [SCHULTES], Johannes (1595–1645). German surgeon, notable for the illustrations of surgical procedures and instruments in his work.

SCULTETUS, J. *Wund-Artzneyisches Zeug-Hauses in zween Theil abgetheilt*. 2 pts. Stuttgart, W.Kohlhammer, 1974. Facsimile of Frankfurt editon of 1666; includes biographical material.

SEACOLE, Mary (1805–1881). Jamaican/British nurse in the Crimea.

SEACOLE, M. The wonderful adventures of Mrs Seacole in many lands. London, X Press, 1999. 188 pp.

SEBRELL, William Henry (1901–1992). American physician.

SEBRELL, W.H. Recollections of a career in nutrition. *Journal of Nutrition*, 1985, **115**, 23–38.

SECHENOV [SETCHENOFF], Ivan Mikhailovich (1829–1905). 'Father of Russian physiology'.

IORASHEVSKII, M.G. Ivan Mikhailovich Sechenov, 1825–1905. Leningrad, Nauka, 1968. 423 pp.
LAROSHEVSKY, M.G. DSB, 1975, **12**, 270–71.
MIRSKII, M.B. Revoliutsioner v nauke, demokrat v zhizni. Ivan Mikhailovich Sechenov. Moskva, Znanie, 1988. 219 pp.
SHATERNIKOV, M.N. The life of I.M.Sechenov. In: Sechenov, I.M. Selected works. Amsterdam, Bonset, 1968, pp. ix–xxxvi.

SEDGWICK, William Thompson (1855–1921). American biologist and epidemiologist.

JORDAN, E.O., WHIPPLE, G.C. & WINSLOW, C.E.A. A pioneer of public health: William Thompson Sedgwick. New Haven, Yale Univ. Press; London, Oxford Univ. Press, 1924. 193 pp.

SÉDILLOT, Charles Emanuel (1804–1883). French surgeon; performed first successful gastrostomy, 1849.

SUNDER, C.H. La vie et les oeuvres de Ch. Emanuel Sédillot, 1804–1883. Strasbourg, Thèse, 1933. 60 pp.

SÉGUIN, Edouard (1812–1880). French psychiatrist; settled in U.S.A.

PELICIER, Y. & THUILLIER, G. Edouard Séguin (1812–1880), "l'instituteur des idiots". Paris, Economica, 1980. 211 pp.

BIBLIOGRAPHY OF MEDICAL AND BIOMEDICAL BIOGRAPHY

PELICIER, Y. & THUILLIER, G. Un pionnier de la psychiatrie de l'enfant. Edouard Séguin (1812–1880). Paris, Association pour l'Étude de l'Histoire de la Sécurité Sociale, 1996. 529 pp.

TALBOT, M. Edouard Séguin: a study of an educational approach to the treatment of mentally defective children. New York, Bureau of Publications, Teachers College, Columbia University, 1964. 150 pp.

SEIDEMANN, Herta (1900–1984). Swiss neuropsychiatrist.

FOCKE, W. Begegnung: Herta Seidemann: Psychiatrin-Neurologin, 1900–1984; eine biographischer Essay. Konstanz, Hartnung-Gorre, 1986. 276 pp.

SELIGMAN, Charles Gabriel (1873–1940). British pathologist and anthropologist; professor of ethnology at London School of Economics.

MYERS, C.S. ONFRS, 1940–41, **3**, 627–46.

SELLARDS, Andrew Watson (1884–1941). American physician.

HARVEY, A.M. Applying the methods of science to the study of tropical diseases – the story of Andrew Watson Sellards. *Johns Hopkins Medical Journal*, 1979, **144**, 45–55.

SELWYN-CLARKE, Selwyn (1893–1976). British army medical officer; served in Hong Kong during its occupation by the Japanese, 1941–45.

SELWYN-CLARKE, S. Footprints; the memoirs of Sir Selwyn Selwyn-Clarke. Hong Kong, Sino-American Publishing Co., 1975, 189 pp.

SELYE, Hans (1907–1982). Austrian/Canadian physician; worked on stress and introduced concept of the 'general adaptation syndrome'.

HALF A CENTURY of stress research: a tribute to Hans Selye by his students. *Experientia*, 1985, **41**, 559–78.
SELYE, H. From dream to discovery: on being a scientist. New York, McGraw Hill, 1964. 419 pp.
SELYE, H. The stress of my life: a scientist's memoirs. 2nd ed. New York, Van Nostrand, c.1979. 267 pp.
YANACOPOULO, A. Hans Selye ou la cathédrale du stress. Montréal, Le Jour, 1992. 430 pp.

SEMMELWEIS, Ignaz Philipp (1818–1865). Hungarian physician; professor of midwifery at Pest; discovered cause and method of prevention of puerperal sepsis.

BENEDEK, I. Semmelweis. Budapest, Gondolat, 1980. 269 pp.
CARTER, K.C. & CARTER, B.R. Childbed fever: a scientific biography of Ignaz Semmelweis. London.
GORTVAY, Gy. & ZOLTÁN, I. Semmelweis: his life and work. Budapest, Akadémiai Kiado, 1968. 287 pp. Translated by E. Rona; first published in Hungarian, 1966.
MURPHY, F.B. Ignaz Philipp Semmelweis (1818–1865); an annotated bibliography. *Bulletin of the History of Medicine*, 1946, **20**, 653–707.
RISSE, G.B. DSB, 1975, **12**, 294–7.
SILLÓ-SEIDEL, G. Die Wahrheit über Semmelweis. Das Werken des grossen Arzt-Forschers un sein tragischer Tod im Licht new endeckter Dokumente. Geneva, Ariston-Verlag, 1978. 214 pp.

SEMON, Felix (1849–1921). German/British laryngologist; Physician Extraordinary to King Edward VII.

BIBLIOGRAPHY OF MEDICAL AND BIOMEDICAL BIOGRAPHY

HARRISON, D. Felix Semon 1849–1921: a Victorian laryngologist. London, Royal Society of Medicine, 2000. 250 pp.
SEMON, F. The autobiography of Sir Felix Semon. Edited by H.C. Semon and Thomas A. McIntyre. London, Jarrolds, 1926, 349 pp.

SENAC, Jean-Baptiste (1693–1770). French physician.

SMEATON, W.A. DSB, 1975, **12**, 302–3.

SENN, Nicholas (1844–1908). Swiss/American surgeon; professor at Rush Medical College, Chicago.

SALMONSEN, E.M. Nicholas Senn, M.D., Ph.D., LL.D. (1844–1908), master surgeon, pathologist and teacher. Biographical sketch with a complete bibliography of his writings. *Bulletin of the Society of Medical History of Chicago*, 1928–35, **4**, 268–94.
WYLIE, S.M. Nicholas Senn: an appreciation. Chicago, Nicholas Senn Club, 1908. 27 pp.

SENNERT, Daniel (1572–1637). German physician; professor at Wittenberg.

KANGRO, H. DSB, 1975, **12**, 310–12.

SEREBROVSKII, Aleksandr Sergeevich (1892–1948). Russian geneticist.

ADAMS, M.B. DSB, 1990, **18**, 803–11.

SERRES, Etienne-Renaud-Augustin (1787–1868). French physician and anatomist; professor of comparative anatomy, Paris.

KRAEGEL-von-der-HEYDEN, M.M. Etienne-Renaud-Augustin Serres (1787–1868), Entdecker der Abdominaltyphus. Zürich, Juris, 1972. 50 pp.

SERTOLI, Enrico (1842–1910). Italian physiologist and histologist.

ZANOBIO, B. DSB, 1975, **12**, 319–20.

SERTÜRNER, Friedrich, Wilhelm Adam Ferdinand (1783–1841). German chemist; isolated morphine.

KROEMKE, F. Friedrich Wilhelm Sertürner, der Entdecker des Morphiums. Lebensbild und Neudruck der Original-Morphiumarbeiten. Jena, G. Fischer, 1925. 93 pp.
MEYER, K. *et al.* F.W. Sertürner: Entdecker des Morphiums. Paderborn, Stadtarchiv, 1983. 84 pp.
SCHMAUDERER, E. D.S.B., 1975, **12**, 320–21.
SCHUMANN, O. Morphium. Ein biographischer Roman über den Entdecker des Morphiums Friedrich Wilhelm Sertürner. Berlin, Deutscher Apotheker-Verlag, 1940. 388 pp.

SERVETUS, Michael [SERVETO, Miguel] (1511–1553). Spanish physician and theologian, described the lesser circulation; martyred in Geneva.

BAINTON, R.H. Hunted heretic: the life and death of Michael Servetus, 1511–1553. Boston, Beacon Press, 1953. 270 pp. Reprinted 1960.
FREIDMAN, J. Michael Servetus: a case study in total heresy. Genève, Librairie Droz, 1978. 149 pp.
FULTON, J.F. Michael Servetus, humanist and martyr ... With a bibliography of his works and census of known copies by Madeline E. Stanton. New York, Reichner, 1953. 98 pp.
PILAPIL, V.R. DSB, 1975, **12**, 322–5.

340

SETCHENOFF, *see* SECHENOV

SEVERINO, Marco Aurelio (1580–1656). Italian anatomist and pathologist.

SCHMITT, C.B. & WEBSTER, C. Harvey and M.A. Severino. A neglected medical relationship. *Bulletin of the History of Medicine*, 1971, **45**, 49–75.

SEVERINUS, Petrus [SØRENSEN, Peder] (1542–1602). Danish physician and Paracelsist.

DEBUS, A.G. DSB, 1975, **12**, 334–6.

SEVERTSOV, Aleksei Nikolayevich (1866–1936). Russian comparative anatomist and evolutionary morphologist.

MIRZOYAN, E. DSB, 1975, **12**, 336–9.

SEWALL, Henry (1855–1936). American physiologist and physician; professor of physiology, University of Michigan.

WEBB, G.B. & POWELL, D.S. Henry Sewall, physiologist and physician. Baltimore, Johns Hopkins Press, 1946. 191 pp.

SHAFFER, Philip Anderson (1881–1960). American biochemist; professor at Washington University, St Louis.

DOISY, E.A. BMNAS, 1969, **40**, 321–36.
Archival material: Washington University Medical School, St Louis.

SHANNON, James Augustine (1904–1994). American physician; Director of National Institutes of Health.

KENNEDY, T.J. Jnr., BMNAS, **75**, 358–78.

SHARPEY, William (1802–1880). British anatomist and physiologist; professor at University College London.

SYKES, A.H. Sharpey's fibres: the life of William Sharpey, the father of modern phsiology. York, William Sessions, *c.* 2001. 164 pp.
TAYLOR, D.W. DSB, 1975, **12**, 354.
TAYLOR, D.W. The life and teaching of William Sharpey (1802–1880), 'Father of modern physiology' in Britain. *Medical History*, 1971, **15**, 126–53, 241–59.
Archival material: University College London; Library, University of Edinburgh; Library, University of Cambridge; Royal Society of London; Arbroath Signal Tower Museum.

SHARPEY-SCHAFER, Edward Albert (1850–1935). British physiologist; professor at University College London, and University of Edinburgh; devised a method of artificial respiration; contributed notable work on cerebral localization of function.

HILL, L.E. ONFRS, 1932, **35**, 401–7.
TAYLOR, D.W. DSB, 1975, **12**, 355–6.
Archival material: CMAC (including material in the collection of I. de Burgh Daly).

SHATTUCK, Frederick Cheever (1847–1929). American physician; Jackson professor of clinical medicine, Harvard Medical School.

SHATTUCK, G.C. Frederick Cheever Shattuck, 1847–1929: a memoir. Boston, privately printed, 1967. 327 pp.
Archival material: Library, Harvard University Medical School.

SHATTUCK, Lemuel (1793–1859). American public hygienist and vital statistician.

KAUFMANN, M. *et al.* Dictionary of American medical biography. Westport, Greenwood Press, 1984, vol. 2, pp. 676–7.

SHAW, Trevor Ian (1928–1972). British physiologist.

DENTON, E.J. BMFRS, 1974, **20**, 359–80.

SHEDLOVSKY, Theodore (1898–1976). American biochemist.

FUOSS, R.M. BMNAS, 1980, **52**, 379–408.

SHEEHAN, Harold Leeming (1900–1988). British pathologist; in 1937 described hypopituitarism due to postpartum pituitary necrosis ('Sheehan's syndrome').

KOVACS, K. Sheehan syndrome. *Lancet*, 2003, **361**, 520–22.

SHEMIAKIN [SHEMIYAKIN], Mikhail Mikhailovich (1908–1970). Russian organic chemist and biochemist.

SHAMIN, A.N. DSB, 1990, **18**, 813–14.

SHEPHERD, Francis John (1851–1929). Canadian surgeon, at Montreal General Hospital and McGill University.

HOWELL, W.B. F.J. Shepherd, surgeon; his life and times. Toronto, Dent, 1934. 251 pp.

SHEPPARD, Philip Macdonald (1921–1976). British geneticist.

CLARKE, C. BMFRS, 1977, **23**, 465–500.
TURNER, J.R.G. DSB, 1990, **18**, 814–16.

SHERMAN, Henry Clapp (1875–1955). American nutritional chemist; professor at Columbia University.

KING, C.G. BMNAS, 1975, **46**, 397–429.

SHERRINGTON, Charles Scott (1857–1952). British physiologist; professor at Oxford University; contributed extensively to the knowledge of neurophysiology; Nobel prizewinner 1932.

COHEN, H. Sherrington: physiologist, philosopher and poet. Liverpool, Liverpool Univ. Press, 1958. 108 pp. Includes bibliography of Sherrington's works.
ECCLES, J.C. & GIBSON, W.C. Sherrington: his life and thought. Berlin, Springer-Verlag, 1979. 269 pp.
FULTON, J.F. Sir Charles Scott Sherrington, O.M. *Journal of Neurophysiology*, 1952, **15**, 167–90.
GRANIT, R. Charles Scott Sherrington: an appraisal. London, Nelson, 1966. 188 pp.
LIDDELL, E.G.T. ONFRS, 1952, **8**, 241–70.
SWAZEY, J.P. DSB, 1975, **12**, 395–403.
Archival material: Fulton papers, Historical Library, Yale University School of Medicine; Liverpool University; Churchill College, Cambridge; Woodward Biomedical Library, University of British Columbia; Physiological Laboratory, University of Oxford.

SHIGA, Kyoshi (1870–1957). Japanese bacteriologist; discovered *Shigella dysenteriae*.

FELSENFELD, O. K. Shiga, bacteriologist. *Science*, 1957, **126**, 113.

BIBLIOGRAPHY OF MEDICAL AND BIOMEDICAL BIOGRAPHY

SHIPPEN, William (1736–1808). American medical teacher, graduated in Edinburgh; professor of anatomy and surgery, College of Medicine, University of Pennsylvania.

CORNER, B.C. William Shippen, Jr., pioneer in American medical education: a biographical essay, with notes and the original text of Shippen's student diary, London, 1759–1760; together with a translation of his Edinburgh dissertation, 1761. Philadelphia, American Philosophical Soc., 1951. 161 pp.
MIDDLETON, W.S. William Shippen, junior. *Annals of Medical History*, 1932, N.S. **4**, 440–52, 538–49.

SHIPWAY, Francis (1875–1968). British anaesthetist; introduced Shipway apparatus, 1916.

COPE, P.A. Sir Francis Shipway (1875–1968): anaesthetist by royal appointment. International Symposium on the History of Anaesthesia (2nd), 1987. History of Anaesthesia, London, 1988, International Congress and Symposium Series, No. 134, pp. 559–64.

SHOPE, Richard Edwin (1901–1966). American virologist; discovered swine influenza virus.

ANDREWES, C.H. BMNAS, 1979, **50**, 353–75.

SHORT, Arthur Rendle (1880–1953). British surgeon; professor at Bristol University.

CAPPER, W.M. & JOHNSON, D. Arthur Rendle Short, surgeon and Christian. London, Inter-Varsity Fellowship, 1954. 208 pp.
CAPPER, W.M. & JOHNSON, D. The faith of a surgeon. Belief and experience in the life of Arthur Rendle Short. Exeter, Paternoster, 1976. 156 pp.

SHORTT, Henry Edward (1887–1987). British protozoologist and medical entomologist; professor of medical protozoology, University of London; elucidated life cycles of Leishmania and malarial parasites.

GARNHAM, P.C. BMFRS, 1988, **34**, 713–51.
Archival material: CMAC.

SHORTT, Thomas (1788–1843).

CHAPLIN, A. Thomas Shortt (Principal Medical Officer in St Helena); with biographies of some other medical men associated with the care of Napoleon from 1815–1821. London, Stanley Paul, 1914. 70 pp.

SHRAPNELL, Henry Jones (1792–1834). British surgeon.

McGOVERN, F.H. The elusive Henry Jones Shrapnell. *Laryngoscope*, 1983, **93**, 903–5.

SHULL, Aaron Franklin (1881–1961). American geneticist.

ALLEN, G. DSB, 1975, **12**, 416–18.
Archival material: Jennings, Blakeslee, Demerec, Dunn and Davenport papers, American Philosophical Society, Philadelphia.

SHUTE, Evan Vere (1905–1978). Canadian physician.

SHUTE, E. The vitamin E story: the medical memoirs of Evan Shute. Edited and introduced by J.C. Shute. Burlington, Ont., Welch Publishing Co., 1985. 222 pp.

343

BIBLIOGRAPHY OF MEDICAL AND BIOMEDICAL BIOGRAPHY

SIBBALD, Robert (1641–1722). Scottish physician; a founder of the Royal College of Physicians of Edinburgh.

THE MEMOIRS of Sir Robert Sibbald (1641–1722). Edited with an introduction and a refutation of the charge against Sir Robert Sibbald of forging Ben Jonson's Conversations. London, Oxford Univ. Press, 1932. 107 pp.

SIMPSON, A.D. Sir Robert Sibbald – the founder of the College. In: Passmore, R. (ed.): Proceedings of the Royal College of Physicians of Edinburgh Tertcentenary Congress, 1981. Edinburgh, Royal College of Physicians of Edinburgh, 1932, pp. 59–91.

SICARD, Jean Marie Athanase (1872–1929). French physician.

ROGER, H. Le professeur J.A.Sicard: sa vie et son oeuvre. *Histoire de la Médecine*, 1955, **5** (11), 61–77.

SICHEL, Julius (1802–1868). German ophthalmologist; published first ophthalmological atlas.

KOEBLING, H.M. Julius Sichel aus Frankfurt (1802–1868), ein Pionier der modernen franzözischen Augenheilkunde. *Ophthalmologica*, 1970, **161**, 83–9.

SIEBOLD FAMILY. Bartel von (1774–1814); **Carl** von (1800–1860); **Carl Caspar** von (1736–1807); **Carl Theodor Ernst** von (1804–1885); **Christoph** von (1767– 1798); **Damien** von (1796–1828); **Eduard** von (1801–1861); **Elias** von (1775–1828); **Philipp Franz Balthasar** von (1796–1866). German family of comparative anatomists, physicians and surgeons.

KÖRNER, H. Die Würzburger Siebold: ein Gelehrtenfamilie des 18. und 19. Jahrhunderts. (Siebold: Beiträge zur Familiengeschichte. Teil 1, Lief.3, 451–1080.) Neustadt, Degener, 1967.

THIEDE, A., HIKI, Y. & KEIL, G. (eds.) Philipp Franz von Siebold and his era: prerequisites, developments, consequences, and perspectives. Berlin, Springer, *c.*2000. 197 pp.

SIGERIST, Henry Ernest (1891–1957). Swiss/American medical historian; director of the Johns Hopkins Institute of the History of Medicine, Baltimore.

BERG-SCHORN, E. Henry E. Sigerist (1891–1957), Medizinhistoriker in Leipzig und Baltimore: Standpunkt und Wirkung. Köln, Institut für Geschichte der Medizin, 1978. 337 pp.

FEE, E. & BROWN, T.M. eds. Making medical history: the life and times of Henry E. Sigerist. Baltimore, Johns Hopkins Univ. Press, 1997. 387 pp.

MILLER, Genevieve (ed.) A bibliography of the writings of Henry E. Sigerist. Montreal, McGill Univ. Press, 1966. 112 pp.

SIGERIST, H.E. Autobiographical writings. Selected and translated by Nora Sigerist Beeson. Montreal, McGill Univ. Press, 1966. 247 pp.

THOM, A. & KARBE, K.-H. Henry Ernest Sigerist (1891–1957). Leipzig, Barth, 1981. 143 pp.

SILBERSCHMIDT, William (1869–1947). Swiss bacteriologist; professor of hygiene and bacteriology.

LORETAN, M. William Silberschmidt (1869–1947), Hygieniker und Bakteriologe. Zürich, Juris, 1988. 124 pp.

SILESIUS, Angelus, *see* **SCHEFFLER, Johannes**

SILLIMAN, Benjamin (1779–1864). American chemist; founder of the Medical Faculty at Yale University.

BROWN, C.M. Benjamin Silliman, a life in the young republic. Princeton, NJ, Princeton Univ. Press, 1989. 377 pp.

FULTON, J.F. & THOMSON, E.H. Benjamin Silliman, pathfinder in American science. New York, Schuman, 1947. 294 pp. Reprinted 1969.

WILSON, L.G. (ed.) Benjamin Silliman and his circle. Studies of the influence of Benjamin Silliman on science in America. New York, Science History Publications, 1979. 227 pp.

Archival material: Library, American Philosophical Society, Philadelphia.

SIMMONDS, Maurice (1855–1925). Danish pathologist.

GRIESBACH, W.E. Maurice Simmonds, pioneer endocrinologist. Some recollections. *Journal of Clinical Endocrinology*, 1965, **25**, 1671–3.

SIMON, Charles Edmund (1866–1927). American virologist.

HARVEY, A.M. Pioneer American virologist – Charles Edmund Simon. *Johns Hopkins Medical Journal*, 1978, **142**, 161–86.

SIMON, Gustav (1824–1876). German surgeon; carried out first sucessful planned nephrectomy; developed cheiloplasty.

GIBSON, T. Gustav Simon (1824–1876): Simonart(s)(z) of the band?? *British Journal of Plastic Surgery*, 1977, **30**, 255–60.

LAURIDSEN, L. Gustav Simon und der Anfang der Nierenchirurgie. *Acta Chirurgica Scandinavica*, 1973, Suppl. 433, 31–41.

SIMON, John (1816–1904). British surgeon and sanitary reformer; first medical officer of health for the City of London.

HALE-WHITE, pp. 189–207 (CB).

LAMBERT, R. Sir John Simon (1816–1904), and English social administration. London, MacGibbon & Kee, 1963. 669 pp.

Archival material: Royal College of Surgeons of England; Bodleian Library, Oxford; Library, University of Manchester.

SIMOND, Paul Louis (1858–1947). French physician at the Institut Pasteur, Paris.

VOELCKEL, J. La vie et l'oeuvre de P.L. Simond (1858–1947). *Médecine Tropicale*, 1969, **29**, 429–41.

SIMPSON, Cedric Keith (1907–1985). British forensic pathologist.

SIMPSON, K. Forty years of murder: an autobiography. London, Harrap, 1978. 327 pp.

SIMPSON, James Young (1811–1870). British obstetrician; professor at Edinburgh; discovered the anaesthetic value of chloroform.

RUSSELL, K.F. & FORSTER, F.M.C. A list of the works of Sir James Young Simpson, 1811–1870. Melbourne, Univ. Melbourne, 1971. 57 pp.

SHEPHERD, J.A. Simpson and Syme of Edinburgh. Edinburgh, London, Livingstone, 1969. 288 pp.

SIMPSON, M. Simpson the obstetrician: a biography. With a foreword by I. Donald. London, Victor Gollancz, 1972. 304 pp.

BIBLIOGRAPHY OF MEDICAL AND BIOMEDICAL BIOGRAPHY

Archival material: Royal College of Physicians, Edinburgh; Royal College of Surgeons, Edinburgh; Medical Library, Duke University, Durham, NC.

SIMS, James Marion (1813–1883). American surgeon and gynaecologist.

HARRIS, S. & BROWIN, F.W. Woman's surgeon: the life story of J. Marion Sims. New York, Macmillan, 1950. 432 pp. Reprinted 1968.
McGREGOR, D.K. Sexual surgery and the origins of gynecology: J. Marrion Sims, his hospital and his patients. New York, Garland Pub., 1989. 431 pp.
SIMS, J.M. The story of my life ... Edited by his son, H. Marion-Sims. New York, Appleton, 1884. 471 pp. Reprinted, New York, Da Capo, 1968. 471 pp.

SINA, ibn, see **AVICENNA**

SINCLAIR, Hugh Macdonald (1904–1990). British physician and nutritionist.

EWIN, J. From wines and fish oil: the life of Hugh Macdonald Sinclair. Oxford, Oxford Univ. Press, 2001. 358 pp.
GALE, M. & LLOYD, B. (eds.) Sinclair: Dr Hugh Macdonald Sinclair D.M., D.Sc., F.R.C.P. Wokingham, McCarrison Society, 1990. 68 pp.

SINGER, Charles Joseph (1876–1960). British physician; historian of science and medicine.

CLARKE, E. Charles Joseph Singer. *Journal of the History of Medicine*, 1961, **16**, 411–19.
SHEPPARD, J. Illustrations from the Wellcome Institute Library. Charles Singer, D.M., D.Litt., D.Sc., F.R.C.P. (1876–1960); papers in the Contemporary Medical Archives Centre. *Medical History*, 1987, **31**, 466–71.
UNDERWOOD, E.A. A bilbiography of the published writings of Charles Singer. In: Underwood, E.A. *Science, medicine and history. Essays...in honour of Charles Singer.* London, Oxford Univ. Press, 1953, vol. 2, pp. 555–81.
Archival material: CMAC.

SINTON, John Alexander (1884–1956). British malariologist; first Director of Malaria Survey of India.

CHRISTOPHERS, S.R. BMFRS, 1956, **2**, 269–90.

SIPPY, Bertram Welton (1866–1924). American gastroenterologist; introduced Sippy diet for treatment of duodenal ulcer.

PALMER, W.L. Dr Bertram W. Sippy's contributions to medicine. *Proceedings of the Institute of the History of Medicine of Chicago*, 1968, **27**, 75–84.
PALMER, W.L. Bertram W. Sippy (1866–1924). Memorial chapters. Aransas Pass, Texas, Biography Press, 1974. 121 ff.

SJÖQVIST, Carl Olof (b. 1901). Swedish neurosurgeon; introduced tractotomy in the treatment of trigeminal neuralgia, 1937.

LIDBERG, L. Olof Sjöqvist. *Svensk Medicinhistorisk Tidskrift*, 1997, **1**, 193–7.

SKINNER, Burrhus Frederic (1904–1990). American psychologist; professor at Harvard University.

BJORK, D.W. B.F. Skinner; a life. New York, Basic Books, 1993. 298 pp.
ŠKODA, Josef (1805–1881). Bohemian physician; remembered chiefly for his work on percussion and auscultation.

346

HORNOF, Z. DSB, 1975, 12, 450–51.

SMITH, L.D. & WOODWARD, W.R. B.F. Skinner and behaviorism in American culture. Bethlehem, Lehigh Univ. Press, 1996. 348 pp.

STERNBERG, M. Josef Škoda. Wien, J. Springer, 1924. 92 pp.

SKRYABIN, Konstantin Ivanovich (1878–1972). Russian helminthologist.

AKADEMIE NAUK SSSR. Konstantin Ivanovich Skriabin 1878–1972. Moskva, Nauka, 1976. 206 pp.

ANTIPIN, D.N. & SHIKHOBALOVA, N.P. Akademik Konstantin Ivanovich Skryabin. Moskva, Selkhozgiz, 1949. 171 pp.

SHIKHOBALOVA, N.P. DSB, 1975, **12**, 452–3.

SLOANE, Hans (1660–1753). British physician, naturalist and collector; his collection formed the nucleus of the British Museum.

BROOKS, E. St J. Sir Hans Sloane, the great collector and his circle. London, Batchworth Press, 1954. 234 pp.

DE BEER, G.R. Sir Hans Sloane and the British Museum. London, Oxford Univ. Press, for British Museum, 1953. 192 pp.

MacGREGOR, A. (ed.) Sir Hans Sloane: collector, scientist, antiquary; founding father of the British Museum. London, British Museum Press, 1995. 308 pp.

SLOAN, W.R. Sir Hans Sloane, founder of the British Museum: legend and lineage. Helen's Bay, Co. Down, W.R. Sloan, 1981. 134 pp.

Archival material: British Library; British Museum (Natural History): Royal Society of London; Library, American Philosophical Society, Philadelphia.

SLUDER, Greenfield (1865–1928). American otologist; first to describe the syndrome of sphenopalatine-ganglion neuralgia, 1901.

BLISS, M.A. *Annals of Otology, Rhinology and Laryngology*, 1928, **37**, 1321–7.

SLYE, Maud (1879–1954). American geneticist.

McCOY, J.J. The cancer lady. Maud Slye and her heredity studies. Nashville, New York, Thomas Nelson, 1977. 191 pp.

SMADEL, Joseph Edwin (1907–1963). American microbiologist; introduced chloramphenicol in treatment of typhus.

WOODWARD, T.E. Joseph E. Smadel, 1907–1963. *Transactions of the Association of American Physicians*, 1964, **77**, 29–32.

SMELLIE, William (1697–1763). British obstetrician; introduced effective forceps and taught safe method of employing them.

JOHNSTONE, R.W. William Smellie, the master of British midwifery. Edinburgh, London, Livingstone, 1952. 139 pp.

KERR, R. Memoirs of the life, writings and correspondence of William Smellie, 2 vols. Edinburgh, 1811. Reprinted in facsimile, Bristol, Thoemmes Press, 1996. 504+488 pp.

SMILES, Samuel (1812–1904). British physician, biographer and social reformer; first published *Self help* in 1859.

JARVIS, A. Samuel Smiles and the construction of Victorian values. Stroud, Sutton, 1997. 176 pp.

SMITH, Andrew (1797–1872). Director-General, Army Medical Department during Crimean War; noted for scientific and anthropological studies in South Africa.

KIRBY, P.R. Sir Andrew Smith, his life, letters and works. Cape Town, Amsterdam, Balkema, 1965. 358 pp.

SMITH, Edward (?1818–1874). British physiologist and nutritionist.

CHAPMAN, C.B. DSB, 1975, **12**, 465–7.

SMITH, Ernest Lester (1904–1992). British chemist.

CUTHBERTSON, W.J.F. & PAGE, J.E. BMFRS, 1994, **40**, 348–65.

SMITH, Grafton Elliot, *see* ELLIOT SMITH, Grafton

SMITH, Hamilton Othanel (b. 1931). American microbial geneticist; shared Nobel Prize (Physiology or Medicine) 1978, for discovery of restriction enzymes and their applications to problems of molecular genetics.

Les Prix Nobel en 1978.

SMITH, Henry (1862–1948). Indian Army surgeon; extracted cataract within the capsule, 1900.

VAIL, D.T. The man with the cigar (Henry Smith). *American Journal of Ophthalmology,* 1973, **75**, 1–10.

SMITH, Herbert Williams (1919–1987). British microbiologist.

DATTA, N. BMFRS, 1988, **34**, 753–86.

SMITH, Homer William (1895–1962). American renal physiologist; professor, New York University College of Medicine.

CHASIS, H. & GOLDRING, W. (eds.) Homer William Smith, Sc.D.: his scientific and literary achievements. New York, NY Univ. Press, 1965. 282 pp.
PITTS, R.F. BMNAS, 1967, **39**, 445–70.

SMITH, Hugh Hollingsworth (b. 1902). American virologist; with Max Theiler developed method of yellow fever immunization without use of immune serum, 1937.

SMITH, H.H. Life's a pleasant institution: the peregrinations of a Rockefeller doctor. Tucson, Ariz., Smith, 1978. 254 pp.

SMITH, James Edward (1759–1828). British physician and botanist; owned collection and library of Linnaeus; founder of Linnean Society of London.

WALKER, M. Sir James Edward Smith, M.D., F.R.S., P.L.S.; first president of the Linnean Society. London, Linnean Society of London, 1988, 60 pp.

SMITH, Kenneth Manley (1892–1981). British virologist.

KASSANIS, B. BMFRS, 1982, **28**, 451–77.
Archival material: Natural Environment Research Council, Institute of Virology. Oxford.

SMITH, Michael (1932–2000). Canadian biochemist; shared Nobel Prize (Chemistry) 1993 for work on mutagenesis and its development for protein studies.

BIBLIOGRAPHY OF MEDICAL AND BIOMEDICAL BIOGRAPHY

ASTELL, C.R. BMFRS, 2001, **47**, 430–41.

SMITH, Nathan (1762–1829). American physician; founder of Dartmouth Medical School and professor of physic, surgery and obstetrics, Yale College, New Haven.

CUSHING, H.W. The medical career and other papers. Boston, Mass., Little, Brown, 1940. 302 pp. [The first address, 'The medical career' deals largely with the career of Nathan Smith.]
HAYWARD, O.S. & PUTNAM, C.E. Improve, perfect, perpetuate. Dr Nathan Smith and early American medical education. Hanover, Univ. Press of New England, 1998. 362 pp.
SMITH, Emily A. The life and letters of Nathan Smith. With an introduction by W.H. Welch. New Haven, Yale Univ. Press, 1914. 185 pp.
Archival material: Library, American Philosophical Society, Philadelphia.

SMITH, Philip Edward (1884–1970). American anatomist and embryologist.

AGATE, F.J. DSB, 1975, **12**, 472–7.

SMITH, Robert William (1807–1873). British physician; described neurofibromatosis ('Recklingshausen's disease') in 1849.

LYONS, J.B. & STAUNTON, H. Neurofibromatosis: why not Smith's disease? *Journal of the History of the Neurosciences*, 1992, **1**, 65–73.

SMITH, Sydney Alfred (1883–1969). British physician; Regius Professor of Forensic Medicine, Edinburgh University.

SMITH, S. Mostly murder. London, Harrap, 1959, 318 pp.

SMITH, Theobald (1859–1934). American microbiologist and comparative pathologist.

DOLMAN, C.E. DSB, 1973, **12**, 480–86.
NUTTALL, G.H.F. ONFRS, 1932–35, **1**, 515–21.
ZINSSER, H. BMNAS, 1936, **17**, 261–303.

SMITH, Thomas Southwood (1788–1861). British physician; pioneer in sanitary reform.

LEWES, C. Dr. Southwood Smith: a retrospect. Edinburgh, London, Blackwood, 1898. 169 pp.
POYNTER, F.N.L. Thomas Southwood Smith – the man (1788–1861). *Proceedings of the Royal Society of Medicine*, 1962, **55**, 381–92.

SMITH, William Robert (1850–1932). British physician; professor of forensic medicine and toxicology, University of London.

LYCETT, C.D.L. Sir William Robert Smith, 1850–1932: a short biography. London, Royal Institute of Public Health and Hygiene, 1989. 131 pp.

SMITH, Wilson (1897–1965). British microbiologist.

EVANS, D.G. BMFRS, 1966, **12**, 479–87.
EVANS, D.G. DSB, 1975, **12**, 492–3.

SMITHERS, David Waldron (1908–1995). British radiologist; professor of radiotherapy, Royal Marsden Hospital.

SMITHERS, D.W. Not a moment to lose: some reminiscences. London, British Medical Journal/Memoir Club, 1989. 112 pp.

SMITHY, Horace Gilbert (1914–1948). American cardiac surgeon.

KILEY, J.C. The heart of a surgeon: the life and writings of Horace Gilbert Smithy, MD, heart surgeon (1914–1948). Berryvale, VA, J.C. Kiley, 1984. 158 pp.

SMOLLETT, Tobias George (1721–1771). British physician and novelist.

BRUCE, D. Radical Doctor Smollett. London, Gollancz, 1964. 240 pp.
KNAPP, L.M. Tobias Smollett, doctor of men and manners. Princeton, NJ, Princeton Univ. Press, 1949. 362 pp.
LEWIS, J. Tobias Smollett. London, Jonathan Cape, 2003. 316 pp.

SMYTH, Andrew Woods (1832–1916). American surgeon; performed the first successful ligation of the innominate artery, 1864.

–. *New Orleans Medical and Surgical Journal*, 1920–1921, **73**, 352–8.

SMYTH, David Henry (1908–1979). British physiologist.

BARCROFT, H. & MATTHEWS, D.M. BMFRS, 1981, **27**, 525–61.

SNELL, George D. (1903–1996). American immunologist.

–. The Nobel Prize for Physiology or Medicine, 1980, awarded to Baruj Benacerraf, Jean Dauset and George D. Snell. *Scandinavian Journal of Immunology*, 1922, **35**, 373–98.
KLEIN, J. In Memoriam: George D. Snell (1903–1996). The last of the just. Immunogenetics, 1996, **44**, 409–18.
McKENZIE, L.F. George Snell: reminiscences 1969–1990. *Immunogenetics*, 1997, **46**, 10–16.

SNELLEN, Herman (1834–1908). Dutch ophthalmologist; introduced ('Snellen's') sight test types.

SISSON, E.O. *Ophthalmic Record*, 1911, **20**, 16–20.

SNOW, John (1813–1858). British physician, epidemiologist and first professional anaesthetist.

BARRETT, N.R. A tribute to John Snow, M.D., London, 1813–1858. *Bulletin of the History of Medicine*, 1946, **19**, 517–35.
SHEPHARD, D.A.C. John Snow: anaesthetist to a Queen and epidemiologist to a nation. Cornwall, PE, Canada, York Point Publishing, 1995. 347 pp.
THOMAS, K.B. DSB, 1975, **12**, 502–3.
Archival material: Clover/Snow Collection, Woodward Biomedical Library, University of British Columbia [annotated by K.B. Thomas in *Anaesthesia*, 1972, **27**, 436–49].

SOCIN, August (1837–1899). Swiss surgeon; professor in Basle.

MEIER, S. August Socin (1837–1899). Leben und Werk des Basler Chirurgen. Basel, Schwabe, 1985. 74 pp.

SOEMMERRING, Samuel Thomas (1755–1830). German anatomist.

BAST, T.H. The life and works of Samuel Thomas Soemmerring. *Annals of Medical History*, 1924, **6**, 369–86.

HINTZSCHE, E. DSB, 1975, **12**, 492–3.

WAGNER, R. Samuel Thomas Soemmerrings Leben uns Verkehr mit seinem Zeitgenossen. Nachdruck der Ausgabe Leipzig, Voss, von 1844. Stuttgart, Fischer, 1986. 712 pp.

WENZEL, M. *et al.* Samuel Thomas Soemmerring: Naturforscher der Goethezeit in Kassel. Kassel, Buch- und Kunstverlag Weber und Weidemayer, *c.*1991. 131 pp.

SOLIS-COHEN, Jacob da Silva (1838–1927). American otorhinolaryngologist; first successfully to remove larynx for cancer, 1867.

KAGAN, S.R. Jacob da Silva Solis-Cohen (1838–1927). *Medical Life*, 1937, **44**, 291–313.

SOLOMONS, Bethel Albert Herbert (1885–1965). Irish gynaecologist; Master of the Rotunda Hospital, Dublin.

SOLOMONS, B. One doctor in his time. London, Christopher Johnson, 1956. 224 pp.

SONDEREGGER, Jakob Laurenz (1825–1896). Swiss physician and medical administrator.

BURKHARDT, R. Arzt und Menschenfreund, Dr St Gall: Doktor Jakob Laurenz Sonderegger. St Gallen, Evangelistische Gesellschaft, 1925. 231 pp.

HAFFTER, E. Dr L. Sonderegger in seinen Selbstbiographie und seine Briefen. Frauenfeld, J. Huber, 1898. 498 pp.

SONES, Frank Mason (1918–1985). American physician; jointly performed cine-coronary arteriography.

SHELDON, W.C. F. Mason Sones, Jr.: stormy petrel of cardiology. *Clinical Cardiology*, 1994, **17**, 405–7.

SONNEBORN, Tracy Morton (1905–1981). American geneticist.

BEALE, G.H. BMFRS, 1982, **29**, 537–74.

SORANUS of Ephesus (*fl.* 2nd century AD). Greek physician in Rome.

MICHLER, M. DSB, 1975, **12**, 538–42.

SØRENSEN, Peder, *see* **SEVERINUS, Petrus**

SOUBEIRAN, Eugène (1793–1858). French chemist; discovered chloroform, 1831.

ROBIQUET, F. Élogue de M. Soubeiran. Paris, Renon & Maude, 1859. 16 pp.

SOURDILLE, Maurice Louis Joseph (1885–1961). French otologist.

LEGENT, F. Maurice Sourdille, précurseur de la chirurgie moderne de l'oreille. *Revue de Laryngologie, Otologie, Rhinologie*, 1984, **105**, 511–14.

SOUTH, John Flint (1797–1882). British surgeon at St Thomas' Hospital, London.

FELTOE, C.L. Memorials of John Flint South, collected by Charles Lett Feltoe. London, Murray, 1884. 220 pp. Facsimile reprint, Fontwell, Centaur Press, 1984. 220 pp. [With new introduction by R.Gittings, who omitted Feltoe's.]

SOUTTAR, Henry Sessions (1875–1964). British surgeon.

ELLIS, R.H. & ADAMS, A.K. Henry Souttar and the surgery of the mitral valve. *Journal of Medical Biography*, 1997, **5**, 8–13, 63–9.

BIBLIOGRAPHY OF MEDICAL AND BIOMEDICAL BIOGRAPHY

SOXHLET, Franz von (1848–1926). Czechoslovak physiologist.

ROMMEL, O. Franz v. Soxhlet (Zum ehrenden Gedächtnis). *Münchener Medizinischer Wochenschrift*, 1926, **73**, 994–5.

SPAHLINGER, Henri (1882–1965). Swiss bacteriologist; introduced Spahlinger method of treatment of tuberculosis.

MACASSEY, L.L. & SALEEBY, C.W. (eds.) Spahlinger contra tuberculosis, 1908–1934; an international tribute. London, John Bale, 1934. 271 pp.
Archival material: CMAC.

SPALDING, Lyman (1775–1821). American physician.

SPALDING, J.A. Dr. Lyman Spalding, the originator of the United States pharmacopoeia, co-labourer with Dr. Nathan Smith in the founding of the Dartmouth medical school, and its first chemical lecturer, president and professor of anatomy and surgery of the College of Physicians and Surgeons of the Western District at Fairfield, NY, by his grandson. Boston, Leonard, 1916. 380 pp.

SPALLANZANI, Lazzaro (1729–1799). Italian physiologist and experimental biologist.

CAPPARONI, P. Lazzaro Spallanzani. Torino, Unione Tip.-Edit. Torinese, 1941. 282 pp. [Corrected reprint, 1948. 310 pp.]
DOLMAN, C.E. DSB, 1975, 12, 553–67.
PIETRO, P. di. Lazzaro Spallanzani. Modena, Aedes Moratorcana, 1979. 324 pp.
PRANDI, D. Bibliografia delle opera di Lazzaro Spallanzani, Firenze, Sonsoni Antiquaristo, 1951. 171 pp.
ROSTAND, J. Les origines de la biologie expérimentale de l'Abbé Spallanzani. Paris, Fasquelle, 1951. 284 pp.

SPATZ, Hugo (1888-1969). German neurologist.

KAHLE, W. Hugo Spatz (1888–1969). *Zeitschrift für Mikroskopisch-Anatomische Forschung*, 1971, **82**, 1–6.

SPEMANN, Hans (1869–1941). German embryologist; professor at Rostock and Freiburg im Breisgau; Nobel prizewinner 1935 for discovery of the organizer in animal development.

HAMBURGER, V. The heritage of experimental embryology: Hans Spemann and the organizer. Oxford, Oxford University Press, 1988. 196 pp.
MANGOLD, O. Hans Spemann; ein Meister der Entwicklungsphysiologie; sein Leben und sein Werk. 2nd edition. Stuttgart, Wissenschaftliche Verlagsgesellschaft, 1982. 254 pp.
SPEMANN, F.W. (ed.) Hans Spemann: Forschung und Leben. Stuttgart, J. Engelhorn, 1943. 344 pp.
WADDINGTON, C.H. DSB, 1975, **12**, 567–9.

SPENS, Thomas (1764–1842). British physician.

LEA, C.E. Dr Thomas Spens: the first describer of the Stokes-Adams syndrome. *Edinburgh Medical Journal*, 1914, n.s. **13**, 51–5.

SPERANSKII, Aleksei Dimitrievivh (1888–1961). Russian pathologist.

PLETSITYI, D.F. A.D.Speranskii. Moskva, Meditsina, 1967, 56 pp.

352

SPERRY, Roger Wolcott (1913–1994). American psychobiologist; shared Nobel Prize (Physiology or Medicine) 1981, for discoveries concerning functional specialization of the cerebral hemispheres.

> VONEIDA, T.J. BMFRS, 1997, **43**, 461–70.
> VONEIDA, T.J. BMNAS, 1997, **71**, 315–31.

SPIEGEL, Ernest Adolf (1895–1985). American surgeon; first to perform stereotactic surgery in humans.

> GOLDENBERG, P.L. A tribute to Ernest A. Spiegel. *Confinia Neurologica*, 1975, **37**, 317–28.
> KRAYENBÜHL, Professor Ernest A. Spiegel's 80th birthday. *Confinia Neurologica*, 1975, **37**, 356–63.

SPIEGHEL, Adriaan van den [SPIGELIUS] (1578–1625). Flemish physician, born Brussels; professor in Venice.

> LINDEBOOM, G.A. DSB, 1975, **12**, 577–8.
> LINDEBOOM, G.A. 'Adriaan van den Spieghel (1578–1625): hoogleraar in de ontleed en heelkunde te Padua'. Amsterdam, Rodopi, 1978. 125 pp.

SPILSBURY, Bernard (1877–1947). British forensic pathologist.

> BROWNE, D.G. & TULLET, E.V. Bernard Spilsbury, his life and cases. London, Harrap, 1951. 422 pp.
> ECKERT, W.G. The writings of Sir Bernard Spilsbury. *American Journal of Forensic Medicine and Pathology*, 1984, **5**, 231–8; 1986, **6**, 31–7.

SPINKS, Alfred (1917–1982). British biochemist and pharmacologist.

> JOHNSON, A.W., ROSE, F.L. & SUCKLING, C.W., BMFRS, 1984, **30**, 567–94.

SPOCK, Benjamin McLane (1903–1998). American paediatrician.

> BLOOM, L. Z. Doctor Spock: biography of a conservative radical. Indianapolis, New York, Bobbs-Merrill Co., 1972. 366 pp.
> MAIER, T. Dr Spock: an American life. New York, Harcourt Brace, *c.*1998. 520 pp.
> SPOCK, B. & MORGAN, M. Spock on Spock: a memoir of growing with the century. New York, Pantheon Books, 1989. 281 pp.

SPRENGEL, Kurt Polycarp Joachim (1766–1833). German botanist and medical historian; professor of medicine and of botany, Halle,.

> KAISER, W. & VÖLKER, A. Kurt Sprengel (1766–1833). Halle, Universität, 1982. 115 pp.
> RISSE, G.B. DSB, 1975, **12**, 591–2.

SPRENGEL, Otto Gerhard (1852–1915). German surgeon; described upward displacement of the scapula ('Sprengel's deformity'), 1891.

> BUZEL, H.F. *Archiv für Klinische Chirurgie*, 1915, **106**, i–x.

SPURZHEIM, Johann Christoph Caspar (1776–1832). German neuroanatomist, psychologist and phrenologist.

> CARMICHAEL, A. A memoir of the life and philosophy of Spurzheim. Dublin, Wakeman, 1833. 96 pp.

WALSH, A.A. Johann Christoph Spurzheim and the rise and fall of scientific phrenology in Boston, 1832–1847 [Thesis, University of New Hampshire.] Ann Arbor, University Microfilms, 1974. 547 pp.
WALSH, A.A. DSB, 1975, **12**, 596–7.
Archival material: Harvard Medical School.

SQUIBB, Edward Robinson (1819–1900). American drug manufacturer; a founder of American medical chemistry.

BLOCHMAN, L.G. Doctor Squibb: the life and times of a rugged idealist. New York, Simon & Schuster, 1958. 371 pp.

STACEY, Maurice (1907–1994). British chemist.

OVEREND, W.G. BMFRS, 1997, **43**, 471–89.

STACKE, Ludwig (1859–1918). German otorhinolaryngologist.

KRETSCHMANN, F. *Zeitschrift für Ohrenheilkunde*, 1919, **77**, 195–7.

STACPOOLE, Henry de Vere (1863–1951). British physician and writer.

British Medical Journal, 1951, **1**, 890.

STADIE, William Christopher (1886–1959). American physician; professor of research medicine, University of Pennsylvania.

STARR, I. BMNAS, 1989, **58**, 513–28.

STAFFORD, Richard Anthony (1810–1854). British physician and surgeon; first (1839) to record sarcoma of the prostate.

PETTIGREW, 1840, **4**, No.12. 8pp. (CB)

STAHL, Georg Ernst (1660–1734). German physician and chemist.

KING, L.S. DSB, 1975, **12**, 599–606.
KAISER, W. & VÖLKER, A. Georg Ernst Stahl (1659–1734). Halle, Martin-Luther-Universität Halle-Willerberg, 1985. 321 pp.
STRUBE, I. Georg Ernst Stahl. Leipzig, Teubner, 1984. 82 pp.

STANIER, Roger Yate (1916–1982). Canadian microbiologist.

CLARKE, P.H. BMFRS, 1986, **32**, 543–68.
STANIER, R.Y. The journey, not the arrival. *Annual Review of Microbiology*, 1980, **34**, 1–48.

STANLEY, Edward (1793–1862). British surgeon; introduced liver puncture as a diagnostic procedure.

PLARR (CB).

STANLEY, Wendell Meredith (1904–1971). American chemist and virologist; shared Nobel Prize 1946 for work on crystallization of enzymes.

COHEN, S.S. DSB, 1990, **18**, 841–8.

STANNIUS, Hermann Friedrich (1808–1883). German comparative anatomist and physiologist.

ROTHSCHUH, K.E. DSB, 1975, **12**, 611–12.

STARE, Frederick John (b.1910). American physician; professor of nutrition at Harvard.

STARE, F.J. Adventures in nutrition: an autobiography. Hanover, Mass., Christopher Publishing House, 1991. 388 pp.

STARLING, Ernest Henry (1866–1927). British physiologist; professor at University College London.

CHAPMAN, C.B. Ernest Henry Starling, the clinician's physiologist. *Annals of Internal Medicine*, 1962, Suppl. 2, 1–43.
CHAPMAN, C.B. DSB, 1975, **12**, 617–19.
EVANS, C.L. Reminiscences of Bayliss and Starling. Cambridge, Cambridge Univ. Press, 1964. 17 pp.
HENDRIKSEN, J.H. Ernest Henry Starling (1866–1927); physician and physiologist – a short biography. Copenhagen, Laegeforeningens Forlag, 2000. 140 pp.
HENDRIKSEN, J.H. Starling, his contemporaries and the Nobel Prize: one hundred years of hormones. *Scandinavian Journal of Clinical Investigation*, 2003, **63**, Suppl. 238.
Archival material: Churchill College, Cambridge; Department of Physiology, University College London.

STARR, Arthur (b. 1926). American thoracic surgeon; with M. Lowell Edwards introduced ball valve prosthesis for replacement of mitral valve, 1961.

MATTHEWS, A.M. The development of the Starr Edwards heart valve. *Texas Heart Institute Journal*, 1998, **25**, 282–93.

STARZL, Thomas Earl (b. 1926). American surgeon; pioneer in liver transplantation.

STARZL, T.E. The puzzle picture: memoirs of a transplant surgeon. Pittsburgh, Univ. Pittsburgh Press, *c*.1992. 364 pp.

STEDMAN, Edgar (1890–1975). British biochemist.

CRUFT, H.J. BMFRS, 1976, **22**, 529–53.

STEENBOCK, Harry (1886–1967). American biochemist and nutritionist; professor of biochemistry, University of Wisconsin.

BORELL, M. DSB, 1990, **18**, 849–51.

STEIN, William Howard (1911–1980). American biochemist at Rockefeller Institute; shared Nobel Prize (chemistry) 1972.

MOORE, S. BMNAS, 1987, **56**, 354–85.
SMITH, E.L. DSB, 1990, **18**, 851–5.

STEINACH, Eugen (1861–1944). Austrian surgeon and gerontologist.

BENJAMIN, A. Eugen Steinach (1861–1944): a life of research. *Scientific Monthly*, 1945, **61**, 427–42.

BIBLIOGRAPHY OF MEDICAL AND BIOMEDICAL BIOGRAPHY

STEINHEIM, Salomon Levi (1789–1866). German physician; accredited with the first account of parathyroid tetany.

SCHOEPS, H.J. *et al.* (eds.) Salomon Ludwig Steinheim zum Gedanken. Ein Sammelband. Leiden, Brill, 1966. 359 pp.

STEINSCHNEIDER, Moritz (1816–1907). German medical and scientific historian and archivist.

GARRISON, F.H. Bibliographie der Arbeiten Moritz Steinschneiders zur Geschichte der Medizin und der Naturwissenschaften. *Archiv für Geschichte der Medizin*, 1932, **25**, 249–78.
PAGEL, J.L. Moritz Steinschneider. *Janus*, 1907, **25**, 110–12.

STEKEL, Wilhelm (1868–1940). Austrian psychoanalyst.

GUTHEIL, E.A. (ed.). The autobiography of Wilhelm Stekel; the life story of a pioneer psychoanalyst. New York, Liveright Publishing Corp., 1950. 293 pp.

STENSEN, Niels [STENO, Nicolaus] (1638–1686). Dutch geologist, physiologist and bishop; discovered the parotid duct.

BIERBAUM, F., FALLER, A. & TRÄGER, J. Niels Stensen. Anatom, Geologe und Bischof. 3rd edn, Münster, Aschendorff, 1989. 208 pp.
MOE, H. Nicolaus Steno: an illustrated biography; his tireless pursuit of knowledge. his genius, his quest for the absolute. Copenhagen, Rhodos, *c.*1994. 180 pp.
POULSEN, J.E. & SNORRASON, E. Nicolaus Steno (1638–1686). A reconsideration by Danish scientists. Gentofte, Denmark, Nordisk Insulin-Laboratorium, 1986. 224 pp.
SCHERZ, G. Vom Wege Niels Stensens. Copenhagen, Munksgaard, 1956. 248 pp. Includes list of his works.
SCHERZ, G. DSB, 1976, **13**, 30–35.
SCHERZ, G. Niels Stensen: eine Biographie. 2 vols. Leipzig, St Benno Verlag, 1987–88. 694 pp.

STEPHENS, John William Watson (1865–1946). British malariologist; professor of tropical medicine, Liverpool University.

CHRISTOPHERS, S.R. ONFRS, 1945–48, **5**, 525–40.

STEPHENSON, Marjory (1888–1948). British biochemist; reader in chemical microbiology, University of Cambridge.

KOHLER, R.E. Jr. DSB, 1990, **18**, 857–60.
ROBERTSON, M. ONFRS, 1948–49, **6**, 563–77.

STEPTOE, Patrick Christopher (1913–1988). British obstetrician and gynaecologist; pioneer of *in vitro* fertilization techniques.

EDWARDS, R.G. BMFRS, 1996, **42**, 434–52.

STERN, Curt (1902–1981). German/American geneticist.

CASPARI, E. DSB, 1990, **18**, 860–66.
NEEL, J.V. BMNAS, 1987, **56**, 443–73.
Archival material: Library, American Philosophical Society, Philadelphia.

STERNBERG, Carl (1872–1936). Austrian pathologist; described lymphogranuloma and lymphogranulomatosis.

356

PALTAUF, R. Carl Sternberg 29.XI.1872–15.VIII.1935. *Verhandlungen der Deutschen Pathologischen Gesellschaft*, 1937, **29**, 417–25.

STERNBERG, George Miller (1838–1915). American bacteriologist; Surgeon General, US Army.

GIBSON, J.M. Soldier in white. The life of General George Miller Sternberg. Durham, NC, Duke Univ.Press, 1958. 277 pp.
STERNBERG, M.L. George Miller Sternberg: a biography by his wife. Chicago, American Medical Association, 1920. 331 pp.

STETTIN, Dewitt (1909–1990). American biochemist.

SIEGMILLER, J.E. BMNAS, 1997, **71**, 333–45.

STEVENS, Edward (1755–1834). American physician and physiologist; isolated human gastric juice.

DAY, S.B. (ed.) Edward Stevens, gastric physiologist, physician, and American statesman, with a complete translation of his inaugural dissertation 'De alimentorum concoctione' and interpretive notes on gastric digestion along with certain other selected and diplomatic papers. Montreal, Cincinnati, Cultural & Educational Productions, 1968. 179 pp.
DAY, S.B. DSB, 1976, **13**, 46–7.

STEVENS, Nettie Maria (1861–1912). American cytologist.

MAIENSCHEIN, J. DSB, 1990, **18**, 867–9.

STEVENS, Stanley Smith (1906–1973). American psychophysicist..

MILLER, G.A. BMNAS, 1974, **47**, 425–59.

STEWART, Alexander Patrick (1813–1883). British physician; differentiated typhus from typhoid, 1840.

SENATOR, H. Selected monographs. London, New Sydenham Society, 1881, pp. 222–6.

STEWART, Alice Mary (1906–2002). British physician; drew attention to the dangers of nuclear radiation.

GREENE, G. The woman who knew too much: Alice Stewart and the secrets of radiation. Ann Arbor, Univ. of Michigan Press, 1999. 321 pp.

STEWART, George Neil (1860–1930). Canadian physiologist.

MILLER, G. DSB, 1976, **13**, 53–4.

STILES, Charles Wardell (1867–1941). American zoologist.

CASSEDY, J.H. DSB, 1976, **13**, 62–3.

STILL, Andrew Taylor (1828–1917). American founder of osteopathy.

HILDRETH, A.G. The lengthening shadow of Dr Andrew Taylor Still. Macon, Miss., A.G. Hildreth, 1938, 464 pp.
SCHNUCKER, R.V. (ed.) Early osteopathy in the words of A.T.Still. Kirksville, Missouri, Thomas Jefferson Univ. Press, 1991. 382 pp.

TROWBIDGE, C. Andrew Taylor Still (1828–1917). Kirksville, Mo., Thomas Jefferson Univ. Press, 1991. 457 pp.

STILL, George Frederic (1868–1941). British paediatrician; first professor of paediatrics in England (at King's College Hospital, London); described chronic arthritic rheumatism in children ('Still's disease').

BYWATERS, E.G.L. George Frederic Still (1868–1941): his life and work. *Journal of Medical Biography*, 1994, **2**, 125–31.

STOERCK, Anton (1731–1803). German physician.

ZUMSTEIN, B. Anton Stoerck (1731–1803) und seine therapeutischen Versuche. Zürich, Juris, 1968. 48 pp.

STOERK, Oskar (1870–1926). Austrian pathologist; gave first clinical and pathological description of bundle-branch block.

WIESNER, R. Oskar Stoerk 27.V.1870–1.II.1926. *Verhandlungen der Deutschen Pathologischen Gesellschaft*, 1937, **29**, 425–8.

STOKES, Adrian (1887–1927). British bacteriologist; professor of bacteriology, Trinity College Dublin; died of yellow fever while investigating the disease.

DUNN, W. Adrian Stokes, DSO, OBE, MD Dubl, FRCSI, MRCP Lond. *Guy's Hospital Reports*, 1928, **78**, 1–17.
HUDSON, N.P. Adrian Stokes and yellow fever research; a tribute. *Transactions of the Royal Society of Tropical Medicine and Hygiene*, 1966, **60**, 170–74.

STOKES, William (1804–1878). Irish physician; Regius Professor of Medicine, Dublin University; pioneer in research on chest diseases.

HALE-WHITE, pp. 124–42 (CB).
STOKES, W. William Stokes, his life and work (1804–1878), by his son. London, Fisher Unwin, 1898. 256 pp.

STOKES, William (1839–1900). British surgeon (Gritti-Stokes amputation).

TAYLOR, W. Selected papers...by William Stokes; with a memoir of the author by Alexander Ogston [pp. xi–xxv]. London, Bailliere, Tindall & Cox, 1902. 484 pp.

STOLL, Arthur (1887–1971). Swiss pharmaceutical chemist.

RUZICKA, L. BMFRS, 1972, **18**, 567–93.
Archival material: Sandoz Ltd., Basel.

STONE, Wilson Stuart (1907–1968). American geneticist; professor at University of Texas.

CROW, J.F. BMNAS, 1980, **52**, 451–68.

STOPES, Marie Charlotte Carmichael (1880–1958). British advocate of birth control.

EATON, P. & WARNICK, M. Marie Stopes; a check list of her writings. London, Croom Helm, 1977. 59 pp.
HALL, R. Marie Stopes. London, Deutsch, 1977. 352 pp.
STOPES-ROE, H.V. & SCOTT, I. Marie Stopes and birth control. Hove, Priory Press, 1974. 96 pp.
Archival material: CMAC; British Library.

BIBLIOGRAPHY OF MEDICAL AND BIOMEDICAL BIOGRAPHY

STOPFORD, John Sebastian Bach (1888–1961). British neurologist.

CLARK, W.E. Le Gros, & COOPER, W.M. BMFRS, 1961, **7**, 271–9.

STORER, Horatio Robinson (1830–1922). American gynaecologist and medical numismatist.

DYER, F.N. Champion of women and the unborn: Horatio Robinson Storer, M.D. Canton, MA, Science History Publications, 1999. 614 pp.

STORM VAN LEEUWEN, Willem (1882–1933). Dutch allergologist.

BEUMER, H.M. Willem Storm van Leeuwen und sein Bedeutung für die Asthmaforschung. Düsseldorf, Michael Triltsch Verlag, 1968. 66 pp.

STOWE, Emily Howard Jennings (1831–1903). Canadian doctor and suffragist; first Canadian woman to practise in Canada.

FRYER, M.B. Emily Stowe: doctor and suffragist. Toronto, Hannah Institute and Dundurn Press, 1990. 150 pp.

STRASBURGER, Eduard Adolf (1844–1912). Polish cytologist and botanist.

ROBINSON, G. DSB, 1976, **13**, 87–90.

STRASSER, Charlot (1884–1950). Swiss psychiatrist.

HEINRICH, D. Dr.med. Charlot Strasser (1884–1950), ein Schweizer Psychiater als Schriftsteller, Sozial- und Kulturpolitiker. Zürich, Juris, 1986. 182 pp.

STRATTON, George Malcolm (1868–1957). American experimental psychologist; professor at University of California.

TOLMAN, E.C. BMNAS, 1961, **35**, 292–306.

STRAZHESKO, Nikolai Dmitrievich (1876–1952). Russian physician; diagnosed coronary thrombosis before death, 1910.

BOBROV, V. [Academician Nikolai Strazhesko.] *Agapit*, 1996–97, No. 5–6, 31–36 [In Russian and English.]

STREETER, George Linius (1873–1948). American anatomist.

CORNER, G.W. BMNAS, 1954, **28**, 261–87.
CORNER, G.W. DSB, 1976, **13**, 96–8.

STREISINGER, George (1927–1984). American geneticist, born Hungary.

STAHL, F.W. BMNAS, 1995, **68**, 353–61.

STROMEYER, Georg Friedrich Louis (1804–1876). German orthopaedic surgeon; founder of modern surgery of the locomotor system; 'Father of military surgery in Germany'.

GEORG FRIEDRICH LOUIS STROMEYER. *Deutsches Archiv für Geschichte der Medizin*, 1884, **7**, 195–261, 273–327.
ROHLFS, H. Geschichte der Deutschen Medizin, 1885, vol. 4, 139–260.
STROMEYER, G.F.L. Erinnerungen eines deutschen Arztes. 2te Aufl. 2 vols. Hannover, Carl Rümpler, 1874. 458+484 pp.

359

BIBLIOGRAPHY OF MEDICAL AND BIOMEDICAL BIOGRAPHY

STRONG, Richard Pearson (1872–1948). American physician for tropical medicine.

CAMPBELL, K.H. Knots in the fabric: Richard Pearson Strong and the Bilibid Prison vaccine trials, 1905–1906. *Bulletin of the History of Medicine*, 1994, **68**, 600–638.
CHERNIN, E. Richard Pearson Strong and the iatrogenic plague disaster in Bilibid Prison, Manila, 1906. *Reviews of Infectious Diseases*, 1989, **11**, 996–1004.

STRÜMPEL, Ernst Adolf Gustav Gottfried (1853–1925). German clinician; wrote important treatise on internal medicine.

STRÜMPEL A. Aus dem Leben eines Deutschen Klinikers. Erinnerungen und Beobachtungen. Leipzig, F.C.W. Vogel, 1925. 294 pp.

STRUENSEE, Johann Friedrich von (1737–1772). German physician and statesman; physician to Christian VII of Denmark; minister of state executed for treason.

ENQUIST, P.O. The visit of the royal physician. London, Harvill, 2002, 308 pp.
NEUMANN, R. The Queen's doctor, being the strange case of the rise and fall of Struensee, dictator, lover, and doctor of medicine. London, Gollancz, 1936. 429 pp.
WINKLE, S. Johann Friedrich Struensee: Arzt, Aufklärer und Staatsmann. 2nd ed. Stuttgart, G. Fischer, 1989. 658 pp. First published 1983.

STUART, Alexander (1673–1742). British physiologist and physician.

BROWN, T.W. DSB, 1976, **13**, 121–3.
Archival material: Hunterian Library, University of Glasgow; Royal College of Physicians, London.

STUART, Thomas Peter Anderson (1856–1920). Scottish-born physiologist; professor at Sydney University.

EPPS, W. Anderson Stuart, M.D. Sydney, Angus & Robertson, 1922. 177 pp.

STUBBS, George (1724–1806). British anatomist and artist.

DOCHERTY, T. The anatomical works of George Stubbs. London, Secker & Warburg, 1974. 345 pp.
EGERTON, J. George Stubbs, anatomist and animal painter. London, Tate Gallery, 1976. 64 pp.

STURTEVANT, Alfred Henry (1891–1970). American zoologist and geneticist.

LEWIS, E.B. DSB, 1976, **13**, 133–8.
Archival material: California Instititue of Technology, Pasadena.

SUDHOFF, Karl Friedrich Jakob (1853–1938). German historian of medicine; Director, Institut für Geschichte der Medizin, Leipzig; professor, University of Leipzig.

GARRISON, F.H. Professor Sudhoff and the Institute of Medical History at Leipzig. *Bulletin of the Society for Medical History of Chicago*, 1923–5, **3**, 1–32.
GARRISON, F.H. & TASKER, A.N. A bibliography of the writings of Professor Karl Sudhoff. *Bulletin of the Society for Medical History of Chicago*, 1923–5, **3**, 33–50.
MANI, N. DSB, 1976, **13**, 141–3.

SUE FAMILY, Family of French surgeons: **Jean** (1699–1792); **Pierre** (1739–1816); **Jean-Baptiste** (1760–1830); **Georges Antoine Thomas** (1792–1865).

360

FOLIE-DESJARDINE, P. Contribution à l'histoire de la médecine; un dynastie médico-chirurgicale, les Suë. Paris, Thèse, 1930. 212 pp.

VALLERY-RADOT, P. Chirurgiens d'autrefois: la famille d'Eugène Suë. Paris, R.G. Ricou, 1944. 214 pp.

SUE, Joseph Marie [Eugène] (1804–1857). French naval surgeon and novelist.

BORY, J.-L. Eugène Sue. Paris, Libraire Hachette, 1973. 448 pp.

SÜSSMILCH, Johann Peter (1707–1767). German statistician; a founder of population statistics.

BURGE, H. (ed.) Ursprunge der Demographie in Deutschland. Leben und Werk Johann Peter Süssmilchs, 1707–1797. Frankfurt, New York, Campus Verlag, 1986. 402 pp.

SULLIVAN, Harry Stack (1892–1949). American psychiatrist.

CHATELAINE, K.L. Harry Stack Sullivan: the formative years. Washington, University Press of America, 1981. 553 pp.

EVANS, F.B. Harry Stack Sullivan: interpersonal theory and psychotherapy. London, Routledge, 1996. 241 pp.

MULLAHY, P. The beginnings of modern American psychiatry. The ideas of Harry Stack Sullivan. Boston, Houghton Mifflin, 1973, 699 pp.

PERRY, H. S. Psychiatrist of America. The life of Harry Stack Sullivan. Cambridge, Mass., Belknap Press of Harvard Univ., 1982. 462 pp.

SULSTON, John (b.1931). British molecular biologist; shared Nobel Prize (Physiology or Medicine), 2002, for discoveries concerning genetic regulation of organ development and programmed cell death.

Les Prix Nobel en 2002.

SULZBERGER, Marion Baldur (1895–1983). American dermatologist; demonstrated atopic dermatitis.

BAER, R.L. *et al.* An appreciation of Marion B. Sulzberger. *International Journal of Dermatology*, 1977, **16**, 306–9 and following.

SUMNER, James Batcheller (1887–1955). American biochemist; shared Nobel Prize for chemistry 1946 for discovery that enzymes can be crystallized.

FRUTON, J.S. DSB, 1976, **13**, 152–3.

MAYNARD, L.A. BMNAS, 1958, **31**, 376–96.

Archival material: Olin Library, Cornell University, New York.

SUN YAT-SEN (1866–1925). Chinese statesman and first president of republican China; graduated in medicine, 1892.

MA, KAN-WEN. Sun Yat-Sen (1866–1925), a man to cure patients and the nation: his early years and medical career. *Journal of Medical Biography*, 1996, **4**, 161–70.

SUTHERLAND, Earl W. (1915–1974). American biochemist; Nobel Prize 1971 for discovery of cyclic AMP.

CORI, C.F. BMNAS, 1978, **49**, 319–50.

SUTTON, Henry Gawen (1837–1891). British pathologist.

ROBERTS, M. H. Gawen Sutton, late physician to the London Hospital: an appreciation. London, Baillière, Tindall & Cox, 1925. 20 pp.

SUTTON, John Bland, *see* **BLAND-SUTTON, John**

SUTTON, Walter Stanborough (1877–1916). American cytologist.

McKUSICK, V.A. DSB, 1976, **13**, 156–8.

SVEDBERG, Theodor (1884–1971). Swedish physical chemist; invented ultracentrifuge and applied it to study of proteins; Nobel Prize for chemistry, 1926.

CLAESSON, S. & PEDERSEN, K.O. BMFRS, 1972, **18**, 595–627.
CLAESSON, S. & PEDERSEN, K.O. DSB, 1976, **13**, 158–64.
KERKER, M. The Svedberg and molecular reality: an autobiographical postscript. *Isis*, 1986, **77**, 278–82.
Archival material: Uppsala University.

SWAIN, Clara A. (1834–1910). American missionary physician in India; founded first hospital for women and children in Asia.

WILSON, D.C. Dr Clara Swain, first woman missionary doctor, and the hospital she founded. New York, McGraw-Hill, 1968. 244 pp.

SWAMMERDAM, Jan (1637–1680). Dutch anatomist, naturalist, and early worker with the microscope; discovered the red blood corpuscles.

PÖHLMANN, O. Jan Swammerdam, Naturforscher und Arzt: biographischer Roman. Zürich, Füssli, 1941. 226 pp.
SCHIERBEEK, A. Jan Swammerdam (12 February 1637– 17 February 1680). His life and works. Amsterdam, Swets & Zeitlinger, 1967. 202 pp. First published in Dutch, Lochem, 1947.
WINSOR, M.P. DSB, 1976, **13**, 168–75.
Archival material: Universitätsbibliothek, Göttingen.

SWAN, Harold Charles James (b. 1922). Irish/American cardiologist; with co-workers introduced flow-guided button-tipped catheter in cardiac catheterization, 1970.

BURKE, K.G. From research lab. to routine procedure; a case study of the Swan-Ganz catheter. Thesis. Pittsburgh, University of Pennsylvania, 2001. Ann Arbor, Univ. Microfilms International, 2003. 207 pp.

SWANN, Michael Meredith (1920–1990). British cell biologist and immunologist.

MITCHISON, J.M. BMFRS, 1991, **37**, 445–60.
Catalogue of the papers and correspondence of Lord Swann. National Cataloguing Unit, Archives of Contemporary Scientists, Bath. 279 pp.
Archival material: Edinburgh University Library.

SWIETEN, Gerard van (1700–1772). Dutch physician; founded Vienna school of medicine.

BRECHKA, F.T. Gerard van Swieten and his world, 1700–1772. The Hague, Nijhoff, 1970. 171 pp.

LESKY, E. & ROHL, E. Schrifttum zu Leben und Werk Gerard van Swietens. In: Lesky, E. & Wandruska, A. Gerard van Swieten und seine Zeit. Internationales Symposium. Wien, H. Böhlaus, 1973, pp. 181–94.
VAN DER PAS, P.W. DSB, 1976, **13**, 181–3.

SWINGLE, Wilbur Willis (b. 1891). American endocrinologist.

COLLINS, E.J. (ed.) W.W. Swingle – teacher and investigator; four decades of American endocrinology, with special reference to the works of W.W. Swingle. Princeton, Princeton Univ. Press, 1959. 70 pp.

SYDENHAM, Thomas (1624–1689). English physician; known as 'the English Hippocrates'.

BATES, D.G. Thomas Sydenham: the development of his thought. Dissertation submitted to the Johns Hopkins University ... for the degree of Doctor of Philosophy, Baltimore, 1975. 371 pp. University Microfilms, Ann Arbor, 1982.
BATES, D.G. DSB, 1976, **13**, 213–15.
DEWHURST, K. (ed.) Dr Thomas Sydenham (1624–1689), his life and original writings. London, Wellcome Historical Medical Library, 1966. 191 pp.
MEYNELL, G.G. Materials for a biography of Dr Thomas Sydenham (1624–1689): a new survey of public and private archives. Folkestone, Winterdown Books, 1988. 107 pp.
MEYNELL, G.G. A bibliography of Dr Thomas Sydenham. Folkestone, Winterdown Books, 1990. 160 pp.
PAYNE, J.F. Thomas Sydenham. London, Fisher Unwin, 1900. 264 pp.
Archival material: Public Record Office, London; Royal College of Physicians, London; Bodleian Library, Oxford.

SYLVIUS, Francis de le Boë (1614–1672). Dutch physician; professor at Leiden.

BAUMANN, E.D. François de la Boe Sylvius. Leiden, E.J. Brill, 1949. 242 pp.
FOSTER, M. Lectures on the history of physiology. Cambridge, Cambridge University Press, 1901, pp. 145–73.
LINDEBOOM, G.A. DSB, 1976, **13**, 222–3.

SYLVIUS, Jacobus, *see* **DUBOIS, Jacques**

SYMCOTTS, John (?1592–1662). English medical practitioner in Huntingdon and Bedfordshire.

POYNTER, F.N.L. & BISHOP W.J. A seventeenth-century doctor and his patients: John Symcotts, 1592?–1662. Streatley, Bedfordshire Historical Record Society, 1951. 126 pp.

SYME, James (1799–1870). British surgeon; professor in Edinburgh; one of the first to adopt ether anaesthesia and a champion of Lister's antiseptic method.

PATERSON, R. Memorials of the life of James Syme, Professor of Clinical Surgery in Edinburgh. Edinburgh, Edmonston & Douglas, 1874. 334 pp.
SHEPHERD, J.A. Simpson and Syme of Edinburgh. Edinburgh, London, Livingstone, 1969. 288 pp.

SYNGE, Richard Laurence Millington (1914–1994). British chemist: shared Nobel Prize (Chemistry) 1952, with A.J.P. Martin, for their invention of partition chromatography.

GORDON, A.H. BMFRS, 1996, **42**, 454–479.
POWELL, T.E. et al. (compilers) Catalogue of the papers of R.L.M. Synge. National Cataloguing Unit for the Archives of Contemporary Scientists. 2 vols. 1998.
Archival material: Trinity College, Cambridge.

SZENT-GYÖRGI, Albert (1893–1987). Hungarian biochemist; Nobel prizewinner 1937.

KAMINER, B. (ed.). Search and discovery: a tribute to Albert Szent-Györgyi. New York, Academic Press, 1977. 344 pp. [Historical chapters by H. Krebs, L. Pauling, H.E. Huxley.]

MOSS, R.W. Free radical: Albert Szent-Györgyi and the battle over vitamin C. New York, Paragon House, 1988. 316 pp.

SZENT-GYÖRGYI, A. Lost in the twentieth century. *Annual Review of Biochemistry*, 1963, **32**, 1–14.

Archival material: US Library of Congress.

TADDEO ALDEROTTI, *see* THADDEUS FLORENTINUS

TAGLIACOZZI, Gaspare (1545–1599). Italian pioneer of plastic surgery.

GNUDI, M.T. & WEBSTER, J.P. The life and times of Gaspare Tagliacozzi, surgeon of Bologna (1545–1599) with a documentary study of the scientific and cultural life of Bologna in the sixteenth century. New York, Reichner, 1950. 538 pp.

TAIT, Robert Lawson (1845–1899). British gynaecologist; pioneered several operative techniques.

FLACK, I.H. Lawson Tait, 1845–1899. London, Heinemann, 1949. 148 pp.

McKAY, W.J.S. Lawson Tait, his life and work: a contribution to the history of abdominal surgery and gynaecology. London, Baillière, Tindall & Cox, 1922. 579 pp.

SHEPHERD, J.A. Lawson Tait, the rebellious surgeon. Lawrence, Kansas, Coronado Press, 1980. 249 pp.

TAKAKI, Kanehiro (1849–1920). Japanese physician; showed beriberi to be of dietary origin.

ITOKAWA, Y. Kanehiro Takaki (1849–1920) – a biographical sketch. *Journal of Nutrition*, 1976, **106**, 581–8.

TAKAMINE, Jokichi (1854–1922). Japanese chemist and pharmacologist; isolated adrenaline.

KAWAKAMI, H.K. Jokichi Takamine: a record of his American achievements. New York, W.E. Rudge, 1928. 73 pp.

TALIAFERRO, William Hay (1895–1973). American parasitologist and microbiologist; professor at University of Chicago.

TALIAFERRO, W.H. The lure of the unknown. *Annual Review of Microbiology*, 1968, **22**, 1–14.

TALMAGE, D.W. BMNAS, 1983, **54**, 373–407.

TANDLER, Julius (1869–1936). German anatomist.

GOETZL, A. & REYNOLDS, R.A. Julius Tandler: a biography. San Francisco, privately printed, 1944. 63 pp.

SABLIK, K. Julius Tandler: Mediziner und Socialreformer; eine Biographie. Wien, A. Schendl, 1983. 389 pp.

TARNIER, Stéphane (1828–1897). French obstetrician; introduced forceps, promoted antisepsis.

BAR, P. Le professeur S. Tarnier, 1828–1897. Paris, Carré & Naud, [1898]. 45 pp.

TASHIRO, Shiro (1883–1963). Japanese/American biochemist.

DAY, S.B. DSB, 1976, **13**, 262–3.

TATUM, Arthur Lawrie (1886–1955). American pharmacologist.

SWANN, J.P. Arthur Tatum, Parke-Davis, and the discovery of mapharsen as an antisyphilitic agent. *Journal of the History of Medicine*, 1985, **40**, 167–87.

TATUM, Edward Lawrie (1909–1975). American biochemist and geneticist; shared Nobel Prize 1958 for work on genetics.

LEDERBERG, J. BMNAS, 1990, **59**, 356–86.
Archival material: Rockefeller University Archive Center.

TAUSSIG, Helen Brooke (1898–1986). American physician: with A. Blalock devised operation for the relief of congenital defects of pulmonary artery.

BALDWIN, J. To heal the heart of a child: Helen Taussig, M.D. New York, Walker, 1992. 128 pp.

TAVEL, Ernst (1858–1912). Swiss bacteriologist and surgeon.

KARAMEHMEDOVIC, O. Ernst Tavel (1852–1912): Bakteriologe und Chirurg in Bern. Bern, H. Huber, *c*.1973. 71 pp.

TAYLOR, Alfred Swaine (1806–1880). British physician; forensic toxicologist.

COLEY, N.G. Alfred Swaine Taylor, MD (1806–1880): forensic toxicologist. *Medical History*, 1991, **35**, 409–27.

TeLINDE, Richard Wesley (b.1904). American gynaecologist; authority on endometriosis.

JONES, H.W., JONES, G.S. & TICKNOR, W.E. Richard Wesley TeLinde. Baltimore, Williams & Wilkins, 1986. 117 pp.

TEMIN, Howard Martin (1934–1994). American virologist; shared Nobel Prize (Physiology or Medicine) 1975 for work on interaction between tumour viruses and genetic material of the cell.

DULBECCO, R. BMFRS, 1995, **41**, 472–80.
SUGDEN, B. BMNAS, 2001, **79**, 337–74.

TENDELOO, Nicolaas Philip (1864–1945). Dutch pathologist, born in Celebes; professor of pathological anatomy, Leiden.

GROTE, 1924, **3**, 245–79 (CB).

TENON, Jacques René (1724–1816). French surgeon; responsible for hospital reforms in Paris.

LAIGNEL-LAVASTINE, M. Mémoires de Tenon sur les hôpitaux de Paris. *Histoire de la Médecine*, 1951, **1**, 38–44.

TERMAN, Lewis Madison (1877–1956). American psychologist; professor at Stanford University; advocate of mental testing.

BORING, E.G. BMNAS, 1959, **33**, 414–61.

BIBLIOGRAPHY OF MEDICAL AND BIOMEDICAL BIOGRAPHY

MINTON, H.L. Lewis M.Terman: pioneer in psychological testing. New York, University Press, 1988. 342 pp.

SEAGOE, M.V. Terman and the gifted. Los Altos, Cal., William Kaufmann Inc., 1975. 258 pp.

TERROINE, Emile Florent (1882–1984). French nutritional biochemist; professor of general physiology, University of Strasbourg.

TERROINE, E.F. Fifty-five years of union between biochemistry and physiology. *Annual Review of Biochemistry*, 1959, **28**, 1–14.

THACHER, James (1754–1844). American physician; compiled American Medical Biography, 1828.

STEINER, W.R. Dr James Thacher of Plymouth, Massachusetts, an erudite physician of Revolutionary and post-Revolutionary fame. *Bulletin of the Institute of the History of Medicine*, 1933, **1**, 157–73.

THACKRAH, Charles Turner (1795–1833). British physician; pioneer in study of industrial diseases; founder of Leeds Medical School.

MEIKLEJOHN, A. The life and times of Charles Turner Thackrah, surgeon and apothecary, Leeds (1795–1833). Edinburgh, Livingstone, 1957. 238 pp. [Includes selective bibliography and reprint of Thackrah's The effects of arts, trades and professions ... on health and longevity. 2nd ed., 1832.] Reprinted, Canton, Mass., Science History Publications, 1985.

THADDEUS FLORENTINUS [TADDEO ALDEROTTI] (1223–c.1295). Italian medical teacher; founder of Bolognese school.

SIRAISI, N.G. Taddeo Alderotti and his pupils. Two generations of Italian medical learning. Princeton, NJ, Princeton Univ. Press, [c.1981] 411 pp.

THAYER, William Sydney (1864–1932). American physician; professor of clinical medicine at Johns Hopkins University.

BERGER. G.H. DSB, 1976. 13, 300–301.

HUNLEY, E.S. Bibliography of the writings of W.S. Thayer. *Bulletin of the Institute of the History of Medicine*, 1936, **4**, 751–81.

REID, E.G. The life and convictions of William Sydney Thayer, physician. London, Oxford Univ. Press, 1936. 243 pp.

THEILER, Max (1899–1972). South African/American virologist; awarded Nobel Prize 1951 for his attenuation of the yellow fever virus.

–. *Journal of the South African Veterinary Association*, 1793, **44**, 460–62.

THEODORIC OF LUCCA, *see* **BORGOGNONI OF LUCCA, Theodoric**

THEOPHRASTUS (*c*.371–287 BC). Greek botanist, physician and philosopher; his classical botanical text contained valuable indications of therapeutic importance.

McDIARMID, T.B. DSB, 1976, **13**, 328–34.

THEORELL, Axel Hugo Theodor (1903–1982). Swedish biochemist; Nobel Prizewinner 1955.

366

DALZIEL, K. BMFRS, 1983, **29**, 585–621.

THIERFELDER, Hans (1858–1930). German physiological chemist.

LESCH, J.E. DSB, 1990, **18**, 904–6.

THIERSCH, Carl (1822–1895). German surgeon; introduced Thiersch skin graft.

THIERSCH, J. Carl Thiersch. Sein Leben. Leipzig, J.A. Barth, 1922. 190 pp.

THOMAS, Edward Donnall (b. 1920). American physicician and oncologist; shared Nobel Prize (Physiology or Medicine) 1990, for discoveries concerning organ and cell transplantation in the treatment of human disease.

Les Prix Nobel en 1990.
BEUTLER, E. Presentation of the George M.Kober Medal to E. Donnall Thomas. *Transactions of the Association of American Physicians*, 1992, **105**, cxl–clii.

THOMAS, Hugh Owen (1834–1891). British manipulative surgeon and founder of orthopaedic surgery in Britain.

AITKEN, D.M. Hugh Owen Thomas, his principles and practice. London, Oxford Univ. Press, 1935. 96 pp.
LeVAY, D. The life of Hugh Owen Thomas. Edinburgh, Livingstone, 1956. 144 pp.
ORR, H.W. On the contributions of Hugh Owen Thomas of Liverpool, Sir Robert Jones of Liverpool and London, John Ridlon, M.D, of New York and Chicago, to modern orthopaedic surgery. Springfield, C.C. Thomas, 1949. 253 pp.
WATSON, F. Hugh Owen Thomas. A personal study. London, Milford, 1934. 94 pp.
WILLIAMS, H. Doctors differ. Five studies in contrast ... London, Cape, 1946, pp. 95–123.

THOMAS, Karl (1883–1969). German nutritional biochemist; member of staff at Max Planck Gesellschaft.

THOMAS, K. Fifty years of biochemistry in Germany. *Annual Review of Biochemistry*, 1954, **23**, 1–16.

THOMAS, Vivien T. (1910–1985). American surgeon.

THOMAS, V.T. Pioneering research in surgical shock. Vivien Thomas and his work with Alfred Blalock: an autobiography. Philadelphia, University of Philadelphia Press, 1985. 245 pp.

THOMPSON, Henry (1820–1904). British genito-urinary surgeon; professor of clinical surgery, University College London.

COPE, Z. The versatile Victorian, being the life of Sir Henry Thompson Bt., 1820–1904. London, Harvey & Blythe, 1951. 179 pp.

THOMPSON, Robert Henry Stewart (1912–1998). British biochemist; co-discover of British anti-lewisite (an anti-arsenical compound, used against was gases and poisoning).

CAMPBELL, P.N. & STOCKEN, L.A. BMFRS, 1998, **44**, 419–31.

THOMSEN, Julius Thomas (1815–1896). Danish physician; gave first full description of myotonia congenita.

Medical History, 1968, **12**, 190–94.

BIBLIOGRAPHY OF MEDICAL AND BIOMEDICAL BIOGRAPHY

THOMSON, John (1856–1926). British paediatrician; physician to the Royal Edinburgh Hospital for Sick Children.

CRAIG, W.S.M. John Thomson: pioneer and father of Scottish paediatrics, 1856–1926. Edinburgh, London, Livingstone, 1968. 96 pp.

THOREK, Max (1880–1960). Hungarian/American plastic surgeon.

THOREK, M. A surgeon's world. An autobiography. Philadelphia, Lippincott, 1943. 333 pp.

THORNDIKE, Edward Lee (1874–1949). American psychologist; professor at Columbia University.

JONÇICH, G. The sane positivist. A biography of Edward L. Thorndike. Middletown, Conn., Wesleyan Univ. Press, 1968. 634 pp.
WOODWORTH, R.B. BMNAS, 1952, **27**, 209–37.

THORNTON, James Howard (1834–1919). British Army surgeon in Indian Medical Department.

THORNTON, J.H. Memories of seven campaigns: a record of thirty-five years' service in the Indian Medical Department in India, China, Egypt, and the Sudan. Westminster, Constable, 1895. 359 pp.

THUDICHUM, Johann Ludwig Wilhelm (1829–1901). German emigré chemist in London; Director of Chemical and Pathological Laboratory, St Thomas' Hospital; discovered cephalin and myelin in brain tissue, and haematoporphyrin.

DRABKIN, D.L. Thudichum: chemist of the brain. Philadelphia, Univ. Pennsylvania Press, 1958. 309 pp.
Archival material: Royal Society of London; National Institute for Medical Research, London.

THUNBERG, Thorsten Ludvig (1873–1952). Swedish physiologist; professor in Lund.

KAHLSON, G. DSB, 1976, **13**, 393–4.

THURNAM, John (1810–1873). British psychiatrist and anthropologist.

CRELLIN, J.K. DSB, 1976, **13**, 396.

TIEDEMANN, Friedrich (1781–1861). German anatomist and physiologist; professor at Heidelberg.

KRUTA, V. DSB, 1976, **13**, 402–4.

TIGERSTEDT, Robert Adolf Armand (1853–1923). Finnish physiologist.

SANTESSON, C.G. Professor Robert Tigerstedt. *Skandinavisches Archiv für Physiologie*, 1924, **45**, 1–6.

TILLETT, William Smith (1892–1974). American physician: discovered streptokinase.

LAWRENCE, H.S. BMNAS, 1993, **62**, 382–412.

TIMOFEEF-RESSOVSKY, Nikolai Vladimirovich (1900–1981). Russian geneticist.

GLASS, B. DSB, 1990, **18**, 919–26.

368

TIMONI, Emanuele (d. *c*.1721). Turkish physician.

TERZIOGLU, A. Emanuele Timonius und die Pockeninokulation. *Episteme*, 1973, **7**, 272–82.

TINBERGEN, Nikolaas (1907–1988). Dutch zoologist; shared Nobel Prize (Physiology or Medicine) 1973 for discoveries in animal behaviour.

HARPER, P. & POWELL, T.F. Catalogue of the papers and correspondence of Nicholaas Tinbergen. Bath, Univ. of Bath, 1991. 90 leaves.
HINDE, R.A. BMFRS, 1990, **36**, 548–65.
RÖLL, D.R. Der wereld van instinct: Niko Tinbergen en het ontstaan van de ethologie in Nederland. Rotterdam, Erasmus, 1996. 289 pp.

TISELIUS, Arne Wilhelm Kaurin (1902–1971). Swedish biochemist; Nobel Prize 1948 for his work on electrophoresis and serum proteins.

KEKWICK, R.A. & PEDERSON, K.O. BMFRS, 1974, **20**, 401–28.
PEDERSON, K.O. DSB, 1976, **13**, 418–22.
TISELIUS, A. Reflections from both sides of the counter. *Annual Review of Biochemistry*, 1968, **37**, 1–24.

TISSOT, Samuel August André (1728–1797). Swiss physician, best remembered for his *Avis au peuple sur la santé* (1761), a work intended for the instruction of the general public; it ran through 18 editions.

BARRAS, V. & LOUIS-COURVOISIER, M. (eds.) La médecine des lumiéres; tout autour de Tissot. Geneva, Georg, 2001. 358 pp.
EMCH-DÉRIAZ, A. Tissot, physician of the Enlightenment. New York, Peter Lang, 1992. 339 pp.

TODD, Alexander Robertus (1907–1997). British chemist and biochemist, professor of chemistry, Cambridge; Nobel Prize (chemistry) 1957.

BROWN, D.M. & KORNBERG, H. BMFRS, 2000, **46**, 515–32.
TODD, A.R. A time to remember: the autobiography of a chemist. Cambridge, Cambridge Univ. Press, 1983. 257 pp.

TODD, Charles (1869–1957). British pathologist and virologist.

ANDREWES, C.H. BMFRS, 1958, **4**, 281–90.

TODD, John Lancelot (1876–1949). Canadian parasitologist.

FALLIS, A.M. John L. Todd: Canada's first professor of parasitology. *Canadian Medical Association Journal*, 1983, **129**, 486, 488–90.

TODD, Robert Bentley (1809–1860). British physician; physician and professor of anatomy at King's College Hospital, London.

McINTYRE, M. Robert Bentley Todd. *King's College Hospital Gazette*, 1956, **35**, 79–91.

TÖNNIS, Wilhelm (1898–1978). German neurosurgeon.

ZÜLCH, K.J. Erinnerungen Wilhelm Tönnis 1898–1978. Jahre der Entwicklung der Neurochirurgie in Deutschland. Berlin, Springer-Verlag, 1984. 117 pp.

BIBLIOGRAPHY OF MEDICAL AND BIOMEDICAL BIOGRAPHY

TOLAND, Hugh Huger (1806–1880). American surgeon.

GOTTLIEB, L.S. Gold-mining surgeon, Hugh Huger Toland, M.D., founder of the University of California, San Francisco, School of Medicine. Manhattan, Kansas, Sunflower Univ. Press, 1985. 108 pp.

TOLMAN, Edward Chace (1886–1959). American psychologist.

RITCHIE, B.F. BMNAS, 1964, **37**, 293–324.

TOMES, John (1815–1895). British dental surgeon.

COPE, Z. Sir John Tomes: pioneer of British dentistry. London, Dawsons, 1961. 108 pp.

TOMMASI, Salvatore (1813–1888). Italian physician.

GIANDOMENICO, M. di. Salvatore Tommasi medico e filosofo. Bari, Adiatrica Editrice, 1965. 243 pp.

TOMMASINI, Giacomo Antonio Domenico (1768–1846). Italian physician.

OBERTI, A. Della dignità della medicina in Italia secondo in concetto di Giacomo Tommasini (1817). Pisa, Casa Editrice Giardini, 1970. 123 pp.

TONEGAWA, Susumi (b. 1939). Japanese molecular biologist; awarded Nobel Prize (Physiology or Medicine) 1987, for discovery of the genetic principle for generation of antibody diversity.

Les Prix Nobel en 1987.

TONKS, Henry (1852–1937). British surgeon and artist; abandoned medicine to teach art; Slade Professor of Fine Art, University College London, 1917.

HONE, J. The life of Henry Tonks. London, Heinemann, 1939. 390 pp.
PLARR, 1931, pp. 775–7.

TOOTH, Howard Henry (1856–1925). British physician; described peroneal progressive muscular atrophy (Charcot-Marie-Tooth type).

–. *St Bartholomew's Hospital Journal*, 1924–25, **32**, 141–4.
–. *St Bartholomew's Hospital Reports*, 1925, **58**, 9–15.

TOPLEY, William Whiteman Carlton (1881–944). British pathologist and immunologist; first professor of bacteriology and immunology, London School of Hygiene and Tropical Medicine.

GREENWOOD, M. ONFRS, 1942–44, **4**, 699–712.
Archival material: CMAC.

TORTI, Francesco (1658–1741). Italian pharmacologist; professor at Modena; popularized use of cinchona bark in treatment of malaria.

JARCHO, S. Quinine's predecessor: Francesco Torti and the early history of cinchona. Baltimore, Johns Hopkins Univ. Press, 1993. 400 pp.
TORTI, F. The clinical consultations of Francesco Torti; translated and with an introduction by Saul Jarcho. Mallabar, Fla., Krieger Publishing Co., 2000. 912 pp.

TOYNBEE, Joseph (1815–1876). British aural surgeon; a founder of aural pathology.

WILSON, T.C. Joseph Toynbee. *Archives of Otolaryngology*, 1966, **83**, 498–500.

TRAMER, Moritz (1882–1963). German child psychiatrist.

JORISCH-WISSINK, E. Der Kinderpsychiater Moritz Tramer (1882–1963). Zürich, Juris, 1986. 75 pp.

TRAUBE, Ludwig (1818–1876). German pathologist; professor in Berlin.

STANGIER, M. Ludwig Traube, sein Leben und Werk. Freiburg i. B., Kehrer, 1935. 91 pp.

TRAUBE, Moritz (1826–1894). German physiological chemist.

RUDOLPH, G. DSB, 1976, **13**, 451–3.
SOURKES, T.L. Moritz Traube (1826–1894): his contribution to biochemistry. *Journal of the History of Medicine*, 1955, **10**, 379–91.

TRAVERS, Benjamin (1783–1858). British ophthalmologist; surgeon to St Thomas' Hospital, London.

PETTIGREW, 1839. 14 pp. (CB).

TRÉFOUËL, Jacques Gustave Marie (1897–1977). French therapeutic chemist; introduced sulphanilamide.

DEBRU, C. DSB, 1990, **18**, 932–3.

TRÉLAT, Ulysse (1828–1893). French surgeon; professor at Hôpital Necker, Paris.

MONOD, C. Éloge de Ulysse Trélat. Paris, G.Masson, 1893. 16 pp.

TREMBLEY, Abraham (1710–1784). Swiss scientist; pioneer of experimental morphology and first to record cell division.

BAKER, J.R. Abraham Trembley of Geneva, scientist and philosopher. London, Arnold, 1952. 259 pp.

TRENDELENBURG, Friedrich (1844–1924). German surgeon; professor at Rostock, Bonn and Leipzig.

SCHWOKOWSKI, C. (ed.) Friedrich Trendelenburg, 1844–1924: zeitloser Glanz seiner Verdienste um die Chirurgie, Darmstadt, Steinkopff, 1994. 125 pp.
TRENDELENBURG, F. From my joyful days of youth: a memoir. Cairo, Al-Ahram Publishing House, [*c*.1983]. 289 pp. Translated from the German edition of 1924.

TREVAN, John William (1887–1956). British pharmacologist; Research Director, Wellcome Foundation.

GADDUM, J.H. BMFRS, 1957, **3**, 273–88.

TREVES, Frederick (1853–1923). British surgeon; surgeon at London Hospital.

TROMBLEY, S. Sir Frederick Treves, the extra-ordinary Edwardian. London, Routledge, 1989. 218 pp.

TRIBOULET, Henri (1864–1920). French paediatrician.

MASSARY, F. de. *Bulletin et Mémoires de la Société Médicale des Hôpitaux de Paris.* 1920, 3 sér., **44**, 1699–704.
–. *Union Médicale du Canada,* 1920, **49**, 259–61.

TRÖLTSCH, Anton Friedrich von (1812–1890). German otologist: invented the modern otoscope.

BAUDACH, R. Anton Friedrich von Tröltsch: Begründer der modernen Ohrenheilkunde auf dem europäischen Festland. Würzburg, Königshausen & Neumann, 1999. 166 pp.

TROJA, Michele (1747–1827). Italian physician.

CASTELLANI, C. DSB, 1976, **13**, 464–5.

TROMMSDORFF, Johann Bartholomäus (1770–1837). German pharmacist; professor of chemistry, Erfurt; a founder of modern scientific pharmacy.

GOETZ, W. Zu Leben und Werk von Johann Bartholomäus Trommsdorff (1770–1837). Würzburg, Jal Verlag, 1977. 354 pp.
GOETZ, W. Bibliographie der Schriften von Johann Bartholomäus Trommsdorff. Stuttgart, Wissenschaftliche Verlagsgesellschaft, 1985.
PATZER, H. *et al.*, Johann Bartholomäs Trommsdorff (1770–1837) und die Begründung der modernen Pharmazie. Leipzig, Barth, 1972. 295 pp.

TRONCHIN, Théodore (1709–1781). Swiss physician in Geneva; wrote important treatise on lead poisoning.

TRONCHIN, H. Un médecin du XVIIIe siècle: Théodore Tronchin. Paris, Plon-Nourrit et Cie., 1906. 417 pp.

TROTTER, Thomas (*c.*1760–1832). British naval surgeon; published 3-volume work on diseases of seamen (1799–1803) and the first printed work on alcoholism (1804).

PORTER, I.A. Thomas Trotter, M.D., naval physician. *Medical History*, 1963, **7**, 155–64.
ROLLESTON, H.D. Thomas Trotter, M.D. *Journal of the Royal Naval Medical Service,* 1919, **5**, 412–19.

TROTTER, Wilfred Batten Lewis (1873–1939). British surgeon; professor at University College Hospital, London.

ELLIOTT, T.R. ONFRS, 1940–41, **3**, 325–44.
PILCHER, R.S. Wilfred Trotter, F.R.S., F.R.C.S. *Annals of the Royal College of Surgeons of England*, 1973, **53**, 71–83.

TROUSSEAU, Armand (1801–1867). French physician; professor, Faculté de Médecine, Paris.

GARRISON, F.H. Armand Trousseau: a master clinician. *International Clinics*, 1916, **26**: ser. 3, 284–303.
GOMEZ, D.M. Trousseau (1801–1867) Paris, Marcel Vigne, 1929. 109 pp.

TRUDEAU, Edward Livingston (1848–1915). American physician; tuberculosis specialist; established Adirondack Cottage Sanitarium at Saranac Lake, NY.

BROWN, L. *et al.* (eds.) Edward Livingston Trudeau: a symposium. Livingston, NY, Livingston Press, 1935. 112 pp.
TAYLOR, R. Saranac: America's magic mountain. Boston, Houghton Mifflin, 1986. 308 pp.

TRUDEAU, E.L. An autobiography. Philadelphia, Lea & Febiger, 1915. 322 pp.

TRUETA RASPAIL, Josep (1897–1977). Spanish surgeon, active in Spanish civil war of 1936–39; devised closed plaster method of treating wounds.

RODRIGO, A. Doctor Trueta: héroe anónimo de dos guerras. Barcelona, Plaza & Janes, 1978. 294 pp.
SOLE, J.R. (ed.) Josep Trueta (1897–1977) en hometage. Barcelona, Fundacâo Barcelona, 1996. 200 pp.
TRUETA RASPAIL, J Trueta: surgeon in war and peace. The memoirs of Josep Trueta. Translated by M. and M. Strubell. London, Gollancz, 1980. 288 pp. Originally published in Catalan.

TSCHERMAK VON SEYSENEGG, Erich (1871–1962). Austrian geneticist.

BIEBI, R. DSB, 1976, **13**, 477–9.

TSCHIRCH, Alexander (1856–1939). German pharmacologist; professor at Berne.

TSCHIRCH, A. Erlebtes und Erstrebtes, Lebenserinnerungen. Bonn, Cohen, 1921. 254 pp.

TÜRCK. Ludwig (1810–1868). Austrian physician, laryngologist and neurologist.

CRAIGIE, E.H. DSB, 1976, **13**, 492–3.

TÜRK, Wilhelm (1871–1916). German haematologist; first to report agranulocytosis.

LEHNDORFF, H. Wilhelm Türk; a prominent hematologist of fifty years ago. *Blood*, 1954, **9**, 642–7.

TUFFIER, Théodore (1857–1929). French cardiovascular surgeon.

LE DOCTEUR Tuffier, 1857–1929; sa vie, ses travaux, ses enseignments. Corbeil, Crété, 1935. 341 pp.
TITRES et travaux scientifiques de Pr. Tuffier. *Cahiers d'Anesthésiologie*, 1984, **32**, 443–6.

TUKE FAMILY. William (1732–1822), British Quaker philanthropist, instrumental in the establishment of The Retreat, York, for the humane treatment of the insane; **Samuel** (1784–1857) advocate of humane treatment of insane; **Daniel Hack** (1827–1895). British physician and chemist.

LANGFITT, M.S. William Tuke's contribution to the humane treatment of the insane. Pittsburgh [n.p.], Thesis, 1937 Typescript.
SÉMELAIGNE, R. Alienistes et philanthropes. Les Pinel et les Tuke. Paris, Steinheil, 1912, 548 pp.
SESSIONS, W.K. & SESSIONS, S.M. The Tukes of York. York, Ebor Press, 1971. 117 pp.

TULP, Nicolaas (1593–1674). Dutch physician; his *Observationees medicae*, 1652, includes one of the earliest accounts of beriberi.

DUDOK VAN HEEL, S.A.C. *et al.* Nicolas Tulp: the life and work of an Amsterdam physician and magistrate in the 17th century. 2nd ed. Amsterdam, Six Art Promotions, *c.*1998. 256 pp.
GOLDWYN, R.W. Nicolaas Tulp (1593–1674). *Medical History*, 1961, **5**, 270–76.
THYSSEN, E.H.M. Nicolaas Tulp. *Medical Life*, 1929, **36**, 394–442; 1932, **39**, 297–330.
VAN DER PAS, P.W. DSB, 1976, **13**, 490–91.

TUPPER, Charles (1821–1915). Canadian physician and politician.

MURRAY, J. & MURRAY, J. Sir Charles Tupper: fighting doctor to doctor of confederation. Toronto, Associated Medical Services Inc., 1999. 155 pp.

TURNBULL, Hubert Maitland (1875–1955). British pathologist; professor, London Hospital Medical College.

BESSON, S. BMFRS, 1957, **3**, 289–304.

TURNER, Daniel (1667–1741). British surgeon and dermatologist: wrote first English textbook on dermatology, 1714, and received first medical degree granted (Hon., Yale, 1723) in English-speaking America.

LANE, J.E. Daniel Turner and the first degree of Doctor of Medicine conferred in the English colonies of North America by Yale College in 1723. *Annals of Medical History*, 1919, **2**, 367–80.
LYELL, A. Daniel Turner...surgeon, physician and pioneer dermatologist, the man seen in the pages of his book on the skin. *International Journal of Dermatology*, 1982, **21**, 162–70.
WILSON, P.K. Surgeon "turned" physician: the career and writings of Daniel Turner (1667–1741). Thesis submitted for the degree of Doctor of Philosophy, University College London. London, University of London, 1992. 441 pp.

TURNER, George Grey (1877–1951). British surgeon; professor at Newcastle upon Tyne.

GEORGE GREY TURNER. Portsmouth, Potts & Horsey, 1986, 68 pp.
Archival material: CMAC.

TURNER, Thomas (1793–1873). British surgeon, Manchester Royal Infirmary, and founder of school of medicine in Manchester.

MEMOIR OF Thomas Turner, Esq., FRCS, FLS. By a relative. London, Simpkin, Marshall; Manchester, Cornish, 1875. 294 pp.

TURNER, William (1508–1568). British physician and herbalist.

JONES, W.R.D. William Turner, Tudor naturalist, physician and divine. London, Routledge, 1988. 223 pp.
WEBSTER, C. DSB, 1976, **13**, 501–3.

TURNER, William (1832–1916). British anatomist; professor at Edinburgh.

THOMAS, K.B. DSB, 1976, **13**, 503–4.
TURNER, A.L. Sir William Turner, K.C.B., F.R.S., professor of anatomy and principal and vice-chancellor of the University of Edinburgh. A chapter in medical history. Edinburgh, London, Blackwood, 1919. 514 pp.

TURQUET DE MAYERNE, Théodore (1573–1655). Swiss physician and chemist; physician to James I and to Charles I & II; involved in production of the first edition of Pharmacopoeia Londinensis, 1618.

GIBSON, T. Sketch of the career of Theodore Turquet de Mayerne. *Annals of Medical History*, 1933, n.s. **5**, 315–26.
HANAWAY, O. DSB, 1976, **13**, 307–8.

NANCE, B. Turquet de Mayerne as baroque physician; the art of medical portraiture. Amsterdam, Rodopi, 2001. 237 pp.

TWITCHELL, Amos (1781–1850). American surgeon; ligated carotid artery (1807) and reported it 35 years later.

FAULKNER, J.M. Amos Twitchell, pioneer American surgeon. Harvard Medical Alumni Bulletin, 1969, **43**, 38–41.

TWORT, Frederick William (1877–1950). British microbiologist; discovered bacteriophage.

CLARKE, E. DSB, 1976, **13**, 519–21.
FILDES, P. ONFRS, 1951, **7**, 505–17.
TWORT, A. In focus, out of step; a biography of Frederick William Twort FRS, 1877–1950. Gloucester, Alan Sutton, 1993. 340 pp.
Archival material: CMAC.

TYNDALL, John (1820–1893). British physicist; showed bacteriostatic action of *Penicillium*, 1876.

EVE, A.S. & CREASEY, C.H. Life and work of John Tyndall. London, Macmillan, 1945. 404 pp.
LANDSBERG, H. Prelude to the discovery of penicillin. *Isis*, 1949, **40**, 225–7.

TYSON, Edward (1650–1708). English physician and pioneer comparative anatomist.

ASHLEY-MONTAGUE, M.F. Edward Tyson, M.D., F.R.S., 1650–1708, and the rise of human and comparative anatomy in England: a study in the history of science. Philadelphia, American Philosophical Society, 1943. 488 pp.
WILLIAMS, W.C. DSB, 1975, **12**, 526–8.

TYZZER, Ernest Edward (1875–1965). American pathologist; professor at Harvard University.

WELLER, T.H. BMNAS, 1978, **49**, 353–73.

UGO BENZI (1376–1439). Italian physician.

LOCKWOOD, D.P. Ugo Benzi, medieval philosopher and physician, 1376–1439. Chicago, Univ. Chicago Press, 1951. 441 pp. Contains list of his MSS and printed works.

UHLENHUTH, Paul Theodor (1870–1957). German bacteriologist.

FROMME, W. Einst und jetzt: Weilsche Krankheit. Zur 50 jährigen Wiederkehr der Entdeckung die Erregers der Weilschen Krankheit. *Münchener Medizinische Wochenschrift*, 1965, **107**, 1204–8.

UKHTOMSKII, Aleksei Aleksevich (1875–1942). Russian neurophysiologist; professor at Petrograd (Leningrad).

BRAZIER, M.A.B. DSB, 1976, **13**, 529–30.
MERKULOV, V.L. Aleksei Aleksevich Ukhtomskii (1875–1942). Moskva, Izdatelvo Akad. Nauk. USSR, 1960. 314 pp.

ULLMANN, Emerich (1861–1937). Hungarian surgeon; carried out successful kidney autotransplantation in dogs.

LESKY, E. Die erste Nierentransplantation: Emerich Ullmann (1861–1937). *Münchener Medizinische Wochenschrift*, 1974, **116**, 1081–4.

ULSENIUS, Theodoricua (*c*.1460–1508). Dutch physician and poet; published one of the earliest coloured prints of a syphilitic (1496) by Albrech Dürer.

SANTING, C. Geneeskunde en humanisme. Een intellectuele bigrafie van Theodoricus Ulsenius (*c*.1460–1508). Rotterdam, Erasmus Publishing, 1992. 312 pp.

UMEZAWA, Hamao (1914–1986). American biochemist; isolated kanamycin.

TSUCHIYA, T. *et al.* Hamao Umezawa (1914–1986). *Advances in Carbohydrate Chemistry and Biochemistry*, 1990, **48**, 1–20.

UNDERWOOD, Michael (1736–1820). British surgeon and paediatrician.

MALONEY, W.J. Michael Underwood, a surgeon practising midwifery from 1764–1784. *Journal of the History of Medicine*, 1950, **5**, 289–314.

UNGER, Ernst (1875–1938). German neurosurgeon; Professor and Director at the Rudolf Virchow Krankenhaus, Berlin.

WINKLER, E.A. Ernst Unger (1875–1938) eine Biobibliographie. Berlin, Internationale Verlags-Anstalt, 1975. 206 pp.

UNNA, Paul Gerson (1850–1929). German dermatologist in Hamburg; published several original descriptions of cutaneous diseases and founded the Dermatologische Wochenschrift.

GROTE, 1929, **8**, 175–219 (CB).
STUMM, H. Der Dermatologe Paul Gerson Unna (1850–1929). Leben, klinische Hauptarbeitsgebiete Ansachauung. [M.D. thesis.] Mainz, Universität Institut für Geschichte der Medizin, 1990. 231 pp.

UNVERRICHT, H. (1853–1912). German neurologist; first described familial myoclonus epilepsy.

WENZEL, –. *Münchener Medizinische Wochenschrift*, 1912, **59**, 116–63.

UNZER, Johann August (1727–1799). German physiologist.

KRUTA, V. DSB, 1976, **13**, 543–44.

URDANG, George (1882–1960). German/American pharmacist and historian of pharmacy.

WOLFE, G. George Urdang, 1882–1960, the man and his work. *Acta Pharmaciae Historica*, 1961, Nr.2, 39–84 [English and German].

UTTER, Merton Franklin (1917–1980). American biochemist; professor, Western Reserve University.

WOOD, H.G. & HANSON, R.W. BMNAS, 1987, **56**, 474–99.

VALENTIN, Gabriel Gustav (1810–1883). German physiologist; professor at Bern.

HINTZSCHE, E. & RITZ, W. Gabriel Gustav Valentin (1810–1883). Versuch einer Bio- und Bibliographie. Bern, F. Haupt, 1953. 92 pp.

HINTZSCHE, E. DSB, 1976, **13**, 555–8.

VALLAMBERT, Simon de (*fl.*1537–1565). French paediatrician; published first French work on paediatrics, 1565.

LANSELLE, M. Un puériculteur oublié: Simon de Vallambert. *Bulletin de la Société Française de Histoire de la Médecine*, 1946, **19**, 455–8.

VALLISNERI, Antonio (1661–1730). Italian biologist and physician; professor at Padua.

MONTALENTI, G. DSB, 1976, **13**, 562–5.

VALSALVA, Anton Maria (1666–1723). Italian anatomist.

CASTIGLIONI, A. Antonio Maria Valsalva. *Medical Life*, 1923, **30**, 466–89; 1932, **39**, 83–107.
LUCCHETTA, P.M. Antonio Vallisneri, medico naturalista: scienze e filosofia nel settecento. Venezia, Cafoscarina, *c.*1984. 167 pp.
PORCIA, G. A. di. Notize della vita e degli studi del Kavalier Antonio Vallisneri. Bologna, Pàtron, 1986. 244 pp.
PREMUDA, L. DSB, 1976, **13**, 566–7.
RAVANELLI, P. A.M.Valsalva (1666–1723) anatomico, chirurgo, primo psichiatra. Imola, Galeati, 1966. 109 pp.

VALVERDE, Juan de (*c.*1520–*c.*1588). Spanish anatomist.

GUERRA, F. DSB, 1976, **13**, 568–9.

VANE, John Robert (b. 1927). British pharmacologist; shared Nobel Prize (Physiology or Medicine), 1982, for discoveries concerning the prostaglandins and related biologically active substances.

Les Prix Nobel en 1982.

VANGHETTI, Giuliano (1861–1940). Italian surgeon; introduced 'kinematisation of stumps' after limb amputation to activate artificial prostheses.

MARIANELLI, V. Giuliano Vanghetti (1861–1940): un medico dalle scarpe grosse ma dal fine ingegno. *Lanternino*, 1995, **18**, 1–5.

VAN GIESON, Ira (1866–1913). American neuropsychiatrist; introduced acid fuschin and picric acid stain for nerve tissue.

ROIZIN, L. Van Gieson: a visionary of psychiatric research. *American Journal of Psychiatry*, 1970, **127**, 180–85.

VANLAIR, Constant (1839–1914). Belgian anatomist, physiologist and physician; suggested the concept of hereditary haemolytic anaemia.

NOLF, P. Notice sur Constant Vanlair. *Annuaire l'Académie Royale des Sciences de Belgique*, 1923, **89**, 125–50.

VAN NIEL, Cornelis Bernardus (1897–1985). Dutch/American microbiologist born Netherlands; professor at Stanford University.

BARKER, H.A. & HUNGATE, R.E. BMNAS, 1990, **59**, 388–423.

VAN SLYKE, Donald Dexter (1883–1971). American biochemist.

HASTINGS, A.B. BMNAS, 1976, **48**, 309–60.
PARASCANDOLA, J. DSB, 1976, **13**, 574–5.
Archival material: Rockefeller Archives Center; Brookhaven National Laboratory, NY.

VAQUEZ, Louis Henri (1860–1936). French physician; first described polycythaemia vera.

LAUBRY, C. Henri Vaquez (1860–1936). *Bulletin de l'Académie Nationale de Médecine*, 1958, **142**, 846–54.

VARMUS, Harold (b. 1939). American virologist; shared Nobel Prize (Physiology or Medicine) 1989, for discovery of the cellular origin of retroviral oncogenes.

Les Prix Nobel en 1989.

VAROLI, Costanzo (1543–1575). Italian anatomist.

O'MALLEY, C.D. DSB, 1976, **13**, 587–8.

VASSALE, Giulio (1862–1913). Italian endocrinologist.

FRANCESCHINI, P. DSB, 1976, **13**, 589–90.

VAUGHAN, Henry (1622–1695). Welsh physician and poet.

HUTCHINSON, F.E. Henry Vaughan: a life and interpretation. Oxford, Clarendon Press, 1947. 260 pp.
RUDRUM, H. (ed.) Essential articles for the study of Henry Vaughan. Hamden, Conn., Archon Books, 1987. 332 pp.

VAUGHAN, Janet Maria (1899–1993). British haematologist and radiobiologist.

ADAMS, P. (ed.) Janet Maria Vaughan 1899–1993: a memorial tribute. Oxford, Somerville College, 1994. 35 pp.
OWEN, M. BMFRS, 1995, **41**, 482–98.
Archival material: CMAC.

VAUGHAN, Victor Clarence (1851–1929). American biochemist, microbiologist and medical administrator.

DAVENPORT, H.W. Victor Vaughan, statesman and scientist. Ann Arbor, Historical Center for the Health Sciences, 1996. 148 pp.
VAUGHAN, V.C. A doctor's memories. Indianapolis, Bobbs-Merrill Co., 1926. 464 pp.

VAUQUELIN, Louis Nicolas (1763–1829). French chemist; discovered asparagine; made first complete chemical analysis of nervous system.

DELAUNAY, A. Louis Nicolas Vauquelin. *Histoire de la Médecine*, 1963, **12**, 2–15.

VEDDER, Edward Bright (1878–1952). American physician and nutritionist.

WILLIAMS, R.R. Edward Bright Vedder – a biographical sketch (28 June 1878–30 January 1952). *Journal of Nutrition*, 1962, **77**, 3–6.

VEIT, Gustav von (1834–1905). German gynaecologist.

ENGEL, C. Leben und Werk des Bonner Frauenarztes Gustav von Veit. Dissertation, Bonn, Rheinische Friedrich-Wilhelm-Universität, 1983. 141 pp.

VELPEAU, Alfred Armand Louis Marie (1795–1867). French surgeon; professor of clinical surgery, Paris.

FRÉZOULS, J. Velpeau (1795–1867). Thèse. Paris, 1932. 63 pp.
LIZAK, G. Bio-bibliographie de Alfred Armand Louis Marie Velpeau (1795–1867). Thèse. Paris-Cochin Port Royal, 1972. 142 pp.

VENEL, Jean André (1740–1791). Swiss orthopaedic surgeon; founded first orthopaedic institute.

BISCHOFBERGER, R. Jean André Venel (1740–1791) ein wichtiger Arzt der Aufklärungszeitalters. Zürich, Juris-Verlag, 1970. 43 pp.
VALENTIN, B. Jean André Venel, der "Vater der Orthopädie" (1740–1791). *Sudhoffs Archiv für Geschichte der Medizin*, 1956, **40**, 305–36.

VERAGUTH, Otto (1870–1944). Swiss neurologist and physical therapist.

SÜSSLI, P. Otto Veraguth, 1870–1944: Neurologe und Professor für physikalische Therapie. Zürich, Juris Druck, 1991. 141 pp.

VERNEUIL, Aristide Auguste Stanislas (1823–1895). French surgeon; introduced forcipressure in treatment of haemorrhage.

BARACCHI, B. Bio-bibliographie d'Aristide Verneuil de Saint-Martin (1823–1895). Paris, Faculté de Médecine, 1975. 33 pp.

VERNEY, Ernest Basil (1894–1967). British pharmacologist.

DALY, I. de B. & PICKFORD, L.M. BMFRS, 1970, **16**, 523–42.
LESCH, J.E. DSB, 1990, **18**, 960–62.
Archival material: Physiology Dept., University of Cambridge; CMAC.

VERWORN, Max (1863–1921). German physiologist.

ROTHSCHUH, K.E. DSB, 1976, **14**, 2–3.
WÜLLENWEBER, R. Der Physiologe Max Verworn. Inaugural-Dissertation ... Rheinischen-Friedrich-Wilhelmsuniversität zu Bonn. Bonn, 1968, 145 pp.

VESALIUS, Andreas (1514–1564). Belgian anatomist; founder of scientific anatomy: taught at Padua.

CUSHING, H. A bio-bibliography of Andreas Vesalius. New York, Schuman's, 1943. 229 pp. 2nd edition. Hamden, Conn., 1962. 264 pp.
O'MALLEY, C.D. Andreas Vesalius of Brussels, 1514–1564. Berkeley, Los Angeles, Univ. California Press, 1964. 480 pp.
O'MALLEY, C.D. DSB, 1976, **14**, 3–12.
ROTH, M. Andreas Vesalius Bruxellensis. Berlin, G.Reimer, 1892. 500 pp.
SAUNDERS, J.B. de C.M. & O'MALLEY, C.D. The illustrations from the works of Andreas Vesalius of Brussels, with annotations and translations ... and a biographical sketch of Vesalius. Cleveland, World Publishing Co., 1950, 248 pp. Reprinted, New York, Bonanza Books, 1982, 252 pp.
SPIELMANN, M.H. The iconography of Andreas Vesalius (André Vesale), anatomist and physician 1514–1564 ... With notes, critical, literary and bibliographical. London, Bale, Sons & Danielsson, 1925. 243 pp.

BIBLIOGRAPHY OF MEDICAL AND BIOMEDICAL BIOGRAPHY

VESLING, Johann (1598–1649). German anatomist and botanist.

HINTZSCHE, E. DSB, 1976, **14**, 12–13.

VICARY, Thomas (*c.*1495–1561). English anatomist.

POWER, D'A. Thomas Vicary 1490–1562. *British Journal of Surgery*, 1918, **4**, 359–62.
POWER, D'A. The education of a surgeon under Thomas Vicary. *British Journal of Surgery*, 1920–21, **8**, 240–58.

VICKERY, Hubert Bradford (1893–1978). American biochemist.

IHDE, A.J. DSB, 1990, **18**, 964–5.

VICQ D'AZYR, Félix (1748–1794). French anatomist and epidemiologist.

HUARD, P. & IMBAULT-HUART, M.J. DSB, 1976, **14**, 14–17.
Archival material: Académie Nationale de Médecine, Paris.

VIDIUS, Vidus, *see* **GUIDI, Guido**

VIERORDT, Karl (1818–1884). German physiologist; invented a sphygmograph.

MAJOR, R.H. Karl Vierordt. *Annals of Medical History*, 1938, n.s., **10**, 463–73.

VIEUSSENS, Raymond (1635–1715). French anatomist and physician.

GRMEK, M.D. DSB, 1976, **14**, 25–6.
KELLETT, C.E. The life and work of Raymond Vieussens. *Annals of Medical History*, 1942, 3 ser., **4**, 31–54.

VIGO, Giovanni da (1450–1525). Italian surgeon.

GIORDANO, D. Giovanni da Vigo (1445–1525) *Rivista di Storia della Scienze Mediche e Naturali*, 1926, **17**, 21–35.
TALBOT, C.H. DSB, 1976, **14**, 27–8.

VILLEMIN, Jean-Antoine (1827–1892). French physician; professor at Val de Grâce; proved the contagiousness of tuberculosis.

DEWALD, R. Un ancien élève de la Faculté de Médecine, Villemin; notice biographique et bibliographique. Strasbourg, 1933. 70 pp.
JAMMET, C. Villemin; sa vie et son oeuvre, 1827–1892. Paris, M.Vigné, 1933. 63 pp.

VILLERMÉ, Louis-René (1782–1863). French physician and statistician.

VALENTIN, M. Louis-René Villermé (1782–1863) et son temps. Paris, Docis, 1993. 311 pp.

VINCENT, Clovis (1879–1947). French neurosurgeon.

GIROIRE, H. Clovis Vincent, 1879–1947; pionnier de la neuro-chirurgie française. Paris, Perrin, 1971. 195 pp.

VINCENT, Jean Hyacinthe (1862–1950). French bacteriologist; described ulcerative stomatitis ("Vincent's angina"); isolated *Actinomyces madurae*.

PICKARD, H.M. Historical aspects of Vincent's disease. *Proceedings of the Royal Society of Medicine*, 1973, **66**, 695–8.

TANON, L. Notice nécrologique sur le professeur Vincent. *Bulletin de l'Academie Nationale de Médecine*, 1951, **135**, 119–27.

VINCI, Leonardo da, *see* **LEONARDO DA VINCI**

VINOGRAD, Jerome Ruben (1913–1976). American biochemist and molecular biologist.

HAGAN, W.J. Jr. DSB, 1990, **18**, 965–6.

VINOGRADSKY, Sergey Nikolaevich (1856–1953). Russian microbiologist.

GUTINA, V. DSB, 1976, **14**, 36–8.
THORNTON, H.G. ONFRS, 1952–53, **8**, 635–44.
WAKSMAN, S.A. Sergei N. Winogradsky: his life and work. The story of a great bacteriologist. New Brunswick, N.J., Rutgers Univ. Press, 1953. 153 pp.

VIRCHOW, Rudolf Ludwig Karl (1821–1902). German pathologist and statesman; founder of 'cellular pathology'.

ACKERKNECHT, E.H. Rudolf Virchow: doctor, statesman and anthropologist. Madison, Univ. Wisconsin Press, 1953. 304 pp.
ANDRÉE, C. Rudolf Virchow als Prähistoriker. 2 vols. Köln, Wien, Bohlau Verlag, 1976. 267+541pp.
BOYD, B.A. Rudolf Virchow: the scientist as citizen. New York, Garland 1991. 269 pp.
DOWD, P.S. Rudolf Virchow and the science of humanity. Thesis: University of Pittsburgh, 1999. Ann Arbor, University Microfilms International, 2001. 194 pp.
McNEELEY, I.F. Medicine on a grand scale: Rudolf Virchow, liberalism and the public health. London, Wellcome Trust, 2002. 97 pp.
RATHER, L.J. A commentary on the medical writings of Rudolf Virchow, based on Schwalbe's Virchow-Bibliographie 1843–1901. San Francisco, Norman Publishing, 1990. 236 pp.
RISSE, G.B. DSB, 1976, 14, 39–44.
SCHIPPERGES, H. Rudolf Virchow. Hamburg, Rohwolt, 1994. 153 pp.
SUDHOFF, K. Rudolf Virchow und die deutschen Naturforscherversammlungen. Leipzig, Akademische Verlagsgesellschaft, 1922. 306 pp.
WINTER, K. Rudolf Virchow. Leipzig, Teubner, 1976. 99 pp.

VIREY, Julien-Joseph (1775–1846). French biologist, physiologist and pharmacist.

BERMAN, A. DSB, 1976, **14**, 44–5.

VIRGILI, Pedro (1699–1776). Spanish surgeon.

FERRER, D. Biografia de Pedro Virgili, fundador, restaurador de la cirugia en España. Barcelona, Pentagono, 1963. 404 pp.

VIRTANEN, Artturi Ilmari (1895–1973). Finnish biochemist; professor at Helsinki University; Nobel prizewiriner (chemistry) 1945 for his method of fodder storage.

LEICESTER, H.M. DSB, 1976, **14**, 45–6.
Archival material: Biochemical Research Institute, Helsinki.

VISSCHER, Maurice Bolks (1901–1983). American physiologist.

DAVENPORT, H.W. BMNAS, 1993, **62**, 414–38.

BIBLIOGRAPHY OF MEDICAL AND BIOMEDICAL BIOGRAPHY

VOELCKER, Friedrich (1872–1955). German surgeon; introduced radiography of the kidney and bladder.

KAISER, W. & STOLZ, M. Pro memoria Friedrich Voelcker (1872–1955), Direktor der Chirurgischen Universitätsklinik Halle in den Jahren 1919–1937. Halle, Martin-Luther-Universität, 1972. 32 pp.

VOGT, Alfred (1879–1943). Swiss ophthalmologist; introduced cyclodiathermy in treatment of glaucoma.

NIEDERER, H.M. Alfred Vogt (1879–1943): seine Zürcher Jahre (1923–1943). Zürich, Juris, 1989. 300 pp.

VOGT FAMILY. Cécile (1875–1962), **Oskar** (1870–1959). German neurologists and neurophysiologists; described disease of the corpora striata ('Vogt's syndrome), 1920.

KLATZO, I. Cécile and Oskar Vogt: the visionaries of modern neuroscience. Wien, Springer-Verlag, 2002. 129 pp. (*Acta Neurochirurgica*, Suppl. 20).
SATZINGER, H. Die Geschichte der genetisch orientierten Hirnforschung von Cécile and Oskar Vogt (1875–1962, 1870–1959) in der Zeit von bis ca. 1927. Stuttgart, Deutsche Apotheker Verlag, 1998. 365 pp.

VOIT, Carl von (1831–1908). German physiologist.

HOLMES, F.L. DSB, 1976, **14**, 63–7.

VOLHARD, Franz (1872–1950). German cardiac surgeon and nephrologist.

BOCK, H.-E. *et al.* Franz Volhard. Stuttgart, New York, Schattauer, 1982. 345 pp. [Text in English and German.]
KRONSCHWITZ, C. Franz Volhard: Leben und Werk. Franfurt a. M., Sinemis-Verlags-Gesellschaft, 1997, 328 pp.

VOLKMANN, Richard von (1830–1889). German surgeon; professor at Halle; pioneer in introduction of antisepsis in Germany.

SCHOBER, K.L. Richard von Volkmann (1830–1889): bedeutende Leistungen eines grossen Chirurgen. Leipzig, Barth, 1986.

VOLTOLINI, Friedrich Eduard Rudolf (1829–1888); German otolaryngologist; introduced galvanocauterization of larynx (1867); performed first laryngeal operation by mouth with external illumination.

GRÜBER, J. Prof. Dr. Rudolf Voltolini. *Monatsschrift für Ohrenkheilkunde*, 1889, **23**, 217–23.

VORONOFF, Serge (1866–1951). Russian/French physiologist; advocate of testicular grafting for rejuvenation.

HAMILTON, D. The monkey gland affair. London, Chatto & Windus, 1986. 155 pp.

VRIES, Hugo Marie de (1848–1935). Dutch geneticist, evolutionist and plant physiologist; professor at Amsterdam; rediscovered Mendel's laws.

HALL, A.D. ONFRS, 1935, **1**, 371–3.
STOMPS, T.G. On the rediscovery of Mendel's work by Hugo de Vries. *Journal of Heredity*, 1955, **45**, 293–394.

VAN DER PAS, P.W. DSB, 1976, **14**, 95–105.
Archival material: Library, American Philosophical Society, Philadelphia.

VULPIAN, Edme Félix Alfred (1826–1887). French physician; discovered adrenalin in adrenal capsules.

EBNER, A. Edme-Félix Alfred Vulpian, 1826–1887. Zürich, Juris-Verlag, 1967. 57 pp.

VVEDENSKY, Nikolai Evgenievich (1852–1922). Russian physiologist.

GRIGORIAN, N.A. DSB, 1976, **14**, 105.

WAAL, *see* **WALE**

WAARDENBURG, Petrus Johannes (1886–1979). Dutch geneticist and ophthalmologist.

OPITZ, J.M. In memoriam: Petrus Johannes Waardenburg (1886–1979). *American Journal of Genetics*, 1980, **7**, 35–9.

WADDINGTON, Conrad Hal (1905–1975). British geneticist.

ROBERTSON, A. BMFRS, 1977, **23**, 575–622. [Correction: *British Medical Journal*, 1978, **1**, 174–5.]
WADDINGTON, C.H. The evolution of an evolutionist. Edinburgh, Edinburgh Univ. Press, 1975. 328 pp.

WAGNER, Ernst Leberecht (1829–1888). German pathologist; his *Gebärmutterkrebs*, 1858 was the first important contribution to the knowledge of the gross pathology of uterine cancer.

STRUMPFELL, A. *Fortschritte der Medizin*, 1886, 202–9.

WAGNER, Rudolph (1805–1864). German anatomist and physiologist; professor at Göttingen.

KRUTA, V. DSB, 1976, **14**, 113–14.

WAGNER, Wilhelm (1848–1900). German neurosurgeon.

BUCHFELDER, M. & LJUNGREN, B. Wilhelm Wagner (1848–1900). *Surgical Neurology*, 1988, **30**, 423–33.

WAGNER-JAUREGG, Julius (1857–1940). Austrian psychiatrist; professor at Vienna; introduced fever therapy in treatment of general paresis (Nobel Prize 1927).

LESKY, E. DSB, 1976, **14**, 114–16.
WAGNER-JAUREGG, J. Lebenserinnerungen, herausgegeben und ergänzt von L. Schönbauer und M. Jantsch. Wien, Springer-Verlag, 1950. 187 pp.
WAGNER–JAUREGG, J. Mein Lebensweg als bioorganischer Chemiker. Stuttgart, Wissenschaftlliche Verlagsgesellschaft, 1985. 94 pp.
WHITROW, M. Julius Wagner-Jauregg (1857–1940). *Journal of Medical Biography*, 1993, **1**, 137–43.
WHITROW, M. Julius Wagner-Jauregg (1857-1940). London, Smith-Gordon, 1993.

WAKLEY, Thomas (1795–1862). British physician and pioneer in medical reform; founder and first editor of the *Lancet*.

BIBLIOGRAPHY OF MEDICAL AND BIOMEDICAL BIOGRAPHY

HOSTETTLER, J. Thomas Wakley: an improbable radical. Chichester, Barry Rose Publishers Ltd., 1993. 158 pp.

SPRIGGE, S.S. The life and times of Thomas Wakley. Facsimile of the 1899 edition. With an introduction by C.G. Roland. Huntingdon, NY, R.E.Krieger, 1974. 509 pp. [The introduction by C.G. Roland includes a reprint of his biographical sketch of Wakley originally published in the *New England Journal of Medicine*.]

WAKSMAN, Selman Abraham (1888–1973). American microbiologist, discoverer of streptomycin; Nobel prizewinner 1952.

WAINWRIGHT, M. Streptomycin: discovery and resultant controversy. *History and Philosophy of the Life Sciences*, 1991, **13**, 97–124.

WAKSMAN, B.H. & LECHEVALIER, H.A. DSB, 1990, **18**, 970–74.

Archival material: Jewish Historical Society, Waltham, Mass.; Library of Congress; Institute of Microbiology, Rutgers State University.

WALAEUS, *see* **WALE**

WALD, George (1906–1997). American biologist; shared Nobel Prize (Physiology or Medicine) 1967 for research on the photosensitive pigments of the visual receptor apparatus.

DOWLING, J. E. BMNAS, 2000, **78**, 298–317.

WALDENSTRÖM, Jan Gösta (b.1906). Swedish physician: worked on myelomas and genetic basic of metabolic disease; described macroglobulinaemia.

BJÖRKMAN. S.E. Jan Waldenström, 17 April 1966 [introduction to Festschrift]. *Acta Medica Scandinavica*, 1966. Suppl. 445, 5–8.

WALDENSTRÖM, J. G. Reflections and recollections from a long life with medicine. Rome, Il Pensiero Scientifico, 1994. 147 pp.

WALDEYER-HARTZ, Heinrich Wilhelm Gottfried von (1836–1921). German anatomist; professor at Strassburg and Berlin.

GLEES, P. DSB, 1976, **14**, 125–7.

WALDEYER, W. Lebenserinnerungen. 2te Aufl. Bonn, F.Cohen, 1921. 419 pp.

WALE, Jan de [WAAL, WALAEUS] (1604–1649). Dutch physician; professor at Leyden; gave proof (1640) of Harvey's discovery of the circulation of the blood.

SCHOUTEN, Johannes Walaeus; zyn betekenis voor de verbreiding van de leer van de bloedomloop. Academische proefschrift der Vrije Universiteit te Amsterdam. Assen, Van Gorcum & Comp., 1972. 260 pp.

WALKER, John (1759–1830). British physician; Director of Royal Jennerian and London Vaccine Institutions.

EPPS, J. The life of John Walker, M.D. London, Whittaker, Treacher, 1832. 342 pp.

WALKER, Mary Edwards (1832–1919). American physician; womens' rights advocate; assistant surgeon Union Army (first woman so commissioned in US Army); Congressional Medal of Honor, 1865 (only woman to receive this award).

HALL M. Quite contrary: Dr Mary Edwards Walker. New York, Funk & Wagnalls, 1970. 160 pp.

BIBLIOGRAPHY OF MEDICAL AND BIOMEDICAL BIOGRAPHY

SNYDER, C.M. Dr Mary Walker: the little lady in pants. New York, Vintage Press, 1962. 166 pp. Reprinted, New York, Arno Press, 1974.

WALL, John (1708–1776). British physician; played important part in the establishment of (Royal) Worcester Porcelain Factory.

SMITH, M. Caduceus: porcelain and palette: John Wall of Worcester. *Journal of the Royal Society of Medicine*, 1999, **92**, 641–5.

WALLACE, Alfred Russel (1823–1913), British naturalist.

BERRY, A. Infinite tropics: an Alfred Russel Wallace anthology. London, Verso, 2002. 430 pp.
GEORGE, W. Biologist philosopher. A study of the life and writings of Alfred Russel Wallace. New York, Abelard-Schuman, 1964. 320 pp.
McKINNEY, H.L. DSB, 1976, **14**, 133–40.
RABY, P. Alfred Russel Wallace. London, Chatto & Windus, 2001. 340 pp.
SHERMER, M. In Darwin's shadow: the life and science of Alfred Russel Wallace. Oxford, Oxford Univ. Press, 2002. 442 pp.
WALLACE, A.R. My life; a record of events and opinions. London, Chapman & Hall, 1905. 2 vols., 435 + 459 pp.; 2nd ed. 1908.
Archival material: British Library; British Museum (Natural History): Linnaean Society; University College London; Zoological Society of London; Central Library, Manchester.

WALLACE, William (1791–1837). British surgeon; first to describe lymphogranuloma venereum; introduced potassium iodide in treatment of syphilis.

MORTON, R.S. Dr William Wallace of Dublin. *Medical History*, 1966, **10**, 38–43.

WALLER, Auguste Desiré (1856–1922). French physiologist; demonstrated action current of the heart, preparing the ground for electrocardiography; obtained first human electrocardiogram, 1887.

COPE, Z. Auguste Desiré Waller (1856–1922). *Medical History*, 1973, **17**, 380–85.

WALLER, Augustus Volney (1816–1870). British neurophysiologist and neurohistologist; demonstrated Wallerian degeneration of nerve.

GERTLER-SAMUEL, R. Augustus Volney Waller (1816–1870) als Experimentalforscher. Zürich, Juris-Verlag, 1965. 33 pp.
SWAZEY, J.P. DSB, 1976, **14**, 142–4.

WALSHE, Francis Martin Rouse (1885–1973). British clinical neurologist.

PHILLIPS, C.G. BMFRS, 1974, **20**, 457–81.
Archival material: Library, University College London.

WALTHARD, Max (1867–1933). Swiss gynaecologist; professor at Berne and Zürich.

RÄBER, D. Der Gynäkologe Max Walthard (1867–1933). Zürich, Juris Druck, 1991. 108 pp.

WALTHER, Philipp Franz von (1782–1849). German ophthalmologist; first to describe corneal opacity.

NEUMANN, W. (ed.) Philipp Franz von Walther als Ophthalmologe. Gräfelring, Demeter, 1986. 268 pp.

WALTON, John Nicholas (b.1922). British neurologist; professor, University of Newcastle upon Tyne.

WALTON, Lord. The spice of life: from Northumberland to world neurology. London, Royal Society of Medicine, 1993. 643 pp.

WANGENSTEEN, Owen Harding (1898–1981). American surgeon; surgeon-in-chief, University of Minnesota Hospitals.

LEONARD, A.S. Reflections of the retiring chief. *Surgical Clinics of North America*, 1967, **60**, 1303–14.
PELTIER, L.F. & AUST, J.B. L'étoile du nord: an account of Owen Harding Wangensteen (1898–1981), surgeon – teacher – scholar. Chicago, American College of Surgeons, c.1994. 158pp.
VISSCHER, M.B. BMNAS, 1991, **60**, 354–65.

WARBURG, Otto Heinrich (1883–1970). German biochemist and cell biologist; Nobel Prize (Physiology or Medicine) 1931.

BURK, D. DSB, 1976, **14**, 172–7.
HÖXTERMANN, E. & SUCKER, U. Otto Warburg. Leipzig, Teubner, 1989. 180 pp.
KREBS, H.A. BMFRS, 1972, **18**, 629–99.
KREBS, H. Otto Warburg: cell physiologist, biochemist and eccentric. Oxford, Clarendon Press, 1981. 141 pp. First published in German, Stuttgart, 1979.
WERNER, P. Otto Warburg. Von der Zellphysiologie zur Krebsforschung: Biographie. Berlin, Verlag Neues Leben, 1988. 355 pp.
WERNER, P. Ein Genie irrt seltener...Otto Heinrich Warburg: ein Lebensbild in Dokumenten. Berlin, Akademie Verlag, 1991. 476 pp.

WARDROP, James (1782–1869). British surgeon; first to classify eye inflammations.

ALBERT, D.M. James Wardrop: a brief review of his life and contributions. *Transactions of the Ophthalmological Society of the United Kingdom*, 1974, **94**, 892–908.

WARREN FAMILY. John (1753–1815); John Collins (1778–1856); John Collins (1842–1927); Jonathan Mason (1811–1867); Joseph (1741–1775).

TRUAX, R. The doctors Warren of Boston: first family of surgery. Boston, Houghton Mifflin, 1968. 369 pp.

WARREN, John Collins (1778–1856). American surgeon; performed first public operation using ether.

WARREN, E. The life of John Collins Warren, M.D., chiefly compiled from his autobiography and journals. 2 vols. Boston, Ticknor & Fields, 1860. 420 + 382 pp.

WARREN, John Collins (1842–1927). American surgeon.

WARREN, J.C. To work in the vineyard of surgery. The reminiscences of John Collins Warren (1842–1927). Edited by E.D. Churchill. Cambridge, Mass., Harvard Univ. Press, 1958. 288 pp.

WARREN, Joseph (1741–1775). American physician and politician.

CARY, J. Joseph Warren: physician, politician, patriot. Urbana, Univ. Illinois Press, 1961. 260 pp.

FROTHINGHAM, R. Life and times of Joseph Warren. New York, Da Capo Press, 1971. 558 pp. Facsimile of 1865 edition.

WARTHIN, Aldred Scott (1866–1931). American pathologist; described "Warthin's sign" in pericarditis.

WELLER, C.V. Aldred Scott Warthin, M.D. 1866–1931. *Archives of Pathology*, 1931, **12**, 276–9.

WASHBURN, Margaret Floy (1871–1939). American psychologist; professor at Vassar College.

WOODWORTH, R.S. BMNAS, 1949, **25**, 275–95.

WASSERMANN, August von (1866–1925). German bacteriologist and immunologist; devised Wassermann test, specific diagnostic for syphilis.

FLECK, L. Genesis and development of a scientific fact. Ed. J. Trenn & R.K. Merton. Chicago, Univ. Chicago Press, 1979. 203 pp.
GILLERT, K.-E. DSB, 1978, **15**, 521–4.
KRAUSE, P. August Wassermann (1866–1925). Leben und Werk unter besonderer Berücksichtigung der Wassermannschen Reaktion. Inauguraldissertation...Johannes Gutenberg-Universität Mainz. Mainz, 1998. 263 pp.
Archival material: Leo Baeck Institute, New York.

WATERHOUSE, Benjamin (1754–1846). American physician; introduced Jennerian vaccination against smallpox into USA.

COHEN, I.B. (ed.) The life and scientific and medical career of Benjamin Waterhouse. 2 vols. New York, Arno Press, 1980. [Reprints of articles by or about Waterhouse, published from 1792 to 1970.]
HAWES, E. Benjamin Waterhouse, M.D.: first professor of the theory and practice of physic at Harvard and introducer of cowpox vaccination in America. Boston, Francis A. Countway Library of Medicine, 1974. 55 pp.

WATERS, Ralph Milton (1883–1979). American anaesthetist.

MORRIS, L.E. The combining influence of Ralph M. Waters in anesthesiology. Anaesthesia: essays in its history. Berlin, *c*.1985.
STEINHAUS, J.E. Ralph M. Waters, M.D., a teacher's teacher. *Anesthesia Association Newsletter*, 1988, **6**, No.1, 1–8.

WATERTON, Charles (1782–1865). British naturalist and traveller; gave early description of the paralysing effects of curare.

ALDINGTON, R. The strange life of Charles Waterton, 1782–1865. London, Evans Bros., [1949]. 200 pp.
EDGINGTON, B.W. Charles Waterton: a biography. Cambridge, Lutterworth Press, *c*.1996. 254 pp.

WATSON, Cecil James (1901–1983). American physician.

SCHMID, R. BMNAS, 1994, **65**, 354–72.

WATSON, James Dewey (b.1928). American molecular biologist; shared Nobel Prize 1962 for discovery of the molecular structure of DNA.

BIBLIOGRAPHY OF MEDICAL AND BIOMEDICAL BIOGRAPHY

WATSON, J.D. The double helix; a personal account of the discovery of the structure of DNA. New ed. London, Weidenfeld & Nicolson, 1981. 298 pp. First published 1968. *Archival material*: Harvard University.

WATSON, John Broadus (1878–1958). American exponent of behaviourist psychology.

BUCKLEY, K.W. Mechanical man: John Broadus Watson and the beginnings of behaviorism. New York, Guilford, 1989. 233 pp.
COHEN, D. J.B. Watson, the founder of behaviourism. London, Routledge & Kegan Paul, 1979. 297 pp.

WATSON, William (1715–1787). British physician and botanist; gave early (1746) account of emphysema.

HEILBRON, J.L. DSB, 1976, **14**, 193–6.

WATT, Robert (1774–1819). British physician at Glasgow Royal Infirmary; bibliographer.

CORDASCO, F. A bibliography of Robert Watt, M.D., author of the 'Bibliotheca Britannica'. With a facsimile edition of his Catalogue of medical books, and with a preliminary essay on his works. New York, Kelleher, 1950. 72 pp. Reprinted Detroit, Gale Research Co., 1968.
FINLAYSON, J. An account of the life and works of Dr Robert Watt, author of the Bibliotheca Britannica. London, Smith, Elder, 1897. 46 pp.

WAUCHOPE, Gladys Mary (1889–1966). British physician.

WAUCHOPE, G.M. The story of a woman physician [an autobiography]. Bristol, John Wright, 1963. 146 pp.

WEBB, Gerald Bertram (1871–1948). American immunologist and historian of tuberculosis.

CLAPESATTLE, H. Dr Webb of Colorado Springs. Boulder, Colorado Associated Univ. Press, 1984. 507 pp.

WEBER, Ernst Heinrich (1795–1878). German anatomist, physiologist and psychophysicist; professor at Leipzig.

BÜCK-RICH, U. Ernst Heinrich Weber (1795–1878) und der Anfang einer Physiologie der Hautsinne. Zürich, Juris-Verlag, 1970. 37 pp.
EISENBERG, W. (ed.) Ernst Heinrich Weber. Leipzig, Verlag im Wissenschaften, 2000. 371 pp.
KRUTA, V. DSB, 1976, **14**, 199–202.

WEBER, Frederick Parkes (1863–1962). British physician.

HALL, L.A. A "remarkable collection"; the papers of Frederick Parkes Weber (1863–1962). *Medical History*, 2001, **45**, 523–32.

WEBER, Hans Hermann Julius Wilhelm (1896–1974). German physiologist and biochemist.

BÜTTNER, J. DSB, 1990, **18**, 978–83.

WEBER, Hermann David (1823–1918). German-born British physician; authority of open-air treatment of tuberculosis, climatology, and numismatics.

388

WEBER, H.D. Autobiographical reminiscences written privately for the family, with annotations and a list of his medical writings by his son Frederick Parkes Weber. London, John Bale, Sons, & Danielsson, 1919. 121 pp.

WEBSTER, Leslie Tillotson (1894–1943). American experimental epidemiologist.

OLITSKY, P.K. Leslie Tillotson Webster, M.D. 1894–1943. *Archives of Pathology*, 1943, **35**, 535–7.

WEDEL, Georg Wolfgang (1645–1721). German physician; professor at Jena.

HUFBAUER, K. DSB, 1976, **14**, 212–13.

WEDENSKY, Nikolai Evgenievich, *see* **VVEDENSKY**

WEEKS, John Elmer (1853–1949). American ophthalmologist.

WEEKS, J.E. Autobiography. Portland, Ore., Metropolitan Print Co., 1954. 115 pp.

WEESE, Hellmut (1897–1954). German anaesthetist.

SCHADEWALDT, H. Hellmut Weese – Gedächtnisvorlesung. Von Galen's 'Narkosis' zur modernen 'Balanced anaesthesia'. *Anaesthesiologie und Intensivmedizin*, 1978, **12**, 589–601.

WEICHSELBAUM, Anton (1845–1920). Austrian pathologist; discovered meningococcus and diplococcus.

CHIARI, H. DSB, 1976, **14**, 218–19.

WEIDMANN, Johann Peter (1751–1819). German surgeon and obstetrician.

WEBER, B. Johann Peter Weidmann (1751–1819) und das Mainzer Accouchement. Mainz, Medizinhistorische Institut und Frauenklinik der Johannes Gutenberg-Universität Mainz, 1986. 165 pp.

WEIGERT, Carl (1845–1904). German pathologist and neurohistologist; inaugurated bacterial and differential tissue staining.

MORRISON, H. Carl Weigert. *Annals of Medical History*, 1924, **6**, 163–77.
RIEDER, R. In WEIGERT, C. Gesammelte Abhandlungen. Berlin, Springer, 1906, pp. 1–132.
RUDOLPH, G. DSB, 1976, **14**, 227–30.

WEIKARD, Melchior Adam (1742–1803). German physician; supporter of the Brunonian doctrine.

MILCHER, M. Melchior Adam Weikard (1742–1803) und sein Weg in den Brownianismus: eine medizinische Biographie. Halle (Saale), Deutsche Akademie der Naturforscher Leopoldina, 1995. 134 pp.
SCHMITT, O.M. Melchior Adam Weikard: Arzt, Philosoph und Aufklärer, Fulda Verlag Prazeller & Co, 1970. 134 pp.

WEIL, Adolf (1848–1916). German physician; gave classic description of leptospirosis icterohaemorrhagica.

SCHULTZE, F. Adolf Weil. *Münchener Medizinische Wochenschrift*, 1916, **63**, 1293–4.

WEIL, Edmund (1879–1922). German bacteriologist; introduced Weil-Felix reaction in disgnosis of typhus.

DIEGPEN, pp. 85–7 (CB).

WEILL-HALLÉ, Benjamin (1875–1958). French phusician; involved in introduction of BCG vaccination.

WEILL, J. Homage à Benjamin Weille-Hallé. Pour le 40e. anniversaire de la vaccination par le B.C.G. *Prophylaxis Sanitaire et Morale*, 1964, **36**, 219–24.

WEINBERG, Michel (1868–1940). Russian/French bacteriologist; isolated *Clostridium oedematiens* and *Clostridium histolyticum*.

RAMON, G. Michel Weinberg (1868–1940). *Annales de l'Institut Pasteur*, 1940, **64**, 461–5.

WEIR, Robert Fulton (1838–1927). American plastic surgeon; introduced reduction rhinoplasty.

McDOWELL, F. On restoring sunken noses without scarring the face. *Plastic and Reconstructive Surgery*, 1970, **45**, 382–92.

WEISMANN, August Friedrich Leopold (1834–1914). German zoologist; advanced the chromosome theory of heredity.

AUGUST WEISMANN (1834–1914) und die theoretische Biologie des 19. Jahrhunderts. Freiberg, Rombach Verlag, 1984. 240 pp.
GAUP, E. August Weismann. Sein Leben und Werk. Jena, Gustav Fischer, 1917. 297 pp.
PETRUNKEVITCH, A. August Weismann, personal reminiscences. *Journal of the History of Medicine*, 1963, **18**, 20–35.
ROBINSON, G. DSB, 1976, **14**, 232–9.

WEISS, Paul Alfred (1898–1989). American neurobiologist.

OVERTON, J. BMNAS, 1997, **72**, 373–86.

WEIZSÄCKER, Viktor von (1886–1957). German neurologist; pioneer in unified psychosomatic medicine.

DRESSLER, S. Viktor von Weizsäcker: medizinische Anthropologie und Philosophie. Wien, Uebereuter Wissenschaft, 1989. 161 pp.
HENKELMANN, T. Viktor von Weizsäcker, 1886–1957. Materialen zu Leben und Werk. Berlin, Springer-Verlag, 1986. 191 pp.

WELCH, William Henry (1850–1934). American pathologist; professor at Johns Hopkins Hospital, Baltimore; director of School of Hygiene and Public Health, Johns Hopkins University.

BRIEGER, G.H. DSB, 1976, **14**, 248–50.
FLEMING, D. William H. Welch and the rise of modern medicine. Boston, Little, Brown, 1954, 216 pp. Reprinted Baltimore, Johns Hopkins Univ. Press, 1987. 240 pp.
FLEXNER, S. BMNAS, 1941–43, **22**, 215–31.

BIBLIOGRAPHY OF MEDICAL AND BIOMEDICAL BIOGRAPHY

FLEXNER, S. & FLEXNER, J.T. William Henry Welch and the heroic age of American medicine. New York, Viking Press, 1941. 539 pp. Reprinted New York, Dover, 1966; London, Constable, 1968; Johns Hopkins University Press, 1993.
Archival material: William H. Welch Memorial Library, Johns Hopkins University Medical School.

WELDON, Walter Frank Raphael (1860–1906). British biometrician.

COWAN, R.S. DSB, 1976, **14**, 251–2.
PEARSON, K. *Biometrika*, 1906, **5**, 1–50.

WELLCOME, Henry Solomon (1853–1936). American/British pharmaceutical chemist and philanthropist.

JAMES, R.R. Henry Wellcome. London, Hodder & Stoughton, 1994. 422 pp.
TURNER, H. Henry Wellcome: the man, his collection and his legacy. London, Wellcome Trust and Heinemann, 1980. 96 pp.
WENYON, C.M. ONFRS, 1936–39, **2**, 229–38.
Archival material: Wellcome Foundation, London; Wellcome Trust, London.

WELLER, Thomas Huckle (b. 1915). American physician; shared Nobel Prize (Physiology or Medicine), 1954, for work on the poliomyelitis virus leading to vaccine production.

Les Prix Nobel en 1954.

WELLS, Harry Gideon (1875–1943). American immunologist and pathologist; professor at the University of Chicago.

LONG, E.R. BMNAS, 1951, **26**, 233–63.

WELLS, Horace (1815–1848). American dentist and pioneer in the use of nitrous oxide as anaesthetic.

ARCHER, W.H. Life and letters of Horace Wells, discoverer of anesthesia. *Journal of the American College of Dentists*, 1944, **11**, no.2, 811–210; 1945, **12**, No.2, 1–16.
GIES, W.J. (ed.) Horace Wells, dentist, father of surgical anesthesia. Proceedings of centenary commemorations of Wells' discovery in 1844: and list of Wells' memorabilia, including bibliographies, memorials and testimonials. Chicago, American Dental Association, 1948. 415 pp.

WELLS, Thomas Spencer (1818–1897). British surgeon; pioneer of abdominal surgery, perfecting the operation of ovariotomy.

SHEPHERD, J.A. Spencer Wells. The life and work of a Victorian surgeon. Edinburgh, London, Livingstone, 1965. 132 pp.

WELLS, William Charles (1757–1817). British physician, physiologist and natural philosopher, born U.S.A.

DOCK, W. DSB, 1976, **14**, 253–4.

WENCKEBACH, Karel Frederik (1864–1940). Dutch cardiologist in Vienna; wrote notable works on the arrhythmias.

LINDEBOOM, G.A. Karel Frederik Wenckebach (1864–1940); een koorte schets van zijn leven en werken. Haarlem, Bohn, 1965. 127 pp.

391

WENYON, Charles Morley (1878–1948). British protozoologist; Director of Research, Wellcome Foundation.

HOARE, C.A. ONFRS, 1948–49, **6**, 627–42.

WEPFER, Johan Jakob (1620–1695). Swiss physician and toxicologist.

EICHENBERGER, P. Johann Jakob Wepfer (1620–1694) als klinischer Praktiker. Basel, Schwabe, 1969. 142 pp.
MAEHLE, A.H. Johann Jakob Wepfer (1620–1695) als Toxikologe. Aarau, Sauerländer, 1987. 222 pp.
PILET, P.E. DSB, 1976, **14**, 255–6.

WERDNIG, Guido (1844–1919). German neeurologist.

GRANGER, H. Guido Werdnig. In Ashwell, S. (ed.) The founders of child neurology. *Medical History*, 2001, **45**, 523–32.

WERKMAN, Chester Hamlin (1893–1962). American bacteriologist; professor at Iowa State University.

BROWN, R.W. BMNAS, 1974, **44**, 329–70.
Archival material: Iowa State University.

WERLHOF. Paul Gottlieb (1699–1767). German physician; gave classical account of purpura haemorrhagica.

BENZENHÖFER, U. Der hannoversche Hof- und Leibarzt Paul Gottlieb Werlhof. (1699–1767). Aachen, Verlag Mainz, 1993. 53 pp.
ROHLFS, H. Geschichte der deutschen Medizin. Stuttgart, F. Enke, 1875, vol.1, 32–81.

WERNHER, Adolf Carl Gustav (1809–1883). German surgeon; professor at Giessen.

BIJOK, H. Adolf Carl Gustav Wernher (1809–1883): sein Leben und Wirken am Giessener Akademischen Hospital. Giessen, W. Schmitz, 1979. 235 pp.

WERNICKE, Carl (1848–1905). German neurologist and psychiatrist.

BYNUM, W.F. DSB, 1976, **14**, 277–8.
LANCZIK, M. Der Breslauer Psychiater Carl Wernicke. Werkenanalyse und Wirkungsgeschichte als Beitrag zur Medizingeschichte Schlesiens. Sigmaringen, Jan Thorbecke Verlag, 1988. 97 pp.

WERTHEIM, Ernst (1864–1920). Austrian gynaecologist; devised operation for cervical cancer.

LESKY, E. DSB, 1976, **14**, 278–9.

WESLEY, John Benjamin (1703–1791). English evangelist; author of *Primitive physick*.

HILL, A.W. John Wesley among the physicians: a study of eighteenth-century medicine. London, Epworth Press, 1958. 135 pp.
WILDER, F. The remarkable world of John Wesley, pioneer in mental health. Hicksville, NY, Exposition Press, 1978. 192 pp.

WEST, Charles (1816–1898). British paediatrician; a founder of the Hospital for Sick Children, London.

BIBLIOGRAPHY OF MEDICAL AND BIOMEDICAL BIOGRAPHY

KOSKY, J. Mutual friends: Charles Dickens and the Great Ormond Street Hospital. London, Weidenfeld & Nicolson, 1989. 245 pp.

LOMAX, E. A mid-nineteenth century British paediatrian's interpretation of the mental peculiarities and disorders of childhood. *Clio Medica*, 1982, **17**, 223–33.

WESTPHAL, Carl Friedrich Otto (1833–1890). German neurologist and psychiatrist.

SEIDEL, M. Carl Westphal – ein fortschrittlicher Hochschullehrer der Neurologie und Psychiatrie im 19. Jahrhundert. *Psychiatrie, Neurologie und Medizinische Psychologie*, 1986, **38**, 733–40.

WEYER, Johann, *see* **WIER, Johann**

WEYGANDT, Wilhelm (1870–1939). German psychiatrist.

WEBER-JASPER, E. Wilhelm Weygandt (1870–1939): Psychiatrie zwischen erkenntnistheoretischen Idealismus und Rassenhygiene. Husum, Matthiesen-Verlag, *c*.1976. 349 pp.

WHARTON, Thomas (1614–1673). English anatomist and endocrinologist; first to describe the duct of the submaxillary gland and to give first adequate account of the thyroid gland.

LeFANU, W. DSB, 1976, **14**, 286–7.

WHERRY, William Buchanan (1875–1936). American bacteriologist.

FISCHER, M.H. William B. Wherry, bacteriologist. Springfield, Ill., C.C. Thomas, 1938. 293 pp.

WHIPPLE, George Hoyt (1878–1976). American pathologist; shared Nobel Prize (Physiology or Medicine) 1934 for raw liver treatment of anaemia.

CORNER, G.W. George Hoyt Whipple and his friends: the life story of a Nobel Prize pathologist. Philadelphia, Montreal, Lippincott, 1963. 335 pp.

MILLER, L.S. BMNAS, 1995, **66**, 371–93.

WHIPPLE, G.H. Autobiographic sketch. *Perspectives in Biology and Medicine*, 1959, **2**, 253–89.

WHISTLER, Daniel (1619–1684). English physician; gave first description of rickets as a definite disease (1645).

COOKE, A.M. Daniel Whistler, P.R.C.P. *Journal of the Royal College of Physicians of London*, 1967, **1**, 221–30.

SMERDON, G.T. Daniel Whistler and the English disease. A translation and biographical note. *Journal of the History of Medicine*, 1950, **5**, 397–415.

WHITBY, Lionel Ernest Howard (1895–1956). British physician and clinical pathologist.

–. Sir Lionel Whitby, C.V.O., M.C., M.D., D.Sc., LL.D., F.R.C.P. *British Medical Journal*, 1956, **2**, 1306–9.

WHITE, Abraham (1908–1980). American biochemist and endocrinologist.

SMITH, E.L. BMNAS, 1985, **55**, 507–36.

393

WHITE, Charles (1728–1813). British obstetrician and surgeon; pioneer in aseptic midwifery.

ADAMI, J.G. Charles White of Manchester (1728–1813) and the arrest of puerperal fever. Liverpool, Liverpool Univ. Press; London, Hodder & Stoughton, 1922. 142 pp. [Includes reprint of White's *Treatise on the management of pregnant and lying-in women*, 1773.]
CLARKE, E. DSB, 1976, **14**, 296–7.
CULLINGWORTH, C.J. Charles White, F.R.S., a great provincial surgeon and obstetrician of the eighteenth century. London, Glaisher, 1904. 56 pp.

WHITE, James Clarke (1833–1916). American dermatologist.

SYNTEX LABORATORIES. James Clark White: the first American professor of dermatology. Palo Alto, Syntex Laboratories, 1968. 62 pp.
WHITE, J.C. Sketches from my life. Cambridge, Mass., Riverside Press, 1914. 326 pp.

WHITE, Michael James Denham (1910–1983). British cytogeneticist.

PEACOCK, W.J. & McCANN, D. BMFRS, 1994, **40**, 402–19.

WHITE, Paul Dudley (1886–1973). American cardiologist; professor of medicine, Harvard.

PAUL, O. Take heart. The life and prescription for living of Dr Paul Dudley White. Cambridge, Mass., Harvard Univ. Press, 1986. 336 pp.
PAUL DUDLEY WHITE, a portrait [and bibliography, 1913–1964]. *American Journal of Cardiology*, 1965, **15**, 433–602.
WHITE, P.D. My life and medicine: an autobiographical memoir. With M. Parton. Boston, Gambit, 1971. 269 pp.

WHITE, Philip Bruce (1891–1949). British bacteriologist; worked on *Salmonella*, gas gangrene and cholera.

SMITH, W. ONFRS, 1950–51, **7**, 279–92.

WHITE, William Alanson (1870–1937). American neurologist.

D'AMORE, A.R. & ECKBURG, A.L. (eds.) William Alanson White, the Washington years, 1903–1937. Washington, U.S. Govt. Printing Office, 1975. 189 pp.
WHITE, W.A. William Alanson White: the autobiography of a purpose. Garden City, NY, Doubleday, Doran, 1938. 293 pp.

WHITTERIDGE, David (1912–1994). British neurophysiologist.

GORDON, G. & IGGO, A. BMFRS, 1996, **42**, 524–38.

WHYTT, Robert (1714–1766). British physician and neurophysiologist; professor at Edinburgh.

BARCLAY, M. The life and work of Robert Whytt: a preliminary survey [part of MD thesis] Edinburgh University, 1922.
FRENCH, R.K. Robert Whytt, the soul and medicine. London, Wellcome Institute, 1969. 182 pp.
RADBILL, S.X. DSB, 1976, **14**, 319–24.

WICHMANN, Johann Ernst (1740–1802). German physician at Hannover; wrote important monograph on scabies and confirmed its parasitic nature.

FRIEDMAN, R. Johan Ernst Wichmann (1740–1892). The one hundred and fiftieth anniversary of his contribution to scabies. *Medical Life*, 1936, **43**, 171–210.

ROHLFS, H. Geschichte der Deutschen Medizin. Stuttgart, F. Enke, 1875, vol. 1, 135–75.

WICKERSHEIMER, Ernest (1880–1965). French historian of medicine; librarian, Académie de Médecine, Paris.

KLEIN, M. DSB, 1976, **14**, 324.

WICKMAN, Otto Ivar (1872–1914). Swedish neurologist; first to confirm the infectious nature of poliomyelitis.

DIEPGEN, pp. 107–10. (CB)

WIDAL, Georges Fernand Isidor (1862–1929). French physician; devised serological test for typhoid.

HUNTER, P.R. Fernand Widal. *Medical History*, 1963, **7**, 56–61.
LEMIERRE, A. Un grand médecin français, Widal. Paris, Expansion Scientifique Française, 1955. 146 pp.

WIDDOWSON, Elsie May (1906–2000). British nutritionist; published (with R.A. McCance) tables on The composition of foods.

ASHWELL, M. (ed.) McCance & Widdowson: a scientific partnership of 60 years. London, British Nutrition Foundation, 1993. 264 pp.
ASHWELL, M. BMFRS, 2002, **48**, 485–506.
Archival material: CMAC.

WIDMANN, Johann (1440–1524). German physician; professor of medicine, Tübingen; early user of mercury for syphilis.

HAEBLER, R.G. Doctor Johannes Widmann. Baden-Baden, Stadtbucherei, 1963. 44 pp.

WIEDERSHEIM, Robert Ernest Edouard (1848–1923). German comparative anatomist and embryologist; professor at Freiburg.

GLEES, P. DSB, 1976, **14**, 331.
GROTE, 1923, **1**, 207–27 (CB).
WIEDERSHEIM, R. Lebenserinnerungen. Tübingen, J.C.B. Mohr, 1919. 207 pp.

WIELAND, Heinrich Otto (1877–1957). German organic chemist; Nobel Prize (Chemistry) 19227, for work on alkaloids and steroids.

KARRER, P. BMFRS, 1958, **4**, 341–52.

WIENER, Alexander Solomon (1907–1976). American immuno-haematologist.

Haematologia (Budapest), 1972, **6**, 11–45 [biography, publications and other biographical papers].

WIENER, Norbert (1894–1964). American mathematician; founder of the science of cybernetics.

ILGAUDS, H.J. Norbert Wiener. Leipzig, Teubner, 1980. 86 pp.

WIER [WEYER], Johann (1515–1588). Dutch physician, considered the first clinical and descriptive psychiatrist; showed that witches were persons affected by mental illness.

BINZ, C. Dr. Johann Weyer. New York, Arno Press, 1976. 189 pp. Reprint of 1896 Berlin edition.
COBBEN, J.J. Jan Wier, devils, witches and magic. Translated by S.A.Prins. Philadelphia, Dorrance, 1976. 218 pp.

WIESCHAUS, Eric F. (b. 1943). American geneticist; shared Nobel Prize (Physiology or Medicine) 1995, for discoveries concerning the genetic control of early embryonic development.

Les Prix Nobel en 1995.

WIESEL, Torsten Nils (b. 1924). Swedish/American neurobiologist; shared Nobel Prize (Physiology or Medicine) 1981, for discoveries conerning information processes in the visual system.

Les Prix Nobel en 1981.

WIGGERS, Carl John (1883–1963). American cardiovascular physiologist; professor at Western Reserve University.

LANDIS, E.M. BMNAS, 1976, **48**, 363–97.
WIGGERS, C.J. Reminiscences and adventures in circulation research. New York, Grune & Stratton, 1958. 484 pp.

WILBRAND, Johann Bernard (1779–1846). German physiologist.

MAAS, C. Johann Bernhard Wilbrand (1779–1846): herausragender Vertreter de romantischer Naturlehre in Giessen. 2 vols. Giessen, Wilhelm Schmitz Verlag, 1994. 855 pp.
RISSE, G.B. DSB, 1976, **14**, 351–2.

WILCOCKS, Charles (1896–1977). British physician; Director, Bureau of Hygiene and Tropical Diseases, London.

Archival material: CMAC.

WILD, John Julian (b. 1911). British/American physicist; introduced ultrasonic investigations in soft tissue, 1957.

KOCH, E.B. In the image of science? Negotiating the development of diagnostic ultrasound in the cultures of surgery and radiology. *Technology and Culture*, 1993, **34**, 858–93.

WILD, Oscar (1870–1932). Swiss otolaryngologist.

MÜLLER, E.R. Der Nasen-Hals-Arzt Oscar Wild (1870–1932). Zürich, Juris, 1986. 85 pp.

WILDBERGER, Johannes (1815–1879). Swiss orthopaedic surgeon.

GROSCH, G. Johannes Wildberger (1815–1879): ein Schweizer Messerschmid und Wegbereiter der Orthopädie. Basel, Schwabe & Co. Verlag, 1969. 55 pp.

WILDE, William Robert Wills (1815–1876). Irish surgeon and antiquary; specialized as oculist and aurist; founded St Mark's Ophthalmic Hospital, Dublin.

WHITE, T. de Vere. The parents of Oscar Wilde. Sir William and Lady Wilde. London, Hodder & Stoughton, 1967. 303 pp.

WILSON, T.G. Victorian doctor: being the life of Sir William Wilde. London, Methuen, 1942. 336 pp.

WILDER, Russell Morse (1885–1959). American physician; chief, Department of Medicine, Mayo Foundation.

WILDER, R.M. Recollections and reflections in education, diabetes, other metabolic diseases, and nutrition in the Mayo Clinic and associated hospitals. *Perspectives in Biology and Medicine*, 1958, **1**, 237–77.

WILKIE, Douglas Robert (1922–1998). British physiologist.

WOLEDGE, R.C. BMFRS, 2001, **47**, 483–95.

WILKINS, Maurice Hugh Frederick (b.1916) New Zealand physicist; shared Nobel Prize (Physiology or Medicine) 1962 for discovery of helical nature of DNA.

WILKINS, M. The third man of the double helix: the autobiography of Maurice Wilkins. Oxford University Press, 2003. 312 pp.

WILKS, Samuel (1824–1911). British physician at Guy's Hospital, London.

HALE-WHITE, pp. 227–45. (CB).
WALE, W. Bibliography of the published writings of Sir Samuel Wilks. London, Adlard, 1911. 28 pp.
WILKS, S. A memoir by Sir Samuel Wilks on the new discoveries or new observations made while he was a teacher at Guy's Hospital. London, Adlard, 1911. 200 pp.

WILLAN, Robert (1757–1812). British physician, the founder of modern dermatology; he classified cutaneous diseases and established a standard nomenclature.

BESWICK, T. Robert Willan. *Journal of the History of Medicine*, 1957, **12**, 349–65.
LANE, J.E. Robert Willan. *Archives of Dermatology and Syphilology*, 1926, **13**, 737–60.

WILLCOX, William Henry (1870–1941). British Home Office analyst and medical adviser.

WILLCOX, P.H.A. The detective-physician. The life and work of Sir William Willcox, 1870–1941. London, Heinemann, 1970, 332 pp.

WILLE, Ludwig (1834–1912). Swiss psychiatrist.

MEYER, C. Ludwig Wille, 1834–1912; sein Leben und sein Werk, insbesondere in seiner Bedeutung für die Entstehung der klinischen Psychiatrie in der Schweiz. Zürich, Juris, 1973. 49 pp.

WILLEBRAND, Erik Adolf von (1870–1949). Finnish physician; first reported pseudohaemophilia type B ('von Willebrand's disease') 1926.

LETHAGEN, S. Mannen bakom syndromet: Erik A. Willebrand fann tidigare okand blodningsrubbning. *Lakartidningen*, 1994, **91**, 3805–9.

WILLIAMS, Charles James Blasius (1805–1889). Physician-extraordinary to Queen Victoria; specialist in tuberculosis and chest diseases.

BISHOP, P.J. The life and work of C.J.B.Williams (1805–1869) an eminent pupil of Laennec. *Tubercle*, 1957, No.378, 278–85.
WILLIAMS, C.J.B. Memoirs of life and work. London, Smith, Elder, 1884. 522 pp.

WILLIAMS, Cicely Delphine (1893–1992). British paediatrician.

CRADDOCK, S. Retired – except on demand: the life of Cicely Williams. Oxford, Green College, 1983. 198 pp.

DALLY, A. Cicely: the story of a doctor. London, Gollancz, 1968. 238 pp.

HUNTER, I. The papers of Cicely Williams (1893–1992) in the Contemporary Medical Archives Centre at the Wellcome Institute. *Social History of Medicine*, 1996, **9**, 109–16. *Archival material*: CMAC.

WILLIAMS, Daniel Hale (1858–1931). American surgeon.

BUCKLER, H. Daniel Hale Williams, Negro surgeon. 2nd ed. New York, Pitman, 1968. [First edition had title *Doctor Dan, pioneer in American surgery* , Boston, Little, Brown, 1954.]

WILLIAMS, Francis Henry (1852–1936). American radiologist; first to apply X rays in cardiology.

SNYDER, C. The Williams brothers and the Roentgen ray. *Archives of Ophthalmology*. 1965, **73**, 749–52.

WILLIAMS, Henry Willard (1821–1895). British ophthalmologist.

SNYDER, C. The single suture of Henry Willard Williams. In Snyder, C. Our ophthalmic heritage. London, Churchill, 1967, pp. 41–44.

WILLIAMS, John (1840–1926). British physician and bibliophile; responsible for the foundation of the National Library of Wales, of which his collection of Welsh books and manuscripts formed an integral part.

EVANS, R. John Williams, 1840–1926. Cardiff, University of Wales Press, 1952. 93 pp. [Text in Welsh and English.]

WILLIAMS, John Whitridge (1866–1931). American obstetrician; professor at Johns Hopkins Hospital.

SLEMONS, J.M. John Whitridge Williams: academic aspects and bibliography. Baltimore, Johns Hopkins Press, 1935. 109 pp.

WILLIAMS, Richard Tecwyn (1909–1979). British biochemist and pharmacologist.

NEUBERGER, A. & SMITH, R.L. BMFRS, 1982, **28**, 685–717.

WILLIAMS, Robert Runnels (1886–1965). American biochemist; discovered vitamin B[inf.c]; synthesized aneurine.

IHDE, A.J. DSB, 1976, 14, 392–4.

WILLIAMS, William Carlos (1883–1963). American physician, poet and novelist.

CRAWFORD, T.H. Modernism, medicine and William Carlos Williams. Norman, Univ. Oklahoma Press, 1993. 195 pp.

MARIANI, P. William Carlos Williams: a new world naked. New York, McGraw Hill, 1981. 874 pp.

WILLIER, Benjamin Harrison (1890–1972). American embryologist and developmental biologist; professor of zoology at Johns Hopkins University.

BIBLIOGRAPHY OF MEDICAL AND BIOMEDICAL BIOGRAPHY

WATTERSON, R.L. BMNAS, 1985, **55**, 539–628.

WILLIS, Robert (1799–1878). British physician, medical historian; Librarian, Royal College of Surgeons of England.

LeFANU, W.R. Robert Willis – physician, librarian, medical historian. Proceedings of the XXIII Congress of the History of Medicine, London, 1972. pp. 1111–15.

WILLIS, Thomas (1621–1675). English physician and anatomist.

FRANK, R.G. DSB, 1976, **14**, 404–9.
HUGHES, J.J. Thomas Willis 1621–1675: his life and work. London, Royal Society of Medicine Services, 1991. 151 pp.
ISLER, H. Thomas Willis, 1621–1675, doctor and scientist. New York, Hafner, 1968. 235 pp. First published in German, Stuttgart, 1965.
Archival material: British Library.

WILLSTÄTTER, Richard (1872–1942). German chemist; Nobel Prize (Chemistry) 1915 for his research on plant pigments, especially chlorophyll. Noted for his work on enzymes and development of partition chromatography.

ROBINSON, R. ONFRS, 1953, **8**, 609–34.
WILLSTÄTTER, R. Aus meinen Leben. Von Arbeit, Muße und Freunden. Herausegegeben und mit einem Nachwort versehen von Arthur Stoll, Basel. München, Verlag, Chemie, 1949. 453 pp.

WILMER, Bradford (1727–1813). British surgeon; introduced the 'Coventry' treatment of goitre, 1779.

LANE, J. Eighteenth-century medical practice: a case study of Bradford Wilmer, surgeon of Coventry. *Social History of Medicine*, 1990, **3**, 369–86.

WILMS, Max (1867–1918). German surgeon; described 'Wilms' tumour', embryoma of the kidney.

ROHL, L. Max Wilms (1867–1916). *Investigative Urology*, 1966, **3**, 194–6.

WILSON, Charles McMoran (1882–1977). British physician; dean of St Mary's Hospital Medical School, London.

LOVELL, R. Churchill's doctor: a biography of Lord Moran. London, Royal Society of Medicine Services, 1993. 457 pp.
Archival material: CMAC.

WILSON, David Wright (1889–1965). American biochemist, involved in early use of radioisotopic tracers; Professor, University of Pennsylvania.

BALL, E.G. & BUCHANAN, J.M. BMNAS, 1973, **43**, 261–84.

WILSON, Edmund Beecher (1856–1939). American cytologist, embryologist and geneticist.

ALLEN, G.E. DSB, 1976, **14**, 423–36.
MORGAN, T.H. BMNAS, 1941, **21**, 315–42.
MORGAN, T.H. ONFRS, 1940–41, **3**, 123–38.
MULLER, H.J. *American Naturalist*, 1943, **77**, 5–37, 142–72.
Archival material: Library, American Philosophical Society, Philadelphia.

WILSON, Edward Adrian (1872–1912). British physician and naturalist; accompanied Captain Scott to the South Pole and died there.

SEAVER, G. Edward Wilson of the Antarctic, naturalist and friend. London, Murray, 1933, 300 pp.

WILSON, Edward Osborne (b.1929). American zoologist and sociobiologist.

WILSON, E.O. Naturalist. Washington DC, Island Press, 1994. 380 pp.

WILSON, Frank Norman (1890–1952). American cardiologist.

SCHWARTZE, D. Frank N. Wilson und seine Bedeutung für die Kardiologie. *Zeitschrift für die gesamte innere Medizin*. 1991, **46**, 160–63.

WILSON, Graham Selby (1895–1987). British bacteriologist; creator and first director, [British] Public Health Laboratory Service.

ANDERSON, E.S. & WILLIAMS, R.E.O. BMFRS, 1988, **34**, 887–919.
Archival material: CMAC.

WILSON, James Thomas (1861–1945). British anatomist; professor at University of Cambridge.

HILL, J.P. ONFRS, 1948–49, **6**, 643–60.
Archival material: Basser Library, Australian Academy of Sciences, Canberra.

WILSON, Percy William (1902–1981). American microbiologist.

BURRIS, R.H. BMNAS, 1992, **61**, 438–67.

WILSON, Samuel Arthur Kinnier (1878–1937). British neurologist.

WESTERMARK, K. The man behind the syndrome: S.A. Kinnier Wilson. *Journal of the History of the Neurosciences*, 1993, **2**, 143–50.

WILSON, William James Erasmus (1809–1884). British dermatologist.

HADLEY, R.M. The life and works of Sir William Erasmus Wilson, 1809–84. *Medical History*, 1959, **3**, 214–47.

WINDAUS, Adolf Otto (1876–1959). German biochemist; synthesized histamine; Nobel Prize (Chemistry) 1928 for work on sterols.

BUTENANDT, A. The Windaus Memorial Lecture. *Proceedings of the Chemical Society*, 1961, pp. 131–8.
LEICESTER, H.M. DSB, 1976, **14**, 443–6.

WINDLE, Bertram Coghill Alan (1858–1929). Irish anatomist and anthropologist.

TAYLOR, M. Sir Bertram Windle: a memoir. London, Longmans Green, 1932. 428 pp.

WINDSOR, Harry Matthew (1914–1987). Australian surgeon.

WINDSOR, H. The heart of a surgeon: the memoirs of Harry Windsor. Kensington, New South Wales Press, [1988?]. 161 pp.

BIBLIOGRAPHY OF MEDICAL AND BIOMEDICAL BIOGRAPHY

WINIWARTER, Felix von (1852–1931). Austrian physician; described thrombo-angiitis obliterans, 1879.

LIE, J.T. *et al.* The brothers Winiwarter, Alexander (1848–1917) and Felix (1852–1931) and thromboangiitis obliterans. *Mayo Clinic Proceedings.* 1979, **54**, 802–7.

WINKLER, Cornelis (1855–1941). Dutch neurologist; professor of neurology and psychiatry at Amsterdam and Utrecht.

WINKLER, C. Herinneringen. Arnhem, Van Loghum Slaterus, 1947. 173 pp. Reprinted with introduction, Utrecht/Antwerpen, Bohn Scheltema & Holkema, 1982. 180 pp.

WINNICOTT, Donald Woods (1896–1971). British psychoanalyst.

KAHR, B. D.W. Winnicott: a biographical portrait. Madison, Conn., International Universities Press, 1996. 189 pp.
RIBAS, D. Donald Woods Winnicott. Paris, Presses Universitaires de France, 2000. 127 pp.

WINOGRADSKY, Sergey Nikolaevich, *see* **VINOGRADSKY**

WINSLOW, Charles-Edward Amory (1877–1957). American epidemiologist.

ACHESON, R.M. The epidemiology of Charles-Edward Amory Winslow. *American Journal of Epidemiology*, 1970, **91**, 1–18.

WINSLOW, Forbes Benignus (1810–1874). British psychiatrist.

WHITTINGTON-EGAN, M. Dr Forbes Winslow: defender of the insane. Great Malvern, Capella Archive, 2000. 280 pp.

WINSLØW, Jacob (Jacques Benigne) (1669–1760). Danish/French anatomist; wrote first treatise on descriptive anatomy.

SNORRASON, E. L'anatomiste J.-B. Winslow 1669–1760. Copenhagen, Bedingske Bogtrykkeri, 1969. 83 pp.
SNORRASON, E. DSB, 1976, **14**, 449–51.

WINTROBE, Maxwell Myer (1901–1986). Canadian/American haematologist; classified the anaemias; introduced method for determination of erythrocyte sedimentation rate.

VALENTINE, W.N. BMNAS, 1990, **59**, 447–72.

WIRTZ [WÜRTZ] Felix (*c.*1510–*c.*1590). Swiss barber-surgeon; wrote book on surgery (1563) containing a section on paediatric surgery and which was later translated into English, French and Dutch.

COURVOISIER, L.G. Felix Wirtz, ein Basler Chirurg des sechszehnten Jahrhunderts. *Correspondenz-Blatt für Schweizer Aertze*, 1880, **10**, 291–312.

WISEMAN, Richard (1622–1676). English surgeon.

LONGMORE, T. Richard Wiseman, surgeon and Sergeant-Surgeon to Charles II. A biographical study. London, Longmans Green, 1891. 210 pp.
SMITH, A.D. Richard Wiseman: his contributions to English surgery. *Bulletin of the New York Academy of Medicine*, 1970, **46**, 167–82.

BIBLIOGRAPHY OF MEDICAL AND BIOMEDICAL BIOGRAPHY

WITEBSKY, Ernest (1901–1989). German/American immunologist.

ROSE, N.R. The discovery of thyroid autoimmunity. *Immunology Today*, 1991, **12**, 167–8.

WITHERING, William (1741–1799). British physician and botanist; described the correct use of foxglove (digitalis) in the treatment of dropsy.

ARONSON, J.K. An account of the foxglove and its medical uses 1785–1985. Oxford, Univ. Press, 1985. 399 pp. [Includes a biography.]

BYNUM, W.F. DSB, 1976, **14**, 463–5.

MANN, R.D. William Withering and the foxglove. A bicentenary selection of letters ...together with a transcription of 'An account of the foxglove' and an introductory essay. Lancaster, MTP Press, 1986. 178 pp.

PECK, T.W. & WILKINSON, K.D. William Withering of Birmingham. Bristol, Wright; London, Simpkin Marshall, 1950. 239 pp.

RODDIS, L.H. William Withering: the introduction of digitalis into medical practice. New York, Hoeber, 1936. 131 pp.

Archival material: Royal College of Surgeons of England; Library, University of Birmingham; Royal Society of Medicine, London; Wellcome Institute, London.

WÖHLER, Friedrich (1800–1882). German chemist; synthesized urea.

KEEN, R. The life and work of Friedrich Wöhler. Ph.D. thesis, University College London, 1976.

KEEN, R. DSB, 1976, **14**, 474–9.

VALENTIN, J. Friedrich Wöhler. Stuttgart, Wissenschaftliche Verlagsgesellschaft, 1949. 178 pp.

Archival material: Bayerische Staatsbibliothek, Munich; Akademie der Wissenschaften, Berlin.

WÖLFLER, Anton (1850–1917). Bohemian surgeon; professor at Graz und Prague; introduced gastro-enterostomy.

HÖFERLIN, A. Der Chirurg Anton Wölfler (1850–1917): sein Leben und Werk mit besonderer Berücksichtigung seiner Arbeiten zur Schilddrüsenchirurgie. Dissertation, Mainz, 1989. 178 pp.

WOLBACH, Simeon Burt (1880–1954). American pathologist; carried out important work on Rocky Mountain spotted fever and on typhus.

JANEWAY, C.A. S. Burt Wolbach. *Transactions of the Association of American Physicians*, 1954, **67**, 30–35.

WOLCOT, John (1738–1819). Physician in Jamaica and Cornwall; satirist and poet under the pseudonym 'Peter Pindar'.

GIRTIN, T. Doctor with two aunts: a biography of Peter Pindar. London, Hutchinson, 1959. 272 pp.

WOLCOTT, Erastus Bradley (1804–1880). American surgeon.

MILLER, W.S. *Wisconsin Medical Journal*, 1931, **30**, 8–12.

WOLF, Alfred P. (1923–1998). American chemist and nuclear physicist; pioneered development of positron emission tomography.

FOWLER, J.S. & WELCH, M.J. BMNAS, 2000, **78**, 354–67.

WOLFF, Caspar Friedrich (1733–1794). German biologist; a founder of the science of embryology.

GAISSINOVITCH, A.E. DSB, 1978, **15**, 524–6.
USCHMANN, G. Caspar Friedrich Wolff: ein Pionier der modernen Embryologie. Leipzig, Jena, Urania Verlag, 1955. 86 pp.

WOLLASTON, William Hyde (1766–1828). British physician and chemist; isolated cystine; discovered palladium and rhodium; invented goniometer for the measurement of crystals.

GOODMAN, D.C. DSB, 1976, **14**, 486–94.
WAYLING, H.G. A short biography of William Hyde Wollaston. *Science Progress*, 1927, **22**, 81–95.

WOOD, Alexander (1817–1884). British physician; introduced hypodermic medication.

BROWN, T. Alexander Wood, M.D., F.R.C.P.E. A sketch of his life and work. Edinburgh, Macniven & Wallace, 1886. 203 pp.

WOOD, Arthur Michael (1919–1987). British plastic surgeon; founder and director-general, African Medical and Research Foundation.

WOOD, M. Different drums: reflections on a changing Africa. London, Century, 1987. 223 pp.

WOOD, Casey Albert (1856–1942). American ophthalmologist.

ASTBURY, E.C. Casey A.Wood (1856–1942), ophthalmologist, bookman, ornithologist. A bio-bibliography. Montreal, Graduate School of Library Science, McGill University, 1981.

WOOD, Francis Clark (1901–1990). American cardiologist.

HOROWITZ, D. Francis Clark Wood. *Transactions of the American Clinical and Climatological Association*, 1992, **103**, lxvi–lxix.
RHOADS, J.E. Memoir of Francis C. Wood, 1901–1940. *Transactions of the American Clinical and Climatological Association*, 1992, **103**, 205–8.
SCHNABEL, T.G. Francis Clark Wood: his contributions to medical science, 1901–1990. *Transactions and Studies of the College of Physicians of Philadelphia*, 1991, 5 ser., **13**, 209–14.

WOOD, Harland Goff (1907–1991). American biochemist, Case Western University Medical School.

GOLDTHWAIT, D.A.& HANSON, A.W. BMNAS, 1996, **69**, 395–428.
WOOD, H.G. Then and now. *Annual Review of Biochemistry*, 1985, **54**, 1–41.

WOOD, Horatio Charles (1841–1920). American pharmacologist and therapist.

ROTH, G.B. BMNAS, 1959, **33**, 462–84.
SONNEDECKER, G. DSB, 1976, **14**, 495–7.

WOOD, Leonard (1860–1927). American surgeon, soldier and politician.

HAGEDORN, H.L. Leonard Wood: a biography. 2 vols. New York, Harper, 1931.

SEARS, J.H. The career of Leonard Wood. London, Appleton, 1919. 273 pp.

WOOD, Paul Hamilton (1907–1962). American cardiologist.

DIMOND, E.G. Paul Hamilton Wood. *American Journal of Cardiology*, 1972, **30**, 121–95.

WOOD, William Barry (1910–1971). American microbiologist; vice-president and professor, Johns Hopkins University.

HIRSCH, J.G. BMNAS, 1980, **51**, 387–418.

WOODALL, John (?1556–1643). English military surgeon and later surgeon to St Bartholomew's Hospital, London. His *The surgions mate*, 1617, was one of the earliest books on naval medicine.

APPLEBY, J.H. New light on John Woodall, surgeon and adventurer. *Medical History*, 1981, **25**, 251–68.
KEYNES, G. John Woodall, surgeon, his place in medical history. *Journal of the Royal College of Physicians of London*, 1967, **2**, 15–33.

WOODHEAD, German Sims (1855–1921). British pathologist; professor at University of Cambridge and editor of *Journal of Pathology.*

WOODHEAD, H.E. In memoriam, Sir German Sims Woodhead, KBE, 1855–1921. Edinburgh, Oliver & Boyd, 1923. 79 pp.

WOOD JONES, Frederic (1879–1954). British anatomist; professor at Royal College of Surgeons of England.

CHRISTOPHERS, B.E. A list of the works of Frederic Wood Jones, 1879–1954. 2nd ed., Melbourne, The author, 1985. 124 pp. Supplement, 1988. Richmond, Vic., Australia, The author.
CLARK, W.E. Le G. BMFRS, 1955, **1**, 19–34.

WOODRUFF, Michael Francis Addison (1911–2001). Australian/British transplant surgeon.

WOODRUFF, M. Nothing venture, nothing win. Edinburgh, Scottish Academic Press, 1996. 234 pp.

WOODS, Donald Devereux (1912–1964). British microbiologist.

GALE, E.F. & FILDES, P.G. BMFRS, 1965, **11**, 203–19.
Archival material: Bodleian Library, Oxford.

WOODWARD, John (1665–1728). English physician and geologist.

EYLES, V.A. DSB, 1976, **14**, 500–503.
EYLES, V.A. John Woodward, F.R.S., F.R.C.P., M.D. (1665–1728): a bio-bibliographical account of his life and work. *Journal of the Society for the Bibliography of Natural History*, 1971–75, **5**, 399–427.

WOODWARD, Joseph Janvier (1833–1864). U.S.Army pathologist; wrote authoritative work on amoebiasis.

BILLINGS, J.S. BMNAS, 1886, **2**, 285–307.
HEATON, L.D. & BLUMBERG, J.M. Lt. Col. Joseph J. Woodward, U.S. Army, pathologist-researcher-photomicroscopist. *Military Medicine*, 1966, **131**, 530–38.

THOMPSON, R.A. BMNAS, 1999, **76**, 360–74.

WOODWARD, Robert Burns (1917–1979). American organic chemist.

TODD, Lord, & CORNFORTH, J. BMFRS, 1981, **27**, 629–95.

WOODWORTH, Robert Sessions (1869–1962). American psychologist; professor at Columbia University.

GRAHAM, C.H. BMNAS, 1967, **39**, 541–72.

WOOLLARD, Herbert Henry (1889–1939). Australian-born anatomist; professor at St Bartholomew's Hospital Medical College, London.

CLARK, W.E. Le G. ONFRS, 1940–41, **3**, 89–95.

WOOLSEY, Clinton Nathan (1904–1993). American neurophysiologist.

THOMPSON, R.A. BMNAS, 1999, **76**, 360–74.

WORCESTER, Noah (1812–1847). American dermatologist; published first American text on skin diseases.

BEUTNER, E.H. & BEUTNER, G.P. Classification of skin diseases by Noah Worcester. *International Journal of Dermatology*, 1991, **30**, 670–73.

WORM, Oläus (1588–1654). Danish anatomist; described Wormian bones in skull.

HOVESEN, E. Laegen Ole Worm, 1588–1654. Aarhus, Universitetsverlag, 1987. 367 pp.

WORMALL, Arthur (1900–1964). British biochemist and immunologist; first professor of biochemistry at St Bartholomew's Hospital Medical College, London.

MORGAN, W.T.J. & FRANCIS, G.E. BMFRS, 1966, **12**, 543–64.

WRIGHT, Almroth Edward (1861–1947). British physician and pathologist; Head of Inoculation Department, St Mary's Hospital, London; pioneer immunologist.

COLEBROOK, L. ONFRS, 1948–49, **6**, 297–314.
COLEBROOK, L. Almroth Wright: provocative doctor and thinker. London, Heinemann, 1954, 286 pp.
COLEBROOK, L. Bibliography of the published writings of Sir Almroth E. Wright. London, Heinemann, 1952. 32 pp.
DUNNILL, M. The Plato of Praed Street: the life and times of Almroth Wright. London, Royal Society of Medicine Press, 2000. 269 pp.
PARKER, DSB, 1976, **14**, 511–13.
Archival material: British Library; CMAC; Royal College of Pathologists; Trinity College, Dublin; British Library.

WRIGHT, Helena Rosa (1887–1982). Pioneer of family planning in Britain.

EVANS, B. Freedom to choose: the life and work of Helena Wright, pioneer of contraception. London, Bodley Head, 1984. 272 pp.
Archival material: CMAC.

WRIGHT, Sewall (1889–1988). American biologist and geneticist.

HILL, W.G. BMFRS, 1990, **36**, 567–79.

PROVINE, W.B. Sewall Wright and evolutionary biology. Chicago, Univ. Chicago Press, 1986. 545 pp.

WRIGHT, William (1735–1819). British physician and botanist; Physician-General Jamaica.

MEMOIR of the late William Wright, M.D. With extracts from his correspondence, and a selection of his papers on medical and botanical subjects. Edinburgh, Blackwood; London, Cadell, 1828.

WRISBERG, Heinrich August (1739–1808). German anatomist; professor at Göttingen.

HEINE, B. Heinrich August Wrisberg's Bedeutung für die Pathologie, 1736–1808. Göttingen, 1935. 23 pp.

WU, Hsien (1893–1959). Chinese biochemist; jointly devised Folin-Wu test for blood sugar.

CARMICHAEL, E.B. Dr Hsien Wu. *Nature*, 1960, 809–10.

WUCHERER, Otto Eduard Heinrich (1820–1873). German physician in Brazil; he saw the filaria worm (1866), later named *Wuchereria bancrofti*, and confirmed hookworm disease as cause of tropical anaemia.

CONI, A.C. Otto Wucherer, sua vida e sua obra. *Revista Brasileira de Malariologia*, 1967, **19**, 91–118.

WÜRTZ, Feliz *see* **WIRTZ, Felix**

WU LIEN-TEH (1879–1960). Chinese physician; public health pioneer.

WU LIEN-TEH. Plague fighter: the autobiography of a modern Chinese physician. Cambridge, Heffer, 1959. 667 pp.
WU YU-LIN. Memories of Dr Wu Lien-teh, plague fighter. Singapore, World Scientific, 1995. 185 pp.

WUNDERLICH, Carl Reinhold August (1815–1877). Germam physician; professor of medicine, Leipzig; laid foundation of modern clinical thermometry.

HEUBNER, O. C.A.Wunderlich, Nekrolog. *Archiv der Heilkunde*, Leipzig, 1878, **19**, 289–320.
ROSER, W. Zur Erinnerung C.A.Wunderlich. *Archiv der Heilkunde*, 1878, **19**, 321–32.

WUNDT, Wilhelm Max (1832–1920). German physiologist and psychologist; founder of experimental psychology.

DIAMOND, S. DSB, 1976, **14**, 526–9.
LAMBERTI, G. Wilhelm Maximilian Wundt (1832–1920). Leben, Werk und Persönlichkeit in Bildern und Texten. Bonn, Deutscher Psychologen Verlag, 1995. 175 pp.
RIEBER, R.W. *et al.* (eds.) Wilhelm Wundt and the making of a scientific psychology. New York, Plenum Press, [*c*.1980]. 252 pp.

WYETH, John Allan (1845–1922). American surgeon and medical educator; founder of New York Polyclinic Medical School.

CARMICHAEL, E.B. John Allan Wyeth. *Bulletin of the History of Medicine*, 1945, **18**, 329–37.
WYETH, J.A. With sabre and scalpel. The autobiography of a soldier and surgeon. New York, Harper Brothers, 1914. 535 pp.

WYNDHAM, Charles (1837–1919). British physician and actor-manager; opened Wyndham's Theatre, London, 1899.

DNB Twentieth Century.

WYNTER, Walter Essex (1860–1945) introduced lumbar puncture (1891) independently of Quincke.

FREDERIKS, J.A.M. & KOEHLER, P.J. The first lumbar puncture. *Journal of the History of the Neurosciences*, 1997, **6**, 147–53.

YALOW, Rosalyn Sussman (b.1921). American immunologist and endocrinologist: shared Nobel Prize (Physiology or Medicine) 1977 for radioimmunossay of a hormone (insulin).

STRAUS, E. Rosalyn Yalow, Nobel Laureate. Her life and work in medicine: a biographical memoir. New York, Plenum, 1988. 277 pp.

YAMAGIWA, Katsusaburo (1863–1930). Japanese physician; produced experimental tar cancer.

HENSCHEN, F. Yamagiwa's tar cancer and its historical significance. From Percivall Pott to Katsusaburo Yamagiwa. *Gann*, 1968. **59**, 447–51.

YERKES, Robert Mearns (1876–1956). American comparative psychologist.

BURNHAM, J.C. DSB, 1976, **14**, 549–51.
HILGARD, E.R. BMNAS, 1965, **38**, 385–425.
Archival material: Yale University Library.

YERSIN, Alexandre Émile Jean (1863–1943). Swiss-born bacteriologist at Institut Pasteur, Paris; discovered plague bacillus and introduced anti-plague vaccine.

BERNARD, N.P.J.L. Yersin: pionnier, savant, explorateur, 1863–1943. Paris, La Colombe, 1955. 190 pp.
MOLLARET, H.H. & BROSSOLET, J. Alexandre Yersin, le vainquer de la peste. Paris, Fayard, 1985. 320 pp.
PILET, P.E. DSB, 1976, **14**, 551–2.

YONGE, James (1647–1721). British surgeon; described use of turpentine for arresting haemorrhage, and the flap operation in amputation.

YONGE, J. The journal of James Yonge (1647–1721), Plymouth surgeon. Edited by F.N.L. Poynter. London, Longmans, 1963, 247 pp. [Includes *A catalogue of the books I have writ and published.*]

YORKE, Warrington (1883–1943). British parasitologist; professor of tropical medicine, Liverpool School of Tropical Medicine.

WENYON, C.M. ONFRS, 1942–44, **4**, 523–45.

YOUNG, Francis Brett (1884–1954). British physician, poet and novelist.

YOUNG, J.B. Francis Brett Young: a biography. London, Heinemann, 1962. 360 pp.

YOUNG, Frank George (1908–1986). British biochemist; professor at St Thomas's Hospital Medical School and Cambridge University.

BIBLIOGRAPHY OF MEDICAL AND BIOMEDICAL BIOGRAPHY

POWELL, T. & HARPER, P. Catalogue of papers and correspondence of Sir Frank George Young (1908–1988). Bath, National Cataloguing Unit for the Archives of Contemporary Scientists, 2001. 134 leaves. NCUACS Cat. No. 97/2/01.
RANDALL, P. BMFRS, 1990, 36, 581–99.
Archival material: Cambridge University Archives.

YOUNG, Hugh Hampton (1870–1945). American urological surgeon.

YOUNG, H.H. A surgeon's autobiography. New York, Harcourt Brace, 1940. 554 pp.

YOUNG, John Richardson (1782–1804). American physiologist; performed early experiments on gastric juice.

YOUNG, J.R. An experimental inquiry into the principles of nutrition and the digestive process. With an introductory essay by W.C. Rose. Urbana, Univ. Illinois Press, 1959. xxvi + 48 pp. [Reprinted inaugural address of J.R. Young and a biographical notice.]

YOUNG, John Zachary (1907–1997). British neurobiologist.

BOYCOTT, B.B. BMFRS, 1998, **44**, 487–509.

YOUNG Thomas (1773–1829). British scientist, physician, and egyptologist; established wave theory of light and Young-Helmholtz theory of colour vision.

MORSE, E.W. DSB, 1976, **14**, 562–72.
OLDHAM, F. Thomas Young, F.R.S., philosopher and physician. London, Arnold, 1933. 159 pp.
WOOD, A. Thomas Young, natural philosopher (1773–1829). Completed by F. Oldham. Cambridge, Cambridge Univ.Press, 1954. 355 pp.
Archival material: Royal Society of London; University College London; British Library.

YPERMAN, Jan (*fl*. 1310). Flemish surgeon; the great authority on surgery in the Low Countries in the 14th century.

DAMME, L. van. De chirurgijn van Ieper. Blankenberg, Saeftinge, 1967. 31 pp.
LEERSUM, E.C. van. Notes concerning the life of Jan Yperman. *Janus*, Amsterdam, 1913, **18**, 1–15.

YUDIN, *see* **IUDIN**

YULE, George Udney (1871–1951). British statistician.

NORTH, J.D. DSB, 1976, **14**, 573–4.
YATES, F. ONFRS, 1952, **8**, 309–23.

ZACCHIAS, Paolo (1584–1659). Italian physician; Papal physician and an early writer on medical jurisprudence.

FOSSEL, V. Studien zur Geschichte der Medizin. Stuttgart, F. Enke, 1909, pp. 46–110.

ZAHORSKY, John (b. 1871). American paediatrician; first to describe herpangina and roseola subitum.

ZAHOSRKY, J. From the hills: the autobiography of a pediatrician. St Louis, Mosby, 1949. 388 pp.

408

ZAHR-WI, Abu al-Qasim, *see* **ABULCASIS**

ZAKRZEWSKA, Maria Elizabeth (1829–1902). German/American physician; qualified USA, 1850; with Elizabeth Blackwell founded first training school for nurses in USA (1873).

VIETOR, A.C. A woman's quest. The life of Marie E. Zakrzewska. New York, Arno Press, 1972, 514 pp. First published New York, Appleton, 1924.

ZAMBECCARI, Giuseppe (1655–1728). Italian physician.

CASTELLANI, C. DSB, 1976, **14**, 586–7.
JARCHO, S. Guiseppe Zambeccari. A 17th century pioneer in experimental physiology and surgery. *Bulletin of the History of Medicine*, 1941, **9**, 144–76.

ZAMENHOF, Lazarus Ludwik (1859–1917). Polish oculist; invented the artificial language Esperanto.

BOULTON, M. Zamenhof, creator of Esperanto. London, Routledge & Kegan Paul, 1960. 223 pp.
PRIVAT, E. The life of Zamenhof, translated from the original Esperanto by R. Eliott. London, Allen & Unwin, 1931. 123 pp.

ZAUFAL, Emanuel (1833–1910). Bohemian otologist; improved the Schwartze-Eysell mastoid operation.

PIFFEL, O. *Archiv für Ohrenheilkunde*, 1910, **82**, 132–41.

ZEIS, Eduard (1807–1868). German plastic surgeon (introduced the term 1838); published a history of plastic surgery with an annotated bibliography of the literature prior to 1860 (1863) which was revised as the *Zeis Index and the History of Plastic Surgery*, 1977.

SEBASTIEN, G. 150 Jahre "Handbuch der plastischen Chirurgie". Erinnerungen an Eduard Zeis (1807–1868). *Hautarzt*, 1989, **40**, 45–52.

ZENKER, Friedrich Albert (1825–1898). German pathologist.

SCHRODER, H. Ein Erinnerungensblatt für Friedrich Albert Zenker. *Münchener Medizinische Wochenschrift*, 1925, **72**, 436–7.

ZERBI, Gabriele (1445–1505). Italian anatomist; published first printed work on geriatrics.

NANNINI, M.C. Processo storico a Jacopo Berengarius ad a Gabriele Zerbi. *Pagine di Storia della Medicina*, 1967, **11**, 78–85.

ZERNIKE, Frits (1888–1966). Dutch physicist; invented phase contrast microscopy; Nobel Prize (Physics) 1953.

TOLANSKY, S. BMFRS, 1967, **13**, 393–402.

ZIEGLER, Ernst (1849–1905). German pathologist; professor at Freiburg and founder of *Beiträge zur pathologischen Anatomie und Physiologie*.

HODEL, C. & BUESS, H. Ernst Ziegler (1849–1905), ein grosser Forscher und Lehrer. *Clio Medica*, 1966, **1**, 303–18.

ZIEHL, Franz (1857–1926). German physician.

BISHOP, P.J. The history of the Ziehl-Neelsen stain. *Tubercle*, 1970, **51**, 196–206.

ZIMMERMANN, Johann Georg (1728–1795). Swiss physician; professor of medicine at Hannover; wrote important work on bacillary dysentery.

ISCHER, R. Johann Georg Zimmermanns Leben und Werke. Bern, K.J. Wyss, 1893. 428 pp.
SCHRAMM, H.P. (ed.) Johann George Zimmermann: königlich grossbrittanischer Leibarzt (1728–1795) Wiesbaden, Harrassowitz, 1998, 366 pp.

ZINKERNAGEL, Rolf M. (b. 1944). Swiss immunologist; shared Nobel Prize (Physiology or Medicine) 1996, for discoveries concerning the specificity of the cell-mediated immune defence.

Les Prix Nobel en 1996.

ZINSSER, Hans (1878–1940). American bacteriologist and immunologist; professor at Stanford and Harvard universities.

BYNUM, W.F. DSB, 1976, **14**, 622–4.
WOLBACH, S.B. BMNAS, 1947, **24**, 323–60.
ZINSSER, H. As I remember him: the biography of H.S. [Hans Zinsser] Boston, Little, Brown, 1940, 443 pp.
Archival material: Library, Harvard University Medical School.

ZIRM, Eduard Konrad (1863–1944). Austrian ophthalmic surgeon; performed first keratoplasty.

LESKY, E. Eduard Konrad Zirm (1863–1944). *Wiener Klinische Wochenschrift*, 1963, **75**, 1–6.

ZOLL, Paul Maurice (b. 1911). American cardiologist; introduced external cardiac pacemaker.

ADELMANN, W.H. Paul M. Zoll and electrical stimulation of the heart, 1952. *Clinical Cardiology*, 1986, **9**, 131–5.

ZOLLINGER, Robert Milton (1904–1992). American physician; with E.H. Ellison described 'Zollinger-Ellison syndrome', 1955.

HANLON, C.R. Former President Robert M. Zollinger died. *Bulletin of the American College of Surgeons*, 1992, **77**, 34–5.

ZONDEK, Bernhard (1891–1966). German/Israeli endocrinologist and gynaecologist.

FINKELSTEIN, M. Professor Bernhard Zondek. An interview. *Journal of Reproduction and Fertility*, 1966, **12**, 3–19.

ZOTTERMAN, Yngve (1898–1982). Swedish physiologist.

ZOTTERMAN, Y. Touch, tickle and pain; an autobiography. 2 vols. Oxford, Pergamon Press, 1969–71. 269 & 293 pp.

ZUCKERMAN, Solly (1904–1993). South African/British zoologist; scientific adviser to British Government.

KROHN, P.L. BMFRS, 1995, **41**, 576–98.
RYTON, J. Solly Zuckerman: a scientist out of the ordinary. London, John Murray, 2001. 252 pp.
ZUCKERMAN, S. From apes to warlords: the autobiography (1904–1946). London, Hamish Hamilton, 1978. 447 pp.

ZUCKERMAN, S. Monkeys, men and missiles; an autobiography 1946–1988. London, Collins, 1988. 498 pp.

ZUELZER, Georg Ludwig (1870–1949). German physician; isolated insulin-containing pancreatic extract, 1908, which was abandoned due to hypoglycaemic reactions.

GEORG LUDWIG ZUELZERS Beitrag zur Insulinforschung. Düsseldorf, Triltsch, 1971. 58 pp.

ZUNTZ, Nathan (1847–1920). German physiologist.

GUNGA, H.C. Leben und Werk des Berliner Physiologen Nathan Zuntz (1847–1920): unter besonderer Berücksichtigung seiner bedeutung für die Frühgeschichte der Höhenphysiologie und Luftfahrtmedizin. Husum, Matthiesen Verlag, 1989. 343 pp.

ZWINGER, Theodor (1658–1724). Swiss physician; professor of medicine in Basel.

BUESS, H., PORTMANN, M.L. & MOLLING, P. Theodor Zwingler III (1658–1724). Ein Basler Arzt und Kinderarzt der Barockzeit. Basel, Helbig & Lichtenhahn, 1962. 246 pp.

COLLECTIVE BIOGRAPHIES

National biographical dictionaries such as American National Biography, Allgemeine Deutsche Biographie, etc. are not listed here but the reader is reminded of their importance.

AMERICAN MEDICAL ASSOCIATION. Directory of deceased American physicians 1804–1929. 2 vols. Chicago, American Medical Association, 1993. 1824 pp.

AMERICAN MEN AND WOMEN OF SCIENCE. Physical and biological sciences. 21st ed., 2003. 8 vols. Thomson Gale, Belmont, CA. Entries for 125,000 active US and Canadian scientists. Published annually.

BEIGHTON, B. & BEIGHTON, G. The person behind the syndrome. London, New York, Springer, 1996. 231 pp.

BINDMAN, L. *et al.* (eds.) Women physiologists: an anniversary celebration of their contributions to British medicine. London, Portland Press, 1992. 166 pp.

BROOKE, E. Medicine women: a pictorial history of women healers. London, Thorsons, 1997. 127 pp.

COAKLEY, D. Irish masters of medicine. Dublin, Town House, 1988. 170 pp.

CRAWFORD, D.G. Roll of the Indian Medical Service, 1615–1930. London, Thacker, 1930. 711 pp.

DEBUS, A.G. (ed.) World who's who in science. A biographical dictionary of notable scientists from antiquity to the present. Chicago, Marquis, 1968. 1855 pp.

DICTIONARY OF NATIONAL BIOGRAPHY. London, Oxford University Press.

DICTIONARY OF SCIENTIFIC BIOGRAPHY. Charles C, Gillispie, Editor-in-Chief. 18 vols. New York, Charles Scribners Sons, 1970–90 Concise biographies of some 5,000 scientists with informed discussions of their work. Most entries conclude with a bibliography listing (I) the subject's principal works and the location of any archival material, and (2) other biographical sources. The one-volume Concise Dictionary of Scientific Biography, Thomson Gale, Belmont, CA. contains abbreviated versions of all the articles in the first 16 vols; the bibliography portions are excluded.

DIEPGEN, P. Unvollendete: vom Leben und Wirken frühverstorbener Forscher und Ärzte. Stuttgart, G. Thieme, 1960. 223 pp. Short biographies of 32 distinguished research workers. Appropriate biographies are recorded in the main sequence of this Bibliography.

DREW, R. Commissioned officers of the medical services of the British Army 1660–1960. [Vol.2. Roll of officers in the Royal Army Medical Corps. 1898–1960]. 2 vols. London, Wellcome Historical Medical Library, 1968. 1126 pp. Vol.1 contains reprints of books by William Johnston and Alfred Peterkin (see below).

ENGELHARDT, D. von (ed.) Biographisches Enzyklopädie deutschsprachiger Mediziner. 2 vols. München, Saur, 2002.

FRIEDENWALD, H. The Jews and medicine. Jewish luminaries in medical history. 2 vols. Baltimore, Johns Hopkins Press, 1944. 817 pp. Vol.1 includes a classified bibliography of ancient Hebrew medicine. Reprinted New York, Ktav, 1967.

FRUTON, J.S. A biobibliography for the history of the biochemical sciences since 1800. Philadelphia, American Philosophical Society, 1992. 425 pp.

FURST, L.R. (ed.) Women healers and physicians; climbing a long hill. Lexington, University Press of Kentucky, *c.*1997. 274 pp.

GOLDSCHMIDT-LEHMANN, R.P. A bibliography of Anglo-Jewish medical biography. Jerusalem, Jerusalem Academy of Medicine, 1985. 108 pp. Lists over 330 doctors, including details of comprehensive biographies, evaluations and obituaries. Supplement to *Koroth*. Vol. 9, 1988.

GREENWOOD, M. The medical dictators, and other biographical studies. London, British Medical Association/Keynes Press, 1986. 137 pp. First published 1936.

GROTE, L.R.R. (ed.) Die Medizin der Gegenwart in Selbstdarstellungen. 8 vols. Leipzig, F. Meiner, 1923–29. 47 autobiographies of distinguished physicians and surgeons. Appropriate biographies are recorded in the main sequence of this Bibliography.

HALE-WHITE, W. Great doctors of the nineteenth century. London, Arnold, 1935. 325 pp. Appropriate biographies are recorded in the main sequence of this Bibliography.

HARPER, P. Guide to the manuscript papers of British scientists catalogued by the Contemporary Scientific Archives Centre and the National Cataloguing Unit for the Archives of Contemporary Scientists 1973–2003. Bath, NCUACS, University of Bath, 1993. 111 pp.

HIRSCH, A. Biographisches Lexikon der hervorragenden Ärzte aller Zeiten und Völker. 2te Aufl.. Wien, Leipzig, Urban & Schwarzenberg, 1929–35. First published 1884–88. Revised edition, containing entries for 1880–1930, published Munich, 1962.

HUTT, M. Medical biography and autobiography in Britain, *c*.1780–1920. Oxford, Oxford University Press, 1995. Abstract in *Index to Theses (1997)*, Vol. 46, pt. 1, ref. G1:46–2096.

INNES SMITH, R.W. English speaking students of medicine at the University of Leyden. Edinburgh, Oliver & Boyd, 1932. 258 pp.

INTERNATIONALER BIOGRAPHISCHER INDEX DER MEDIZIN… World biographical index of medicine. Physicians, homeopaths, veterinarians. 3 vols. München, Saur, 1996. An alphabetically-arranged index of 81,000 biographies contained in six national biographical archives.

IRELAND, N.O. Index to scientists of the world from ancient to modern times: biographies and portraits. Boston, MA, Faxon, 1962. 662 pp. Indexes 338 books published between 1877 and 1961 (some 7,500 scientists). Supplement by P.A. Pelletier (see below).

ISIS CUMULATIVE BIBLIOGRAPHY. A bibliography of the history of science formed from *Isis* Critical Bibliographies 1–90, 1913–65. London (Boston), History of Science Society. 3 supplements, 1966–1975; 1976–1985; 1986–1995.

JOHNSTON, W. Roll of commissioned officers in the medical service of the British Army who served on full pay within the period … 20 June, 1727 to 23 June, 1898. Aberdeen, University Press, 1917. 638 pp, Reprinted in Vol. 1. of Drew (see above).

KAUFMAN, M. *et al.* (eds.). Dictionary of American medical biography. 2 vols. Westport, Conn,, Greenwood Press, 1984. 1027 pp.

KELLY, H.A. A cyclopedia of American medical biography, comprising the lives of eminent deceased physicians and surgeons from 1610–1910. 2 vols. Philadelphia, Saunders, 1912. 969 pp.

KELLY, H.A. & BURRAGE, W.L. Dictionary of American medical biography. New York, Appleton, 1928. 1364 pp. Reprinted New York, Milford House, 1971.

KOREN, N. Jewish physicians: a biographical index. Jerusalem, Israel Universities Press, 1973. 275 pp.

LINDEBOOM, G.A. Dutch medical biography. A biographical dictionary of Dutch physicians and surgeons 1475–1975. Amsterdam, Rodopi, 1984. 2244 cols.

MACMICHAEL, W. The gold-headed cane. London, Murray, 1827. 179 pp. Biographies of the holders of the famous gold-headed cane now in the library of the Royal College of Physicians of London (Radcliffe, Mead, Askew, William and David Pitcairn, and Matthew Baillie). Later editions up to the 7th (1953) have appeared. A facsimile of the author's 1827 copy, illustrated and interleaved with his own amendments and additions, was published by the College in 1968. 26 + 279 pp.

MONRO, T.K. The physician as man of letters, science and action. 2nd ed. Edinburgh, Livingstone, 1951. 212 pp. Arranged under subjects, e.g., poetry, drama, saints, soldiers, explorers, piracy, crime, etc.

MUNK, W. *see* ROYAL COLLEGE OF PHYSICIANS OF LONDON.

NATIONAL ACADEMY OF SCIENCES. Biographical memoirs of the National Academy of Sciences. Vol. 1–, Washington, 1877– . Includes portraits and lists of publications. Appropriate biographies listed in the above are recorded in the main sequence of this Bibliography.

NEW YORK ACADEMY OF MEDICINE. Catalog of biographies in the library. Boston, Mass., Hall, 1960. 165 pp.

NOBEL PRIZE

    FOX, D.M. *et al.* Nobel laureates in physiology or medicine: a biographical dictionary. New York, Garland, 1990. 595 pp.

    LES PRIZ NOBEL EN 1901–. Stockholm, P.A.Norstedt (Almqvist & Wiksell), 1904– . Published annually; contains prizes, presentations, biographies, lectures, portraits.

    RAJU, T.N.K. The Nobel chronicles:a handbook of Nobel Prizes in physiology or medicine, 1901–2000. Bloomington, Ind., 1st Books, 2002. 508 pp.

OGILVIE, M.B. Women in science: antiquity through the nineteenth century. A biographical dictionary with annotated bibliography. Cambridge, MA, MIT Press, 1986. 254 pp.

PELLETIER, P.A. (ed.) Prominent scientists; an index to collective biographies. New York, Neal-Schuman, 1994. 353 pp. Some 10,000 scientists in English-language books. Supplements N.O. Ireland (see above).

PETERKIN, A. A list of commissioned medical officers of the Army, Charles II to accession of George II, 1660–1727. Aberdeen, University Press, 11925. 38 pp. Reprinted in vol. 1 of work by Drew (see above).

PETTIGREW, T.J. Medical portrait gallery. Biographical memoirs of the most eminent physicians, surgeons, etc., who have contributed to the advancement of medical science. 4 vols. London, Fisher Whittaker, 1838–40. 766 pp. 59 biographies, with engraved portraits.

PLARR. *See* ROYAL COLLEGE OF SURGEONS OF ENGLAND.

ROYAL COLLEGE OF OBSTETRICIANS AND GYNAECOLOGISTS.

    PEEL, J.H. The lives of the Fellows of the Royal College of Obstetricians and Gynaecologists 1929–1969. London, Heinemann Medical, 1976. 390 pp.

ROYAL COLLEGE OF PHYSICIANS OF LONDON.

    MUNK, W. The roll of the Royal College of Physicians of London, comprising biographical sketches 2nd ed. 3 vols, London, Royal College of Physicians, 1878. Covers the period 1518–1825. Continued in further volumes containing biographies of Fellows who died during the period 1826–1997.

ROYAL COLLEGE OF SURGEONS OF ENGLAND.

    PLARR, V.G. Plarr's Lives of the Fellows of the Royal College of Surgeons of England. 2 vols. Bristol, for the Royal College of Surgeons, 1930. Continued in further volumes containing biographies of Fellows who died to 1996.

ROYAL SOCIETY OF LONDON Obituaries of deceased Fellows chiefly for the period 1898–1904 with a generaL index to previous obituary notices contained in the Proceedings, vols. X to LXIV, 1860–1899. London, Harrison & Sons, 1903. 351 pp. (*Proceedings, vol. LXXV*). Obituary notices of Fellows of the Royal Society. Vol. 1–19. London, 1932–54. Include portraits and lists of publications. Appropriate biographies are recorded in the main sequence of this Bibliography. Continued as: Biographical Memoirs of Fellows of the Royal Society. Vol. 1– , London, 1955– . Includes portraits and lists of publications. Appropriate biographies are recorded in the main sequence of this Bibliography.

BIBLIOGRAPHY OF MEDICAL AND BIOMEDICAL BIOGRAPHY

WELLCOME LIBRARY FOR THE HISTORY AND UNDERSTANDING OF MEDICINE. Subject catalogue of the history of medicine and related sciences. Biographical section. 5 vols. Munich, Kraus International Publications, 1980. 2824 pp. Includes cumulation of biographical entries in nos. 1–173 (1954–1977) of *Current Work in the History of Medicine*.

WHO'S WHO IN SCIENCE IN EUROPE. 9th ed. 2 vols. London, Cartermill, 1985. 2083 pp.

WORLD WHO'S WHO IN SCIENCE.. A biographical dictionary of notable scientists from antiquity to the present. Chicago, Marquis, 1968. 1855 pp.

WÜSTENFELD, H.F. Geschichte der arabischen Ärzte und Naturforscher. Göttingen., Vandenhoek & Ruprecht, 1840. 167 pp. Reprinted Hildesheim, Olms, 1963.

JOURNAL OF MEDICAL BIOGRAPHY. London, Royal Society of Medicine. Vol. 1– , 1993–

SHORT LIST OF PUBLICATIONS ON THE HISTORY OF MEDICINE AND RELATED SUBJECTS

GENERAL WORKS   417
ANESTHESIOLOGY   417
ANATOMY   418
ART IN MEDICINE   418
BACTERIOLOGY, *SEE* MICROBIOLOGY
BIOCHEMISTRY   418
BIOLOGICAL STANDARDS   418
CANCER   418
CARDIOLOGY   418   *SEE ALSO* SURGERY, cardiothoracic   424
COMMUNICABLE DISEASES   419   *SEE ALSO* SEXUALLY-TRANSMITTED DISEASES   423
CONTRACEPTION   419
DENTISTRY   419
DERMATOLOGY   419
DIAGNOSIS   419   *SEE ALSO* RADIOLOGY   423
EMBRYOLOGY   419
ENDOCRINOLOGY   419
GASTROENTEROLOGY   420
GENETICS, HEREDITY   420
GYNAECOLOGY & OBSTETRICS   420
HAEMATOLOGY   420
HOSPITALS   420
HYGIENE, *SEE* PUBLIC HEALTH, HYGIENE
IMMUNOLOGY   420
MICROBIOLOGY   421
MOLECULAR BIOLOGY   421
NEUROLOGY, NEUROPHYSIOLOGY   421
NUTRITION   421
OBSTETRICS, *SEE* GYNAECOLOGY & OBSTETRICS
OCCUPATIONAL MEDICINE   421
OPHTHALMOLOGY   422
PAEDIATRICS   422
PARASITOLOGY   422
PATHOLOGY   422
PHARMACOLOGY, THERAPEUTICS   422
PHYSIOLOGY   422
PSYCHIATRY   423
PUBLIC HEALTH, HYGIENE   423
RADIOLOGY, IMAGING   423   *SEE ALSO* DIAGNOSIS   419
RHEUMATOLOGY   423
SEXUALLY-TRANSMITTED DISEASES   423
SURGERY, general   423
SURGERY, cardiothoracic   424

SURGERY, neurological   424
SURGERY, orthopaedic   424
SURGERY, plastic and cosmetic   424
TROPICAL MEDICINE   424   *SEE ALSO* PARASITOLOGY   422
TUBERCULOSIS   424
UROLOGY   424
VENEREAL DISEASES, *SEE* SEXUALLY-TRANSMITTED DISEASES
VIROLOGY, *SEE* MICROBIOLOGY

GENERAL WORKS

ACKERKNECHT, E.G. A short history of medicine. Baltimore, Johns Hopkins University Press, 1982. 277 pp.

ASIMOV, I. Asimov's chronology of science and discovery. New York, HarperCollins, 1994. 790 pp.

BYNUM, W.F. & PORTER, R. (eds.) Companion encyclopaedia of the history of medicine. 2 vols. London, Routledge, 1993. 1792 pp.

CASTIGLIONI, A. A history of medicine. New York, Aronson, 1969. 1013 pp.

CURRENT WORK IN THE HISTORY OF MEDICINE, London. No, 1–184, 1954–2000. Subsequently available on INTERNET: http://library.wellcome.ac.uk. No. 1–173 reprinted in WELLCOME INSTITUTE FOR THE HISTORY OF MEDICINE: Subject catalogue of the library of medicine and related sciences. (CB)

GARRETT, L. The coming plague: newly emerging diseases of the world out of balance. New York, Farrar, Strauss & Giroux, 1994. 750 pp.

GARRISON, F.H. An introduction to the history of medicine. 4th ed. Philadelphia, Saunders, 1929. 996 pp.

LAIGNEL-LAVASTINE, M.P.M. (ed.) Histoire générale de la médecine, de la pharmacie, de l'art dentaire et de l'art vétérinaire. 3 vols. Paris, Michel, 1936–49.

LeFANU, J. The rise and fall of modern medicine. London, Little, Brown, 1999. 490 pp.

MORTON, L.T. & MOORE, R.J. A chronology of medicine and related subjects. Aldershot, Scolar Press, 1997. 784 pp.

NORMAN, J.M. (ed.) Morton's medical bibliography: an annotated check-list of texts illustrating the history of medicine (Garrison and Morton). 5th ed. Aldershot, Scolar Press, 1991. 1243 pp.

PORTER, R. The greatest benefit to mankind. A medical history of humanity from antiquity to the present. London, HarperCollins, 1997. 831 pp.

PORTER, R. (ed.) The Cambridge illustrated history of medicine. Cambridge, Cambridge Univ. Press, 1996. 400 pp.

RHODES, P. An outline history of medicine. London, Butterworth, 1985. 219 pp.

SEBASTIAN, A. A dictionary of the history or medicine. Pearl River NY, Parthenon, 1999. 781 pp.

SINGER, C. & UNDERWOOD, E.A. A short history of medicine. 2nd ed. Oxford, Clarendon Press, 1962. 854 pp.

ANAESTHESIOLOGY

ATKINSON, R.B.. & BOULTON, T.B. (eds.) The history of anaesthesia. London, Royal Society of Medicine Press, 1988. 649 pp.

BERGMAN, N.A. The genesis of surgical anesthesia. Park Ridge, Ill., Wood Library, Museum of Anesthesiology, c.1998. 448 pp.

DUNCUM, B. The development of inhalation anaesthesia, with special reference to the years 1846–1900. London, Oxford Univ. Press, 1947. 640 pp.

MALTBY, R.J. Notable names in anaesthesia. London, Royal Society of Medicine Press, 2002. 200 pp.

BIBLIOGRAPHY OF MEDICAL AND BIOMEDICAL BIOGRAPHY

ANATOMY

CHOULANT, L. History and bibliography of anatomic illustration in its relations to anatomic science. New York, Schuman, 1945. 435 pp. Reprinted New York, Hafner, 1962.

DOBSON, J. Anatomical eponyms. Edinburgh. Livingstone, 1962. 235 pp. Very brief biographies but useful references.

PERSAUD, T.V.N. Early history of human anatomy from antiquity to the beginning of the modern era. Springfield, Ill., C.C. Thomas, 1984. 200 pp.

PERSAUD, T.V.N. A history of anatomy: the post-Vesalian era. Springfield, Ill., C.C. Thomas, 1997. 357 pp.

SINGER, C. The evolution of anatomy. A short history of anatomical and physiological discovery to Harvey. London, Kegan, Paul, 1935. 200 pp. Reprinted Dover, 1957.

ART IN MEDICINE

WHEELER, S. (ed.) Five hundred years of medicine in art. Aldershot, Ashgate, 2001. 3391 pp. Reproductions of 1668 prints and drawings from the Harvey Cushing/John Hay Whitney Medical Library, Yale University.

BACTERIOLOGY, SEE MICROBIOLOGY

BIOCHEMISTRY

FLORKIN, M. A history of biochemistry. 5 vols. Amsterdam, Elsevier, 1972–79. Continued in SEMENZA, G. A history of biochemistry. Vol. 1– . Amsterdam, Elsevier, 1983– .

FRUTON, J.S. A biobibliography for the history of the biochemical sciences since 1800. Philadelphia, American Philosophical Society, 1982. 885 pp. Supplement, 1985, 262 pp.

TEICH, M. & NEEDHAM, D. A documentary history of biochemistry, 1770–1940. Leicester, Leicester Univ. Press, 1992. 579 pp.

BIOLOGICAL STANDARDS

BANGHAM, D.R. A history of biological standardization: the characterization and measurement of complex molecules important in clinical and research medicine. Contributions from the U.K. 1900–1995. London, The Author, 4 Crown Close NW7 4HN. 1999. 293 pp.

CANCER

OLSON, J.S. The history of cancer: an annotated bibliography. New York, Greenwood Press, 1989. 426 pp.

RATHER, L.T. The genesis of cancer; a study of the history of ideas. Baltimore, Johns Hopkins Univ. Press, 1978. 262 pp.

RAVEN, R.W. The theory and practice of oncology: historical evolution and present principles. Carnforth, Parthenon, 1990. 366 pp.

CARDIOLOGY, SEE ALSO SURGERY, cardiothoracic

ACIERNO, L.J. A history of cardiology. Carnforth, Parthenon, 1994. 758 pp.

BING, R.J. (ed.) Cardiology: the evolution of the science and the art. 2nd ed. New Brunswick, Rutgers Univ. Press, 1999. 360 pp.

BURCH, G.E. & DePASQUALE, N.P. A history of electrocardiography. Chicago, Year Book Publishing, 1964. 309 pp. Reprinted, San Francisco, Norman Publishing, 1990.

BYNUM, W.F. et al. (eds.) The emergence of modern cardiology. Medical History, 1985, Supplement 5. 165 pp.

FLEMING, P.R. A short history of cardiology. Amsterdam, Rodopi, 1997. 241 pp.

MURPHY, W.B. The healing heart: an illustrated history of cardiology. Greenwich, Conn., Greenwich Press, 1998. 206 pp.

SILVERMAN, M.E. (ed.) British cardiology in the 20th century London, Springer, 2000. 390 pp.

COMMUNICABLE DISEASES, *SEE ALSO* SEXUALLY-TRANSMITTED DISEASES

FELDMAN, D.A. & MILLER, J.W. (eds.) AIDS crisis: a documentary history. Westport, Greenwood Press, 1998. 266 pp.

GOULD, T. A summer plague: polio and its survivors. New Haven, Yale Univ. Press, 1995. 366 pp.

GRMEK, M. History of AIDS. Princeton, Princeton Univ. Press, 1994. 279 pp.

MOHR, N. Malaria: evolution of a killer. Seattle, Seril & Pixel, 2001. 544 pp.

MONTAGNIER, L. Virus: the co-discoverer of HIV tracks its rampage and charts the future. New York, W.W. Norton, 2000. 256 pp.

SCOTT, S. & DUNCAN, C.J. Biology of plagues. New York, Cambridge Univ. Press, 2001. 420 pp.

SPINK, W.W. Infectious diseases. Prevention and treatment in the nineteenth and twentieth centuries. Minneapolis, Univ. Minnestoa Press, 1978. 577 pp.

CONTRACEPTION

McLAREN, A. A history of contraception from antiquity to the present time. Oxford, Blackwell, 1990. 275 pp.

MARKS, L.V. Sexual chemistry: a history of the contraceptive pill. New Haven, Yale Univ. Press, 2001. 372 pp.

ROBERTSON, W.H. An illustrated history of contraception. Carnforth, Parthenon, 1990. 152 pp.

DENTISTRY

HOFFMANN-AXTHELM, W. History of dentistry. Chicago, Quintessence Books, 1981. 435 pp.

RING, M.E. Dentistry; an illustrated history. New York, Abrams, 1985. 319 pp.

DERMATOLOGY

CRISSEY, J.T. *et al.* Historical atlas of dermatology and dermatologists. London, Parthenon, 2002. 234 pp.

PUSEY, W.A. The history of dermatology. Springfield, C.C. Thomas, 1933. 233 pp. Reprinted 1976.

DIAGNOSIS, *SEE ALSO* RADIOLOGY

BLAUFOX, M.D. An ear to the chest: an illustrated history of the evolution of the stethoscope. London, Parthenon, 2002. 149 pp.

MATTSON, J. & SIMON, M. The pioneers of NMR and magnetic resonance in medicine: the story of MRI. Ramat Gan, Israel, Bar-Ilan Univ. Press, 1996. 838 pp.

EMBRYOLOGY

GILBERT, S.F. A conceptual history of modern embryology. New York, Plenum Press, 1991. 250 pp. (Vol. 7 of *Developmental Biology*.)

HORDER, T.J. *et al.* (eds.) A history of embryology. Cambridge, Cambridge Univ. Press, 1986. 477 pp.

ENDOCRINOLOGY

MEDVEI, V.C. The history of clinical endocrinology. 2nd ed. Carnforth, Parthenon, 1993. 551 pp.

WELBOURN, R.B. (ed.) The history of endocrine surgery. New York, Praeger, 1990. 385 pp.

GASTROENTEROLOGY

CHEN, T.S. & CHEN, P.S. The history of gastroenterology. London, Parthenon, 1995. 313 pp.

DAVENPORT, H.W. A history of gastric secretion and digestion. Experimental studies to 1975. New York, Oxford Univ. Press, 1992. 414 pp.

KIRSNER, J.A. (ed.) The growth of gastroenterologic knowledge during the twentieth century. Philadelphia, Lea & Febiger, 1994. 522 pp.

GENETICS, HEREDITY

BODMER, W.F. & McKIE, R. The book of man: the quest to discover our genetic heritage. London, Little, Brown, 1994, 259 pp.

CAVALLI-SFORZA, L.L. *et al.* The history and geography of human genes. Princeton, NJ, Princeton Univ. Press, 1994. 518 pp.

HARRIS, H. The cells of the body: a history of somatic cell genetics. Cold Spring Harbor, Cold Spring Harbor Laboratory Press, 1995. 263 pp.

KAY, L.E. Who wrote the book of life? A history of the genetic code. Stanford, Stanford Univ. Press, 2000. 441 pp.

KELLER, E.F. The century of the gene. Cambridge, MA, Harvard Univ. Press, 2000. 186 pp.

SHANNON, T.A. (ed.) Genetic engineering: a documentary history. Westport, Conn., Greenwood Press, 1999. 282 pp.

GYNAECOLOGY & OBSTETRICS

GÉLIS, J. History of childbirth. Oxford, Polity Press, 1991. 326 pp.

MOSCUCCI, O. The science of woman: gynaecology and gender in England, 1800–1929. Cambridge, Cambridge Univ. Press, 1990. 270 pp. Reprinted 2000.

O'DOWD, M.J. & PHILIPP, E.E. History of obstetrics and gynaecology. Lonmdon, Parthenon, 1994. 710 pp.

HAEMATOLOGY

WINTROBE, M.M. Hematology: the blossoming of a science. Philadelphia, Lea & Febiger, 1985. 563 pp.

HOSPITALS

RISSE, G. Mending bodies, saving souls: a history of hospitals. New York, Oxford Univ. Press, 1999. 716 pp.

HYGIENE, SEE PUBLIC HEALTH, HYGIENE

HYPNOTISM

FORREST, D. Hypnotism: a history. London, Penguin, 2000. 334 pp. First published as *Evolution of hypnotism*, 1999.

IMMUNOLOGY

BRENT, L. A history of transplantation immunology. San Diego, CA, Academic Press, *c.*1997. 482 pp.

GALLAGHER. R.B. *et al.* (eds.) Immunology; the making of a modern science. London, Academic Press, 1995. 246 pp.

MAZUMDAR, P.M.H. Species and specificity: on the interpretation of the history of immunology. Cambridge, Cambridge Univ. Press, 1995. 457 pp.

PARISH, H.J. A history of immunization. Edinburgh, Churchill, 1965. 355 pp.

Microbiology

BECK, R.W. A chronology of microbiology in historical perspective. Washington, ASM Press, 2000. 391 pp.

BULLOCH, W. The history of bacteriology. London, Oxford Univ. Press, 1979. 422 pp.

CALISHER, H. & HORZINEK, M.E. (eds.) 100 years of virology: the birth and growth of a discipline, Vienna, Springer, 1999. 220 pp.

FENNER, F. & GIBBS, J.J. (eds.) Portraits of viruses: a history of virology. Basel, Karger, 1988. 344 pp.

GRAFE, A. A history of experimental virology. Berlin, Springer, 1991. 343 pp.

KOPROWSKI, H. & OLDSTONE, M.B.A. (eds.) Microbe hunters: then and now. Bloomington, Medi-Ed Press, 1996. 456 pp.

Molecular Biology

CAIRNS, J. *et al.* (eds.) Phage and the origins of molecular biology Cold Spring Harbor, Cold Spring Harbor Laboratory, 1966. 340 pp.

ECHOLS, H. Operators and promoters: the story of molecular biology and its creators. Berkeley, Univ. of California Press, 2001. 466 pp.

JUDSON, H.F. The eighth day of creation: makers of the revolution in biology. Plainview, CSHL Press, 1996. 714 pp.

LAGERKVIST, U. DNA pioneers and their legacy. New Haven, Yale Univ. Press, 1998. 156 pp.

MORANGE, M. A history of molecular biology. Cambridge, MA, Harvard Univ. Press, 1998. 336 pp.

Neurology, Neurophysiology

CLARKE, E. & O'MALLEY, C.D. The human brain and spinal cord: a historical study. 2nd ed. San Francisco, Norman Publishing, 1995. 951 pp.

FINGER, S. Minds behind the brain: a history of brain pioneers and their discoveries. New York, Oxford Univ. Press, 2000. 364 pp.

HAYMAKER, W.E. & SCHILLER, F. The founders of neurology. 2nd ed. Springfield, C.C. Thomas, 1970. 616 pp.

ROSE, F.C. Twentieth century neurology: the British contribution. London, Imperial College Press, 2001. 313 pp.

Nutrition

DARLEY, W.J. & JUKES, T.H. (eds.) Founders of nutrition science: biographical articles from the *Journal of Nutrition*. 2 vols. Bethesda, MD, American Institute of Nutrition, 1992. 1182 pp.

GOLDBLITH, S.A. & JOSLYN, M.A. Milestones in nutrition. Westport, CT Publishing Co., 1964. 797 pp.

KAMMINGA. H. & CUNNINGHAM, A. (eds.) The science and culture of nutrition, 1840–1940. Amsterdam, Rodopi, 1995. 344 pp.

McCOLLUM, E.V. A history of nutrition. Boston, Houghton Mifflin, 1956, 451 pp.

Obstetrics, *see* Gynaecology & Obstetrics

Occupational Medicine

HUNTER, D. Diseases of occupations. 6th ed. London, Hodder & Stoughton, 1978. 1257 pp.

SELLERS, C.C. Hazards of the job: from industrial disease to environmental health science. Chapel Hill, NC, Univ. of North Carolina, 1997. 227 pp.

BIBLIOGRAPHY OF MEDICAL AND BIOMEDICAL BIOGRAPHY

WEINDLING. P. (ed.) The social history of occupational health. London, Croom Helm, 1985. 267 pp.

OPHTHALMOLOGY

ALBERT, D.M. & EDWARDS, D.H. (eds.) The history of ophthalmlogy. Oxford, Blackwell Science, 1996. 394 pp.

GORIN, G. History of ophthalmology. Wilmington, Publish or Perish, 1982. 630 pp.

HIRSCHBERG, J. Geschichte der Augenheilkunde. 10 vols. Leipzig, W. Engelmann. 1899–1918. Reprinted, Hildesheim, G. Olms, 1977. 7 vols. English translaton by F.C. Blodi. Nonn, J.P. Watenborg, 1982. 11 vols. & supplements.

OTORHINOLARYNGOLOGY

PAPPAS, D. & KENT, L. Otolpogy's great moments: illustrations and annotations. [s.l., s.n.] 2000. 176 pp.

PIRSIG, W. & WILLEMOT, J. (eds.) Ear, nose and throat in culture. Oostende, G. Schmidt, 2001. 332 pp.

STEVENSON, R.S. & GUTHRIE, D.J. A history of oto-laryngology. Edinburgh, E. & S. Livingstone, 1945. 155 pp. Reprinted, San Francisco, Norman Publishing, 1991.

PAEDIATRICS

ABT, A.F. & GARRISON, F.H. History of pediatrics. Philadelphia, Saunders, 1965. 316 pp.

STILL, G.F. History of paediatrics: the progress of the study of diseases of children up to the end of the XVIII century. London, Dawsons, 1965. 526 pp.

PARASITOLOGY

FOSTER, W.D. A history of parasitology. Edinburgh, Livingstone, 1965. 202 pp.

GROVE, D.I. A history of human helminthology. Wallingford, CAB International, 1990. 848 pp.

POSER, C.M. & BRUYN, G.W. An illustrated history of malaria. New York, Parthenon, 1999. 165 pp.

PATHOLOGY

CUNNINGHAM, G. The history of British pathology. Bristol, White Tree Books, 1992. 375 pp.

DHOM, G. Geschichte der Histopathologie. Berlin, Springer, 2001. 812 pp.

FOSTER, W.D. A short history of clinical pathology. Edinburgh, Livingstone, 1961. 154 pp.

PHARMACOLOGY, THERAPEUTICS

ACKERKNECHT, E.H. Therapeutics from the primitives to the 20th century. New York, Hafner Press, 1973. 194 pp.

DAVENPORT-HINES, R. The pursuit of oblivion: a history of narcotics . London, Weidenfeld & Nicolson, 2001. 466 pp.

HIGBY, G.J. & STROUD, E.C. The history of pharmacy. New York, Garland Publications, 1995. 321 pp.

LEAKE, C.D. An historical account of pharmacology to the 20th century. Springfield, C.C. Thomas, 1975. 321 pp.

PHYSIOLOGY

FOSTER, M. Lectures on the history of physiology from the sixteenth, seventeenth and eighteenth centuries. Cambridge, Cambridge Univ. Press, 1901. 310 pp. Reprinted, Dover Publications, 1970.

HALL, T.S. History of general physiology 600 BC to AD 1900. 2 vols. Chicago, Chicago Univ. Press, 1975. 1513 pp.

HODGKIN, A.L. *et al.* The pursuit of nature: informal essays on the history of physiology. Cambridge, Cambridge Univ. Press, 1977. 180 pp.

ROTHSCHUH, K.E. History of physiology. Huntington, NY, Krieger, 1973. 379 pp.

PSYCHIATRY

ALEXANDER, F.G. & SLESNICK, S.T. History of psychiatry. Northville, NJ, J. Aronson, 1995. 471 pp.

BERRIOS. G.E. & PORTER, R. A history of clinical psychiatry: the origin and history of psychiatric disorders. London, Athlone Press, 1995. 684 pp.

FREEMAN, H. (ed.) A century of psychiatry. London, Mosby-Wolfe, 1999. 360 pp.

HOWELLS, J.G. (ed.) World history of psychiatry, New York, Brunner Maizels, 1975. 770 pp.

HUNTER, R.A. & MACALPINE, I.. Three hundred years of psychiatry, 1535–1860. London, Oxford Univ. Press, 1963. 1107 pp. Reprinted, Hartsdale, NY, 1982.

SHORTER, E. A history of psychiatry: from the era of the asylum to the age of Prozac. Chichester, Wiley, 1997. 436 pp.

PUBLIC HEALTH, HYGIENE

PORTER, D. (ed.) The history of public health and the modern state. Amsterdam, Rodopi, 1994. 439 pp.

ROSEN, G. A history of public health. Baltimore, Johns Hopkins Univ. Press, 1993. 535 pp.

RADIOLOGY, IMAGING, *SEE ALSO* DIAGNOSIS

GAGLIARDI, R.A. A history of the radiological sciences: diagnosis. Reston, VA, Radiological Centennial Inc., *c.*1996. 673 pp.

KEVLES, B. Naked to the bone: medical imaging in the twentieth century. New Brunswick, NJ, Rutgers Univ. Press, 1997. 378 pp.

MICHETTE, A. & PFAUNTSCH, S. (eds.) X-rays: the first hundred years. Chichester, John Wiley, *c.*1996. 262 pp.

ROSENBUSCH, G. Radiology in medical diagnostics: evolution of X-ray applications, 1895–1995. Oxford, Blackwell Science, 1995. 534 pp.

RHEUMATOLOGY

KERSLEY, G.D. & GLYN, J. A concise history of rheumatology and rehabilitation. London, Royal Society of Medicine Press, 1990. 150 pp.

SEXUALLY-TRANSMITTED DISEASES

ORIEL, J.D. The scars of Venus: a history of venereology. London, Springer, 1994. 248 pp.

QUETEL, C. History of syphilis. London, Polity Press, 1992. 342 pp.

SURGERY, general

ELLIS, H. A history of surgery. London, Greenwich Medical Media, 2001. 264 pp.

RAVITSCH, M.M. A century of surgery 1880–1980. 2 vols. Philadelphia, Lippincott, 1981. 1737 pp.

RUTKOW, I.M. Surgery: an illustrated history. St Louis, Mosby-Yearbook Inc., 1990. 550 pp.

WANGENSTEEN, O.H. & WANGENSTEEN, S.D. The rise of surgery, from empiric craft to scientific discipline. Minneapolis, Univ. Minnesota Press, 1978. 785 pp.

SURGERY, cardiothoracic

HURT, R. History of cardiothoracic surgery from early times. New York, Parthenon, 1996. 514 pp.

RICHARDSON, R. Heart and scalpel: a history of cardiac surgery. London, Quiller, 2001. 312 pp.

WESTABY, S. Landmarks in cardiac surgery. Isis Medical Media, 1997, 683 pp.

SURGERY, neurological

GREENBLATT, S.H. *et al.* (eds.) A history of neurosurgery in its scientific and professional contexts. Park Ridge, American Association of Neurological Surgeons, 1997. 625 pp.

SURGERY, orthopaedic

KLENERMAN, L. (ed.) The evolution of orthopaedic surgery. London, Royal Society of Medicine Press, 2001. 246 pp.

PELTIER, L.F. Orthopaedics: a history and iconography. San Francisco, Norman Publishing, 1993. 304 pp.

SURGERY, plastic and cosmetic

GILMAN, S.L. Making the body beautiful: a cultural history of aesthetic surgery. Princeton, NJ, Princeton Univ. Press, 1999. 396 pp.

HAIKEN, E. Venus envy: a history of cosmetic surgery. Baltimore, Johns Hopkins Univ. Press, 1997. 370 pp.

HOFFMANN-AXTHELM, W. Die Geschichte der Mund-, Kiefer-, und Gesichtschirurgie. Berlin, Quintessenz, *c.*1995. 351 pp.

TRANSPLANTATION

BRENT, L. A history of transplantation immunology. San Diego, CA, Academic Press, *c.*1997. 482 pp.

HAKIM, N.S. & PAPALOIS, V.E. History of organ and cell transplantation. London, Imperial College Press, 2003. 464 pp.

TROPICAL MEDICINE, *SEE ALSO* PARASITOLOGY

COX, F.E.G. (ed.) Illustrated history of tropical diseases. London, Wellcome Trust, 1996. 454 pp.

DESOWITZ, E.S. Tropical diseases from 50,000 BC to AD 2500. London, HarperCollins, 1997. 256 pp.

SCOTT, H.H. A history of tropical medicine. London, Arnold, 1939. 1165 pp.

TUBERCULOSIS

DORMANDY, T. The white death: a history of tuberculosis. London, Hambledon Press, 1999. 433 pp

OTT, K. Fevered lives. Cambridge, Harvard Univ. Press, 1999. 252 pp.

ROSENKRANTZ, B.G. (ed.) From consuption to tuberculosis: a documentary history. New York, Garland, 1994. 623 pp.

UROLOGY

Lee, H.S.J. (ed.) Dates in urology. New York, Parthenon, 2000. 116 pp.

MURPHY, L.J.T. The history of urology. Springfield, C.C. Thomas, 1971. 531 pp.

VENEREAL DISEASES, *SEE* **SEXUALLY-TRANSMITTED DISEASES**

VIROLOGY, *SEE* **MICROBIOLOGY**